VHDL Design Representation and Synthesis

Second Edition

James R. Armstrong
F. Gail Gray

Prentice Hall PTR
Upper Saddle River, NJ 07458
www.phptr.com

Library of Congress Cataloging-in-Publication Data

Armstrong, James R., 1939-
    VHDL design representation and synthesis / James R. Armstrong, F. Gail. Gray.
      p. cm.
    Includes index.
    ISBN 0-13-021670-4
    1. VHDL (Computer hardware description language) I. Gray, F. Gail.

TK7885.7 .A7623 2000
621.39'2--dc21

                                      99-056533

Editorial/Production Supervision: *Vincent J. Janoski*
Acquisitions Editor: *Bernard Goodwin*
Manufacturing Manager: *Alexis Heydt*
Marketing Manager: *Lisa Konzelmann*
Editorial Assistant: *Diane Spina*
Cover Design Director: *Jerry Volta*
Composition: *Aurelia Scharnhorst*

© 2000 by Prentice Hall PTR
Prentice-Hall, Inc.
Upper Saddle River, New Jersey 07458

Prentice Hall books are widely used by corporations and government agencies for training, marketing, and resale.
The publisher offers discounts on this book when ordered in bulk quantities. For more information, contact Corporate Sales
Department. Phone: 800-382-3419; Fax: 201-236-7141; E-mail: corpsales@prenhall.com; Or write: Prentice Hall PTR, Corp.
Sales Dept., One Lake Street, Upper Saddle River, NJ 07458

First edition originally entitled *Structured Logic Design With VHDL.*

Printed in the United States of America
10 9 8 7 6 5 4 3 2

ISBN 0-13-021670-4

Prentice-Hall International (UK) Limited, *London*
Prentice-Hall of Australia Pty. Limited, *Sydney*
Prentice-Hall Canada Inc., *Toronto*
Prentice-Hall Hispanoamericana, S.A., *Mexico*
Prentice-Hall of India Private Limited, *New Delhi*
Prentice-Hall of Japan, Inc., *Tokyo*
Pearson Education Asia Pte. Ltd.
Editora Prentice-Hall do Brasil, Ltda., *Rio de Janeiro*

Contents

3   Basic Features of VHDL   **41**

7   Gate Level and ASIC Library Modeling   261

8   HDL-Based Design Techniques   315

Preface

The purpose of this book is to integrate hardware description languages into the digital design process at all levels of abstraction. There are two main steps in this process: (1) development of a hardware description language model and (2) synthesis of the model into an ASIC logic circuit or FPGAs. In teaching this process, we use VHDL, the VHSIC Hardware Description Language. VHDL, whose development began in 1983 under DOD sponsorship, was further developed by the IEEE and released as IEEE Standard 1076 in 1987. Further improvements were incorporated since then and the language was re-released as an updated standard in 1993. Since that time, VHDL has evolved into a de facto industry standard for hardware description languages. In the opinion of the authors, it has the most comprehensive set of modeling constructs available in any hardware description language. For these reasons, VHDL was chosen as the base language for this book. We explore the language in an in-depth, unified manner.

Most books currently on the market that treat hardware description languages, particularly VHDL, are either: (1) language texts that cover the VHDL language thoroughly, but do not show how to integrate the language into the digital design process, or (2) logic design books that primarily use VHDL models as simulation tools to validate designs that are produced in the classical manner. This book fully integrates VHDL into the design process starting with a high-level executable model that provides an unambiguous, executable version of the specification, and concluding with a gate-level implementation.

In this book, synthesis is viewed as a multistep process, beginning with an English description which is transformed first into VHDL and then from VHDL into a circuit schematic. We discuss synthesis from two viewpoints: 1) the mappings: emphasis is placed on understanding the relationship between VHDL language constructs and the implied logic circuit. A full chapter is devoted to correct modeling style for synthesis; 2) the tools: we illustrate the synthesis process using two very popular tool sets, the Synopsys Design Analyzer and Compiler (for ASICs) and the Xilinx Foundation Series (for FPGAs). Since ASICs and FPGAs are the targets, a chapter is devoted to these technologies. The book also contains a chapter illustrating the complete top-design design process from specification to logic synthesis.

This book is written for three main educational purposes: (1) for a second course in logic design for undergraduate students in Electrical Engineering, Computer Engineering, and Computer Science; (2) for a graduate course dealing with hardware description languages and other design aids; and (3) for practicing engineers who wish to learn about design with hardware description languages. Thus the assumed background for the book is (1) a basic course in computer organization and logic design and (2) some knowledge of high-level languages, such as C, C++, or JAVA.

The authors use the text in a course, which is the second course in a logic design sequence. The students are either juniors in Computer Engineering, for whom the course is required, or Electrical Engineering seniors, for whom the course is an elective. In this semester length course we cover Chapters 1, 2, 3, 4, 5, 9, 10, and 11. The emphasis is on developing VHDL models in a conservative algorithmic style that can be synthesized. To support this in the laboratory, we use a PC version of ViewLogic, Inc.'s Workview for VHDL modeling and simulation and schematic capture. Xilinx software and XS40/XTEND boards are used for FPGA synthesis. We also employ System View from Elanix to provide for high-level design of digital filters. Workstation-based Synopsys tools are used for ASIC synthesis. All students in our department have their own PCs, so the use of a PC-based system such as Workview has been effective in being able to serve the large number of students we normally teach in our second digital design course. For this same reason, we use telnet and dc_shell scripts for Synopsys synthesis. Typical assignments include:

1. An introductory assignment to familiarize students with Workview's VHDL modeling, simulation and schematic capture environment.
2. An assignment to develop and simulate a single VHDL behavioral model.
3. An assignment to develop a model of a counter, or some similar circuit. VHDL behavioral models are developed for counter flip-flops and gates, and the schematic capture capability of Workview is used to construct the structural model of the counter.
4. An assignment to translate a system description is first translated into a VHDL behavioral model which is simulated. This is typically a state machine such as an interface protocol, a vending machine, or a traffic light controller.
5. An introductory tutorial to the Xilinx Foundation Series Software.
6. An assignment to implement a small circuit in both Synopsys ASIC logic and FPGAs and compare speed of execution.
7. A fairly complicated FPGA project such as a booth multiplier, calculator, small processor, digital filter, or graphics display. For the filter, the codec on the XSTEND board is used for A/D and D/A comversion. The Xilinx filter code is developed using System View. The graphics display displays a pattern stored in RAM on a VGA monitor.

If used for a graduate course, the entire book can be covered in one semester. In such a course, one can cover the broad range of constructs in the language and examine in detail the language semantics for both simulation and synthesis. In our graduate course at Virginia Tech, we synthesize with Synopsys and validate synthesized models. We study ways to control the synthesis to achieve optimum circuits in a delay or area sense. High-level modeling tools such as

Express VHDL, SPW, and System View are also covered. A comparison is done between VHDL and Verilog.

For this course, the student's laboratory assignments include:

1. An assignment to develop and simulate a single VHDL behavioral model.
2. An assignment to develop a model of a counter or some similar circuit. VHDL behavioral models are developed for counter flip-flops and gates, and then a VHDL structural model is developed for the whole system.
3. An assignment involving complex data types, e.g., using array aggregates and record types to implement a tabular representation of a finite state machine.
4. A system modeling assignment that involves the use of bus resolution and bus protocols. This system employs the IEEE 9 valued logic system. Examples include the URISC processor system in Chapter 6 or a histogram construction system for image processing.
5. An assignment where a model is written, simulated, and synthesized using both VHDL and Verilog and comparison's made
6. A semester project where the students model a system of their choice. One can choose projects, which stretch the language, i.e., involve applications that are not typical, such as modeling parallel processing systems or modeling systems which are not digital.

The book contains hundreds of VHDL models and code fragments. All code has been analyzed, and simulated, and synthesized (where required), using the Synopsys VHDL system. The only exception to this is the VHDL 93 code. In addition, the text contains over 300 homework problems with a wide range of difficulty. Types of problems include short answer questions, simple design problems, complex system design problems involving design, modeling, and simulation, and problems that require a study of a design or design tool issue. Some problems in this latter category would make good thesis projects!

Accompanying the book is a CD-ROM. On the CD are: 1) source files for all VHDL code in the book, 2) a set of projects accompanied by supporting data command files, and 3) packages to support common design paradigms. Problem and project solutions and Power Point lecture slides are available to instructors who adopt our text for classroom use.

Writing a book of this nature is a large undertaking. In doing so we have received the help and assistance of a number of individuals and organizations. We would like to thank:

1. Viewlogic, Synopsys, Inc., and Xilinx, for their support in providing us with the VHDL software to check out the VHDL code in the book and for use in our courses on hardware description languages.
2. Dave Barton of Intermetrics for his review of the manuscript.
3. Our production editor at Prentice Hall, Vincent Janoski.
4. Our book editor at Prentice Hall, Bernard Goodwin, for his enthusiastic support of the project.

We would also like to thank our wives, Marie and Caryl, for their encouragement and support in spite of the long hours we spent in front of our computers and their tolerance of laptops accompanying us on trips and vacations over a period of two years.

Structured Design Concepts

In this chapter we present basic definitions that relate to the design process. It is necessary to introduce them now so that other concepts can be explained. The reader should study them carefully in order to comprehend material introduced later. It will also be useful to revisit this chapter as one proceeds through the text since the full meaning of the terms will only become clear through use and example.

1.1 THE ABSTRACTION HIERARCHY

In this section we present the abstraction hierarchy employed by digital designers. Abstraction can be expressed in the following two domains:

Structural domain.   A domain in which a component is described in terms of an interconnection of more primitive components.

Behavioral domain.   A domain in which a component is described by defining its input/output response.

Figure 1.1 shows structural and behavioral descriptions for a logic circuit, which detects two or more consecutive 1's or two or more consecutive 0's on its input X. The structural description is an interconnection of gate and flip-flop primitives. The behavioral description is expressed textually in a hardware description language (HDL).

An abstraction hierarchy can be defined as follows:

Abstraction hierarchy.   A set of interrelated representation levels that allow a system to be represented in varying amounts of detail.

Figure 1.2 shows a picture of a typical abstraction hierarchy. For each level i in the hierarchy there exists a transformation to level i+1. The level of detail usually increases monotonically as one moves down in the hierarchy.

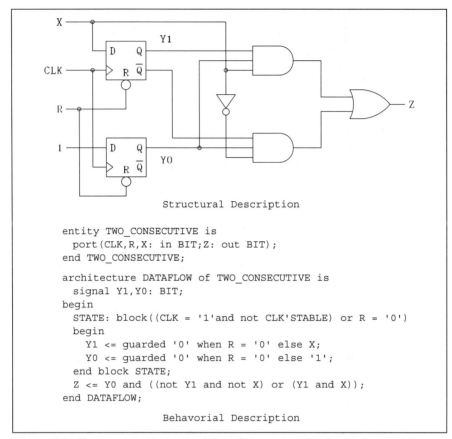

```
entity TWO_CONSECUTIVE is
  port(CLK,R,X: in BIT;Z: out BIT);
end TWO_CONSECUTIVE;

architecture DATAFLOW of TWO_CONSECUTIVE is
  signal Y1,Y0: BIT;
begin
  STATE: block((CLK = '1'and not CLK'STABLE) or R = '0')
  begin
    Y1 <= guarded '0' when R = '0' else X;
    Y0 <= guarded '0' when R = '0' else '1';
  end block STATE;
  Z <= Y0 and ((not Y1 and not X) or (Y1 and X));
end DATAFLOW;
```

Figure 1.1   Structural and behavioral descriptions of a sample circuit.

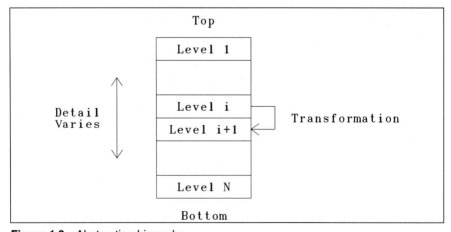

Figure 1.2   Abstraction hierarchy.

Table 1.1 Design abstraction hierarchy.

Level of Detail	Behavioral Domain Representation	Structural Domain Primitive
System	Performance specifications (English)	Computer, disk, unit, radar
Chip	Algorithm	Microprocessor, RAM, ROM, UART, parallel port
Register	Data flow	Register, ALU, COUNTER, MUX, ROM
Gate	Boolean equations	AND, OR, XOR, FF
Circuit	Differential equations	Transistor, R, L, C
Layout/silicon	Equations of electron and hole motion	Geometrical shapes

This book uses a design hierarchy that has six levels: silicon, circuit, gate, register, chip, and system. Table 1.1 illustrates this hierarchy. The silicon level is the lowest level in the hierarchy, the system level the highest. One can represent a design at any of these levels. As one moves down in the hierarchy, one is closer to a physical implementation and the design representation is less abstract. Correspondingly, the amount of detail required to represent a design increases as one descends in the hierarchy. As will be emphasized throughout the text, it is important that a particular design activity be carried out at a level which has sufficient but not excessive detail. A level with insufficient detail yields inaccurate results; whereas, a level with excessive detail can make the design activity too expensive.

Table 1.1 shows the nature of this hierarchy in terms of the structural primitives and behavioral representation for each level. Structural primitives are interconnected to form a structural model at a given level. Figure 1.3 shows examples of the structural primitives at each level. The behavioral representation is the textual or pictorial form of a device's I/O response at that level.

At the lowest level, the silicon level, the basic primitives are geometric shapes that represent areas of diffusion, polysilicon, and metal on the silicon surface. The interconnection of these patterns models the fabrication process from the designer's point of view. Behavioral description at this level are the physical equations that describe electron and hole motion in electrical materials. At the next level up, the circuit level, the representation is that of an interconnection of traditional passive and active electrical circuit elements: resistors, capacitors, and bipolar and MOS transistors. The interconnection of components is used to model circuit behavior in terms of voltage and current. The behavioral content at this level can be expressed in terms of differential equations.

The third level up, the gate level, has traditionally been the major design level for digital devices. The basic primitives are the AND, OR, and INVERT operators and various types of flip-flops. Interconnection of these primitives forms combinational and sequential logic circuits. Boolean equations define the behavior at this level.

The level above the gate level is the register level. Here the basic primitives are such things as registers, counters, multiplexers, and ALUs. These primitives are sometimes referred to as functional blocks. They also correspond to VLSI design macros. Thus, this level is also referred to as the functional or macro level. Although the register-level primitives can be expressed in

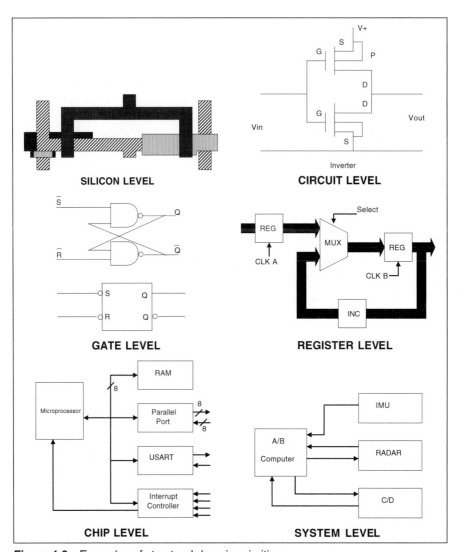

Figure 1.3   Examples of structural domain primitives.

terms of an interconnection of gates, when working at this level, one does not take this viewpoint. Register-level primitives are expressed in terms of truth tables and state tables; thus, these two forms can be used to represent the behavioral content at this level. Behavioral descriptions at this level are termed *data flow*, i.e., they reflect the way data is actually distributed in a real implementation. In this book, we will show how these data flow descriptions can be implemented in a hardware description language.

The level above the register level is the chip level. At this level, the structural primitives are such things as miroprocessors, memories, serial ports, parallel ports, and interrupt controllers. Although chip boundaries are typically the model boundaries, other situations are possible.

For example, collections of chips, which together form a single functional unit, can be modeled as a single entity. Or alternatively, sections of a chip design could be modeled as separate entities during this design phase. The key aspect is that a large block of logic is to be represented in which long and frequently convergent data paths from inputs to outputs must be modeled. In the behavioral domain, at each level in the hierarchy, the primitives are behavioral descriptions which are not structural models that are created from more basic primitives. Each primitive is a distinct model entity. Thus, if a serial I/O port (UART) is to be modeled, the model is not created by interconnecting simpler functional models of such things as registers and counters—the UART itself is the basic model entity. Behavioral domain models are important to system manufacturers who buy a chip from another manufacturer but have no knowledge of its proprietary gate-level structure. Chip level models of complicated circuits are viewed as Intellectual Property (IP) and are frequently sold by one company to another. The behavioral content of a chip-level model is defined in terms of the I/O response of the device—the algorithm that the chip implements. In this book, a hardware description language is used to code these algorithmic descriptions.

The top level in the structural hierarchy is the system level. The primitive elements of this level are computers, bus interface units, disk units, radar units, etc. The behavioral content of this level is frequently expressed in terms of performance specifications, which give, for example, the MIPS rating (million instructions per second) of a processor or the bandwidth in bits per second of a bus, or use a statistical model to determine the percent utilization of a part of the system. If a deterministic model is used at this level, it employs very high-level data types to represent information being passed between systems. For example, if a radar system were modeled, information of type "frequency" would be passed between units in the system.

1.2 TEXTUAL VS. PICTORIAL REPRESENTATIONS

Design representations can be either pictorial or textual. Figure 1.1 shows a pictorial logic schematic for the circuit, which detects two consecutive 1's or two consecutive 0's (referred to as TWO_CON). Figure 1.4 shows a block diagram, a state diagram, a timing diagram, a state table, state assignment, and truth tables (Kmaps) for the same circuit.

These are all examples of pictorial forms. Common textual methods of representation are natural languages (e.g., English), equations (e.g., Boolean or differential), and computer languages. In this text we use a specialized computer language called a *hardware description language*, which can be defined as follows:

Hardware description language.   A high-level programming language with specialized constructs for modeling hardware.

The behavioral description in Figure 1.1 illustrates a hardware description language textual description for TWO_CON.

An important consideration in developing a design process is whether to use pictures or text. Historically, pictures have been the preferred representation for digital design, i.e., block, timing, and logic diagrams (schematics) were the principal forms of representation. However, with the advent of hardware description languages, textual design descriptions have gained in popularity. From inspecting Figures 1.1, 1.3, 1.4, and Table 1.1, one can see that structural

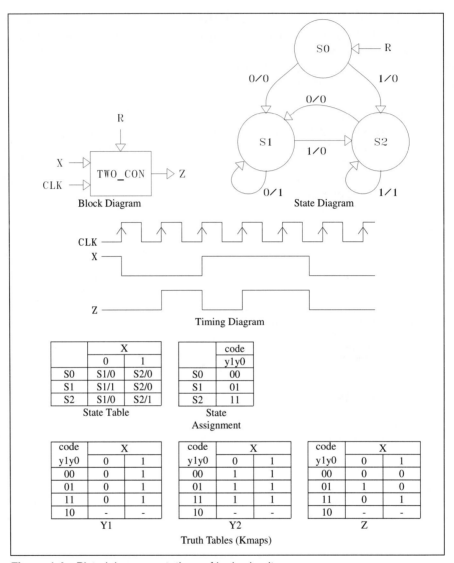

Figure 1.4 Pictorial representations of logic circuits.

descriptions are primarily pictorial, and that behavioral representations are primarily textual. Some exceptions to this classification scheme are state tables, state diagrams, and timing diagrams, which are pictorial but represent behavior. As to the general question of when pictures should be employed or when text should be used, one can make the following generalization: *Text is better for representing complex behavior; pictures are better for illustrating interrelationships.* Excessive use of either text or pictures results in a loss of perspective, i.e., "one cannot see the forest for the trees." Thus, real design systems balance the use of text and pictures—this is the approach we use in this text.

```
Architecture ALGORITHMIC of TWO_CONSECUTIVE is
  type STATE is (S0,S1,S2);
  signal Q: STATE := S0;
begin
  process(R,X,CLK,Q)
  begin
    if (R'EVENT and R = '0') then --reset event
      Q <= S0;
    elsif (CLK'EVENT and CLK = '1') then --clock event
      if X = '0' then
        Q <= S1;
      else
        Q <= S2;
      end if;
    end if;
    If Q'EVENT or X'EVENT then --output function
      if (Q=S1 and X='0') or (Q=S2 and X='1') then
        A <= '1';
      else
        z <= '0';
      end if;
    end if;
  end process;
and ALGORITHMIC;
```

Figure 1.5 Algorithmic description of the example circuit in Figure 1.1.

1.3 TYPES OF BEHAVIORAL DESCRIPTIONS

Behavioral descriptions in hardware description languages are frequently divided into two types: *algorithmic* and *data flow.*

Algorithmic. A behavioral description in which the procedure defining the I/O response is not meant to imply any particular physical implementation.

Thus, an algorithmic description is merely a procedure or program written to model the behavior of a device, to check that it is performing the correct function, without worrying about how it is to be built.

Data flow. A behavioral description in which the data dependencies in the description match those in a real implementation.

Data flow descriptions show how data moves between registers. Figure 1.1 gave a data flow description for the circuit which detects two or more consecutive 1's or two or more consecutive 0's. Figure 1.5 shows an algorithmic description for the same circuit.

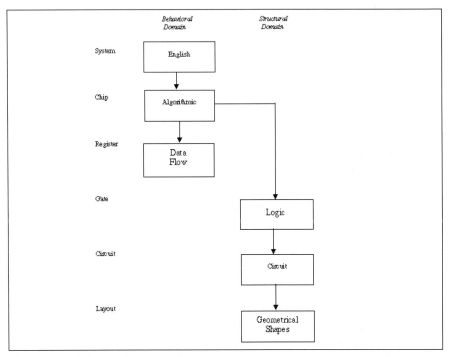

Figure 1.6   A typical design track.

1.4   DESIGN PROCESS

We will describe a structured design process in this text beginning with the following definitions.

> **Design.**   A series of transformations from one representation of a system to another until a representation exists that can be fabricated.

Our approach to design involves synthesis, which the dictionary defines as "the combining of abstract entities into a single or unified entity." Thus, to synthesize is "to put something together." For our context we make a more specialized definition.

> **Synthesis.**   The process of transforming one representation in the design abstraction hierarchy into another representation.

Each step in the design process is referred to as a *synthesis step*. Since the design process usually starts in the behavioral domain at a high level and ends in the structural domain at a low level, each synthesis step is a transformation from level i to level j with $i \leq j$. Therefore, what we are "putting together" is a representation of a design at level j. The representation at level i is used as a guide in the synthesis process, in that the implementation at level j must implement the same function as that at level i. In most cases, j=i+1, or j=i, i.e., either the levels are adjacent, or the transformation is from the behavioral domain to the structural domain at the same level. However, there are situations where levels are skipped in the synthesis process.

The *design cycle* consists of a series of transformations (synthesis steps). Common synthesis steps include:

1. Transformation from English to an algorithmic representation (*natural language synthesis*).
2. Translation from an algorithmic representation to a data flow representation (*algorithmic synthesis*) or to a gate level representation (referred to in the industry as *behavioral synthesis*). Note: In this second case one moves from the behavioral to the structural domain.
3. Translation from data flow representation to a structural logic gate representation (*logic synthesis*). Note: In this process one also moves from the behavioral to the structural domain and skips the register level.
4. Translation from logic gate representation to layout representation (*layout synthesis*). In this synthesis step the circuit level is skipped. This completes the synthesis process since the layout information can be fabricated.

The complete design cycle is sometimes referred to as *design synthesis*.

Figure 1.6 shows a typical design track through the design hierarchy. The track begins in the behavioral domain and descends through the system and chip levels to the register level. At the register level the transformation is made from a behavioral data flow representation at this level to a structural gate level description. Or, the transformation can be made directly from an algorithmic representation to a structural gate level description. This gate-level description is then transformed to the circuit level or perhaps directly to the layout or silicon level.

The design cycle steps can be carried out automatically in all stages in descending order, except the first, which is currently an active area of research. It is the purpose of this book to show students and design engineers how to carry out the steps themselves, and therefore, understand which functions are actually performed by automatic synthesis programs. Also, in some engineering organizations, manual transformations are still the only means employed.

1.5 STRUCTURAL DESIGN DECOMPOSITION

The structural form of the design hierarchy implies a design decomposition process. This is because at any level we choose, the system model is composed by interconnection of the "primitives" defined for that level. In the structural domain, primitives are defined in terms of interconnections of primitives at the next lower level. Thus, as shown in Figure 1.7 a design can be represented as a tree, with the different levels of the tree corresponding to levels in the abstraction hierarchy. Eventually, even in structural models, primitives at the leaves of the tree must be represented by behavioral models.

As defined above, a behavioral model is a primitive model in which the operation of the model is specified by a procedure as opposed to redefining it in terms of other components. Since a behavioral model can exist at any level in the design hierarchy, different parts of the design can have behavior specified at different levels. In Figure 1.7(a) the design tree is "full," and all behavior is, therefore, specified at the same level. In Figure 1.7(b) a design that has the form of a partial tree is shown, where behavior is specified at different levels. This situation is

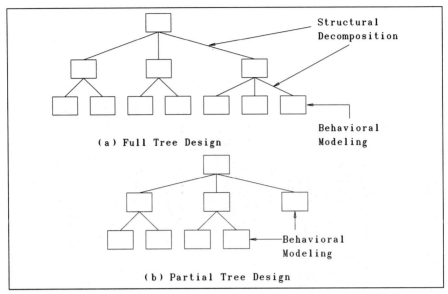

Figure 1.7   Structural decomposition.

encountered because one frequently wants to evaluate the relationships between system components before they have all been completely designed. For example, *multilevel simulation* can be
used to evaluate the interaction of one component whose design has reached the gate level with
other components whose design is currently still at the system level. Thus, it is not required that
all system components be specified at the gate-level in order to evaluate the gate level design of
a specific component. The checking is done by employing a simulation in which the behavioral
content of the component models occurs at different levels in the hierarchy. In fact, it may be
impossible to fully simulate a large system with all components modeled at the gate level. Such
a simulation might take months to execute using current gate-level simulators. Instead, one
would do multiple simulations, with different sets of components modeled at the gate level in
each simulation. The other components would be modeled at the systems level. These simulations would take much less time to execute because system level simulations are more efficient.
A few hundred simulations with each simulation taking several hours is preferable to one giant
simulation that takes several months.

  Two concepts related to the design tree are those of top-down and bottom-up design. Here
the word "top" refers to the root of the tree; whereas, "bottom" refers to the leaves. In *top-down
design*, the designer begins with knowledge of only the function of the root. He or she then partitions the root into a set of lower-level primitives. Each of these lower-level primitives is then
partitioned into an interconnection of primitives at still lower levels. This process continues until
the leaf nodes of the design are reached. At the leaf nodes, the models are always behavioral. An
important point to make about top-down design is that the partitioning is optimized at each level
according to some objective criterion, e.g., cost, speed, and chip area. The partitioning is not
constrained by "what's available."

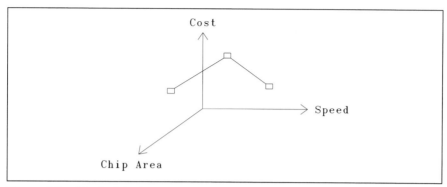

Figure 1.8 A typical design space.

The term *bottom-up design* is somewhat of a misnomer in that the process of design still begins with the definition of the root, but in this case the partitioning is conditioned by what is available. Lower parts of the tree will have been designed previously, perhaps on another project, and the designer is constrained (perhaps ordered!) to use them. Top-down design usage appears to be the most ideal situation, but its disadvantage is that it produces components that are not "standard," thus, increasing the cost of the design. Bottom-up designs are more economical, but they may not meet the objective performance criterion as well as top-down designs. Most real designs are a combination of top-down and bottom-up techniques.

A final concept related to the hierarchy is that of a design window. By this we mean a range of levels over which the designer works in developing a design-tree structure. The VLSI chip designer's window extends over the range of silicon, circuit, gate, register, and chip levels. The computer system designer, on the other hand, is currently concerned with a window consisting of the gate, register, chip, and system levels. Thus, the designer selects the window which is appropriate to his or her design activity and works at those levels of abstraction, which provide the necessary information for that activity, without involving unnecessary detail. CAD systems used to support design should allow easy movement between levels in the design window.

1.6 THE DIGITAL DESIGN SPACE

In the preceding discussion on top-down design, we said that partitioning was carried out in order to meet some objective criteria. These criteria are the major factors one has to consider in arriving at a design. These factors can be considered to be dimensions in a space. Some useful dimensions for the Digital Design Space are: speed, chip area, and cost. Figure 1.8 illustrates such a design space. Various designs have different tracks through the space as they evolve. The designer trades off one factor for another. For example, in his quest for speed he may increase the chip area and cost of a design. Figure 1.9 gives a concrete example of this. Circuits A and B implement the same logic function. Circuit A uses fewer gates (area) than Circuit B but is slower. Circuit B is faster than Circuit A but requires more gates (area). Thus we are trading off area for speed.

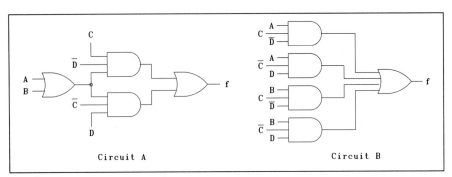

Circuit A   Circuit B

Figure 1.9   An example of a design space trade-off.

PROBLEMS

1.1 Draw a structural model of a 16-bit ripple carry adder as an interconnection of full adders. Develop a behavioral description of the 16-bit adder in C, C^{++}, or JAVA.

1.2 Draw a gate-level diagram of a full adder in terms of AND, OR, and INVERT primitives. Consult books on electronics and VLSI design and develop a CMOS circuit for the adder in terms of transistors, resistors, and capacitors. Compare the two circuits as follows:

    a. Count the number of primitives and wires required for each representation.

    b. Simulate both circuits on the same computer using a logic simulator and SPICE. Drive both circuits with all possible 8-input patterns. Compare simulation times.

1.3 Track the representation of a digital system through the abstraction hierarchy. Use the small computer example in Morris Mano's book, *Computer System Architecture,* 2nd Edition (Englewood Cliffs: Prentice Hall, 1982). Track the description of this system through the hierarchy as follows:

    a. Develop a one page, single-spaced, English, system-level description of the machine.

    b. Using your favorite graphics package, draw a chip-level diagram of a system that uses the Mano machine, RAM memory, and an I/O logic section.

    c. Draw a complete register-level diagram of the Mano machine.

    d. Using gates and J-K flip-flop primitives, draw a complete gate-level diagram of the Program Counter.

    e. Select a gate from the Program Counter, and draw its CMOS circuit equivalent.

    f. Draw the layout for the gate you described in the previous step.

    Which of the representations you developed are structural and which are behavioral?

1.4 Figure 1.10 shows a structural model of a simple RC series circuit. Write a behavioral description which involves current, *i(t)*. Plot *i(t)*.

1.5 Shown is a behavioral description for a gate-level circuit.

$$F = AB \text{ or } C\overline{D} \text{ or } \overline{E}F \tag{1.1}$$

Develop two different, structural gate-level models that implement this function.

1.6 A textual description of the fabled family relationship from N. Wirth's *Algorithms + Data Structures = Programs*, (Englewood Cliffs: Prentice Hall, 1976) follows:

> I married a widow (let's call her W) who had a grown-up daughter (call her D). My father (F), who visited us quite often, fell in love with my stepdaughter and married her. Hence, my father became my son-in-law and my stepdaughter became my

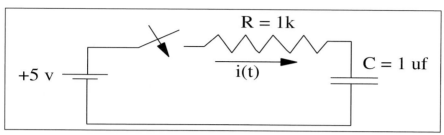

Figure 1.10 RC circuit.

mother. Some months later, my wife gave birth to a son (S1), who became the brother-in-law of my father, as well as my uncle. The wife of my father, that is, my stepdaughter, also had a son (S2). Question: Am I my own grandfather?

Draw a pictorial representation of the relationship by means of a directed graph in which the nodes are the persons' names and the arcs are the relationships between persons, e.g., "father of." Which representation is easier to understand? Can you answer the question?

1.7   Figure 1.11 is a logic diagram of a counter circuit from Morris Mano's *Computer System Architecture,* 2nd Edition (Englewood Cliffs: Prentice Hall, 1982). Outline a behavioral description in either C or C^{++} which describes the behavior. Which description gives a more readily understandable description of the circuit function?

1.8   Figure 1.12 shows a system interface between two devices: a sending device (Device 1) and a receiving device (Device 2). The interface between the two devices consists of a DATA line and three control signals READY, VALID, and ACCEPT. A communications protocol between two asynchronous devices functions as follows:

  a. Device 2 asserts READY.
  b. Device 1 detects the positive going change on READY, and places data on the DATA line and asserts VALID.
  c. Device 2 detects the positive going change on VALID, copies the data and resets READY, and asserts ACCEPT.

Draw a timing diagram for DATA, READY, VALID, and ACCEPT. Use arrows between signals to show how one signal transition triggers another.

1.9   Given the following algorithmic description:

```
for I=1 to 3 loop
  A(I) = B(I) + C(I)
  D(I) = E(I) * A(I)
end for;
```

Draw a data-flow graph in which the nodes are operations (+,*) and the arcs are inputs or computed values.

1.10 From your own experience, describe a design synthesis process that you are familiar with. Identify distinct synthesis steps, abstraction levels, and whether a representation is behavioral or structural. Draw a design track similar to Figure 1.6.

1.11 Design decomposition implies what kind of VHDL model?

1.12 Compare top-down and bottom-up design in two areas (a) cost and (b) whether the design is optimum or not.

1.13 Explain the difference between top-down and bottom-up design. What is the main advantage and disadvantage of each approach?

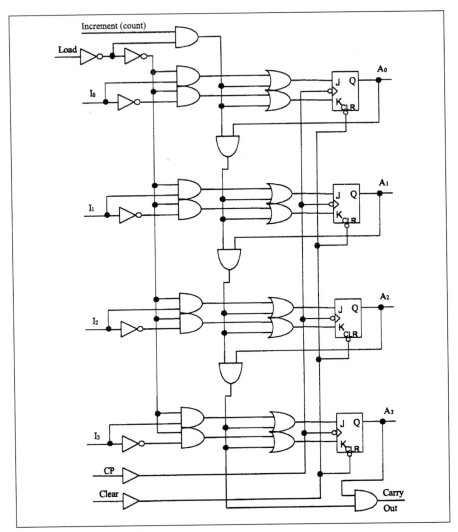

Figure 1.11   Counter circuit.

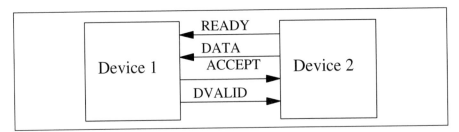

Figure 1.12   Interface protocol.

1.14 Characterize microprocessor system design and design with application-specific integrated circuits (ASICs) as being top-down or bottom-up. Explain your answer.

1.15 Explain how top-down and bottom-up design concepts can be used for software design.

1.16 Use the design of a 16-bit adder to illustrate trade-offs in the design space.

     a. Design a 16-bit carry ripple adder.

     b. Design a 16-bit carry look-ahead adder.

     c. Compare gate counts and delay for the two adder implementations.

     d. Develop a general expression for the gate count and delay for the two adder implementations that is a function of n, the input word size to the adder.

Design Tools

Ⅰn this chapter we describe the CAD tools used in the design process, where the word *tool* has a specialized meaning:

CAD tool.   A software program that assists in performing or automating a particular design function.

2.1 CAD TOOL TAXONOMY

The past 30 years a large number of CAD tools have been developed to aid in the design process. The terminology used by manufacturers to describe a tool's function is not consistent. In this text we adopt the *taxonomy* shown in Figure 2.1, which organizes tools into classes and subclasses. The figure also indicates at what levels in the abstraction hierarchy the tool is employed.

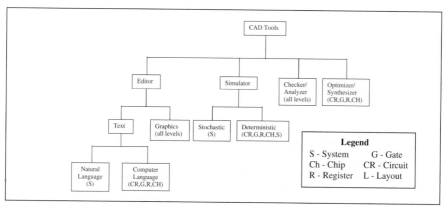

Figure 2.1   CAD tool taxonomy.

2.1.1 Editors

Editors are either textual or graphic. Textual editors can be used to edit the natural language of the specification at the system level, or to edit hardware description language text at the circuit through chip levels. SPICE is an example of an HDL that is appropriate for use at the circuit level. In this text, VHDL is used at the gate, register, and chip levels.

Graphics editors are used at all levels. At the silicon level they are used to create a pattern of geometric shapes that imply silicon treatment. At all the higher levels, the editors are used to create block diagrams and schematics. In fact, the process is sometimes referred to as *schematic capture*. The graphical schematic represents a structural model. It has two interpretations: (1) a *netlist*, which can be used to wire the interconnect between system components and (2) a *simulation model* that can be used to determine system response to inputs.

2.1.2 Simulators

Simulators are either *stochastic* or *deterministic*. Stochastic simulation is carried out at the system level to determine, for example, the percentage of time that a particular unit is busy. Deterministic simulation is carried out at all levels above the silicon level. The units employed, e.g., voltage, BITs, or tokens, depend on the level of abstraction. In Chapter 3, when we discuss data types, we explore this issue further.

2.1.3 Checkers and Analyzers

Checkers and analyzers are employed at all levels. At the silicon level, *design rule checkers* are used to insure that the layout implies a circuit that can be fabricated reliably. Rule checkers are used at other levels to determine if connection rules or fanout rules have been violated. *Timing analyzers* can be used to check for the longest path through a logic circuit or system. Wherever a hardware description language is employed at levels (CR,G,R,CH) (See Figure 2.1), analyzers can be used to check for errors that violate the structure and meaning of the language.

2.1.4 Optimizers and Synthesizers

Optimizers and Synthesizers change the form of the design representation to a new form, which is regarded as "improved" in some fashion. For example, at the gate level one can use a minimization program that yields a simpler Boolean expression. At the register level, optimizers are used to determine the best combination of control sequence and data paths. And, as pointed out in the previous chapter, various forms of synthesis programs are employed to move the representation to a lower level in the hierarchy and thus, closer to fabrication.

2.1.5 Cad Systems

The emphasis in this section has been on classifying CAD tools. In most cases *CAD Systems* are developed by commercial vendors which combine a number of tools. For example, one might combine a text editor, schematic editor, rule checker, analyzer, and simulator in one system. We now describe the more important tools in detail.

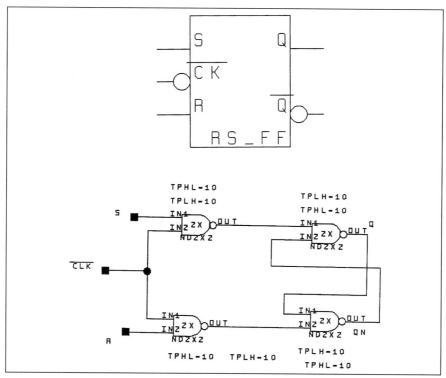

Figure 2.2 RS flip-flop.

2.2 SCHEMATIC EDITORS

At various points in the design process, schematics are created. This is done with a schematic editor, which we define as:

Schematic editor. An editor which can be used to create and display an interconnected set of graphic tokens.

The graphic tokens correspond to structural primitives, and thus, the interconnected net corresponds to a structural model. In fact, the correspondence is frequently strong, i.e., the creation of the graphical network of tokens also creates a simulation model.

A typical schematic capture system has the following features:

1. A library of primitive symbols, including a simulation model corresponding to each primitive. The library primitives are either *native* or correspond to *standard parts* families (e.g., TTL, CMOS, ECL, etc.) ASIC library components, or synthesis tool macros. Native primitives are the most fundamental logic elements, such as ANDs and ORs, and are frequently built into the simulator. The standard parts families are expressed in terms of the native primitives. Figure 2.2 shows the symbol for a clocked RS flip-flop and its corresponding representation in native primitives in the WORKVIEW system

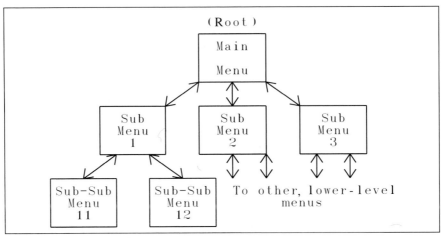

Figure 2.3   Example of a menu tree.

developed by VIEWLOGIC, Inc. Most schematic capture systems allow the user to add new symbols and new simulation models to existing libraries. In the context of this book, simulation models are created by writing VHDL source code for the model.

2. A system of graphic windows that can be used to create an interconnect of graphic tokens. The window system is usually organized as a tree (See Figure 2.3.) One starts at the main menu (the root of the tree) and drops down into windows at lower levels of the tree as more detailed activities are carried out. Figure 2.4(a) shows a menu sequence (one path in the tree) for creating a component in the VIEWLOGIC WORKVIEW System. The menu token sequence is CREATE, COMPONENT. One is then prompted to type in a component name. In response to this, the system places a graphic token corresponding to the library component at the selected spot on the screen. Figure 2.4(b) shows the result of creating two components, a 3-input NAND gate and a D flip-flop. To form the interconnect between components, one would use the menu token sequence CREATE, NET, and then "drag a line" from a source pin to a destination pin. Figure 2.4(c) shows the interconnect net between the NAND gate output and the D input of the flip-flop. Connections to external inputs, which were made by the same method, are also shown. Signals have also been labeled by using the menu token sequence CREATE, LABEL. The menus also allow panning to make a designated point the center of the image, and zooming to change the size of the displayed image. Figure 2.4(d) illustrates a pan followed by a zoom in the NAND gate-D flip-flop net. This would be done with the menu sequence VIEW,IN. The menus also allow moving and copying of graphic elements.

3. Commands for creating wirelists. The wirelist has two applications. First, it can be used to physically build the circuit, but it also implies a structural model that can be simulated to test circuit response. Thus, the use of schematic capture is frequently followed by an invocation of the simulator.

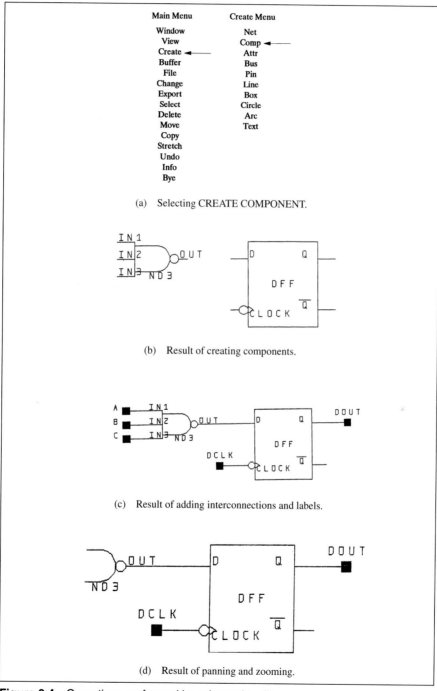

Main Menu   Create Menu

Window   Net
View   Comp
Create   Attr
Buffer   Bus
File   Pin
Change   Line
Export   Box
Select   Circle
Delete   Arc
Move   Text
Copy
Stretch
Undo
Info
Bye

(a)   Selecting CREATE COMPONENT.

(b)   Result of creating components.

(c)   Result of adding interconnections and labels.

(d)   Result of panning and zooming.

Figure 2.4   Operations performed by schematic editors.

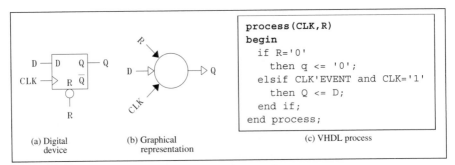

Figure 2.5   Correspondence between devices and processes.

2.3 SIMULATORS

The simulator is a major tool used in the development of digital systems.

Simulator.   A program which models the response of a system to input stimuli.

In our case the system to be modeled will be an interconnected net of digital logic elements. Our modeling approach will map the function of a digital logic element onto one or more processes using the following definition:

Process.   A computational entity that models the function and delay of the digital device.

Figure 2.5 gives an example of a digital device, graphical representation of the corresponding process, and corresponding VHDL source code for the process.

A network of digital devices is modeled by its corresponding network of processes. An example of this is illustrated in Figure 2.6. The correspondence between devices and processes is as follows:

Devices	Processes
D1	P1
D2	P2,P3
D3,D4	P4

This illustrates that (1) sometimes a single device maps to a single process, (2) sometimes a single device maps to multiple processes, and (3) sometimes a set of devices maps to a single process. A wire between devices in the digital device network is modeled by a *signal* in the process network. The wire to signal mapping is as follows:

Wire	Signal
A	SA
B	SB
C	SC
D	SD
E	SE
F	SF
G	SG
H	SH
I, J	*
*	SI, SJ

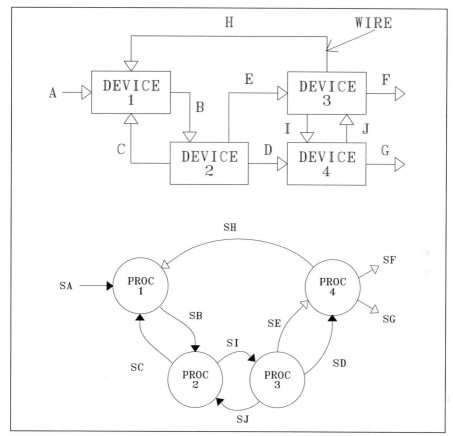

Figure 2.6 Modeling a network of devices by a network of processes.

Wires A through H have corresponding signals SA through SH. Wires I and J have no corresponding signals because devices 3 and 4 are represented by single process 4. Signals SI and SJ have no corresponding wires since they result from the fact that device 2 is represented by processes 2 and 3.

Just as a device computes new outputs when one of its inputs changes, a process is activated to compute new outputs when one of its signal inputs changes. Signals that cause a process to execute are *triggering signals*. They are indicated on the graphical notation for a process with a solid arrow head. Figure 2.5(b) shows that CLK and R are triggering signals. Some signals do not cause process execution but are merely sampled by the process. These are represented graphically by using an open-headed arrow. Figure 2.5(b) shows that D is a *sampled signal*.

In our approach to simulation, a logic circuit is modeled as a network of processes. The simulator operates on the network of processes as shown in Figure 2.7. A process responds to *signal transactions* on its inputs. These transactions are kept in the simulator time queue. Time queue entries are represented as a two-tuple of the form (SN,V), where SN is a signal name and V is the value the signal is scheduled to assume at the scheduled time. Each time queue entry is

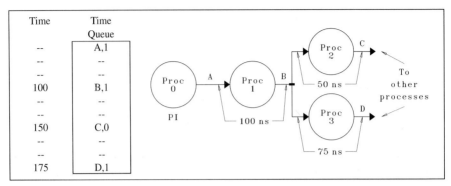

Figure 2.7 Simulator operation.

called a signal transaction. If the entry represents a change in signal value, it is called a *signal event*.

The operation of our example simulation proceeds as follows: The processes are numbered 0 through 3. Suppose that process 0 represents a primary input (PI). The initial entry in the time queue (A,1) specifies that output signal A of process 0 should take on the logic value 1 at time 0. When this occurs, a signal tracer program determines that signal A is tied to an input of process 1. Assume that this value on signal A is a new one (an event has occurred), and that this is a triggering input. Process 1 will be invoked. Assume that it computes a new value 1 for output signal B and that due to internal propagation delay output, signal B will change to a logic 1 after a delay of 100 nanoseconds (ns) from the present simulation time. A new entry is then inserted in the time queue. After all input changes at time 0 are processed, the simulation time is advanced to the time of the next entry in the queue. The queue processing continues until time 100 is reached. Here the queue entry (B,1) is processed: Signal B is set equal to the logic value 1. Again, assume this is an event. The signal tracer program then determines that this event on signal B should trigger processes 2 and 3 to execute. The execution of process 2 results in a new scheduled value for signal C, 50 ns later. The execution of process 3 produces a new scheduled value on signal D, 75 ns later. Signals C and D are presumed to drive other processes in the network, which are not shown. Simulation continues until the time queue is empty or some externally controlled time limit is reached.

2.3.1 Simulation Cycle

Every time simulation time advances, a simulation cycle occurs, which we now define more formally. The execution of a model consists of an initialization phase followed by the repetitive execution of processes in the process network. Each repetition is said to be a *simulation cycle*. In each cycle, it is determined which signals have events on them. If an event occurs on a given signal, the processes that are triggered by that signal will resume and will be executed as part of the simulation cycle. Thus, the simulation cycle consists of these steps:

1. Advance the simulation time to the time of the next entry in the queue. If there are no new time queue entries, stop.
2. For all signals that have events at that time, activate the processes triggered by them.

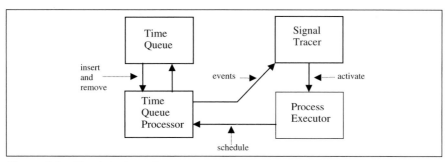

Figure 2.8 Simulator kernel.

3. Execute the activated processes and (possibly) schedule new time queue entries. Go to step 1.

2.3.2 Simulator Organization

Figure 2.8 shows the organization of the kernel of a simulator, which has four basic parts:

1. Time Queue
2. Time Queue Processor
3. Signal Tracer
4. Process Executor

As implied in the previous example, the Time Queue Processor handles the insertion and removal of entries from the Time Queue, and passes signal events to the Signal Tracer. The Signal Tracer determines the fanout of a signal and marks processes for execution. The Process Executor controls the execution of the marked processes and schedules new time queue entries by passing them to the Time Queue Processor.

2.3.3 Language Scheduling Mechanism

The preceding discussion of simulator operation illustrates the need for a scheduling mechanism in the language used to create the device model. In this book we are using VHDL, the VHSIC hardware description language.

In VHDL one might write:

```
X <= Y after 100 ns;
```

and the effect would be to have X take on the value of Y at a simulation time 100 ns from the present time. Thus, a future value for X has been scheduled.

2.3.4 Simulation Efficiency

When simulating models for large systems, simulation efficiency is very important. For logic simulation, *simulation efficiency (E)* is defined as:

$$E = \frac{real\ logic\ time}{host\ CPU\ time}$$

Real logic time is the actual time required to complete an activity sequence in a real logic circuit. *Host CPU time* is the time required to simulate the same activity sequence using a logic simulator running on a host CPU. To make the figures meaningful, the host CPU time must be normalized against the speed of the host. For example, were the simulations run on a 100 MIP machine or a 500 MIP machine?

Another possible measure of efficiency is the number of events that a simulator can process per unit time. For example, gate-level simulators are frequently rated on the number of gate evaluations per host CPU second. For higher-level modeling, the definition of an event is defined in terms of a higher-level activity. For example, when modeling microprocessor systems, one might measure simulation efficiency in terms of the number of microprocessor clock pulses simulated per host CPU second. An even higher-level measure is used when modeling the execution of one CPU on another, that is, the number of (modeled) machine instructions simulated per host CPU second.

Simulation efficiency is determined by programming technique, the computer architecture employed, and the modeling level. By *programming technique*, we mean whether the simulator is compiled or interpreted. In the interpreted approach, the system interconnect is stored in tables, and the models of the individual devices are called at run time as dictated by the interdevice signal flow. In this case the simulation model is "interpreted" by the simulator at run time. In compiled simulators, the entire model is compiled into machine code before simulation, and thus no run-time interpretation is necessary. Compiled simulators are faster than interpreted, but they suffer from portability problems if they are compiled into host machine assembly language. However, the VHDL simulators discussed in this book compile the simulation model into C, solving the portability problem.

The previous discussion of simulators implicitly assumed the use of a uniprocessor host to run the simulator program. Recently, however, simulation engines have been developed which employ parallel architectures to speed up gate-level simulations. For example, IBM's Yorktown Simulation Engine (YSE) and Zycad's simulation accelerator are capable of speeding up gate-level simulation by a factor of 1000.

Therefore, programming technique and host machine architecture can significantly affect simulation efficiency. However, another fundamental property affecting simulation efficiency is the level in the representation hierarchy at which the simulation is performed. For example, circuit-level simulations employing SPICE, although yielding accurate results, are very inefficient. That is, only several hundred circuit nodes can realistically be simulated in a given circuit simulation run. Gate-level simulations can also be very inefficient. If accurate timing is modeled, efficiencies as low as 10^{-7} have been experienced. Unit-delay or zero-delay simulations at the gate level can achieve better efficiency than this.

Simulations at the chip level can achieve simulation efficiencies that are 100 to 10,000 times better than those at the gate level. Modelers should always choose the highest possible level in the hierarchy, so they can perform their simulations and still achieve the desired accuracy.

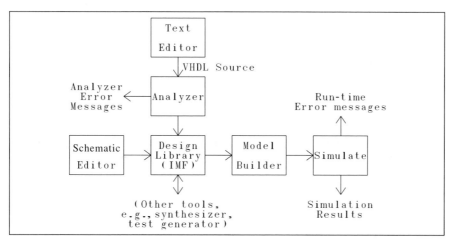

Figure 2.9 Simulation system.

2.4 THE SIMULATION SYSTEM

A simulator program is only part of a *simulation system*. An example of a complete simulation system is shown in Figure 2.9. The source code for VHDL models is prepared in a text editor. The source code is then submitted to the *Analyzer*. The Analyzer checks the source for the following types of errors:

Lexical. The source is checked to see that the correct character set is used.

Syntactic. The source is checked to see that the grammar of the language has been observed. For example, VHDL language syntax specifies that every "if statement" be terminated with an "end if." Not doing this will result in a syntax error.

Semantic. Any language attaches specialized meaning to the constructs used in the language. This specialized meaning is termed the *language semantics*.

As an example, consider the following VHDL code:

```
variable X: BIT;        -- X and Y have
variable Y: INTEGER;    -- different types
---------------
---------------
X:=Y;                   --correct syntax, but incorrect semantics
---------------
---------------
```

Because X and Y are different types, trying to assign the value of Y to X violates the semantics of the language, i.e., the meaning of having different types—in this case types BIT and INTEGER. The syntax however is correct. The semantic checks run by the analyzer are *static*, i.e., independent of simulation results.

After the model code has successfully passed analysis, it is stored in the *design library* for use in subsequent simulations and other design activities. The code is stored in the design library in an *intermediate form* (IMF). The IMF is a data structure which allows for efficient storage and access of the code.

To prepare for an actual simulation, the *Model Builder* links models together to form a system simulation module. This linking process is analogous to the subroutine linking that is carried out in software systems. After the model is totally prepared for simulation, it is said to be *elaborated*. Simulation is then carried out by the *simulator*. It is possible that run-time errors can occur. These are termed *dynamic* semantic errors. An example of this type of error is given in the section on Model Debugging.

Figure 2.9 shows two other interfaces to the Design Library. First, a schematic editor can produce the IMF for VHDL code, which is guaranteed by construction to be lexically, syntactically, and semantically correct. Thus, it can be inserted directly into the Design Library. Second, other tools, such as synthesizers and test generation tools, can also access the Design Library. It is quite possible that the simulator, schematic editor, synthesizer, and test generation tool will be developed by different vendors. Therefore, it is important that the intermediate form (IMF) be standardized.

2.5 SIMULATION AIDS

For those involved with simulation on a day-to-day basis, the preparation of models and input vectors, the debugging of simulation models, and the interpretation of simulation results can be a very tedious process. As a result, simulation companies have developed simulation aids to make these jobs easier.

2.5.1 Model Preparation

The schematic editor discussed previously is a great aid to model preparation. It creates a structural model of a digital system. Figure 2.4 showed an example of this. However, a structural model is made up of an interconnection of primitives for which behavioral models need to be defined. The creation of behavioral models can be eased by the use of an interactive program called the Modeler's Assistant. This is an interactive system that enables a designer to specify the behavior of a model using a graphical notation. The system produces VHDL source code describing the models. The graphical notation is a process network graph like the one described in the section on simulators (see Figure 2.6). In this application it is referred to as a process model graph (PMG) because of the model it implies. The program uses the PMG as the graphical input and produces the VHDL source code interactively.

Specifically, the system operates as follows:

1. The process model graph is entered graphically, defining the shell of the behavioral model.
2. Constituent processes are selected primarily from a library of process primitives which can be specialized for a particular application.
3. Some textual input is required, e.g., to specify delays and data types.
4. The result is the complete VHDL source description of the model.

The code can be viewed as it develops. The tool allows the user to use predefined processes or create new processes and place them in the library.

The coding of behavioral models for complex devices is a labor-intensive task. With the benefit of a graphical representation, a menu-driven system and reusable code, the Modeler's Assistant can improve the productivity of VHDL modelers. It also results in behavioral models that have a well-defined structure. Chapter 10 gives a more detailed discussion of this tool.

2.5.2   Model Test Vector Development

To test models it is necessary to develop input test vectors for them. The following VHDL code was developed to validate the consecutive 1's and 0's detector described in Chapter 1. Signals CLK, X, and R are the three inputs to the model.

```
CLK <='0','1'after 10 ns,'0'after 20 ns,
       '1'after 30 ns,'0'after 40ns,
       '1'after 50ns,'0'after 60 ns,
       '1'after 70 ns,'0'after 80 ns,
       '1'after 90 ns,'0'after 100 ns,
       '1'after 110 ns,'0'after 120 ns,
       '1'after 130 ns,'0'after 140 ns;
X <= '0','1'after 15 ns,'0' after 55 ns;
R <= '1','0' after 125 ns,'1' after 127 ns;
```

To generate this code manually is a tedious and error-prone process. What is required is a Timing Diagram Processor that translates the graphical description of model inputs to the equivalent VHDL statements. Such a capability is described in the Chapter 4 section on test bench development.

2.5.3   Model Debugging

As with the development of software programs, debugging aids are very important in model development. Debuggers can detect errors during *analysis* and *simulation*.

2.5.3.1   *Textual Debugging*

Consider the following VHDL code. Note that signal X is of type INT_RANGE, which consists of the integer range 1 to 4, yet the value 5 is assigned to X.

```
entity NUMS is
end NUMS;

architecture RINT of NUMS is
   type INT_RANGE is range 1 to 4;
signal X: INT_RANGE;
begin
   X <= 5;
end RINT;
```

Analysis of this file on the Synopsys Simulator causes the following error message to print at the terminal:

```
X <= 5;
    *
Warning: vhdlan,2 nums3.vhd(7) : Constraint error.
```

The message indicates that a "constraint error" has been detected in line 7 of the source code file. Also, if the user selects the listing option, the following file will be produced:

```
-- Synopsys 1076 VHDL Analyzer Version 2.1c
--
--              Copyright (c) 1990 by Synopsys, Inc.
--              ALL RIGHTS RESERVED
-- This program is proprietary and confidential information
-- of  Synopsys, Inc. and may be used and disclosed only as
-- authorized in a  license agreement  controlling such use
-- and disclosure.
--

--
-- Source File:  nums3.vhd
-- Thu Jan 30 12:55:17 1992

    1 entity NUMS is
    2 end NUMS;
    3 architecture RINT of NUMS is
    4   type INT_RANGE is range 1 to 4;
    5   signal X: INT_RANGE;
    6 begin
    7   X <= 5;
        *
Warning: vhdlan,2 nums3.vhd(7) :
    Constraint error.
    8 end RINT;
    9
-- "nums3.vhd": errors: 0; warnings: 1.
```

For an example of debugging a run-time error during simulation, consider this example:

```
entity NUMS is
end NUMS;

architecture RINT of NUMS is
   type INT_RANGE is range 1 to 4;
   signal X: INT_RANGE:= 2;
   signal Y: INT_RANGE:= 3;
   signal Z: INT_RANGE;
begin
   Z <= X + Y;
end RINT;
```

When this code is simulated, the value 5 will be assigned to Z, which is again out of range. This error will not be detected during analysis because the analyzer doesn't know what the result of the computation will be. However, during simulation, the Synopsys Simulator detects the error and prints the following message.

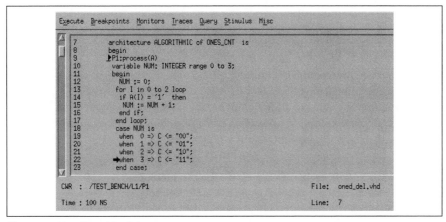

Figure 2.10   Graphic debugger.

	100	0	10	20	30	40	50	60	70	80	90	100
► /TEST_BENCH/A(2:0)	7		0	1	2	3	4	5	6		7	
► /TEST_BENCH/C(1:0)	3		0	1	1	2	1		2		3	

Figure 2.11   Waveform display.

```
**Error: vhdlsim,2:
     Constraint error.
          nums2.vhd(9)  :
```

The message indicates that a constraint error was detected during simulation of file nums2.vhd at line 9, executing the statement: $Z <= X + Y;$.

2.5.3.2   Graphic Debuggers

Graphic debuggers are also very useful. They allow single stepping and the setting of break points to follow the flow of execution. Figure 2.10 shows a window from the Synopsys debugger executing the ALGORITMIC ones count model. This window is used in conjunction with waveform display to validate the model.

2.5.4   Results Interpretation

An extremely important feature of simulation systems is the mechanism for interpretation of results. Figure 2.11 gives a timing diagram result for the simulation of the consecutive 1's and 0's detector on the Synopsys Simulator. Timing diagrams are best for showing the timing relationships for key system signals. However, for complicated models and for detailed analysis of results, textual output displays are better. Figure 2.12 shows a textual display from the Synopsys Simulator which displays the same simulation results as the timing diagram in Figure 2.11. Note: The textual display shows the details of output logic hazards, while the waveform display in the full view mode represents them as double bars.

```
 /
     1 NS
        SMON1:     ACTIVE /TEST_BENCH/C (value = X"0")
    10 NS
         SMON:     ACTIVE /TEST_BENCH/A (value = X"0")
    20 NS
         SMON:     ACTIVE /TEST_BENCH/A (value = X"1")
    22 NS
        SMON1:     ACTIVE /TEST_BENCH/C (value = X"1")
    30 NS
         SMON:     ACTIVE /TEST_BENCH/A (value = X"2")
    32 NS
        SMON1:     ACTIVE /TEST_BENCH/C (value = X"0")
    33 NS
        SMON1:     ACTIVE /TEST_BENCH/C (value = X"1")
    40 NS
         SMON:     ACTIVE /TEST_BENCH/A (value = X"3")
    42 NS
        SMON1:     ACTIVE /TEST_BENCH/C (value = X"3")
    43 NS
        SMON1:     ACTIVE /TEST_BENCH/C (value = X"2")
    50 NS
         SMON:     ACTIVE /TEST_BENCH/A (value = X"4")
    52 NS
        SMON1:     ACTIVE /TEST_BENCH/C (value = X"0")
    53 NS
        SMON1:     ACTIVE /TEST_BENCH/C (value = X"1")
    60 NS
         SMON:     ACTIVE /TEST_BENCH/A (value = X"5")
    62 NS
        SMON1:     ACTIVE /TEST_BENCH/C (value = X"3")
    63 NS
        SMON1:     ACTIVE /TEST_BENCH/C (value = X"2")
    70 NS
         SMON:     ACTIVE /TEST_BENCH/A (value = X"6")
    72 NS
        SMON1:     ACTIVE /TEST_BENCH/C (value = X"2")
    80 NS
         SMON:     ACTIVE /TEST_BENCH/A (value = X"7")
    82 NS
        SMON1:     ACTIVE /TEST_BENCH/C (value = X"3")
```

Figure 2.12   Textual results display.

2.6 APPLICATIONS OF SIMULATION

Simulation is employed in two major application areas: *system validation* and *fault simulation*. In system validation, one wants to demonstrate that the system conforms to a specification. However, at different points in the design process, the system is represented by a model. Based on the design cycle described in Chapter 1, a system can have models at the chip, register, gate, and circuit levels. One uses simulation to validate that these models conform to the specification.

As one moves down in the hierarchy, the model has a closer correspondence to the physical design, and its validation will, therefore, approximate the validation of the system itself. In the past, hardware prototypes were built to complete the validation process, i.e., simulation was not trusted as a complete method of system validation. However, today's simulation tools allow accurate modeling of large systems. Thus, it is possible to insure correct designs using simulation, without the need for building a hardware prototype. Very complicated VLSI devices can be fabricated correctly the first time because the design can be completely checked through simulation. Of course, complete checking of a design through simulation can be very time-consuming and expensive, i.e., simulation efficiency is very important. Thus, the development of simulation engines that employ parallel processing has been critical. Without these simulation engines, complete simulation of complicated VLSI systems on conventional processors can literally require months (and even years) and thus is not practical.

There are a number of issues related to validation through simulation. The first question is, how much should one simulate? A second question is, where does one get the system input vectors to drive the simulation? A simple answer to the first question is that one should use all possible input combinations, and simulate the system exhaustively. However, for complicated systems, particularly those involving sequential circuits, this is not practical. Thus, the designer must select a subset of test cases to simulate.

Test generation programs can be used to develop the test vectors for simulation at the gate level. For higher level models, the designer or model developer frequently creates the tests manually.

Another problem is that as one descends in the hierarchy, the models become increasingly detailed. An issue then becomes how to compare the outputs of models at different levels of the hierarchy. A related problem is the *back annotation problem*, i.e., how to reflect actual layout information back into simulation models. In Chapter 7 we give an example of this.

The other application of simulation is fault simulation. Here, faults are injected into the system model, and the system is simulated to observe the response. This can be done to *validate* tests, i.e., to establish that a test detects a fault. Fault simulation can also be used to create *fault dictionaries*, which relate output signal states (symptoms) to faults, therefore allowing one to diagnose which fault occurred.

2.7 SYNTHESIS TOOLS

A synthesis tool is used either to automate a design step or to provide assistance during the performance of a design step. In Chapter 1 we introduced a hierarchy that defined representations of designs at different levels of abstraction in both behavioral and structural domains. The design process was then defined as a series of transformations starting with a representation at a high level of abstraction and terminating with a low-level representation that can be fabricated. Such a model of the design process leads to the following definition:

Synthesizer. A computer program that automatically performs a translation from one design representation to another or a program that assists a human in making the translation.

In Chapters 9 and 10 we show how to synthesis ASICs with the Synopsys Design Compiler and FPGAs with Xilinx software. These tools work primarily at the register level of abstraction.

Automation at higher levels of abstraction is still the subject of current research, although *high-level synthesizers* are beginning to be used.

A synthesizer can translate a high-level representation to a lower-level representation or can translate from the behavioral domain to the structural domain, or both. The translation task performed by a synthesizer is analogous to the task performed by a language compiler. For example, a C-compiler translates a high-level language representation of an algorithm (a C language program) into a machine-level language representation that can be run on a particular machine. In a similar manner, a *high-level synthesizer* translates a representation of a circuit at a high level of abstraction (perhaps the algorithmic level) into a lower-level representation (perhaps the gate level) that can be implemented within a particular technology (say, CMOS). In fact the analogy is appropriate enough that the term *silicon compiler* is given to a synthesis tool that translates a medium-level representation (often the gate level) into a silicon layout (silicon level in structural domain) that can be implemented directly in some specific technology.

The advantages and disadvantages that accrue to the designer when he designs at a high level of abstraction and uses a synthesis tool to perform a translation to a lower level instead of designing directly at the lower level are exactly the same as those that accrue to a programmer who writes a program in a high-level language and uses a language compiler to translate the program to machine language instead of writing the program directly in assembly language. The advantage in both cases are that the designer or the programmer can concentrate on system issues and is removed from the additional detail needed to design or program directly at low levels of abstraction. Design, or programming, at high levels of abstraction requires less time and results in fewer errors because of the removal of all detail needed to design or program at lower levels. However, because the language compiler or the synthesizer must allow for the most general case in all translation situations, it might not be able to take advantage of optimizations that are possible in the current situation. Therefore, compiled programs and synthesized designs, are usually not optimum. Optimizing compilers, or optimizing synthesizers, can sometimes address specific optimizing opportunities but can never handle every possible situation. In general, programs written directly in assembly language by experienced programmers, and designs done at the lowest possible level by experienced designers, will perform better and cost less than those done at higher levels and translated by computer programs.

Figure 2.13 shows a VHDL description at the register transfer level of a circuit that produces a '1' output when register R contains an 8-bit binary number with an odd number of 1's. The VHDL description uses a FOR loop to compute the function. A FOR loop can be directly translated into an iterative network of XOR gates as shown in Figure 2.14(a). A basic synthesizer would be required to make the translation to an iterative network because such a translation would work for any VHDL FOR loop. However, just as some language compilers can look for certain kinds of optimizations, a hardware synthesizer can also look for certain kinds of optimizations. The process of optimization is usually performed after a general translation has been made. In this example, a synthesizer might be able to determine that a tree of XOR gates as shown in Figure 2.14(b) would operate faster and perform the same function. It might also know that the first XOR gate can be eliminated by using a Boolean algebra theorem (X xor 0 = X). Although this alternative works fine for XOR gates, not all iterative networks can be simplified in this way. The optimizing synthesizer must have sufficient knowledge to know when this is possible. Similarly, if 3-input XOR gates are available, the circuit shown in Figure 2.14(c) could

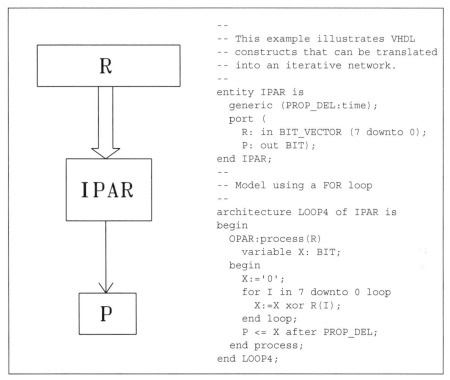

```
                              --
                              -- This example illustrates VHDL
                              -- constructs that can be translated
   ┌─────────────────┐        -- into an iterative network.
   │                 │        --
   │        R        │        entity IPAR is
   │                 │          generic (PROP_DEL:time);
   └─────────────────┘          port (
            │                      R: in BIT_VECTOR (7 downto 0);
            │                      P: out BIT);
            ▼                    end IPAR;
   ┌─────────────────┐          --
   │                 │          -- Model using a FOR loop
   │      IPAR       │          --
   │                 │          architecture LOOP4 of IPAR is
   └─────────────────┘          begin
            │                      OPAR:process(R)
            │                        variable X: BIT;
            ▼                        begin
   ┌─────────────┐                   X:='0';
   │             │                   for I in 7 downto 0 loop
   │      P      │                     X:=X xor R(I);
   │             │                   end loop;
   └─────────────┘                   P <= X after PROP_DEL;
                                    end process;
                                  end LOOP4;
```

Figure 2.13 VHDL description of ODD PARITY component: Behavioral domain description of a circuit that determines whether register R contains an 8-bit binary number with an odd number of 1's.

be designed with both fewer gates and smaller delays. This illustrates that optimization often needs to be an interactive process in which the user can specify certain parameters, such as availability of gate types, maximum number of gate levels, etc.

This discussion illustrates that translation may be a two-step process consisting of a general translation followed by an optimizing step. It is obvious that the synthesizer could never know all possible alternatives for design, and therefore, automated design will usually not be optimum. In many design situations lack of optimality is not as important as low design cost and high reliability of the design process. For this reason, many companies employ some kind of automated design methodology often requiring human interaction for best results.

In general, more highly-trained personnel are required to write programs in assembly language or to do designs at lower levels of detail, which adds to the cost of design. Frequently, the best approach is dictated by expected volume. A low-volume product should be designed at a high level and automatically translated to avoid the high costs of design. A high-volume product should be designed at a low level because the higher cost of design can be spread over the expected volume and the cost savings (or faster circuit) obtained in production will more than offset the additional design cost. Choice of design level will also affect the time required to get the product to market. Designs done at a higher level and translated by computer program will

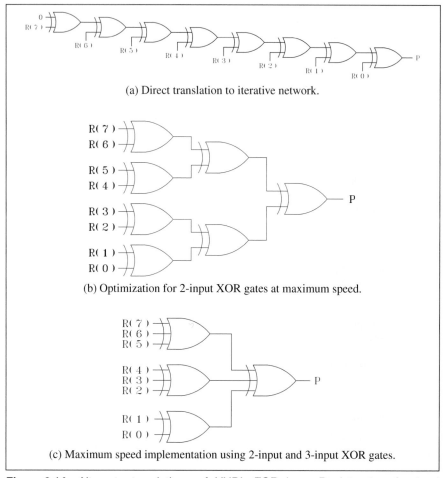

(a) Direct translation to iterative network.

(b) Optimization for 2-input XOR gates at maximum speed.

(c) Maximum speed implementation using 2-input and 3-input XOR gates.

Figure 2.14 Alternate translations of VHDL FOR loop: Register transfer level (behavioral domain) to gate level (structural domain).

take less time to complete. Designs done at lower levels will take more time to complete, but may cost more to produce.

PROBLEMS

2.1   Identify specific tools that match the CAD tool taxonomy in Figure 2.1. From your own experience, through literature searches, and vendor documentation, identify specific tools that fit each slot in the CAD tool taxonomy.

2.2   Use your favorite graphics editor to create a drawing of a gate-level description of a full adder. For what purpose can you use the drawing? Compare this with the drawing obtained in a digital design capture system.

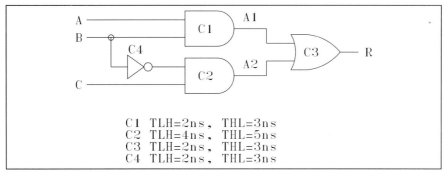

Figure 2.15 Circuit with timing.

2.3 The following VHDL description contains errors. Using VHDL software, analyze and simulate the description. Classify errors as being lexical, syntactic, or semantic. Which errors are discovered during analysis and simulation? Do you get the same result on different VHDL systems?

```
entity ERROR is
  port(X: in  BIT; C: in BIT; Z1: out INTEGER; Z2: out BIT);
end ERROR;
architecture HAS of ERROR is
  signal S^D: BIT;
begin
  Z1 <= X;
  process(X)
  begin
    if C= '1' then Z2 <= X;
  end process;
end HAS;
```

2.4 What is the difference between a transaction and an event?

2.5 In the Synopsys Simulator, one can set up a control file to monitor all signals that are "active." What does active mean?

2.6 Consider the logic circuit shown in Figure 2.15. For each gate in the circuit, its output rise (TLH) and output fall delay (THL) is given, i.e., the delay through the circuit is different depending on whether the output is rising or falling. In Chapter 7, we discuss this in more detail. Assume that the inputs to the circuit (A,B,C) are initially all equal to '1' and that B switches to '0' at t = 5 ns. Draw a time queue which shows how the signals A, B, C, $\overline{B}$, A1, A2, and R are updated. Also, draw a timing diagram.

2.7 Figure 2.16 is the same circuit as the previous problem but with different time delays. In this case the initial state of the inputs is A = B = '0', C = '1'. Shown below are a series of time queue entries for the circuit. Draw a timing diagram that shows all transactions and events.

2.8 Study the operation of a simple logic simulator. For example, consult *Digital Logic Testing and Simulation* by Alexander Miczo (New York: Harper & Row, 1986, pp. 174-177). This simulator is written in BASIC. Study the listing. Run the simulator on your personal computer and observe its operation. Simulate some simple gate circuits with the simulator. Give a word statement of the simulation cycle for this simulator.

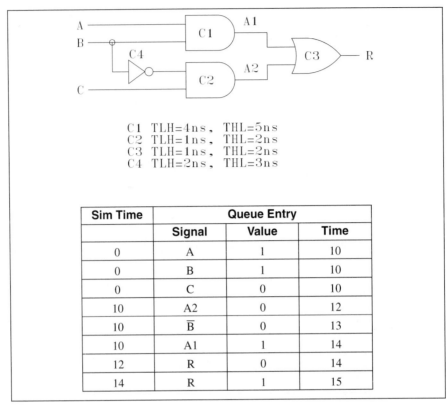

C1 TLH=4ns, THL=5ns
C2 TLH=1ns, THL=2ns
C3 TLH=1ns, THL=2ns
C4 TLH=2ns, THL=3ns

Sim Time	Queue Entry		
	Signal	Value	Time
0	A	1	10
0	B	1	10
0	C	0	10
10	A2	0	12
10	$\overline{B}$	0	13
10	A1	1	14
12	R	0	14
14	R	1	15

Figure 2.16   Time queue problem.

2.9   Analyze and simulate the following VHDL oscillator model and observe its response using both tabular and waveform output.

```
entity OSC is
end OSC;
architecture WAIT_DEL of OSC is
  signal CLOCK: BIT := '0';
begin
  process
    variable NUM: INTEGER := 0;
  begin
    while NUM < 10 loop
      CLOCK <= '1';
      wait for 75 ns;
      CLOCK <= '0';
      wait for 25 ns;
      NUM := NUM + 1;
    end loop;
    wait;
  end process;
end WAIT_DEL;
```

2.10 This problem has two parts:
    a. Perform the following parts of the WORKVIEW DIGITAL TUTORIALS. These tutorials will acquaint you with the operation of the WORKVIEW schematic capture and simulation system. (1) Design Capture Tutorial and (2) Simulation Tutorial.
    b. Prepare a model of the positive edge triggered D flip-flop shown in Figure 5-15 (page 360) of *Digital Design Principles and Practices* by John F. Wakerly (Englewood Cliffs: Prentice Hall, 1990). Note that the two D symbols in that diagram are D latches shown in Figure 5-12 (page 359). Simulate the model. Select an appropriate input sequence that will fully test the model. The selection of an appropriate input sequence is necessary for full credit.

An informal report is required for this problem. At a minimum, the report must include the following items for both the the tutorial and the D flip-flop:
    a. A printout of the schematic created by WORKVIEW, including any subschematics.
    b. Simulation printouts. 1) Using tabular output and 2) Using waveform output.

2.11 Demonstrate the difference between deterministic and stochastic phenomena by testing a 32-bit adder using both approaches.
    a. Using WORKVIEW or similar design software, develop a gate-level schematic of a 16-bit, ripple carry adder.
    b. Write a high-level language program, e.g., C or C^{++}, which generates all possible input patterns. Simulate the adder using these patterns in an input data file. Record the wall clock time for the simulation.
    c. Using a random number routine, write a high-level language program that generates 100 random input patterns for the adder and places them in a data file. Simulate the adder using these patterns and record wall clock time.
    d. Examine the 100 random patterns. Are there some additional patterns you would add to more thoroughly exercise the adder?
    e. Comment on the relative wall clock times in the two approaches vs. how thoroughly the adder was exercised.

2.12 In this chapter we defined simulation efficiency as the ratio of real logic time and host CPU time. Many operating systems have mechanisms for measuring elapsed CPU time. Using such a system, measure the simulation efficiency of simulations:
    a. A circuit-level model of a full adder using SPICE.
    b. A gate-level model of a full adder using a gate-level simulator.
    c. A behavioral model of a full adder. This can be written in a hardware description language, such as VHDL, or a high-level language, such as PASCAL or C. Plot simulation efficiency as a function of abstraction level.

2.13 Simulation can be accelerated through the use of parallel processing. Research the literature and prepare a report on this approach.

2.14 In the VLSI design process, it is desirable to have chips that work the first time. In light of this requirement, discuss how simulation efficiency controls the complexity of chips that can be built.

2.15 Timing analyzers are used to identify the longest delay paths in a circuit. Research timing analyzers in the literature. Write a report which compares timing analyzers to simulation as a means for discovering timing problems.

Basic Features of VHDL

This chapter describes the basic features of the VHDL language. An example first demonstrates the general nature of VHDL. This is followed by detailed discussion that describes the syntax and semantics of individual language statements.

In the VLSI era, a structured design process is required. In response to this need, considerable effort is being expended in the development of design aids. Hardware description languages are a specific example of this, and a great deal of effort is being expended in their development. Actually, the use of these languages is not new. Languages, such as CDL, ISP, and AHPL, have been used since the 1970s. However, their primary application has been the verification of a design's architecture. They do not have the capability to model designs with a high degree of accuracy. That is, their timing model is not precise, and/or their language constructs imply a certain hardware structure. Newer languages, such as Verilog and VHDL, have more universal timing models and imply no particular hardware structure.

Hardware description languages have two main applications: documenting a design and modeling it. Good documentation of a design helps to ensure design accuracy. It is also important in ensuring design portability, that is, solving the "Tower of Babel" problem that can exist among manufacturers.

All useful hardware description languages are supported by a simulator. Thus, the model inherent in an HDL description can be used to validate a design. Prototyping of complicated systems is extremely expensive, and the goal of those concerned with the development of hardware languages is to replace this prototyping process with validation through simulation. Another important use of HDL descriptions is in logic synthesis, that is, the automatic generation of gate-level circuits from the HDL model. HDL models are also used for test generation.

The use of hardware description languages in the design process implies a different approach to design than was used in the past. Formerly, digital designers were restricted to relying on a combination of word descriptions, block diagrams, timing diagrams, and logic schematics to describe their designs. Also, for a designer to be personally involved in the simulation process was considered to be "anathema." The attitude was, "Those who can't design, simulate." The situation today is markedly different. Designers have a great deal more training in software. They prepare their designs in a computer environment, where the design is entered as an HDL source file, as a schematic diagram, or as some combination of the two approaches. Simulation is a tool frequently employed by the designers themselves to verify the correctness of the design as it evolves. Synthesis tools translate the HDL models into gate-level designs. Designers trained in the old school of design can have difficulty adjusting to this new approach. They may consider the use of a hardware description language unnecessary overhead. Hopefully, attitudes will change as they see the results from using these languages and supporting tools—better documented, error-free, and more quickly generated designs.

To illustrate the concepts of structured design, a specific hardware description language must be used. Until recently, there has not been a standard hardware description language, as there are standard programming languages, such as C, C++, and JAVA. However, beginning in 1983, the U.S. Department of Defense sponsored the development of the VHSIC hardware description language (VHDL). The original intent of the language was to serve as a means of communicating designs from one contractor to another in the very high speed integrated circuit (VHSIC) program. The design of the language has received input from many individuals in the computer industry and therefore, reflects a consensus of opinion as to what characteristics a hardware description language should have.

In August 1985 version 7.2 of the language was released, representing the completion of the first major stage of the language development. Version 7.2 was a complete language in that it comprehensively provided constructs for structural and behavioral modeling, as well as the means to document designs. After the release of version 7.2 by the Department of Defense, IEEE sponsored the further development of VHDL, with the ultimate goal being the development of an improved standard version of the language. The review process was completed by May 1987 and the language reference manual (LRM) released for industrial review. In June 1987 eligible IEEE members voted to accept this version of VHDL as the standard and in December 1987 it became official. From 1988 through 1992, VHDL users suggested minor changes to the language. The proposed changes were balloted in 1993, some of which were incorporated into the language. In this book most of the VHDL conforms to the 1987 standard for reasons we discuss below.

Here, we use VHDL as the working language to illustrate the concepts of structured design. This choice was made, not only because of the popularity of VHDL, but because the authors feel that the set of constructs available in VHDL are very effective for representing designs.

This chapter constitutes a brief introduction to VHDL. In subsequent chapters more advanced features of VHDL are introduced "online" as the need arises. The authors have found this to be an effective approach to the teaching of computer languages. The reader is also encouraged to consult the references for more language details.

```
entity ONES_CNT is
port (A: in BIT_VECTOR(2 downto 0);
        C: out BIT_VECTOR(1 downto 0));

------ Truth Table:
---
----------------------------
---|A2    A1    A0   |  C1   C0  |
----------------------------
-- |0     0     0    |  0    0   |
-- |0     0     1    |  0    1   |
-- |0     1     0    |  0    1   |
-- |0     1     1    |  1    0   |
-- |1     0     0    |  0    1   |
-- |1     0     1    |  1    0   |
-- |1     1     0    |  1    0   |
-- |1     1     1    |  1    1   |
end ONES_CNT;
```

Figure 3.1 Commented interface description.

3.1 MAJOR LANGUAGE CONSTRUCTS

To introduce the language, this section illustrates the major language constructs with some simple examples. The language syntax will be covered in detail in the next section.

3.1.1 Design Entities

In VHDL a given logic circuit is represented as a *design entity*. The logic circuit represented can be as complicated as a microprocessor or as simple as an AND gate. A design entity, in turn, consists of two different types of descriptions: the *interface description* and one or more *architectural bodies*. We illustrate this concept with an example. Consider the interface description for a circuit that counts the number of 1's in an input vector of length 3:

```
entity ONES_CNT  is
port (A: in BIT_VECTOR(2 downto 0);
        C: out BIT_VECTOR(1 downto 0));
end ONES_CNT;
```

One can see that the interface description declares the entity and describes its inputs and outputs. The description of interface signals includes the mode of the signal (i.e., *in* or *out*) and the type of signal. In this case, A is a 3-bit input array of type BIT_VECTOR with an index range (2,1,0). C is a 2-bit output array of type BIT_VECTOR with an index range (1,0). More details about data types are provided later in this chapter.

The interface description is also a place where documentation information about the nature of the entity can be recorded. For example, let us rewrite the interface description above and include, in comment form, the truth table of the entity. Figure 3.1 shows how the truth table can be inserted as a comment in the description. On any line, any text after two dashes is interpreted as a comment. This is only one example of the type of information that can be added to an interface description. In later chapters, we give other examples.

```
architecture ALGORITHMIC of ONES_CNT is
begin
  process(A)
    variable NUM: INTERGER range 0 to 3;
  begin
    NUM := 0;
    for I in 0 to 2 loop
      if A(I) = '1'  then
        NIM := NUM + 1;
      end if;
    end loop;
    case NUM is
      when 0 => C <= "00";
      when 1 => C <= "01";
      when 2 => C <= "10";
      when 3 => C <= "11";
    end case;
  end process;
end ALGORITHMIC;
```

Figure 3.2   Algorithmic description of the 1's counter.

3.1.2 Architectural Bodies

The interface description defines only the inputs and outputs of the design entity. An architectural body specifies either the behavior of the entity or specifies a structural decomposition of the entity using more primitive components.

At the beginning of the design process, designers usually have an algorithm in mind that they would like to implement. Initially, however, they want to check the accuracy of the algorithm without specifying the detailed implementation. Thus, the first architectural body that would be implemented is one that is algorithmic. Figure 3.2 shows such an architectural body for the 1's counter.

Note in the architectural body shown in Figure 3.2, the loop scans the inputs A(0) through A(2) and increments the integer variable NUM whenever a particular bit is set. Based on the final value of NUM, a *case* statement then selects which bit pattern to transfer to the output.

This behavioral body describes the operation of the algorithm perfectly, but its correspondence to real hardware is weak. Specifically, one might ask such questions as: What logic does the looping construct correspond to? How easy would it be to synthesize? What is the delay through the circuit? This architectural body provides no answers to these questions. However, at the early stages of design, one is usually not concerned about these detailed considerations.

Continuing the example, suppose that the design of the 1's counter advances to the logic design stage. Figure 3.3 shows the Karnaugh maps for the two outputs, C1 and C0. From these maps one can determine that the Boolean equations for C1 and C0 are:

$$C1 = (A1)(A0) + (A2)(A0) + (A2)(A1)$$

$$C0 = (A2)(\overline{A1})(\overline{A0}) + (\overline{A2})(\overline{A1})(A0) + (A2)(A1)(A0) + (\overline{A2})(A1)(\overline{A0})$$

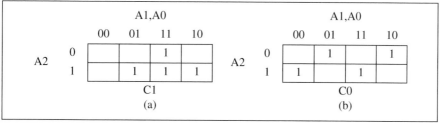

Figure 3.3 K-maps for the 1's counter.

```
architecture DATA_FLOW of ONES_CNT is
  begin
    C(1) <= (A(1) and A(0)) or (A(2) and A(0))
          or (A(2) and A(1));
    C(0) <= (A(2) and not A(1) and not A(0))
          or (not A(2) and not A(1) and A(0))
          or (A(2) and A(1) and A(0))
          or (not A(2) and A(1) and not A(0));
  end DATA_FLOW;
```

Figure 3.4 Data flow model of the 1's counter.

It should also be noted for later use that C1 is the majority function of three inputs A2, A1, and A0 [MAJ3(A2,A1,A0)], and that C0 is the odd-parity function of three inputs [OPAR3 (A2,A1,A0)]. At this point the designer could replace the algorithmic architecture body with the data flow architecture body shown in Figure 3.4.

In the statements implementing the "sum of products" equations for C(1) and C(0), the AND terms are enclosed in parentheses. This is because in VHDL, the *and* and the *or* operators have equal precedence.

The two-level logic mechanization shown in Figure 3.4 implies a standard gate structure utilizing AND gates, OR gates, and Inverters. However, since C1 can be computed by the MAJ3 function and C0 by the OPAR3 function, an even simpler macro level architectural body would be:

```
architecture MACRO of ONES_CNT is

begin

    C(1) <= MAJ3(A);
    C(0) <= OPAR3(A);

end MACRO;
```

This architectural body implies the existence of MAJ and OPAR gates at the hardware level. In terms of a VHDL description, it requires that the functions MAJ3 and OPAR3 must have been declared and defined previously, and that they are visible. In a subsequent section we show how to do this.

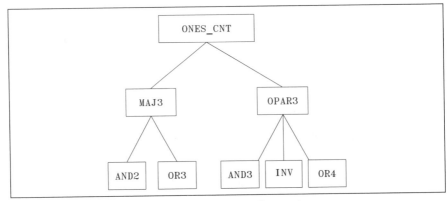

Figure 3.5 Structural design hierarchy for the 1's counter.

The foregoing three architectural bodies for the 1's counter are behavioral in that they specify the input/output response of the entity without exactly specifying the internal structure. We now give a structural architectural body for the 1's counter. In the approach shown here, we use a design hierarchy of the form shown in Figure 3.5; that is, the 1's counter is first decomposed into MAJ3 and OPAR3 gates, which are in turn decomposed into AND and OR gates and Inverters. Figure 3.6 shows the interface descriptions and architectural bodies for the MAJ3 gate, AND2 gates, and OR3 gates.

We will explain architecture AND_OR of entity MAJ3 in detail. A logic diagram of its gate structure is shown in Figure 3.7. Referring again to Figure 3.6(a), note that in the declaration section of architecture AND_OR, two components (AND2C and OR3C) are declared. The port specification used is identical to the port specification used in the entity definitions for these components (see Figures 3.6(b) and 3.6(c)). Next, three signals (A1,A2,A3) are declared. The outputs of the three AND gates must be connected to the inputs of the OR gate, and the signal definitions allow us to represent these connections. After the key word *begin*, four components are *instantiatea*; that is, a specific instance of a general component entity is created. Note that each instantiation has a unique label associated with it (i.e., G1, G2, G3, or G4) as well as a port map. The port map creates an association between signal (wire) names and the inputs and outputs of the component declaration. In this case the association is "by position." In another approach, association can be performed that is position independent. For example, the port map for instantiation labeled G1 could be written as follows:

```
G1: AND2C
    port map(O => A1, I1 => X(0), I2 => X(1));
```

In this case the arrows associate the *formal signals* in the component declaration with *actual signals* declared in the structural architecture. Since this association is position independent, it is less prone to error and more readable.

To define the operation of entity MAJ3 completely, there must exist interface descriptions and architectural bodies for all components instantiated within the architectural body for MAJ3. Figures 3.6(b) and 3.6(c) gives these descriptions. The OPAR3 structure can be similarly defined. We leave it as an exercise to the reader to do this (see Problem 3.2.). The component declarations for AND2C and OR3C are like connectors with no chips inserted. They must be

```
use work.all;
entity MAJ3 is
  port (X: in BIT_VECTOR(2 downto 0); Z: out BIT);
end MAJ3;
architecture AND_OR of MAJ3 is
  component AND2C
    port (I1,I2: in BIT; O: out BIT);
  end component;
  component OR3C
    port (I1,I2,I3: in BIT; O: out BIT);
  end component;
  for all: AND2C use entity AND2(BEHAVIOR);
  for all: OR3C use entity OR3(BEHAVIOR);
  signal A1,A2,A3: BIT;
begin
  G1: AND2C
    port map (X(0),X(1),A1);
  G2: AND2C
    port map (X(0),X(2),A2);
  G3: AND2C
    port map (X(1),X(2),A3);
  G4: OR3C
    port map (A1,A2,A3,Z);
end AND_OR;
```

(a) MAJ3 Description

```
entity AND2 is
  port (I1,I2: in BIT; O: out BIT);
end AND2;
architecture BEHAVIOR of AND2 is
begin
  O <= I1 and I2;
end BEHAVIOR;
```

(b) AND2 Description

```
entity OR3 is
  port (I1,I2,I3: in BIT; O: out BIT);
end OR3;
architecture BEHAVIOR of OR3 is
begin
  O <= I1 or I2 or I3;
end BEHAVIOR;
```

(c) OR3 Description

Figure 3.6 Structural description of MAJ3.

bound to library models (analagous to inserting the chips). The following statements from Figure 3.6 do this:

```
for all: AND2C use entity AND2(BEHAVIOR);
for all: OR3C use entity OR3(BEHAVIOR);
```

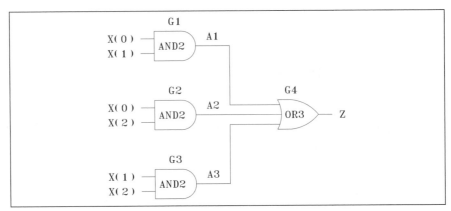

Figure 3.7   Majority function gate structure.

```
use work.all;
architecture STRUCTURAL of ONES_CNT is
  component MAJ3C
    port (X: in BIT_VECTOR(2 downto 0); Z: out BIT);
  end component;
  component OPAR3C
    port (X: in BIT_VECTOR(2 downto 0); Z: out BIT);
  end component;
  for all: MAJ3C use entity MAJ3(AND_OR);
  for all: OPAR3C use entity OPAR3(AND_OR);
begin
  COMPONENT_1: MAJ3C
    port map (A,C(1));
  COMPONENT_2: OPAR3C
    port map (A,C(0));
end STRUCTURAL;
```

Figure 3.8   Structural architectural body for the 1's counter.

After the MAJ3 and OPAR3 components have been defined, a structural architectural description for the 1's counter can be given as shown in Figure 3.8.

3.1.3   Model Testing

As with other forms of programming, after VHDL models are written they must be tested. We do this by forming a top-level entity called a *test bench*. As an example, Figure 3.9 shows the code for a test bench to test the entity ONES_CNT that we developed previously. Note: The entity declaration for TEST_BENCH contains no port statement because the test signals are generated internally to the test bench. Within the architecture ONES_CNT1, the test input (A) and test output (C) for the entity ONES_CNT are declared as signals.

Next, a component declaration is made for a component ONES_CNTA, which is then bound to entity ONES_CNT, architecture ALGORITHMIC in the design library. After the key-

```
entity TEST_BENCH is
end TEST_BENCH;

use WORK.all;
architecture ONES_CNT1 of TEST_BENCH is
  signal A: BIT_VECTOR(2 downto 0); ----Declare signals
  signal C: BIT_VECTOR(1 downto 0);
  component ONES_CNTA              ----Declare component
    port (A: in BIT_VECTOR(2 downto 0);
          C: out BIT_VECTOR(1 downto 0));
  end component;
  for L1: ONES_CNTA use entity ONES_CNT(ALGORITHMIC);
begin
  L1: ONES_CNTA
    port map(A, C);
  process
  begin
    A <= "000" after 1 ns,
         "001" after 2 ns,
         "010" after 3 ns,
         "011" after 4 ns,
         "100" after 5 ns,
         "101" after 6 ns,
         "110" after 7 ns,
         "111" after 8 ns;
    wait;
  end process;
end ONES_CNT1;
```

Figure 3.9 Test bench for entity ONES_CNT.

word *begin*, the component ONES_CNTA is instantiated and its port signals mapped to A and C with the port map statement. Finally, the process statement contains a sequence of test vectors to drive the entity ONES_CNT. The process runs just once at the start of simulation and then suspends because of the wait statement. All eight test vectors are scheduled to occur at nanosecond intervals.

The result of simulating this test bench entity is shown in Figure 3.10. The left-hand column specifies the time (in nanoseconds) at which some signal changes value. The rest of the line specifies the new values for each signal that changes value at this time. Dashes in a column mean that the signal does not change value at this time. The input A is in the middle column and the corresponding response of C is in the right-hand column. The response on C exhibits *delta* delay, e.g., the value "001" is impressed on A at time equal to 2 ns. The response on C ("01") appears at time (2 ns + 1 delta). For now, we can regard delta as a unit delay. In Chapter 4, we will define delta delay more precisely.

3.1.4 Block Statements

A basic element of a VHDL description is the *block*. A block is a bounded region of text that contains a declaration section and an executable section. Thus, the architectural body itself is a

```
VHDL Report Generator

  TIME     |------------------SIGNAL NAMES------------------|

  (ns)               A(2 DOWNTO 0)             C(1 DOWNTO 0)

    0                    "000"                     "00"
    2                    "001"                     ----
   +1                    -----                     "01"
    3                    "010"                     ----
    4                    "011"                     ----
   +1                    -----                     "10"
    5                    "100"                     ----
   +1                    -----                     "01"
    6                    "101"                     ----
   +1                    -----                     "10"
    7                    "110"                     ----
    8                    "111"                     ----
   +1                    -----                     "11"
```

Figure 3.10   Simulation results for entity ONES_CNT.

block. However, within an architectural body, internal blocks can exist. Consider the example shown in Figure 3.11. Here, blocks A and B are nested within the outer block of the architectural body. In general, any number of nesting levels is possible; for example, blocks A and B could also be decomposed into subblocks. There are two main reasons why this structure is employed. First, it supports a natural form of design decomposition, and second, a "guard" condition can be associated with a block. When a guard condition is TRUE, it enables certain types of statements inside the block. Figure 3.12 shows an example of a guarded block. In the block header "block (CON = '1')" implies that "CON = '1'" is a "guard" expression. The guard is implicitly defined to be of type Boolean. Any guarded statement will be executed when (1) the guard is TRUE and a signal on the right-hand side of the guarded statement changes, or (2) the guard changes from FALSE to TRUE. When the guard is false, the signal on the left-hand side retains its old value. For the example in Figure 3.12, when CON is a "1," the output O1 will follow the input I1. When CON changes from '1' to '0', the last value stored in O1 is retained. Note: The signal assignment to O2 is independent of the guard and thus will be activated whenever I2 changes, regardless of the state of the guard.

Guarded blocks are very useful for modeling register primitives. We'll pursue this more in Chapter 4.

3.1.5 Processes

Another major modeling element in VHDL is the process. Figure 3.2 gave an example of a process in which the behavioral description of the ONES_CNT circuit was given. Note: The process in this case began with the keyword *process(A)*. The vector A in this instance is said to make up the *sensitivity list* of the process. Whenever a signal in the sensitivity list changes, the process is activated and the statements within the process block are executed. A typical application of the process block is to implement algorithms at an abstract level, which, of course, is the case in Figure 3.2.

```
architecture  BLOCK_STRUCTURED of SYSTEM is
--------------        Outer Block Declaration Section
--------------
--------------
begin
-----------        Outer Block Executable Statements
----------
A: block
--------        Inner Block A Declaration Section
--------
begin
--------        Inner Block A Executable Statements
--------
end block A;
B: block
--------        Inner Block B Declaration Section
--------
begin
--------        Inner Block B Executable Statements
--------
end block B;
end  BLOCK_STRUCTURED;
```

Figure 3.11   Example of nested blocks.

```
entity GUARD_EXAMP is
    port(I1,I2,CON: in BIT; O1,O2: out BIT);
end GUARD_EXAMP;

architecture DF of GUARD_EXAMP is
  begin
   B:block(CON = '1')
    begin
     O1 <= guarded I1;
     O2 <= I2 ;
    end block B;
end DF;
```

Figure 3.12   Guarded block example.

We discuss the implications of the process construct in a later section and discover that it represents the fundamental method by which concurrent activities in digital circuits are modeled.

3.2   LEXICAL DESCRIPTION

The text of a VHDL description consists of one or more design files. Each design file must be prepared using only character codes from the standard 7-bit ASCII character code set. If the file is prepared using a word processing package, such as WORD, it is necessary to request the

```
" quotation mark      # sharp, pound sign    & ampersand
' apostrophe          ( left parenthesis     ) right parenthesis
* star, multiply      + plus                 , comma
- hyphen, minus       . dot,period,point     / slash, divide
: colon               ; semi-colon           < less than
= equal               > greater than         _ underline
| vertical bar        ! exclamation point    $ dollar
% percent             @ commercial at        [ left square bracket
\ back-slash          ] right square bracket ^ circumflex
` grave accent        { left brace           } right brace
~ tilde
```

Figure 3.13 Special characters.

ASCII option to create a file of pure ASCII character codes. The normal file produced by such word processing packages contains non-standard ASCII character codes that control the formatting of the file. These nonstandard character codes will cause compiler errors. Therefore, use of an editor that produces pure ASCII files (e.g., DOS editor or Windows NotePad) is recommended.

3.2.1 Character Set

In the 1987 version of VHDL, only the following characters are allowed in a design file. Nonprintable characters, such as carriage return, are represented by the ASCII codes for the characters, expressed in hexadecimal and enclosed within parentheses. For example, the carriage return character is (0D) = 00001101.

1. Upper case letters: A B C D E F G H I J K L M N O P Q R S T U V W X Y Z
2. Digits: 0 1 2 3 4 5 6 7 8 9
3. Special characters: " # & ' ( ) * + , - . / : ; < = > _ |
4. Space character: (20)
5. Format effectors:
   a. Carriage return: (0D)
   b. Line feed: (0A)
   c. Form feed: (0C)
   d. Horizontal tabulation: (09)
   e. Vertical tabulation: (0B)
6. Lower case letters: a b c d e f g h i j k l m n o p q r s t u v w x y z
7. Other special characters: ! $ % @ ? [\ ] ^ ` { } ~

Figure 3.13 shows the definitions of the special characters.

3.2.2 Lexical Elements

A *lexical element* is a sequence of characters that makes up a fundamental element that cannot be divided into smaller elements. The text of a design file is a sequence of lexical elements and separator characters. The types of lexical elements are:

1. Delimiter
2. Identifier
3. Comment
4. Character literal
5. String literal
6. Bit string literal
7. Abstract literal

Any number of *separators* are allowed between adjacent lexical elements, before the first lexical element, and after the last lexical element in a line or file. In some cases, an explicit separator is required between two lexical elements when the concatenation of the lexical elements without the separator could be interpreted as a single lexical element.

The *separators* are:

1. Space character
2. Format effectors
3. End of line

The language does not specify what constitutes an "end of line." This is implementation dependent. However, the end of line must be defined using only format effector characters.

3.2.3 Delimiters

A *delimiter* is a character that is used to separate lexical elements and has a specific meaning in the language. The following characters are classified as *delimiters*.

$$\& \quad ' \quad (\quad) \quad * \quad + \quad , \quad - \quad . \quad / \quad : \quad ; \quad < \quad = \quad > \quad |$$

All VHDL statements are terminated by a ; character. The meanings of the other characters are defined in later sections as they are needed.

The term *compound delimiter* refers to a sequence of two delimiters that have special meaning. The *valid compound delimiters* are

$$=> \quad ** \quad := \quad /= \quad >= \quad <= \quad <> \quad --$$

The uses of compound delimiters will be defined as needed in context. For example, the compound delimiter <= means "less than or equal to" in one context but means "assign to a signal" in another context.

In most cases, the use of a delimiter or compound delimiter removes the necessity for inserting specific separators between lexical elements. For example, (A * B) is equivalent to (A*B).

3.2.4 Identifiers

An *identifier* is the name of an object or is a *reserved word*. Reserved words are identifiers that have special significance in the language. Reserved words may not be explicitly declared as an identifier by the user. Figure 3.14 shows the reserved words defined in 1987 standard VHDL.

User-defined identifiers are used to name variables, blocks, procedures, etc. An *identifier* is any sequence of characters that starts with a letter and includes only letters, digits, and isolated

abs	disconnect	label	package	units
access	downto	library	port	until
after		linkage	procedure	use
alias	else	loop	process	
all	elsif			variable
and	end	map	range	
architecture	entity	mod	record	wait
array	exit		register	when
assert		nand	rem	while
attribute	file	new	report	with
	for	next	return	
begin	function	nor		xor
block		not	select	
body	generate	null	severity	
buffer	generic		signal	
bus	guarded	of	subtype	
		on		
case	if	open	then	
component	in	or	to	
configuration	inout	others	transport	
constant	is	out	type	

Figure 3.14   Reserved words in 1987 standard VHDL.

underline characters. There may not be two consecutive underline characters. The last character must be either a letter or digit. The following are *valid* user-defined identifiers:

```
COUNT    cout    c_out    AB2_5C
VHSIC    X1      FFT      Decoder
A_B_C    xyZ     h333     STORE_NEXT_ITEM
```

The following are *invalid* user-defined identifiers:

```
2CA      My-name   H$B       LOOP      _ABC
A__B     Decode_   alpha 2   END       AB AC
N#3      BEGIN     MAP       REGISTER
```

All user-defined identifiers must be distinct from each other and also distinct from the reserved word identifiers. Because VHDL is case insensitive, two user-defined identifiers that differ only in the upper and lower case letters are considered to be the same identifier. For example, STATE = state, ABC = AbC, and A23_D = a23_d. Underline characters are significant in identifiers. By convention, in this book, user-defined identifiers are set in all capital letters (COLOR, TRISTATE, GREEN, S1, etc.). Reserved words are set in lower case italics (*type, is*). Built-in data types are set with only the initial letters capitalized. This is done only to improve readability.

3.2.5 Comments

A *comment* is a lexical element that begins with the delimiter -- and ends at the end of the line. Clearly, a comment must be the last lexical element on a line. Comments can start anywhere on the line. A comment can follow a legal statement on the same line, or can be the only lexical element on the line. Comments have no influence on either the compiler or the simulator. Their only purpose is for program documentation. Comments can contain any valid characters, including all special characters, except the end of line marker. Some examples follow.

```
-- This line consists of only a comment.

C:=A*B; -- This line contains a VHDL statement followed by a comment.
        -- Long comments can be continued on the next line. The statement
        -- contains three user-defined identifiers (A B C) and three
        -- delimiters (:=  *  ;).

---------- Extra hyphens can enhance readability.
```

3.2.6 Character Literal

A *character literal* is a lexical element formed by inserting exactly one character between two apostrophe delimiters. It is used to define constant values for initialization of scalar objects. Here are some examples:

```
'A'    ' '    '$'    '''    'b'    '.'    '1'
```

3.2.7 String Literal

A *string literal* is a lexical element formed by inserting a sequence of graphical characters between two quotation delimiters. The sequence may contain no characters at all. If a quotation character is needed in the sequence, use two consecutive quotation characters. Note: This convention excludes the possibility of writing two string literals on the same line without a separator character between the strings. The length of a string literal is the number of characters in the sequence (counting doubled quotation characters as single characters). Here are some examples:

```
"A simple string literal."  -- length=24
""                          -- length=0
"A"                         -- length=1,  This is different from 'A'.
""""                        -- length=1, a single quotation character.
"Special characters, $, #, and | are allowed."
```

A string literal must be typed on one line since it is a lexical element. Strings with lengths exceeding the capacity of a line must be formed by concatenating shorter strings. The *concatenation operator* is the "&" character.

```
"This is a long string literal that will not fit on one"&
"line which requires using the concatenation operator."
```

3.2.8 Bit String Literal

A *bit string literal* is a lexical element that consists of a string of digits enclosed by quotation character delimiters, preceded by a base specifier. The base specifier is one of the letters B (binary), O (octal), or X (hexadecimal). The digits in the string must be appropriate for the base, i.e., B(0,1), O(0-7), X(0-9,A-F). Regardless of the base specifier, the value of a bit string literal is taken to be the equivalent string of bits. For example, the value of X"A" is "1010". In any case, a bit string literal is just a string of bits, and the VHDL language treats it as such. Any association with a numerical value is specific to the user. For example, if "1010" is loaded into a register, the user may choose to interpret this as (+10) if he interprets the register contents as a positive integer, or the user may choose to interpret this as (-6) if he interprets the register contents as a twos complement number. The VHDL language always interprets the contents as the bit string "1010". Bit string literals are used to specify initial conditions for the contents of binary registers. Bit string literals may also be directly specified as a string of bits without any base specifier.

Underline characters may be used to improve the readability of bit string literals without affecting the value of the literal. The length of a bit string literal is the number of equivalent bits in the string, regardless of the base specifier. Underline characters do not affect the string length. However, underline characters are not allowed if the bit string literal is specified without a base specifier. Here are some examples:

```
B"11011110"    -- Length 8
"11011110"     -- Length 8, no base specifier
B"1101_1110"   -- Also length 8 and equivalent to the previous two examples.
X"DE"          -- Also length 8 and equivalent to the previous examples.
X"27"          -- Length 8, equivalent to B"0010_0111".
O"742"         -- Length 9, equivalent to B"111_100_010".
"1101_1110"    -- ILLEGAL SYNTAX, underscore is not a valid element value
               -- of type Bit.
```

3.2.9 Abstract Literal

An *abstract literal* is a lexical element that has a numerical value. There are two classes of abstract literal, *real* and *integer*. Real literals have a base point; integer literals do not. There are two types of abstract literals: decimal literals and based literals.

3.2.10 Decimal Literal

A *decimal literal* is an abstract literal expressed in the standard decimal notation used in many high-level computer languages. As in bit string literals, underline characters may be used to improve the readability of decimal literals without affecting the value of the literal. The exponent part of the literal is preceded by the letter E. Negative exponents are only allowed in real literals. The first character in a decimal literal must be a decimal digit. Here are some examples of integer decimal literals:

```
5       85_257   3E3      0          015         23e0
```

The following are examples of real decimal literals:

```
2.5     0.0      0.25     256.857   25.834_233 0.000_000_25
2.0e5   2.0E-5   2.0E+5   15.67e25  1.94E-15   03.14159
```

No spaces are allowed anywhere in an abstract literal. The following are *invalid* decimal literals:

```
2e-2    2.0e0    2,354    2.5 e-2   2.5e -2    .25
```

3.2.11 Based Literal

A *based literal* is an abstract literal that includes a base specification between 2 and 16. The base specification is always written in base 10. Any exponent is also written in base 10. The significant digits in the based literal are written in the specified base with A-F representing 10-15 in bases greater than 10. Upper and lower case letters are equivalent. The based literal consists of a base, sequence of significant digits, and optional exponent in that order. The significant digits must be preceded and followed by a # character. Again, no spaces are allowed within a based literal.

```
-- Integer-based literals with value 4095

16#FfF#   2#1111_1111_1111#  10#4095#   8#7777#   4#333333#

-- Integer-based literals with value 480

16#1E#E1  2#1_1110#e4  4#132#E2  16#1e0#  2#0001_1110_0000#

-- Real-based literals with value 4095.0

16#0.fff#e3  2#1.1111_1111_111#E11  4#3.33333#e5
```

3.3 VHDL Source File

A *VHDL source file* consists of a sequence of lexical elements delineated by separator characters or delimiter characters. No lexical element may be split between two lines. The lexical elements (delimiters, identifiers, comments, character literals, string literals, bit string literals, and abstract literals) constitute the only valid constructs that may appear in the file. Separators and format effectors define lexical element boundaries and line spacing.

Throughout the sections on VHDL language syntax, a set of running examples are defined. Examples of data types defined in the following section are used many times in later sections to define variables and signals.

3.4 Data Types

A *data type* is a named set of values. The data type of an object specifies what values the object may have and limits the kinds of operations that may be performed on objects of that type. In VHDL the constraints imposed by data types are strictly enforced. It is a strongly typed language. For example, an object of type integer must have values that are integers (... -2, -1, 0, 1, 2, ...). Operations on that object are restricted to those operations that are defined for integer data types, such as addition, multiplication, etc.

A *subtype* is a type along with a *constraint*. A value belongs to the subtype if it is a legitimate value for the type and satisfies the constraint. For example, the type natural is a subtype of type integer with the constraint that the values be natural numbers (0, 1, 2, ...). The *base type* of a subtype is the type from which it was constructed. For example, the base type of subtype natural is integer. A type is a subtype of itself, known as an *unconstrained subtype*. The base type of a type is the type itself.

```
1. Scalar
      Enumeration - discrete
      Integer - discrete, numeric
      Physical - numeric
      Floating point (or real) - numeric
2. Composite
      Array - All elements have the same type.
      Record - Elements may have different types.
3. Access
4. File
```

Figure 3.15   Classification of data types.

3.4.1   Classification of Types

Figure 3.15 shows the complete classification system for data types in VHDL. *Scalar types* have values consisting of a single element. Examples include integer and natural. *Composite types,* such as arrays and records, have complex values that include multiple elements. *Access types* are types that provide access to other types. *File types* provide access to files. Enumeration and integer types are classified as *discrete types.* Integer, Real, and Physical types are classified as *numeric types.*

3.4.2   Scalar Data Types

Scalar data types have simple, single values. For example, an integer data type may have the value 15. Scalar is contrasted with composite data types, such as vectors, which may have a complex value, such as (15, 25, 30). Scalar data types include enumeration data types, numeric data types (integer and real), and physical data types.

3.4.2.1   Enumeration Types

An *enumeration type* is a scalar type whose values are defined by simply listing them in an ordered list. The elements in the ordered list may be either identifiers or character literals. Figure 3.16 shows the *predefined enumeration data types* included in the VHDL language.

   The two values of type Bit are the character literals '0' and '1'. The VHDL language defines built-in logical operators *not, and, or, nand, nor,* and *xor* that apply to variables and signals of type Bit.

   Built-in type Boolean has values FALSE and TRUE. Variables and signals of type Boolean are used to control conditional execution of code segments. Built-in operators *not, or, and, nand, nor,* and *xor* also apply to variables and signals of type Boolean.

   Built-in type Character defines the standard ASCII character code set of 128 characters. This type is used for character manipulation. Built-in type Severity_Level is discussed later in conjunction with *assert* statements.

   Examples of user-defined enumeration data types include:

```
type COLOR is (RED, ORANGE, YELLOW, GREEN, BLUE, INDIGO, VIOLET);
type Bit is ('0', '1');
type TRISTATE is ('Z', '0', '1');
```

```
     type Bit is ('0', '1');
     type Boolean is (FALSE, TRUE);
     type Severity_Level is (NOTE, WARNING, ERROR, FAILURE);
     type Character is (
          NUL, SOH, STX, ETX, EOT, ENQ, ACK, BEL,
          BS,  HT,  LF,  VT,  FF,  CR,  SO,  SI,
          DLE, DC1, DC2, DC3, DC4, NAK, SYN, ETB,
          CAN, EM,  SUB, ESC, FSP, GSP, RSP, USP,
          ' ', '!', '"', '#', '$', '%', '&', ''',
          '(', ')', '*', '+', ',', '-', '.', '/',
          '0', '1', '2', '3', '4', '5', '6', '7',
          '8', '9', ':', ';', '<', '=', '>', '?',
          '@', 'A', 'B', 'C', 'D', 'E', 'F', 'G',
          'H', 'I', 'J', 'K', 'L', 'M', 'N', 'O',
          'P', 'Q', 'R', 'S', 'T', 'U', 'V', 'W',
          'X', 'Y', 'Z', '[', '\', ']', '^', '_',
          '`', 'a', 'b', 'c', 'd', 'e', 'f', 'g',
          'h', 'i', 'j', 'k', 'l', 'm', 'n', 'o',
          'p', 'q', 'r', 's', 't', 'u', 'v', 'w',
          'x', 'y', 'z', '{', '|', '}', '~', DEL);
```

Figure 3.16 Predefined enumeration data types.

```
type STATE is (S0, S1, S2, S3, S4);
type STD_ULOGIC is ('U','X','0','1','Z','W','L','H','-');
```

Type STD_ULOGIC is the nine-valued type defined in IEEE standard 1164, which is important for synthesis. A subtype of this type, STD_LOGIC, with an associated resolution function, is the type actually used for signal and variable declarations. More details on this are included later.

The values of a data type are listed in ascending order from left to right, separated by commas, and enclosed in parentheses. As indicated earlier, each VHDL statement ends with a semicolon.

If the same identifier or character literal is declared for more than one enumeration type, it is said to be *overloaded*. For example, the character literals '0' and '1' would be overloaded if both type Bit and type TRISTATE were declared.

Since the values of the data type are considered to be an ascending ordered list, it is convenient to associate a *position number* with each element in the list. Position numbers begin with 0 for the left element in the list and proceed by successive positive integers toward the right. The position numbers for the values of type COLOR are shown below.

RED	ORANGE	YELLOW	GREEN	BLUE	INDIGO	VIOLET
0	1	2	3	4	5	6

Since the position numbers define a numerical ordering for the values, expressions, such as "GREEN < BLUE", are TRUE. As described in the next definition, attributes provide a mechanism to refer to the position numbers.

Predefined attributes for enumeration data types. A predefined *attribute* is a value, function, type, or range that may be associated with various kinds of constructs. There are a number of predefined attributes associated with enumeration data types. Most of these have some relationship to the implicitly defined position numbers.

Attribute '*pos* is a predefined function that defines the position number of a specific value in the list. Attribute '*val* is a predefined function that defines the list value that corresponds to a specified position number. Attribute names are attached to the type name using the single quote character (sometimes called the *tick* character), as shown in these examples:

```
COLOR'pos(GREEN) = 3          COLOR'val(3) = GREEN
STATE'pos(S2) = 2             STATE'val(0) = S0
TRISTATE'pos('Z') = 0         TRISTATE'val(1) = '0'
Bit'pos('0') = 0              Bit'val(0) = '0'
```

For built-in type Character, the position number of a character is the normal ASCII code for the character, as shown in these examples:

```
Character'pos(NUL) = 0        (00 in hexadecimal)
Character'pos(CR)  = 13       (0D in hexadecimal)
Character'pos('0') = 48       (30 in hexadecimal)
```

The attributes '*left* and '*right* are predefined values associated with enumeration data types that define the left element in the list and the right element in the list respectively, as shown in these examples:

```
COLOR'left = RED             COLOR'right = VIOLET
STATE'left = S0              STATE'right = S4
TRISTATE'left = 'Z'          TRISTATE'right = '1'
Bit'left = '0'               Bit'right = '1'
```

Attributes '*high* and '*low* are predefined values associated with enumeration data types that define upper and lower bounds, respectively, for values in the list. Since the only values associated with enumeration types are the position numbers, attribute '*high* is equivalent to '*right* and attribute '*low* is equivalent to '*left*. However, as we will see later, this equivalence does not necessarily hold for other data types and for subtypes of enumeration data types. The following are examples of these values:

```
COLOR'low = RED              COLOR'high = VIOLET
STATE'low = S0               STATE'high = S4
TRISTATE'low = 'Z'           TRISTATE'high = '1'
Bit'low = '0'                Bit'high = '1'
```

Attribute '*leftof* is a predefined function associated with enumeration data types that defines the list value that occurs immediately to the left of a specified list value. Similarly, attribute '*rightof* is a predefined function associated with the data type that defines the list element that is immediately to the right of a specified list value. Some example values are:

```
COLOR'leftof(GREEN) = YELLOW    COLOR'rightof(BLUE) = INDIGO
STATE'leftof(S2) = S1           STATE'rightof(S3) = S4
TRISTATE'leftof('0') = 'Z'      TRISTATE'rightof('0') = '1'
```

Attribute '*pred* is a predefined function associated with enumeration data types that defines the list value that immediately precedes a given list value. Similarly, attribute '*succ* is a predefined function associated with the data type that defines the list element that immediately follows a specified list value. Attributes '*pred* and '*succ* refer to numerical values, not to relative position. Attribute '*pred* implies less than and attribute '*succ* implies greater than. Examples include:

```
COLOR'pred(GREEN) = YELLOW        COLOR'succ(BLUE) = INDIGO
STATE'pred(S2) = S1               STATE'succ(S3) = S4
TRISTATE'pred('0') = 'Z'          TRISTATE'succ('0') = '1'
```

For enumeration data types, *'leftof* and *'pred* are equivalent and *'rightof* and *'succ* are equivalent. However, for other classes of data types, this equivalence does not hold.

The concept of an ordered list also has implications on variable and signal initialization. If a variable or signal is declared without an explicit initial value, the default initial value will be the left element in the list.

Subtype declarations do not define new types. They simply constrain the values that objects of the specified subtype are allowed to possess. For example:

```
subtype LONGWAVE is COLOR range COLOR'left to YELLOW;
subtype ODDSTATE is STATE range S1 to S3;
subtype SHORTWAVE is COLOR range COLOR'right downto GREEN;
```

Signals declared to have subtype LONGWAVE and subtype SHORTWAVE will all have base type COLOR. Signals with the same base type are considered to be the same type. Signals declared to have subtype LONGWAVE and SHORTWAVE will have base type COLOR. Default initialization goes by subtype. The default initialization of signals of subtype LONGWAVE will be value RED. The default initialization of signals of subtype SHORTWAVE will be value VIOLET. Some attributes of type SHORTWAVE are:

```
SHORTWAVE'right = GREEN
SHORTWAVE'high = VIOLET
SHORTWAVE'low = GREEN
SHORTWAVE'val(1) = ORANGE
SHORTWAVE'pos(RED) = 0
```

Study these attributes carefully. Attributes SHORTWAVE'right, SHORTWAVE'left SHORTWAVE'low, and SHORTWAVE'high refer to the subtype ranges. However, attributes SHORTWAVE'val and SHORTWAVE'pos refer to the basetype COLOR. Position numbers always refer to the base type.

User-defined attributes for enumeration data types. Users may define their own attributes. In most cases, user-defined attributes do not affect simulation. More often, user-defined attributes are employed to provide design documentation or to provide information to other design tools. For example, using the following code, a designer can specify that a synthesis tool must use a one-hot state assignment to implement a state machine. The synthesis tool will extract the state assignment information from the user defined STATE_ASSIGNMENT attribute.

```
type STATE_TYPE is (S1, S2, S3, S4, S5, S6, S7);
attribute STATE_ASSIGNMENT: String;
attribute STATE_ASSIGNMENT of STATE_TYPE: type is
"0000001 0000010 0000100 0001000 0010000 0100000 1000000 ";
```

The general form to declare a user-defined attribute is:

```
attribute ATTRIBUTE_NAME: ATTRIBUTE_SUBTYPE;
```

The keyword *attribute* is followed by the user-selected name for the attribute. The ATTRIBUTE_SUBTYPE may be any previously declared subtype (or type) that is neither an

```
type Integer is range _____ ;
  -- Range is implementation dependent.
type Real is range _____ ;
  -- Range is implementation dependent.
subtype Natural is Integer range 0 to Integer'high;
subtype Positive is Integer range 1 to Integer'high;
```

Figure 3.17   Predefined numeric data types.

access type nor a file type. In the previous example, STATE_ASSIGNMENT is the attribute name, and String (a built-in data type) is the attribute type.

Attributes are attached to specific entities by using a statement of the following form:

```
attribute ATTRIBUTE_NAME of ENTITY_NAME: entity_class is expression;
```

The keyword *attribute* is followed by the name of the attribute that was previously declared. The keyword *of* separates the attribute name from the name of the entity that is to be associated with the attribute. The colon delimiter separates the name of the entity from the entity class. In the STATE_ASSIGNMENT example, the attribute STATE_ASSIGNMENT was associated with the data type STATE_TYPE. The entity class was *type*. Attributes may be associated with the following entity_classes: *entity, procedure, type, signal, label, architecture, function, subtype, variable, configuration, package, constant,* or *component*. Finally, the keyword *is* separates the entity_type from an expression that evaluates to the type ATTRIBUTE_SUBTYPE.

In the example, STATE_ASSIGNMENT is the attribute name, STATE_TYPE is the entity name, and *type* is the entity class.

3.4.2.2   Numeric Data Types—Integer and Real

The VHDL language includes built-in numeric data types Integer and Real with associated operations of addition, subtraction, multiplication, and division.

Figure 3.17 shows the predefined integer and real data types and subtypes. The ranges of types Integer and Real are implementation dependent. The ranges are usually related to the register size in the target computer. The language requires that the range include -2147483647 to $+2147483647$. Some machines may have an extended range. The user may obtain the range by using the predefined attributes Integer'high and Integer'low. Positive and Natural are predefined subtypes of type Integer that are of use in many applications. The declarations for these subtypes use the predefined attribute Integer'high. Therefore, the declarations will work on any machine regardless of the implementation dependent range for data type Integer. This demonstrates the use of an attribute to make a declaration more general than would otherwise be possible. Now, consider the following user-defined numeric types. They are constrained types because the declaration constrains the range. (When we discuss modeling for synthesis, we will see that it important to use constrained integer types as opposed to type Integer.)

The following are examples of user-defined constrained numeric data types:

```
type COUNTER is range 0 to 100;
subtype LOW_RANGE is COUNTER range 0 to 50;
subtype MID_RANGE is COUNTER range 25 to 75;
subtype HIGH_RANGE is COUNTER range 50 to 100;
type REG is range 0 to 100;
```

```
type INDEXA is range 0 to 7;
type INDEXD is range 7 downto 0;
type PROBABILITY is range 0.0 to 1.0;
type ANGLE is range -90.0 to 90.0;
type TESTSCORE is range 100.0 downto 0.0;
```

The VHDL analyzer distinguishes Integer from Real by searching for the decimal point. The previous examples define only seven different data types, COUNTER, REG, INDEXA, INDEXD, PROBABILITY, ANGLE, and TESTSCORE. COUNTER and REG are different data types, even though they have the same range declaration. Variables with different data types cannot be combined in the same expression. For example, if variable A has data type COUNTER and variable B has data type REG, it is an error to write:

$$A := A + B$$

The ranges for numeric data types can be either ascending or descending. For example, the range of INDEXA is ascending, and the range of INDEXD is descending. Whereas, certain predefined data type attributes (such as 'left and 'low) were indistinguishable for enumeration data types, this is no longer the case for numeric data types with descending ranges. The following examples illustrate several similar concepts. The reader should study these examples carefully:

```
INDEXD'left = 7        INDEXD'low = 0
INDEXD'right = 0       INDEXD'high = 7
INDEXA'left = 0        INDEXA'low = 0
INDEXA'right = 7       INDEXA'high = 7
INDEXD'pred(6)  = 5    INDEXD'leftof(6)  = 7
INDEXD'succ(6)  = 7    INDEXD'rightof(6) = 5
INDEXA'pred(6)  = 5    INDEXA'leftof(6)  = 5
INDEXA'succ(6)  = 7    INDEXA'rightof(6) = 7
```

It should be clear that for data types with ascending ranges (this includes all enumeration data types), attributes *'left* and *'low* are equivalent, attributes *'right* and *'high* are equivalent, attributes *'pred* and *'leftof* are equivalent, and attributes *'succ* and *'rightof* are equivalent. Similarly, for data types with descending ranges, attributes *'left* and *'high* are equivalent, attributes *'right* and *'low* are equivalent, attributes *'pred* and *'rightof* are equivalent, and attributes *'succ* and *'leftof* are equivalent.

3.4.2.3 Physical Data Types

A *physical data type* is a scalar numeric data type with an associated system of units that is used to represent entities subject to physical measurements, such as time, length, voltage, current, etc. The only predefined physical data type is type Time, which is defined as follows:

```
type Time is range _____
   units
     fs;                -- femptosecond
     ps = 1000fs;       -- picosecond
     ns = 1000ps;       -- nanosecond
     us = 1000ns;       -- microsecond
     ms = 1000us;       -- millisecond
     sec= 1000ms;       -- second
     min= 60 sec;       -- minute
     hr = 60 min;       -- hour
   end units;
```

The range is implementation dependent. The first unit given (fs) is called the *base unit*. All other units must be integer multiples of the base unit with the integer multiple being in the range specified on the "type" line. Also, the actual value computed for any variable of type Time must also be an integer multiple of the base unit. In effect, the base unit is an indivisible unit. There cannot be fractional numbers of base units. There may, however, be fractional numbers of other units as long as the result is still an integer number of base units. For example, 2.5 ps is valid because 2.5 ps = 2500 fs. However, 2.250326 ps is invalid because this is not an integer number of femtoseconds (2.250326 ps = 2250.326 fs).

Implementations may restrict the ranges of physical data types. However, an implementation must allow the declaration of any physical data type whose range is wholly contained within the bounds –2,147,483,647 to +2,147,483,647. If the range is restricted to these limits, any user-defined physical data type may have at most three units that are 1000 multiples of the next lower unit.

Here are some examples of user-defined physical data types:

```
-- Electrical Resistance
  type RESISTANCE is range 0 to Integer'high
    units
       NOHM;                       -- Base unit is nano-ohm.
       UOHM   = 1000 NOHM;    -- Microohm
       MOHM   = 1000 UOHM;    -- Milliohm
       OHM    = 1000 MOHM;    -- Ohm
       KOHM   = 1000 OHM;     -- Kilohm
       MEGOHM = 1000 KOHM;    -- Megohm
    end units;
-- Electrical Power
  type POWER is range 0 to 1e9
    units pW;                  -- Pico-watt
          nw = 1000 pW;        -- Nano-watt
          uW = 1000 nW;        -- Micro-watt
    end units;
-- Frequency
  type FREQUENCY is range 0 to 1E9
    units Hz;                  -- Hertz
          KHz = 1000 Hz;       -- Kilohertz
          MHz = 1000 KHz;      -- Megahertz
    end units;
```

In the examples above, the physical data type RESISTANCE would not work in an implementation that was restricted to the minimum range mentioned above. A workable range would include ohms, Kohms, and Mohms.

3.4.3 Composite Data Types

Composite data types have complex values. For example, a one-dimensional array of integers could have the value (15, 35, 2, 956). The complex value has several components and represents an array of values that have some relationship to one another.

3.4.3.1 Arrays

An *array* is a composite data type in which each element has the same subtype. The array elements are said to be *homogeneous*. For example, there are two predefined array data types. Data type String is an array of Character elements. It is used to create text for printing and for other program uses. Data type Bit_Vector is an array of bits. It is used to describe the data stored in a register or the data being transmitted over a parallel bus. The definitions are:

```
type String is array (Positive range <>) of Character;
type Bit_Vector is array (Natural range <>) of Bit;
```

Note: The *index range* is placed within parentheses following the reserved word *array*. The index range for data type String is of data type Positive. The notation <> means that the range is *unconstrained*. The effect is that a user must specify the index range when declaring an object of type String, subject to the restriction that the range must be positive integers. A great degree of freedom results. Similarly, the index range for data type Bit_Vector is unconstrained (except the range must be natural numbers). Objects of type Bit_Vector may have index number 0, but objects of type String cannot.

A variety of user-defined array data types follow:

```
type STD_LOGIC_VECTOR is array (Natural range <>) of STD_LOGIC;
type REGISTER_32_Bit is array (31 downto 0) of Bit;
       -- Descending index range.
type COUNTER_16_Bit is array (0 to 15) of Bit;
       -- Ascending index range.
subtype BITVECT3 is Bit_Vector(0 to 2);
type COLOR_COUNT is array (COLOR range <>) of Natural;
type ARRAY_OF_COLORS is array (Natural range <>) of COLOR;
type ROM_TYPE is array(Natural range <>)of REGISTER_32_Bit;
type ARRAY_2D is array (Natural range <>, Natural range <>) of Bit;
```

Type STD_LOGIC_VECTOR is an unconstrained data type used to represent registers and buses when the IEEE 9 valued logic system is required, e.g., in models that are to be synthesized. Array type REGISTER_32_BIT is a constrained one-dimensional array of 32 BITS numbered from 31 down to 0 (left to right). For example, if a data item of type REGISTER_32_BIT has value:

B"1111_1101_0101_1110_0000_0101_1100_0010"

then bit 31 (the left bit) has value '1', bit 25 has value '0', bit 2 has value '0', and bit 0 (the right bit) has value '0'. The index range for type REGISTER_32_BIT is said to be *descending*.

Similarly, array type COUNTER_16_BIT is a constrained 16-bit one-dimensional array with the bit positions numbered in *ascending* order from 0 to 15 (left to right). If an object of type COUNTER_16_BIT has value:

B"1011_0010_1011_0010"

then bit 0 (the left bit) has value '1', bit 4 has value '0', and bit 15 (the right bit) has value '0'.

Since Bit_Vector is an unconstrained data type, one can declare subtypes with specific index ranges. BITVECT3 is a subtype of type Bit_Vector with an ascending index range from 0 to 2.

Array type COLOR_COUNT is an array of natural numbers. The index range is unconstrained of type COLOR, which permits the user to select any range he wishes when declaring an object to have type COLOR_COUNT. Suppose the range RED to YELLOW is selected for some object. Then, the array consists of three natural numbers. The index sequence is RED, ORANGE, YELLOW. For example, the value of the array might be (24, 35, 0). In this case, the element with index RED is 24; the element with index ORANGE is 35, and the element with index YELLOW is 0.

Similarly, array type ARRAY_OF_COLORS is an array of colors. The index range is unconstrained but must be natural numbers. Suppose the index range is declared as 3 to 5. If the value of the array is (BLUE, ORANGE, VIOLET), the element with index 3 has value BLUE; the element with index 4 has value ORANGE, and the element with index 5 has value VIOLET.

There are two ways to declare a two-dimensional array. Array type ROM_TYPE is an array of arrays. Each element in the array has type REGISTER_32_BIT. In other words, ROM_TYPE is a one-dimensional array of 32-bit words. A typical value for an entity of type ROM_TYPE that consists of four words is:

```
(X"2F3C_5456", X"FF22_A5B9", X"9900_AD51", X"FFFF_FFFF") .
```

In other words, each element of the array is a 32-bit quantity.

In contrast, ARRAY_2D is a two dimensional array in which each element is of type Bit. If an object is declared to have five rows and ten columns, then a typical value for the entity is:

```
(1, 0, 1, 1, 0, 0, 0, 1, 1, 1,
 0, 0, 1, 1, 1, 0, 1, 1, 0, 1,
 1, 1, 1, 0, 1, 0, 1, 0, 0, 1,
 1, 1, 1, 1, 0, 0, 0, 1, 1, 0,
 0, 0, 0, 0, 1, 1, 1, 0, 0, 1)
```

In this case, each element of the array is a single bit. Manipulating elements of the array is discussed in the entity declaration section.

3.4.3.2   *Array Attributes*

Arrays have many useful built-in attributes. Attribute A'range is a function that returns the subscript range for a one-dimensional array named A. This allows the creation of a process that examines the elements of an array without worrying about what the actual range of the array might be. For example, the following process counts the number of 1's in the one-dimensional array DBUS with type Bit_Vector. The range can be either ascending or descending. The process works regardless of the number of elements in the index range.

```
PROCESS_RANGE: process (DBUS)
    variable COUNT3: Integer := 0;
  begin
    COUNT3 := 0;
    for I in DBUS'range loop
      if DBUS(I) = '1' then
        COUNT3 := COUNT3 + 1;
      end if;
    end loop;
  end process;
```

Attribute A'high is a function that returns the greatest value in the index range of array A. Attribute A'low is a function that returns the least value in the index range. Similarly, attribute A'right and A'left are functions that return the right-most value and left-most value, respectively, from the index range of A. Attribute A'length is a function that returns the number of elements in the index range for array A. For example, if DBUS is declared as:

```
signal DBUS: Bit_Vector (15 downto 0);
```

the array attributes have the following values:

```
DBUS'right = 0
DBUS'left = 15
DBUS'high = 15
DBUS'low = 0
DBUS'length = 16
```

The following process uses two of these attributes to compute the number of 1's in signal DBUS. This process works for any signal of type Bit_Vector, regardless of the range of index values. The range can be either ascending or descending.

```
PROCESS_LENGTH: process (DBUS)
    variable COUNT2: Integer := 0;
  begin
    COUNT2 := 0;
    for I in 0 to DBUS'length-1 loop
      if DBUS(DBUS'low+I) = '1' then
        COUNT2 := COUNT2+1;
      end if;
    end loop;
  end process;
```

All the array attributes can be applied to multidimensional arrays. The general form of the attribute for a multidimensional array is A'attribute(n), where the attribute applies to the *n*th index range.

3.4.3.3 *Records*

Record types are composite types in which the elements are *heterogeneous*; that is, elements may have different types. As in array types, the elements form an ordered list. For example, consider the following declaration for a record used to hold a date:

```
type MONTH_NAME is (JANUARY, FEBRUARY, MARCH, APRIL, MAY, JUNE, JULY,
AUGUST, SEPTEMBER, OCTOBER, NOVEMBER, DECEMBER);

type DATE is record
   DAY  : Integer range 1 to 31;
   MONTH: MONTH_NAME;
   YEAR : Integer range 0 to 3000;
end record;
```

A typical value for an object of type DATE is:

```
(16, AUGUST, 1943)
```

where DAY=16, MONTH=AUGUST, and YEAR=1943.

3.4.4   Access Types

Access types are types used in conjunction with allocation and deallocation of storage in dynamic storage applications, such as linked lists and trees. Access types are not discussed in this book.

3.4.5   File Types

File types are used to define objects that represent the contents of files in the host system environment. Files may be used to hold data used by the VHDL program. Test vectors for the VHDL model are frequently stored in files. These types are discussed in a later section.

3.4.6   Type Marks

Sometimes it is impossible for VHDL to correctly decide what type is intended by the programmer when overloading occurs. For example, consider the following situation:

```
type TRISTATE is ('Z', '0', '1');
type MVL is ('0', '1', 'Z');
```

Is the expression ('Z' < '0') TRUE or FALSE? Since the characters 'Z' and '0' are overloaded, the simulator has no way to tell whether to use type TRISTATE (where the expression is TRUE) or type MVL (where the expression is FALSE). The problem is solved by writing the expression using *type marks* as follows:

```
(MVL'('Z') < MVL'('0'))
```

When the type is not evident from the context of the statement, use the type name followed by the "tick" mark (called the type mark in this context) with the value name enclosed in parentheses. There is no ambiguity in the above expression—it evaluates to FALSE.

3.5   DATA OBJECTS

A *data object* is a container in which values of a stated type may be stored.

3.5.1   Classes of Objects

Table 3.1 shows the three classes of data objects.

Table 3.1   Classes of data objects.

Class	Description
Constant	An object whose value is specified at compile time and cannot be changed by VHDL statements.
Variable	A data object whose current value can be changed by VHDL statements.
Signal	A data object that has a time dimension. VHDL statement may assign future values to the data object without affecting the current value. Using waveforms, a series of future values may be assigned. The current value cannot be changed.

Variables in VHDL are similar to variables in other high-level programming languages, such as C, C++, and Java. Variables can be declared only within processes or subprograms. A variable declared within a process is local to the process. Within a process, a variable is *static*, i.e., a value assigned in one process call is held until the next process call. Process variables are initialized only once at the beginning of simulation. However, variables declared within subprograms are *dynamic*. Such values are not held from one call to the next. Subprogram variables are initialized at the beginning of each call to the subprogram.

Signals are used to represent data values on actual physical data lines in circuits. In a physical system, a signal, such as a voltage signal, cannot change value instantaneously. Therefore, to correctly represent physical quantities, VHDL signals do not change instantaneously either. Thus, a new value assigned to a signal does not take effect until some future time. The future time at which the change is to take affect can be explicitly stated. If no time is specified for a signal change, the default future time is an infinitesimally small time called *delta time*. The important concept is that a scheduled change for a VHDL signal never occurs at the present time but is always delayed until some future time. Keeping this principle in mind aids the understanding of signal properties and reduces errors. A series of changes can also be specified for a VHDL signal using a waveform-like structure.

3.5.2 Declaration of Data Objects

Data objects must be declared before they can be referenced in an assignment statement. Examples of declared objects in this section use types defined in previous sections. These declared objects are subsequently used in later sections to write various types of statements.

3.5.2.1 Declaration of Constants

Normally, the value of constants must be specified when they are declared. The following constant values are specified using data types defined in an earlier section of this chapter.

```
constant ALPHA_LEVEL: PROBABILITY := 0.75;
constant INITIAL_STATE: STATE := S0;
constant MID_COLOR: COLOR := GREEN;
constant HIGH_IMPEDANCE: TRISTATE := 'Z';
constant END_MARKER: Character := DEL;
constant PI: REAL := 3.14159;
constant C1K: Integer := 2 ** 10;
   -- Operator ** is exponentiation.
constant MESSAGE1: STRING := "Happy Birthday";
```

Note: The constant name is separated from the type name by the : delimiter, and the type name is separated from the constant value by the := delimiter. As usual, every VHDL statement is terminated by the ; delimiter.

3.5.2.2 Declaration of Variables

Variable names must be declared before they are used in expressions. Variables can only be declared within processes or subprograms. The variable declaration always specifies a data type for the variable. Initial values for variables can be explicitly specified in the declaration. If no initial value is specified for a variable, the default initial value is the left element in the type range specified in the declaration. The following are examples of scalar variable declarations:

```
variable STAR_COLOR, HAT_COLOR: COLOR;
variable BETA_LEVEL: PROBABILITY := 0.0;
variable CURRENT_STATE: STATE := INITIAL_STATE;
variable A, B, C, D: Bit;
variable R1, R2, R3, R4: REG;
variable A_COUNT, B_COUNT: COUNTER := 10;
variable TRANSMIT_FREQUENCY: FREQUENCY;
variable EVENT_Timer: Time := 1 ns;
variable MIDDLE_INITIAL: Character;
variable Z: Integer := 0;
variable RADIUS: Real := 0.0;
```

To declare more than one variable of the same type, separate the variable names by the , delimiter. Separate the list of variable names from the data type by the : delimiter. If an initial value is specified, separate the initial value from the data type by the := delimiter.

By default, the initial value of variable STAR_COLOR is

```
COLOR'left = RED
```

Similarly, by default, the initial values of A, R1, TRANSMIT_FREQUENCY, and MIDDLE_INITIAL are respectively '0', 0, 0 Hz, and NUL.

Default initial values can be overridden by stating the initial value explicitly. For example, the initial value of variable BETA_LEVEL is explicitly declared to be 0.0 even though the default value would have been the same value. The initial value of CURRENT_STATE is defined explicitly to be the value of constant INITIAL_STATE which is S0. The initial value of A_COUNT is explicitly declared to be 10 and the initial value of EVENT_Timer is declared to be 1 ns.

Here are some declarations for array variables:

```
variable REG1: Bit_Vector (15 downto 0) := X"F5A2";
variable REG2: COUNTER_16_Bit := B"1111_0101_1010_0010";
variable COLOR_STRENGTH: COLOR_COUNT(ORANGE to INDIGO) :=
   (GREEN=> 15, BLUE | YELLOW => 20, others => 10);
variable MAP_COLOR:
   ARRAY_OF_COLORS(5 to 7) := (VIOLET, ORANGE, BLUE);
variable ROM_A: ROM_TYPE(0 to 7) :=
       (0=> X"FFFF_FFFF",
        5=> X"2222_CCCC",
    others=> X"0000_0000");
variable MATRIX_3X4: ARRAY_2D(0 to 2, 0 to 3) :=
       ( ('0','1','0','0'),
         ('1','1','0','0'),
         ('1','0','1','0'));
```

When a variable is associated with a data type that has an unconstrained index range, the actual index range for the variable is enclosed within parentheses and placed immediately after the type name. For example, variable REG1 is an array with elements of type Bit that has a descending index range. Variable REG2 is an array with elements of type Bit that has an ascending range (0 to 15) that is pre-specified in the constrained data type declaration for type COUNTER_16_BIT. Both REG1 and REG2 are initialized by placing the desired initial value after the delimiter := . For example, some initial values are REG1(15)='1', REG1(0)='0', REG2(0)='1', and REG2(15)='0'.

Variable COLOR_STRENGTH is an array with elements of type Natural and index range of type COLOR. Again, since type COLOR_COUNT has an unconstrained index range, the actual range of COLOR_STRENGTH is specified by placing the range in parentheses immediately after the type name. COLOR_STRENGTH is initialized by using a *named association list*. The index values are listed in any order and separated from the assigned value by the delimiter => . The assignments are separated by the , delimiter. Multiple elements can be assigned the same value by separating the index names by the | delimiter. For example, both the BLUE and YELLOW index are assigned the value 20. The reserved word *others* is used to refer to all index values not yet assigned values. In this example, index names ORANGE and INDIGO are assigned the value 10 by the *others* clause. Finally, the entire list of initial value assignments is enclosed in parentheses. Variable MAP_COLOR is an array with element type COLOR and index type Natural. The initial value is specified by using a *positional association list*. In this method, the initial values are listed in position order from left to right and enclosed within parentheses. For example, the initial value of MAP_COLOR(5) is VIOLET.

An *aggregate* is an element association list that is used to create a value of composite type. It may be either a named association list, or a positional association list. In the previous examples, variables ROM_A and COLOR_STRENGTH were initialized with an aggregate that uses a named association list. Variables MATRIX_3X4 and MAP_COLOR were initialized with an aggregate that uses a positional association list.

Elements of an array are referenced by placing the index name in parentheses following the array name. Using this convention, the initial values for some array elements are:

```
MAP_COLOR(5)  = VIOLET
MAP_COLOR(6)  = ORANGE
MAP_COLOR(7)  = BLUE
COLOR_STRENGTH(YELLOW) = 20
COLOR_STRENGTH(ORANGE) = 10
REG2(0) = '1'
REG1(15) = '1'
ROM_A(0) = X"FFFF_FFFF"
MATRIX_3X4(0,0) = '0'
MATRIX_3X4(1,0) = '1'
MATRIX_3X4(2,3) = '0'
```

Examples of record declarations include:

```
constant BIRTH_DATE: DATE := (16, AUGUST, 1943);
variable CURRENT_DATE: DATE := (MONTH => JANUARY, DAY => 1, YEAR => 2000);
```

Constant BIRTH_DATE uses the positional association list to define the initial value. Variable CURRENT_DATE uses the named association list to define the initial value.

3.5.2.3 *Declaration of Signals*

Signal declaration is similar to variable declaration. Signals may not be declared within processes or subprograms. Signals may be declared as ports in entity declarations or in the declarative region of architecture declarations. Examples of signal declarations include:

```
signal X1, X2, X3, X4, X5: Bit;
signal SR1, SR2, SR3, SR4: REG;
signal DOWN_COUNT: COUNTER := COUNTER'right;
```

Sequential Statements	Concurrent Statements
Assertion	Assertion
Signal assignment	Signal assignment
Procedure call	Procedure call
Variable assignment	Process
If…then…else	Block
Case	Component instantiation
Loop	Generate
Wait	
Null	
Next	
Exit	
Return	

Figure 3.18   Sequential and concurrent statements.

```
signal ROM_B: ROM_TYPE(0 to 3) :=
   (X"FFFF_FFFF", X"2222_CCCC", X"A03B_0020", X"ABCC_506C");
signal DATE_REGISTER: DATE := (1, JANUARY, 1900);
```

The initial value of X1 is '0' by default. Similarly, the initial value of SR1 is '0' by default. However, the initial value of DOWN_COUNT is 100. The initial value of ROM_B(0) is X"FFFF_FFFF".

All ports must be signals.

3.6   LANGUAGE STATEMENTS

This section describes various kinds of language statements. The VHDL language has two classes of language statements. Sequential statements are used in processes and subprograms to describe algorithms. They are executed sequentially, except when a branch statement is encountered, similar to statements in other procedural languages, such as C. Concurrent statements are used in architecture bodies to model signals. Unlike typical programming languages, these statements are not executed in the order written. Instead, concurrent statements are only executed when signals change that trigger the statement. In addition, all concurrent statements are executed once at the beginning of simulation. The order in which concurrent statements appear in the architecture body is unimportant. Some statement types are both concurrent and sequential. Figure 3.18 shows the classification of statements.

3.6.1   Assignment Statements

The values of both variables and signals may be changed by using assignment statements.

3.6.1.1   *Variable Assignment Statements*

A *variable assignment statement* causes replacement of the current value of the variable by a new value determined by an expression. The variable assumes its new value instantaneously. The compound delimiter := is used to define a variable assignment. Some simple examples of variable assignment statements follow using variables declared in a previous section:

```
STAR_COLOR := GREEN;
HAT_COLOR := MID_COLOR;                  -- Assigns value GREEN to HAT_COLOR.
BETA_LEVEL := 0.5;
MAP_COLOR(5) := YELLOW;
COLOR_STRENGTH(BLUE) := 100;
ROM_A(6) := X"2AA6_344F";
ROM_A(5) := ROM_A(0);
MATRIX_3X4(1,2) := '1';
MATRIX_3X4(2,2) := MATRIX_3X4(0,1);
CURRENT_STATE := S4;
A := '1';
R1 := REG'(50);                          -- Need type mark because R1 is type
                                         -- REG not type Integer.
MIDDLE_INITIAL := Character'val(68); -- Assigns value 'D'
                                         -- to MIDDLE_INITIAL
CURRENT_DATE := BIRTH_DATE;
CURRENT_DATE := (10, JANUARY, 1992);
CURRENT_DATE.YEAR := 1993;               -- Note that the "period" is used to
                                         -- separate the variable name from the
                                         -- element name.
Z := CURRENT_DATE.DAY;
R1 := CURRENT_DATE.DAY;                   -- ILLEGAL because R1 is not type
                                         -- Integer.
```

3.6.1.2 *Signal Assignment Statements*

A *signal assignment statement* schedules a new value for a signal to assume at some future time. The current value of the signal is never changed by a signal assignment statement. If no specific value of time is specified, the default value is an infinitesimally small value of time into the future known as *delta time*. To differentiate signal assignment from variable assignment, the delimiter <= is used instead of :=. Here are examples of simple signal assignment statements:

```
X1 <= '1' after 10 ns;
SR1 <= 5 after 5 ns;
X2 <= '0' after 10 ns, '1' after 20 ns, '0' after 30 ns;
X5 <= '1';
```

Figure 3.19 shows the timing implied by the signal assignment statements shown above, assuming that all statements are executed at time *t*. The current value of X1 at time *t* is assumed to be '0', the value of SR1 at time *t* is assumed to be 10, the value of X2 at time *t* is assumed to be '1', and the value of X5 at time *t* is assumed to be '0'. The first statement causes signal X1 to change to '1' at *t* + 10 ns. Similarly, the second statement causes signal SR1 to change from 10 to 5 at time *t* + 5 ns. The statement that assigns values to signal X2 is a waveform type that causes three future changes in signal X2. Since there is not an explicit specification of the time to schedule the change in X5, the change will occur at time *t* + delta.

The base type of the value assigned to a signal must be the same as the base type declared for the signal. For example, the signal assignment statement:

```
DOWN_COUNT <= 50.5 after 5 ns;
```

produces an error condition because the base type of DOWN_C0UNT is Integer but the value 50.5 is Real.

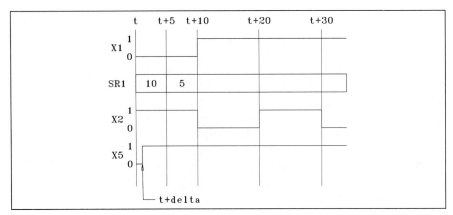

Figure 3.19   Timing diagrams for selected signal assignment statements.

3.6.1.3   Signal Drivers

If a process contains one or more signal assignment statements that schedule future values for some signal X, the VHDL simulator creates a single value holder called a *signal driver* for signal X that is associated with the process. The driver maintains an ordered list of scheduled value assignments for signal X. Each scheduled signal assignment is called a *transaction*. The new value assigned to signal X by a transaction may or may not be different from the current value of signal X. If signal X undergoes a change in value due to a transaction, an *event* occurs. The driver maintains the value of the signal between transactions. The VHDL simulator reads the current value of the signal from the driver.

   If more than one process contains signal assignment statements for the same signal X, the simulator creates a separate driver for signal X for each such process. Therefore, at any given time, there may be more than one driver, each associated with a different process, scheduling values for signal X. Obviously, there must be some way to determine the correct value for signal X if different drivers specify different values for signal X at the same time. The problem is solved by specifying a *resolution function* for signal X that computes the *resolved value* for all possible pairs of values that two different drivers might attempt to assign to signal X at a given time. Signal X is then called a *resolved signal*. The resolution function may be associated with the data type or the signal. These examples show how to specify a resolved signal:

```
signal X1: WIRED_OR Bit;        --Example 1

subtype RESOLVED_Bit is WIRED_OR Bit;
signal X2: RESOLVED_BIT;

signal Y1: RESOLVED STD_ULOGIC; --Example 2

subtype STD_LOGIC is RESOLVED STD_ULOGIC;
signal Y2: STD_LOGIC;
```

The first statement in the first example directly declares that signal X1 is a resolved signal of type Bit with resolution function WIRED_OR. The resolution function name is separated from the type name by a space separator character. The second statement associates the resolution function WIRED_OR with the data subtype RESOLVED_Bit which has base type Bit. Any signal declared to have subtype RESOLVED_BIT will automatically be a resolved signal with resolution function WIRED_OR. The third statement declares SIGNAL X2 to have subtype RESOLVED_BIT.

The second example repeats the process for IEEE standard logic. First, signal Y1 is directly declared to be a resolved signal of type STD_ULOGIC that uses resolution function RESOLVED. Then, the resolution function RESOLVED is associated with type STD_ULOGIC to form the subtype STD_LOGIC. Signal Y2 is declared to have subtype STD_LOGIC and, therefore, inherits the resolution function RESOLVED. In models for synthesis, most scalar signals and variables are declared to be of subtype STD_LOGIC.

When different signal drivers for signal X specify values for X at a given time, the simulator executes the resolution function to determine the correct value for signal X. Requirements for writing resolution functions are described in Chapter 5.

3.6.1.4   Signal Attributes

The VHDL language predefines several attributes for signals. S'*active* is a Boolean function. If S is a scalar signal, S'active returns the value TRUE if there is a transaction on the indicated signal during the current simulation cycle. If S is a composite signal, S'active returns TRUE if any scalar component of S has a transaction during the current simulation cycle. Similarly, if S is a scalar signal, S'*event* is a Boolean function that returns a value TRUE if there is an event on scalar signal S during the current simulation cycle. If S is a composite signal, S'event is TRUE if any scalar component of S has an event during the current simulation cycle.

```
X2'active         -- An expression that is TRUE when a transaction occurs on
                  -- signal X2 during the current simulation cycle.
X3'event          -- An expression that is TRUE when an event occurs on
                  -- signal X3 during the current simulation cycle.
```

S'*stable(n)* defines a Boolean signal that is TRUE at simulation time *t* only if there have been no events on signal S during the previous *n* units of time. Similarly, S'*quiet(n)* defines a Boolean signal that is TRUE at simulation time *t* only when there have been no transactions on signal S for the previous *n* units of time.

```
X1'stable(5 ns)   -- An expression that is TRUE when there have been no
                  -- events on signal X1 during the past 5 ns.

X4'quiet(5 ns)    -- An expression that is TRUE when there have been no
                  -- transactions on signal X4 during the past 5 ns.

X2'stable         -- Either expression is TRUE unless there
X2'stable(0 ns)   -- is an event on X2 during the current simulation cycle.

X3'quiet          -- Either expression is TRUE unless there
X3'quiet(0 ns)    -- is a transaction on signal X3 during the current
                  -- simulation cycle.
```

Note: The default time is 0 ns for both attributes '*stable* and '*quiet*.

S'*transaction* defines a signal of type Bit that toggles in each simulation cycle in which there is a transaction on signal S. An event on signal S'*transaction* indicates a transaction on signal S.

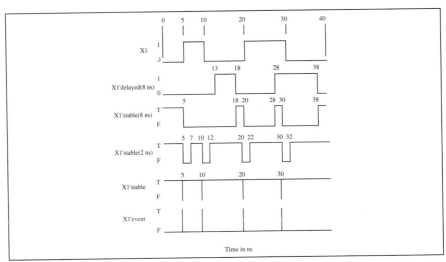

Figure 3.20 Timing diagram for signal X1 and its attributes.

S'*last_active* is a function that returns the amount of time that has elapsed since the last transaction on signal S. Similarly, S'*last_event* is a function that returns the amount of time that has elapsed since the last event on signal S.

```
X1'last_event    -- An expression that has a value equal to the amount of
                 -- time that has elapsed since the last event on signal X1.
X2'last_active   -- An expression that has a value equal to the amount of
                 -- time that has elapsed since the last transaction
                 -- on signal X2.
```

S'*last_value* is a function that returns the value of a signal S immediately before the last change of the signal.

```
X4'last_value    -- An expression that has a value equal to the value of
                 -- signal X4 immediately before the last event on
                 -- signal X4.
```

S'*delayed* defines a signal that is equivalent to a given signal delayed by a specified amount of time.

```
X1'delayed(5 ns) -- An expression that defines the current value of a
                 -- signal that is equal to signal X1 delayed by 5 ns.
```

Figure 3.20 shows a timing diagram for a signal X1 of type Bit and some of its signal attributes. There are events on signal X1 at times 5, 10, 20, and 30 ns. Signal X1'*delayed*(8 ns), which is also of type Bit, is just a copy of signal X1 that is shifted 8 ns forward in time. Therefore, there are events on signal X1'*delayed*(8 ns) at times 13, 18, 28, and 38 ns. Signal X1'*stable*(8 ns) is type Boolean. Its initial value is TRUE. It changes to FALSE at t=5 ns due to the event on X1. It changes from FALSE to TRUE at t=18 ns because signal X1 has been stable for 8 consecutive ns (had no events for the previous 8 ns). Signal X1'*stable*(8 ns) becomes FALSE again at t=20 ns due to the event on signal X1 at that time. At t=28 ns, signal X1 has been stable for the previous 8 ns; therefore signal X1'*stable*(8 ns) becomes TRUE again. The reader should verify the

Miscellaneous Operators	** abs not
Multiplying Operators	* / mod rem
Signing Operators	+ -
Adding Operators	+ - &
Relational Operators	= /= < <= > >=
Logical Operators	and or nand nor xor

Operators on the same line have equal precedence. For example, all logical operators have equal precedence. All operators on a given line have higher precedence than all operators below that line. Miscellaneous operators have the highest priority and logical operators have the lowest priority.

Figure 3.21 Precedence among VHDL operators.

remaining events on signal X1'*stable*(8 ns). Signal X1'*stable*(8 ns) has events at times 18, 20, 28, 30, and 38 ns.

Similarly, signal X1'*stable*(2 ns) is initialized to TRUE and becomes FALSE at t=5 ns to the event on X1. Signal X1'*stable*(2 ns) changes from FALSE to TRUE at t=7 ns because signal X1 has been stable for the previous 2 ns. The reader should verify all events on signal X1'*stable*(2 ns). Signal X1'*stable* is TRUE most of the time. It is only FALSE at times 5, 10, 20, and 30 ns which corresponds to events on signal X1.

Attribute X1'*event* is unlike the other attributes in Figure 3.20 because it is not a signal. Attribute X1'*event* is a function attribute that evaluates to TRUE only if there is an event on signal X1 during the current simulation cycle. Note: X1'*event* and X1'*stable* have complement values at all instances of time.

The following expressions illustrate various aspects of signal attributes. Note: Since X1'*stable* and X1'*delayed* are a signals, they also have attributes.

```
X1'delayed(8 ns)              -- this expression has value '1' at t=15ns
X1'stable(8 ns)               -- this expression has value FALSE at t=15ns
X1'stable(2 ns)               -- this expression has value TRUE at t=15ns
X1'event                      -- this expression has value FALSE at t=15ns
X1'last_event                 -- this expression has value 5ns at t=15ns
X1'last_value                 -- this expression has value '1' at t=15ns
X1'stable'last_event          -- Value is 5ns at t=15ns
X1'stable(2 ns)'last_event    -- Value is 3ns at t=15ns
X1'delayed(8 ns)'event        -- Has value TRUE only at times
                              -- 13,18,28, and 38 ns
```

3.6.2   Operators and Expressions

The VHDL language has a rich collection of built-in operators. Figure 3.21 shows the VHDL operators and the precedence relations among them. The set of operators in each line have equal precedence. The set of operators on a given line have higher precedence than all operators that appear below that line and lower precedence than all operators that appear above that line. For example, the relational operators have higher precedence than the logical operators and lower precedence than the adding operators.

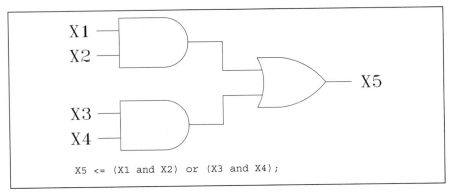

Figure 3.22   Logic circuit for VHDL statement.

3.6.2.1 *Logical Operators*

The *logical operators* define logical operations on variables or signals of type Bit or Boolean. The built-in logical operators cannot be applied to any user-defined data types. Logical operators may also define operations on arrays whose elements are Bit or Boolean. In this case, the logical operations are performed on each component of the array. Note: All logical operators have equal precedence. Parentheses must be used to force desired order of operator execution. For example, the signal assignment statement:

```
X5 <= X1 and X2 or X3 and X4;
```

is illegal and produces an analyzer error. The expression:

```
X5 <= (X1 and X2) or (X3 and X4);
```

is valid. Figure 3.22 shows the corresponding circuit.

The logical operators *and*, *or*, and *xor* are *associative*. For example:

```
(X1 and X2) and X3 = X1 and (X2 and X3).
```

Therefore, an assignment of the following type without parentheses:

```
X5 <= X1 and X2 and X3
```

is valid and correctly performs the function of a 3-input AND gate. However, the *nand* and *nor* operators are not associative. Therefore, a three input NAND operation *cannot* be written as:

```
X5 <= X1 nand X2 nand X3.   -- Erroneous format.
```

The previous expression produces an analyzer error. In fact, although the expression:

```
(X1 nand X2) nand X3
```

is a valid expression, it does not correctly represent a 3-input NAND gate because:

```
(X1 nand X2) nand X3 = (X1 and X2) or (not X3)
```

which is not the desired result. A correct representation for a 3-input NAND gate is:

```
X5 <= not (X1 and X2 and X3).
```

3.6.2.2 Relational Operators

The relational operators are defined in Table 3.2.

Table 3.2 Relational operators.

Symbol	Operation
=	Equal to
/=	Not equal to
>	Greater than
>=	Greater than or equal to
<	Less than
<=	Less than or equal to

Relational operators are used in expressions when it is necessary to compare the values of objects. The value of any expression that includes a relational operator is of type Boolean. Relational operators may be applied to any built-in or user-defined scalar data object. Objects that are being compared must have the same base type. For enumeration data types, the position numbers of the values are compared. The following example contains some typical relational expressions that involve constants, variables, and signals, defined earlier in this chapter.

```
BETA_LEVEL := 0.5;                       -- Variable assignment statement.
ALPHA_LEVEL := 0.75;
(BETA_LEVEL < ALPHA_LEVEL)               -- Expression that is TRUE.
STAR_COLOR := GREEN; HAT_COLOR := VIOLET;
(STAR_COLOR > HAT_COLOR)                 -- Expression that is FALSE.
R1 := REG'(3);
(COLOR'val(R1) = STAR_COLOR)             -- Expression that is TRUE.
CURRENT_STATE := S3;
(CURRENT_STATE /= INITIAL_STATE)         -- Expression that is TRUE.
(CURRENT_STATE > INITIAL_STATE)          -- Expression that is TRUE.
(STATE'right < CURRENT_STATE)            -- Expression that is FALSE.
(STATE'left = INITIAL_STATE)             -- Expression that is TRUE.
MIDDLE_INITIAL := 'G';
(MIDDLE_INITIAL >= '4')                  -- Expression that is TRUE.
(MIDDLE_INITIAL <= 'g')                  -- Expression that is TRUE.
(Character'pos(MIDDLE_INITIAL) <= 40) -- Expression that is FALSE.
SR1 <= 10;                               -- Signal assignment statement.
SR2 <= 25;                               -- Signal assignment statement.
DOWN_COUNT <= 50;                        -- Signal assignment statement.

(DOWN_COUNT <= 50)                       -- Expression that is TRUE.
B1 <= (DOWN_COUNT <= 50);                -- Assigns the value TRUE to a
                                         -- signal, B1, of type Boolean.
(SR2 <= SR1)                             -- Expression that is FALSE.
(DOWN_COUNT <= SR2)                      -- Illegal expression because objects
                                         -- have different base types.
```

Notice that the symbol "<=" has two valid meanings in the language. It can mean "less than or equal to" or it can denote a signal assignment. The context in which the symbol occurs determines its meaning.

3.6.2.3 Adding Operators

The *adding operators* are defined in Table 3.3.

Table 3.3   Adding operators.

Symbol	Operation
+	Arithmetic addition for numeric operands of the same base type.
-	Arithmetic subtraction for numeric operands of the same base type.
&	Concatenation of one-dimensional arrays of the same base type.

Arithmetic addition and subtraction apply only to types Integer or Real unless they are overloaded. Data types are strictly enforced. It is an error to attempt to add or subtract two operands that have different base types—even if the ranges of the base types are identical. Some examples follow:

```
BETA_LEVEL := 0.2;                          -- Variable assignment statement.
SR1 <= 10;                                  -- Register assignment statement.
SR2 <= 25;                                  -- Register assignment statement.
DOWN_COUNT <= 50;                           -- Register assignment statement.
SR3 <= SR2 - SR1;                           -- SR3 is assigned the value 15.
BETA_LEVEL := BETA_LEVEL + BETA_LEVEL;      -- BETA_LEVEL is assigned
                                            -- the value 0.4
DOWN_COUNT <= DOWN_COUNT - SR1;             -- Illegal expression because the
                                            -- operands do not have the same
                                            -- base type.
```

Concatenation produces a new one-dimensional array using two, one-dimensional arrays of the same base type. The length of the result is the sum of the lengths of the two operands. The elements of the new array consist of the elements of the left operand (in left to right order) followed by the elements of the right operand (also, in left to right order). The left bound of the result is the left bound of the left operand unless the left operand is the null array, in which case, the result of the concatenation is just the right operand. The direction of the result is the direction of the left operand unless the left operand is the null array, in which case, the direction of the result is the direction of the right operand. Some examples follow:

```
variable FIRST_NAME: STRING (1 to 4);
variable MIDDLE_INITIAL: Character;
variable LAST_NAME: STRING (7 downto 1);
variable NAME: STRING(1 to 15);
FIRST_NAME:= "JOHN";
MIDDLE_INITIAL := 'Q';
LAST_NAME:= "CITIZEN";
NAME:= FIRST_NAME & ' ' & MIDDLE_INITIAL & ". " & LAST_NAME;
        -- The value of NAME is "JOHN Q. CITIZEN".
```

```
   A     B   A/B   A rem B

   11    4    2       3
  -11    4   -2      -3
   11   -4   -2       3
  -11   -4    2      -3
```

Figure 3.23 Results of integer division.

Note that an object whose type is the array element type is a valid operand. Specifically, the Character 'Q' is a valid operand for the concatenation operation with a second operand of type STRING.

3.6.2.4 Sign Operators

The sign operators are the *unary* operators + (for keep the same sign) and – (for change the sign). They are only valid for numeric operands. Due to the precedence structure, neither of these unary operators can immediately follow a multiplying or miscellaneous operator. For example, A/–B and A**-B are illegal expressions. The use of parentheses is required. Expressions A/(–B) and A**(–B) are legal versions of the desired operations.

```
type SMALL_Integer is Integer range -100 to 100;
variable A: SMALL_Integer := 10;
signal B, C: SMALL_Integer;
C <= -A;    -- Signal C has value -10.
A := -B;    -- Variable A has value +100.
```

3.6.2.5 Multiplying Operators

Table 3.4 shows the multiplying operators.

Table 3.4 Multiplying operators.

Symbol	Operation
*	Arithmetic multiplication for Integer and Real types.
/	Arithmetic division for Integer and Real types.
Mod	Modulus operation for Integer types.
Rem	Remainder operation for Integer types.

Integer division and remainder are defined by this statement:

$$A = (A/B)*B + (A \text{ rem } B)$$

where (A rem B) has the sign of A and an absolute value less than the absolute value of B. Figure 3.23 shows the results of integer division for representative numbers.

The result of the modulus operation is such that (A mod B) has the sign of B and an absolute value less than the absolute value of B; in addition, for some integer N, this result must satisfy:

$$A = B*N + (A \text{ mod } B).$$

A	B	A mod B
11	4	3
-11	4	1
11	-4	-1
-11	-4	-3

Figure 3.24 Results of modulus operation.

Figure 3.24 shows the results of the *mod* operation for representative integer values. Other examples include:

```
variable A, B, C, D, E: Integer;
A := 5; B := 10; C := 15;  -- Initialize variables.
D := A * (-B) / 15;        -- Variable D has value -3.
E := (A *(-B)) rem 15;     -- Variable E has value -5.
E := (A * (-B)) mod 15;    -- Variable E has value 10.
```

Any object having a physical data type can also be multiplied or divided by an entity of type Integer or Real. The result has the same physical type as the first object. Two objects having the same physical data type may also be divided. The result is *universal integer*, an Integer data type that is not constrained by the implementation. For example,

```
variable RES1, RES2, RES3: RESISTANCE := 1 KOHM;
variable T1, T2: Time := 10 ns;
variable Z1, Z2: Integer := 10;
RES1 := Z1 * RES2;      -- RES1 has value 10 kohm.
T1 := 2.5 * T2;         -- T1 has value 25.0 ns.
Z1 := RES1/RES2;        -- Z1 has value 10.
T2 := T2/10;            -- T2 has value 1 ns.
Z1 := RES1/Z2;          -- Invalid statement. RES1 is type RESISTANCE,
                        -- Z1 is type Integer.
RES1 := RES1 * RES2;    -- Invalid statement.
                        -- Units do not match.
```

3.6.2.6 Miscellaneous Operators

The miscellaneous operators have the highest precedence of all operators and are shown in Table 3.5.

Table 3.5 Miscellaneous operators.

Symbol	Operation
**	Exponentiation. The left operator must be of type Integer or Real. The right operator must be of type Integer. The result will be of the same type as the left operator.
abs	Unary operator that computes the absolute value of a numeric data type.
not	Logical NOT operator for types Bit or Boolean.

Exponentiation with a negative exponent is only allowed if the left operand is Real. Some examples of these operations are:

```
variable A, B, C: Integer := 5;
variable D: REAL := 2.5;
signal X1, X2, X3, X4: Bit := '1';
C := 2 ** A;              -- C is assigned the value 32.
D := 2.5 ** 2;           -- D is assigned the value 6.25
D := D ** (-1);          -- D is assigned the value 0.16
C := A ** (-2);          -- Invalid expression.
C := A ** (D);           -- Invalid expression.
C := abs (A - (3 * B));  -- C is assigned the value 10
X4 <= X1 and not X2;     -- X4 is assigned the value '0'.
```

3.6.3   Sequential Control Statements

The control statements described in this section are used in processes and subprograms to define algorithms. They are executed in order of appearance.

3.6.3.1   *Wait Statement*

The full *wait* statement format is illustrated as follows:

```
wait on X,Y until Z=0 for 100 ns;
```

This statement suspends execution of the process for a maximum of 100 ns. If an event occurs on X or Y prior to 100 ns, the condition "Z=0" is evaluated. If "Z=0" is TRUE when the event on X or Y occurs, the process will resume at that time; otherwise, suspension continues. In effect, the process resumes after 100 ns has passed or when an event occurs on X or Y with Z=0 — whichever comes first. Other valid forms of the *wait* statement are:

```
wait for 100 ns;          -- Always resume after 100 ns.
wait on A,B,C;            -- Resume when an event occurs on signal
                         -- A, B, or C.
wait until Z=0;           -- Resume when the expression "Z=0" becomes TRUE
                         -- due to an event on signal Z.
wait on A,B,C for 100 ns; -- Wait for a maximum of 100 ns.
                         -- Resume earlier than 100 ns if there is an
                         -- event on A,B, or C.
wait on X,Y until Z=0;    -- Wait for an event on X or Y with Z=0.
wait until A=B for 100 ns; -- Wait for the expression "A=B" to become TRUE
                         -- as the result of an event on A or B.
                         -- Maximum wait is 100 ns.
wait;                    -- wait forever, i.e., permanently suspend
                         -- the process.
```

3.6.3.2   *If Statement*

The format for the *if* statement is as follows.

```
if CONDITION1 then
   -- sequence_of_statements_1
elsif CONDITION2 then
   -- sequence_of_statements_2
        .
   -- any number of elsif clauses
        .
else
   -- last_sequence_of_statements
end if;
```

The conditions are evaluated in order from top to bottom. The first condition that evaluates to TRUE causes the corresponding {sequence_of_statements} to be executed. If all conditions evaluate to FALSE, the sequence associated with the *else* clause is executed. Note: At most one {sequence_of_statements} will be executed, even if more than one condition is true. Here is an example:

```
if A < 0 then
   LEVEL := 1;
elsif A > 1000 then
   LEVEL := 3;
else
   LEVEL :=2;
end if;
```

3.6.3.3   Case *Statement*

The format for the *case* statement is as follows:

```
case EXPRESSION is
   when CHOICES1 => --sequence_of_statements_1
   when CHOICES2 => --sequence_of_statements_2
               .
               .
               .
   when others => --last_sequence_of_statements
end case;
```

It is required that all possible choices for values for the expression be included exactly once in the set of choices. The sets of choices must be mutually exclusive, and the union of all sets must equal the set of all possible values for the expression. The expression must be a discrete type or a one-dimensional character array type. Each choice must be of the same type as the expression. The choices may be a list, for example:

```
when 0|1|2 => (AB <= "LOW");
```

The keyword *others* is used to cover all possible values for the expression not included in any of the specific choices. This is a convenient way to insure that all possible values for the expression are covered.

If A and B are of type Integer with range (-10 to +10) and X is type STRING, consider this example:

```
case A+B is
   when 0       => X <= "ZERO";
   when (1 to 20)=> X <= "Positive";
   when others  => X <= "NEGATIVE";
end case;
```

3.6.3.4   Loop *Statement*

There are three types of *loop* statements as shown:

```
-- FOR loop
for NAME in RANGE loop
   -- sequence_of_statements
end loop;
```

```
-- WHILE loop
while CONDITION loop
   -- sequence_of_statements
end loop;

-- Simple loop
loop
   -- sequence_of_statements
end loop;
```

In the *for* loop, NAME must have the same base type as the RANGE values. The loop is executed multiple times with NAME assuming a different value in RANGE each time.

In the *while* loop, the CONDITION is evaluated prior to execution of the loop statements. Therefore, the sequence of statements will never be executed if the CONDITION is initially FALSE. The sequence of statements is continually executed for as long as the CONDITION is TRUE. An infinite loop is possible if the CONDITION is always TRUE.

In the *simple* loop, the sequence of statements is continuously executed. To prevent an infinite loop, the sequence of statements must contain a *wait* statement or an *exit* statement.

Some examples follow:

```
while A<B loop
   -- sequence_of_statements
   A := A + 1;
end loop;
--
for I in 1 to 10 loop
   A(I) := A(I) + 1;
end loop;
--
loop
   compute (x);
   exit when x < 10;-- Exit statement is defined below.
end loop;
```

3.6.3.5 Next *Statement*

The *next* statement is used to terminate the current loop iteration if a given condition is true. When the iteration is terminated, the sequence of statements begins again with the next iteration value. This is the format:

```
next [loop_label] [when CONDITION];
```

If the loop_label is present, the current iteration of the loop with the given label is terminated when the CONDITION is TRUE. If the loop_label is absent, the innermost loop is terminated when the CONDITION is TRUE. If the CONDITION is omitted, the appropriate loop iteration is always terminated.

3.6.3.6 Exit *Statement*

The *exit* statement behaves in a similar manner to the *next* statement except that the entire loop statement is terminated. The format for the *exit* statement is:

```
exit [loop_label] [when CONDITION];
```

If loop_label is present, the loop with the given label is terminated when the CONDITION is TRUE even though all values in the iteration range may not have been processed. If loop_label is absent, the innermost loop is terminated when the CONDITION is TRUE even though all values in the iteration range may not have been processed. If the CONDITION is omitted, the appropriate loop is always terminated.

3.6.3.7 Null *Statement*

The *null* statement does nothing. It is useful in *case* statements, for example, if no action is desired for certain choices. Recall that all choices must be covered. Therefore, the *null* statement is mandatory for this situation.

3.6.4 Architecture Declarations and Concurrent Statements

Concurrent statements are used in architecture bodies to describe the behavior of signals. The placement of concurrent statement is illustrated as follows:

```
architecture ARCHITECTURE_NAME of ENTITY_NAME is
   -- Architecture declaration section
   -- Signals are declared here.
   -- Variables may NOT be declared here.
begin
   --
   -- Concurrent statements that describe signal behavior
   --
end ARCHITECTURE_NAME;
```

In this section, we discuss process statements, assert statements, and signal assignment statements. Other concurrent statements are introduced as needed.

3.6.4.1 Process *Statement*

The *process* statements we have been using as examples are concurrent statements. They are the fundamental statement type used in architectures. In fact, all other concurrent statements can be written as an equivalent *process* statement. As we discuss various concurrent statements, we will take advantage of this equivalence to aid in the description of other concurrent statements. *Process* statements come in two forms. One form contains a sensitivity list. This is the form used in most examples up to this point in the book.

```
LABEL:process (SENSITIVITY_SIGNAL_LIST)
           -- constant_declarations
           -- variable_declarations
           -- subprogram declarations
           -- signal declarations are NOT permitted here
           -- other, more specialized, items may also be declared here
       begin
           -- sequential_statements
       end process LABEL;
```

For example, the architecture ALGORITHMIC for the 1's counter of Figure 3.2 consists of just one concurrent process statement. Note: The label is optional. In this example, there is no label for the *process* statement. The sensitivity list consists of the single signal A. This process is executed once at the beginning of simulation, and thereafter, only when there is an event on signal

A. In general, every *process* statement within an architecture is executed once at the beginning of simulation, and thereafter, only when a signal in its sensitivity list changes value (i.e., when there is an event on one or more of the signals in the sensitivity list). Notice that constants and variables can be declared within processes, but not signals. The statements within processes must be sequential statements. Each time that the process is activated, the sequential statements are executed, as in more familiar procedural languages, such as C or C++.

Variables declared within processes are *static*. They are initialized only once at the beginning of simulation and retain their values between process activations.

The other form of *process* statement has no sensitivity list.

```
LABEL: process
          -- constant_declarations
          -- variable_declarations
          -- subprogram declarations
          -- signals may NOT be declared here.
          -- other, more specialized, items may also be declared here
       begin
          -- sequential_statements
       end process LABEL;
```

This kind of process is executed once at the beginning of simulation and continues to execute until a *wait* statement is encountered. When the last sequential statement is executed, the process automatically begins again with the first sequential statement. Therefore, at least one of the sequential statements must be a *wait* statement in order to prevent an infinite loop. A *process* statement with a sensitivity list is equivalent to a process without a sensitivity list that has a *wait* statement as the last sequential statement in the process. For example, the process:

```
S_LIST: process (S1, S2)
           -- constant_declarations
           -- variable_declarations
        begin
           -- sequential_statements
        end process S_LIST;
```

is equivalent to the process:

```
NO_LIST: process
            -- constant_declarations    -- Same as in S_LIST
            -- variable_declarations     -- Same as in S_LIST
         begin
            -- sequential_statements     -- Same as in S_LIST
            wait on S1, S2
         end process NO_LIST;
```

assuming that all declarations and sequential statements are the same except for the last wait statement. In either case, the process executes once at the beginning of simulation, and thereafter, whenever an event occurs on signal S1 or S2.

Clearly, a process without a sensitivity list must have a *wait* statement and a process with a sensitivity list may not have a *wait* statement. The first requirement prevents infinite loops, and the second requirement is necessary to prevent conflicting process activation.

3.6.4.2   Concurrent Assert Statement

The *assert* statement may be either sequential or concurrent. This is the format for a concurrent *assert* statement:

```
LABEL: assert Boolean_EXPRESSION
        report "Message_string"
        severity SEVERITY_LEVEL;
```

When there is an event on any signal in the Boolean_EXPRESSION, the Boolean_EXPRESSION is evaluated. If the expression evaluates to FALSE, the Message_string is written to the standard output device. SEVERITY_LEVEL is a predefined enumeration data type in the standard VHDL language. The default level is WARNING. Interpretation of SEVERITY_LEVEL is implementation dependent. For example, some implementations abort simulation if the SEVERITY_LEVEL is too high. See Figure 3.16 for the definition of severity levels. Here is an example:

```
LABEL: assert (A or B) = C
        report "WARNING: C is NOT equal to (A or B)"
        severity WARNING;
```

It is assumed that A, B, and C are signals of type Bit. Under normal operating conditions, C is always supposed to be equal to (A or B). We want to detect any violation of this condition. The concurrent assert statement is activated once at the beginning of simulation, and thereafter, any time there is an event on either of the three signals (A, B, or C). When activated, the condition in the first line is evaluated. The condition is "(A or B) is equal to C." When the condition is FALSE, i.e., when (A or B) is NOT equal to C, the warning message will be printed at the standard output device. This output can be redirected to a file in most computer systems.

This concurrent assertion statement is equivalent to the following process statement.

```
LABEL: process (A, B, C)
        begin
           assert (A or B) = C
           report "WARNING: C is NOT equal to (A or B)"
           severity WARNING;
        end process LABEL;
```

3.6.4.3   Concurrent Signal Assignment Statement

Signal assignment statements can be either sequential or concurrent. If they are sequential, they are only executed when the algorithm reaches the statements. When they are concurrent, they are executed any time there is an event on any signal on the right hand-side of the assignment statement. For example, the concurrent signal assignment statement:

```
LABEL: C <= A or B;
```

is activated once at the beginning of simulation, and thereafter, when an event occurs on signal A or signal B. This concurrent signal assignment is equivalent to the following process:

```
LABEL: process (A,B)
          begin
            C <= A or B;
          end process LABEL;
```

Concurrent signal assignment statements can also be conditional. The format for a conditional concurrent signal assignment statement is:

```
LABEL: SIGNAL_NAME <= [transport]
        WAVEFORM1 when CONDITION1 else
        WAVEFORM2 when CONDITION2 else
                        .
                        .
                        .
        WAVEFORMn when CONDITIONn else
        WAVEFORMq;
```

The concurrent statement is activated once at the beginning of simulation, and thereafter, when there is an event on any signal in any of the WAVEFORMs or on any signal in any of the CONDITIONs. If CONDITION1 is TRUE, WAVEFORM1 is assigned to SIGNAL_NAME. If CONDITION1 is FALSE and CONDITION2 is TRUE, WAVEFORM2 is assigned to SIGNAL_NAME. WAVEFORMq is assigned to SIGNAL_NAME only if CONDITION1 through CONDITIONn are all FALSE. Only one of the WAVEFORMs is assigned to the signal, even if more than one of the CONDITIONs happen to be TRUE. The WAVEFORM that goes with the first CONDITION that is TRUE is assigned. An example follows:

```
LL1: S <= A or B when XX=1 else
         A and B when XX=2 else
         A xor B;
```

In this example, it is assumed that A, B, and S are signals of type Bit and that XX is a signal of type Integer. The statement is executed once at the beginning of simulation, and is executed again any time there is an event on signal A, B, or XX. This concurrent signal assignment statement is equivalent to the following *process* statement:

```
LL1: process (A,B,XX) begin
         if XX=1 then S <= A or B;
         elsif XX=2 then S <= A and B;
         else S <= A xor B;
         end if;
     end process LL1;
```

The format for a selected concurrent signal assignment statement is:

```
LABEL: with EXPRESSION select
        SIGNAL_NAME <= [transport]
        WAVEFORM1 when CHOICES1,
        WAVEFORM2 when CHOICES2,
                        .
                        .
                        .
        WAVEFORMn when CHOICES,
        WAVEFORMq when others;
```

Every possible value for EXPRESSION must be included in exactly one of the CHOICES. The sets of CHOICES must be mutually exclusive, and the union of all sets of CHOICES must equal the set of all possible values for EXPRESSION. Note: *Others* is a keyword that includes all values for EXPRESSION not included in any of the CHOICES. It is, therefore, a convenient way to insure that all possible values are covered. The concurrent statement is activated once at the beginning of simulation, and thereafter, when an event occurs on any signal in EXPRESSION or on any signal in any of the WAVEFORMS. For example, the statement:

```
LL2: with (S1 + S2) select
       C <= A after 5 ns when 0,
            B after 10 ns when 1 to Integer'high,
            D after 15 ns when others;
```

activates at t=0 and whenever there is an event on any of the signals S1, S2, A, B, or D. All signals are assumed to have type Integer. This selected concurrent signal assignment statement is equivalent to the following *process* statement:

```
LL2: process (S1, S2, A, B, D)
       begin
         case (S1 + S2) is
            when 0 => C <= A after 5 ns;
            when 1 to Integer'high => C <= B after 10 ns;
            when others => C <= D after 15 ns;
         end case;
       end process LL2;
```

3.6.5   Subprograms

In any computer language one augments the basic operations built into the language by writing functions and procedures. In VHDL, functions and procedures fall under the general class of programming constructs known as *subprograms*. Functions compute and return a value to the invoking expression. A function does not modify any of its arguments and may be invoked only in an expression. Procedures are both sequential and concurrent statements. They do not return a value to the invoking program, but may modify their arguments. The following architecture shows where subprograms are declared and where they are invoked:

```
architecture WITH_SUBPROGRAMS of ENTITY_NAME is
   -- Subprograms may be declared here
begin
   -- Invoke subprograms here
end WITH_SUBPROGRAMS;
```

Subprograms declared in the architecture declaration region are visible only within that architecture. They may also be declared in packages and made visible to any architecture via a *use* statement. (See the section of this chapter on packages for more details.) Subprograms may also be declared in the declaration regions of processes, blocks, and even other subprograms. A subprogram is only directly visible within the construct in which it is declared. (See the section of this chapter on visibility for further details.)

We will now explain how subprograms are *declared* and how they are *invoked*.

3.6.5.1   *Functions*

Functions can be declared in VHDL by specifying:

1. The name of the function.
2. The input parameters, if any.

3. The type of the returned value and function.

4. Any declarations required by the function itself.

5. An algorithm for the computation of the returned value.

This is the general format for a function declaration:

```
function FUNCTION_NAME (FORMAL_PARAMETER_DECLARATIONS)
    return RETURN_TYPE is
        -- constant and variable declarations
        -- no signal declarations are allowed here
begin
    -- sequential statements
    --
    return (RETURN_VALUE);
end FUNCTION_NAME;
```

Formal parameters in functions must have mode *in*. Therefore, it is not necessary to specify the mode of function formal parameters. The only allowed object classes are *constant* and *signal*. The default object class is *constant*. Since the formal parameters are always mode *in*, functions do not have any side effects on the invoking environment. Functions return a value to the invoking expression, but no actual parameter in the invoking environment is modified.

Variables and constants may be declared in the declaration region of functions. Variables in functions are dynamic. Variable values are initialized every time the function is called. The values are not held between invocations of the function. Since a function specifies an algorithm, all statements in the function body must be sequential statements. Since a value must be immediately returned to the invoking expression, *wait* statements are not permitted in the function body. This restriction also applies to any function or procedure called by the function.

At the beginning of this chapter, we presented an architectural body MACRO for the entity ONES_CNT, where the functions MAJ3(X) and OPAR3(X) were employed. To use these functions they must first be declared as follows:

```
function MAJ3(X: Bit_Vector(0 to 2)) return Bit is
    begin
        return (X(0) and X(1)) or (X(0) and X(2)) or (X(1) and X(2));
    end MAJ3;
```

Parameter X has default class *constant*. Note that the value returned is of type Bit. The value returned can be specified by an expression, as in the case above, or can be computed by executing a sequence of statements. As stated, local declarations can be made, but they are not required for this simple case. Functions are used as part of expressions. Thus, the MAJ3 function was invoked in architecture MACRO of ONES_CNT as follows:

```
C(1) <= MAJ3(A);
```

Figure 3.25 illustrates the use of a function to convert an Integer to a Bit_Vector of length *N*. It is declared within architecture ALG of entity PULSE_GEN. The function is passed the integer INPUT and the number of bit positions *N*. Since no class is specified for parameters INPUT and N, they are both assumed to be class *constant*. It is assumed that *N* is large enough that the resultant bit vector can hold the converted value. The function implements the standard conversion algorithm employing successive integer division by decreasing powers of two. At

```
entity PULSE_GEN is
  generic(N: INTEGER; PER: TIME);
  port(START:in BIT; PGOUT:out BIT_VECTOR(N-1 downto 0);
       SYNC:inout BIT);
end PULSE_GEN;
architecture ALG of PULSE_GEN is
  function INT_TO_BIN (INPUT : INTEGER;N : POSITIVE)
    return BIT_VECTOR is
    variable FOUT: BIT_VECTOR(0 to N-1);
    variable TEMP_A: INTEGER:= 0;
    variable TEMP_B: INTEGER:= 0;
  begin                 -- Begin function code.
    TEMP_A := INPUT;
    for I in N-1 downto 0 loop
      TEMP_B := TEMP_A/(2**I);
      TEMP_A := TEMP_A rem (2**I);
      if (TEMP_B = 1) then
        FOUT(N-1-I) := '1';
      else
        FOUT(N-1-I) := '0';
      end if;
    end loop;
    return FOUT;
  end INT_TO_BIN;
begin                   -- Begin architecture body.
  process(START,SYNC)
    variable CNT: INTEGER:= 0;
  begin                 -- Begin process
    if START'EVENT and START='1' then
      CNT := 2**N-1;
    end if;
    PGOUT <=  INT_TO_BIN(CNT,N) after PER;
    if CNT /= -1 and START ='1' then
      SYNC <= not SYNC after PER;
      CNT := CNT -1;
    end if;
  end process;
end ALG;
```

Figure 3.25   Function to convert an Integer type to type Bit_Vector.

each step, the bit of the result is determined by the quotient of the integer division operation. The remainder from the integer division becomes the dividend for the next iteration.

The purpose of entity PULSE_GEN is to generate all possible binary patterns of length N. In architecture ALG of PULSE_GEN, the function INT_BIN is used to convert the value of counter (CNT) to a bit vector of length N. CNT cycles through values of 0 through 2**N - 1. Thus, all bit vector combinations are produced. Each bit vector persists for period PER. A synchronization signal is toggled every time the bit vector value changes.

The parameters declared in the function declaration are called *formals*. In this example, the formals for function INT_TO_BIN are INPUT and N. The formal parameters are only visible inside the function declaration. They are used to specify the algorithm performed by the function. When a function is invoked, the parameters in the invoking statement are called *actuals*. In this example, the actuals in the invoking statement are CNT and N. This example uses the positional association list to match the actuals to the formals. CNT is associated with INPUT, and N is associated with N. Therefore, in the architecture body, variable CNT is converted from Integer to Bit_Vector form.

3.6.5.2 *Procedures*

Procedures are subprograms that can modify one or more of the input parameters. A procedure is declared by specifying:

1. The name of the procedure.
2. The input and output parameters, if any.
3. Any declarations required by the procedure itself.
4. An algorithm.

Here is an outline for a procedure declaration:

```
procedure PROCEDURE_NAME (FORMAL_PARAMETER_DECLARATIONS)
    -- Procedure declaration part
    -- Variables and constants may be declared
    -- Signals may NOT be declared here
begin
    -- Sequential statements
end PROCEDURE_NAME;
```

Formal parameters may have mode *in*, *out*, or *inout*. If the mode is unspecified, it is assumed to be *in*. The allowable object classes are *constant*, *variable*, and *signal*. If the mode is *in* and no object class is specified, then *constant* is assumed. If the mode is *out* or *inout*, the default object class is *variable*.

Variables and constants may be declared in the declaration section, but not signals. Variables in procedures are dynamic; they are initialized every time the procedure is invoked and do not hold values between invocations. Sequential statements are used in the procedure body after the keyword *begin* to specify the algorithm to be performed by the procedure. Unlike functions, procedures may contain *wait* statements. However, if a procedure is called by a function, it may not contain any *wait* statements. The procedure does not return a value to the invoking expression. Instead, a procedure modifies one or more of its formal parameters. This causes the actual parameter that is matched to the formal parameter in the invoking statement to be modified in the invoking environment. As usual, if the formal parameter is class *signal*, the signal is not updated during the current simulation cycle, but is instead scheduled for update in a future simulation cycle.

Figure 3.26 shows the declaration of a procedure ADD that performs addition of bit vectors. Formal parameters A, B, and CIN are mode *in*; therefore, they can only be read by the procedure. The procedure cannot assign new values to formal parameters of mode *in*. Since no object class is specified for formal parameters A, B, and CIN, they are assumed to be *constant*. Formal parameters SUM and COUT are mode *out*; therefore, they can only be assigned values by the procedure. The procedure cannot read the current value of formal parameters of mode *out*. Since no object class is specified, parameters SUM and COUT have default class *variable*.

```
procedure ADD(A,B: in BIT_VECTOR; CIN: in BIT;
              SUM: out BIT_VECTOR; COUT: out BIT) is
   variable SUMV,AV,BV: BIT_VECTOR(A'LENGTH-1 downto 0);
   variable CARRY: BIT;
begin
   AV := A;
   BV := B;
   CARRY := CIN;
   for I in 0 to SUMV'HIGH loop
     SUMV(I) := AV(I) xor BV(I) xor CARRY;
     CARRY := (AV(I) and BV(I)) or (AV(I) and CARRY)
              or (BV(i) and CARRY);
   end loop;
   COUT := CARRY;
   SUM := SUMV;
end ADD;
```

Figure 3.26   Procedure to add bit vectors.

Note: The statements that assign new values to SUM and COUT use the := operator. The new values assigned to SUM and COUT take effect immediately during the current simulation cycle. Several internal procedure variables are declared that are visible only inside the procedure. These variables are dynamic and are initialized every time the procedure is called. Since no specific initial values are specified, the default initial values are '0' for variables of type Bit and a vector of all zeros for variables of type Bit_Vector. This implementation of the add operation works for bit vectors having either descending or ascending ranges and works for bit vectors of any length. This is accomplished by using the internal variables of type Bit_Vector (A'LENGTH-1 downto 0) and using SUMV'high in the loop range specification. It is assumed, however, that the least significant bit is on the right. Each pass through the loop implements the full adder equations. Repeated application of the loop implements a mutltibit addition.

In general, procedures can be invoked as concurrent statements in architecture bodies or as sequential statements in processes, functions, or other procedures. However, procedure ADD can only be used as a sequential statement in processes, functions, or other procedures because the outputs, SUM and COUT, have object class *variable* and must, therefore, be matched to actuals of class *variable*. Since variables cannot be declared in architecture declaration regions, procedure ADD cannot be invoked as a concurrent statement in an architecture body.

Similar to functions, when a procedure is invoked, actual parameters from the invoking environment are associated with the formal parameters using either a positional association list or a named association list. A formal and its associated actual must have the same base type. Object class matching is discussed in the next section.

3.6.5.3   *Matching Object Classes of Formals and Actuals in Subprograms*

Table 3.6 shows the required matching of object classes during the invocation of subprograms. The fact that a formal object of class *constant* can be matched with an actual *expression* is an interesting case. When a formal object has class *constant*, that implies the subprogram cannot alter it. When the subprogram is called, the value of the parameter takes on the value of the expression. In its primitive forms, an expression can be a signal or a variable; therefore, these

Table 3.6   Object class matching for formals and actuals.

Formals	Actuals
Signal	Signal
Variable	Variable
Constant	Expression

objects can also be matched with the formal when the formal has class *constant*. Again, the MAJ3 function provides an example. No object class is specified for the parameter X. Thus, by default, its object class is *constant*. In the architecture MACRO of ONES_CNT, the actual used in the call is the signal A, which is a valid expression, and can, therefore, be matched with the *constant* formal parameter X.

For procedure ADD in Figure 3.26, formal parameters A, B, and CIN have default class *constant*. Therefore, either formal parameter can be matched to an actual expression in the invoking statement. In the following invocation, process P1 uses procedure ADD to select one of two possible sums to assign to signal R5. Note that both procedure calls match formal parameters A, B, and CIN (mode *in* and object class *constant*) to signals. This is valid since a signal name is a valid expression. Each procedure call matches formal parameters SUM and COUT (mode *out* and object class *variable*) to actual variables R6 and COUT_TEMP, respectively. This is necessary since formal parameters of class *variable* must be matched to actuals of class *variable*. Signal assignment statements at the end of the process assign the desired value to signals R5 and COUT.

```
architecture PROCEDURE_INVOCATION of PROCEDURE_TEST is
   -- Procedure declaration goes here.
   signal R1,R2,R3,R4,R5: Bit_Vector (7 downto 0);
   signal CIN: Bit := '0';
   signal COUT: Bit;
   signal S1: Boolean;
begin
   -- Other concurrent statements.
P1:  process (S1)
      variable R6: Bit_Vector (7 downto 0);
      variable COUT_TEMP: Bit;
   begin
     if S1 then
        ADD(R1, R2, CIN, R6, COUT_TEMP);
     else
        ADD(R3, R4, CIN, R6, COUT_TEMP);
     end if;
     R5 <= R6;
     COUT <= COUT_TEMP;
   end process P1;
   -- Other concurrent statements.
end PROCEDURE_INVOCATION;
```

```
function "and" (L, R: MVL4) return MVL4 is
  -- Declare a two-dimensional table type.
  type MVL4_TABLE is array (MVL4, MVL4) of MVL4;
  -- truth table for "and" function
  constant table_AND: MVL4_TABLE :=
--  ------------------------------
--  |  X    0    1    Z  |
--  ------------------------------
    (('X', '0', 'X', 'X'),    --  |  X  |
     ('0', '0', '0', '0'),    --  |  0  |
     ('X', '0', '1', 'X'),    --  |  1  |
     ('X', '0', 'X', 'X'));   --  |  Z  |
begin
  return table_AND(L, R);
end "and";
```

Figure 3.27 Overloading the *and* operator.

3.7 ADVANCED FEATURES OF VHDL

This section covers advanced features of the VHDL language.

3.7.1 Overloading

In VHDL one has the capability to change the meaning of literals, and the names of operators, functions, and procedures by redeclaration. This process is called *overloading*. Consider the following signal assignment statement:

```
F <= (A and B) or (C and D);
```

Suppose that A, B, C, D, and F were originally declared to be of type Bit, i.e., two-valued logic was used. Later, it becomes desirable to model the same logic equation using a four-valued logic type known as MVL4, which has the values 'X', '0', '1', and 'Z'. We can redeclare A, B, C, D, and F to be of type MVL4. Note that in doing this, we have overloaded the literal values '0' and '1' because these two literals are members of the type set for type Bit as well. However, having declared F, A, B, C, and D to be of type MVL4, the built-in AND and OR operators do not apply since they are defined for types Bit and Boolean only. One could define two functions, MVL4_AND(X;Y) and MVL4_OR(X;Y), to perform the indicated operations, but this would require rewriting the equation using the following function call notation:

```
F <= MVL4_OR(MVL4_AND(A,B),MVL4_AND(C,D));
```

There are a number of problems with this approach. For complicated Boolean expressions, this functional form is difficult to read, and the writing of Boolean equations in this form is error prone. A better approach is to overload the meaning of the *and* and *or* operators. One can declare a new *and* operator as shown in Figure 3.27.

In the declaration area of the function, a two-dimensional array type is declared. The elements of arrays of this type are of type MVL4, as are the two array indices. The operation table for the *and* operation is declared next as a constant. The function returns only the value accessed from the table. Given the declaration of this function, one can then write expressions, such as:

```
F <= A and B;
```

```
function INTVAL (VAL: MVL4_VECTOR)
    return INTEGER is
  variable SUM: INTEGER:= 0;
begin
  for  N in VAL'LOW to VAL'HIGH loop
    assert not(VAL(N) = 'X' or VAL(N) = 'Z')
    report "INTVAL inputs not 0 or 1"
    severity WARNING;
    if  VAL(N) = '1' then
      SUM := SUM + (2**N);
    end if;
  end loop;
  return SUM;
end INTVAL;
```

Figure 3.28   Declaration of a function to convert type MVL4_VECTOR to type Integer.

provided that F, A, and B are of type MVL4. If for some reason one wishes to use functional call notation, the same expression would be written as:

```
F <= "and"(A,B);
```

An immediate question might arise: Is the built-in *and* operator still available for use with objects of type Bit and Boolean? The answer is yes. The VHDL analyzer can infer which *and* operator is required in an expression by examining the parameter profiles and result profiles for the operations having the same name. The parameter profile specifies the number, order, and type of operand parameters. The result profile specifies the type of the returned result. Using these profiles, the analyzer can differentiate between the different operators. For example, the types of the two parameters for the built-in *and* function are of type Bit or Boolean. For the MVL4 *and* operator, the two operands are of type MVL4.

Subprogram names can also be overloaded. For example consider the function shown in Figure 3.28 which converts an MVL4_VECTOR to an integer. The function first checks to see that bit positions are not equal to X or Z. Assuming that the vector is made up of all 0's or 1's, the function adds in the appropriate power of two for each detected value of '1'. Note: The function assumes that the least significant bit of the word to be converted has the lowest index.

Suppose another integer conversion function shown in Figure 3.29 is then declared that converts a parameter of type Bit_Vector to an Integer: Since the names of the two functions are the same, the type of the calling parameter determines which function is invoked.

Another question about overloading might arise: If two subprograms are declared with identical names and identical parameter and result profiles, which one will be visible? The answer is that the most recent declaration hides the previous one.

The power of overloading is best illustrated by showing how the IEEE packages (discussed in detail in the next section) can be used to overload operations. First, all Boolean operations are overloaded for type STD_LOGIC and STD_LOGIC_VECTOR. Second, arithmetic operators can be overloaded. In Figure 3.30, the three inputs to the entity are type STD_LOGIC_VECTOR. However, because package STD_LOGIC_SIGNED is used, the overloaded + operator in this package interprets the inputs as signed numbers and performs twos complement arithmetic and returns a result of type STD_LOGIC. If package STD_LOGIC

```
function INTVAL(VAL:BIT_VECTOR) return INTEGER is
  variable SUM: INTEGER:=0;
begin
  for N in VAL'LOW to VAL'HIGH loop
    if VAL(N)='1' then
      SUM := SUM + (2 ** N);
    end if;
  end loop;
  return SUM;
end INTVAL;
```

Figure 3.29   Declaration of function to convert type BIT_VECTOR to type Integer.

```
use IEEE.STD_LOGIC_1164.all;
use IEEE.STD_LOGIC_SIGNED.all;.
entity ADD_OVERLOAD is
  port ( A, B, C: in  STD_LOGIC_VECTOR (7 downto 0);
         SUM: out STD_LOGIC_VECTOR (7 downto 0));
end ADD_OVERLOAD;
architecture PACK_SIGNED of ADD_OVERLOAD is
begin
  SUM <=  A + B + C; --This is now 2's-
                     --complement addition
end PACK_SIGNED;
```

Figure 3.30   Overloading ADD with signed package.

_UNSIGNED is used instead, the overloaded + operator in this package performs unsigned addition.

In summary, overloading gives one the ability to alter the meaning of a name without having to change the VHDL code in which it is invoked.

Overloaded subprogram names must either have a different number of parameters, or at least one of the formal parameters must have a different data type. For example, one can have three overloaded versions of a D flip-flop as illustrated by the following three procedure calls, where all formal parameters are assumed to have type Bit:

```
DFF (CLK, D, Q);
DFF (CLK, D, Q, QBAR);
DFF (CLK, D, CLEAR, PRESET, Q);
```

However, one cannot have two overloaded versions of the D flip-flop, illustrated by the following two procedure calls, where all formal parameters have type Bit:

```
DFF (CLK, D, CLEAR, Q, QBAR);
DFF (CLK, D, PRESET, Q, QBAR);
```

The preceding version of overloaded DFF procedures are invalid because the number of formal parameters is the same for each, and all parameters have the same type.

```
    package HANDY is
      subtype BITVECT3 is BIT_VECTOR(0 to 2);
      subtype BITVECT2 is BIT_VECTOR(0 to 1);
      function MAJ3(X: BIT_VECTOR(0 to 2)) return BIT;


    ------ Other Declarations---------------------------

    end HANDY;
    package body HANDY is
      function MAJ3(X: BIT_VECTOR(0 to 2))
        return BIT is
      begin
        return (X(0) and X(1)) or (X(0) and X(2))
               or (X(1) and X(2));
      end MAJ3;

    ------ Other Subprogram Declarations ----------------

    end HANDY;
```

Figure 3.31 Definition of a package.

3.7.2 Packages

It becomes tedious to repeat declarations each time they are needed. VHDL uses a package to hold frequently used declarations, and each package has a name associated with it. Declarations in the package may be made visible by referring to the package name. First, however, the package must be declared. Figure 3.31 gives an example of such a declaration. Note: The package declaration for package HANDY declares subtypes and the interface for a function. The code for the function is given in the package body. If a package contains no subprograms, a package body is not required.

Given the definition of package HANDY, suppose one wants to give an entity named LOGSYS access to the package. One can do this by placing a *use* clause before the interface description for entity LOGSYS:

```
use work.HANDY.all;
entity LOGSYS is
   port(X: in BITVECT3;Y: out BITVECT2);
end LOGSYS;
```

All the declarations contained in HANDY are then *visible* within the entity LOGSYS, including any of its architecture bodies. We discuss visibility in greater detail in the next section.

The general form for a *use* statement is:

```
use LIBRARY_NAME.PACKAGE_NAME.ELEMENT_NAME;
```

where LIBRARY_NAME is the name of a library that contains the package, PACKAGE_NAME is the name of the package, and ELEMENT_NAME is the name of a specific item in the package. For example:

```
use MY_LIBRARY.MY_PACKAGE.DFF; -- Get one component.
use MY_LIBRARY.MY_PACKAGE.all; -- Get everything in package.
```

In the HANDY example, the keyword *work* refers to the default library that is associated with the current project. This library contains all packages, procedures, components, etc., that have been analyzed as part of the current project.

Packages are a very useful language feature. Design groups can use standard packages that contain the type declarations and subprograms related to their systems. As shown in Chapter 5, when we discuss system modeling, the amount of code in these declarations and subprograms is substantial. The package mechanism relieves the modeler from having to repeatedly enter this code. If shared by different designers in a design group, it also ensures consistency across the design.

The VHDL language defines a package STANDARD that can be used by all entities. Among other things, this package contains the definitions for types Bit, Bit_Vector, Boolean, Integer, Real, Character, String, and Time, as well as subtypes Positive and Natural.

IEEE has developed Standard 1164 as a standard nine-valued logic system, with particular application to logic synthesis. The IEEE committee developed two packages (1) STD_LOGIC_1164, which defines the basic value system and associated functions (it is used as is by vendors) and (2) NUMERIC_STD, which provides overloaded arithmetic and other operators for synthesis. Vendors have developed their own versions of this package. In particular, Synopsys has developed:

1. STD_LOGIC_ARITH, which defines two new standard logic vector types SIGNED and UNSIGNED and associated overloaded arithmetic and relational operators
2. STD_LOGIC_UNSIGNED, which provides overloaded arithmetic and relational operators for objects of type STD_LOGIC_VECTOR that are interpreted as unsigned numbers, and
3. STD_LOGIC_SIGNED, which provides overloaded arithmetic and relational operators for objects of type STD_LOGIC_VECTOR that are interpreted as signed (twos complement) numbers.

All of the above three packages allow one of the operands to be of type Integer. Synopsys package STD_LOGIC_MISC includes Boolean reduction operators, e.g., the *and* reduction operator computes the AND of all bits of a standard logic vector to obtain a scalar result. Synopsys package STD_LOGIC_TEXTIO provides text I/O for objects of type STD_LOGIC and STD_LOGIC_VECTOR.

IEEE has also developed packages MATH_REAL and MATH_COMPLEX, which contain operators and functions to support the evaluation of mathematical expressions, such as MATH_REAL, which contains code to evaluate transcendental functions.

3.7.3 Visibility

In describing the VHDL language, we have given numerous examples of the declaration of VHDL types, objects, and subprograms, but have not given an exact definition of where they are declared. We will do so now:

Region. A logically continuous, bounded portion of text.

Declaration region. A region in which a name can be used to unambiguously refer to a declared item.

After an item has been declared at some point in the declaration region, its name is then said to be *visible* to the end of the declaration region. The practical effect of this is that an item must be declared before use. Names are normally visible only within the region where they are declared, i.e., they are *directly visible*. However, as we will see, names can be made visible outside their declaration region through *library* and *use* clauses. This second type of visibility is called *visibility by selection*.

We now list the following constructs, which establish declaration regions. In doing so, we divide them into two categories: *general declaration regions* that allow a wide range of declarations and *specialized declaration regions* that allow only a restricted set of declaration types.

Constructs which establish a general region are:

1. An entity declaration and an associated architecture.
2. A process.
3. A block.
4. A subprogram.
5. A package.

Figure 3.32 gives an example with signals and variables declared with the name X in five different declarative regions: a package SIG, an entity Y and its architecture Z, a function R, a nested block B, and a process within architecture Z. Since each of these declarations is local to the given declaration region, they can be made without conflict. In each case, the name X is directly visible to the end of the declaration region in which X is declared.

Signal X (X=1) declared in package SIG is directly visible throughout package SIG. Signal X (X=2) declared in entity Y is directly visible throughout entity Y and all its architectures. Therefore, Z2 is assigned the value 2. The *return X* statement in function R returns the value of X=3 that is directly visible to function R. Therefore, Z3 is assigned the value 3. The value of Z5 is 5 because it is assigned the value of X (X=5) that is directly visible to process P1.

In three cases the name X is made visible by selection outside the declaration region in which it was declared:

1. The package name is made visible by the "use work.SIG" statement. (More is said about this in the next section.) Then, the selected name SIG.X is used to refer to the value of X within package SIG. The value of this signal (X=1) is assigned to Z1.
2. The selected name B.X is used to refer to the value of X within block B. The value of this signal (X=4) is assigned to Z4.
3. Inside block B, the selected name Y.X is used to refer to the value of X declared in entity Y. The value of this signal (X=2) is added to the local value of X (X=4) that is directly visible to block B and the result (6) is assigned to Z6.

Note: In each case, the selected name is of the form P.S where P is the prefix denoting the construct in which the name declaration occurs, and S is the suffix, which is the declared name. Name selection can be carried to an unlimited number of levels of nesting. For example, a selected name of the form P1.P2.P3.S denotes a declared name S inside P3 that is nested in P2, which is in turn nested in P1.

```
package SIG is
  signal X: INTEGER:= 1;
end SIG;

use work.SIG.all;
entity Y is
  signal X: INTEGER:= 2;
end Y;

architecture Z of Y is
  signal Z1,Z2,Z3,Z4,Z5: INTEGER:= 0;
  function R return INTEGER is
    variable X: INTEGER := 3;
  begin
    return X; -- Returns value of 3.
  end R;
begin
  B: block
    signal X: INTEGER := 4;
    signal Z6: INTEGER := 0;
  begin
    Z6 <= X + Y.X;   -- Z6 = 6
  end block B;

P1: process
    variable X: INTEGER :=5;
  begin
    Z5 <= X;         -- Z5=5
    wait;
  end process;

  Z1 <= work.SIG.X; -- Z1=1
  Z2 <= X;           -- Z2=2
  Z3 <= R;           -- Z3=3
  Z4 <= B.X;         -- Z4=4
end Z;
```

Figure 3.32 Declarative regions.

If one simulates the architecture Z of entity Y, one finds that signal Z_i is assigned the value i. The following constructs establish specialized declaration regions:

1. Record types: The names of the record fields are implicitly declared.
2. Loops: The loop index control variable is implicitly declared.
3. Component declarations: The component name and port and generic names are implicitly declared.
4. Configuration: Configuration declarations have a declaration region for attribute specifications and *use* clauses.

3.7.4 Libraries

When VHDL models are analyzed with no errors, the result is stored in a library. The various libraries that design groups create maintain the state of the designs that the group is developing. The existence of these libraries allows the existing VHDL models to be used in future VHDL descriptions. As will be shown below, the models resident in a design library can be pointed to, so it is not necessary to have the actual code resident in the model using the design library models.

There are two types of VHDL libraries, the *work* library, and *resource* libraries. All current analysis results are stored in the work library. Resource libraries can be referenced during analysis and simulation, but cannot be written into. However, at some time a resource library must be designated the work library so that models can be analyzed into it. Below, we illustrate how this is done.

Libraries contain both *primary* and *secondary* units. Primary units are entity, package, or configuration declarations. Secondary units are architectures and package bodies. A primary unit must be analyzed before any of its corresponding secondaries. A secondary unit must be analyzed into the same library as its corresponding primary.

Libraries have both *logical* and *physical* names. The logical name is used in the VHDL description. This name is "portable," and is used to refer to the library regardless of the system in which the library is installed. The physical name is the name used by the host operating system to refer to the library. In general, it changes depending on the system in which the library is installed. There must be some method of mapping logical names to physical names. One approach follows.

Logical library names must be declared to make them visible. This is done by means of a library clause. Suppose, there exists a library whose logical name is DESIGN1. This name is made visible by means of the following library clause:

```
library DESIGN1;
```

One can then refer to this library in other statements. For example, suppose that package SIG presented in the previous section is in library DESIGN1. The signal X could be made visible by the following two statements:

```
use DESIGN1.SIG;
use SIG.X;
```

Or the single statement:

```
use DESIGN.SIG.X;
```

However, it is interesting that, in terms of their total effect, the two approaches are not equivalent. Using the two statements makes both the package name SIG and the signal named X visible. Using the single statement makes the signal name X visible but not the package name SIG.

For the logical library names WORK and STD, no library clause is necessary, i.e., the library names WORK and STD are always visible. The library STD contains the packages STANDARD and TEXTIO. All names within package STANDARD are visible, i.e., the clause "use STD.STANDARD.all" is implicitly assumed. However, a *use* clause is required to access items within package TEXTIO.

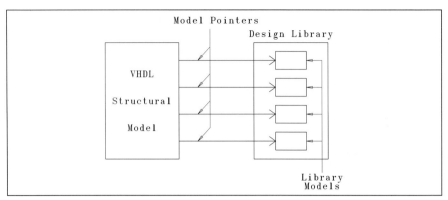

Figure 3.33   Pointers to library models.

To illustrate the mapping of logical library names to physical library names, we explain the approach used in the Synopsys system. The system employs a User Options File to hold the mappings. For example, the logical to physical name mapping for logical library name DESIGN1 can be mapped as:

```
DESIGN1 > LIB1
LIB1 : /VTVHDL/USERS/JRA/LIB1
```

The first statement maps the logical library name DESIGN1 to the host library name LIB1. The second statement maps LIB1 to the directory /VTVHDL/USERS/JRA/LIB1.

As discussed, the library whose logical name is WORK is always the library which receives the results of analysis. In the Synopsys system, one can cause analyzed files to be added to library LIB2 by placing the following statements in the User Options File:

```
WORK  > LIB2
LIB2: /VTVHDL/USERS/JRA/LIB2
```

and host library LIB2 will receive the results of analysis. If one later wishes to make LIB1 the work library, the User Options File file can be changed, or one can temporarily make LIB1 the WORK library by invoking the analyzer:

```
VHDLAN -W LIB1  MODEL.VHD
```

As soon as this analysis is complete, LIB2 becomes the WORK library again.

3.7.5   Configurations

VHDL structural architectures are developed by instantiating components that are declared in the declaration region of the architecture. Before a structural model can be simulated, each instantiated component must be bound to a library model. Figure 3.33 shows how this is accomplished. A VHDL structural model contains a set of pointers to models in a design library of the type discussed in the previous section. One way to specify the pointers is to place configuration specification statements directly in the structural model. Figure 3.34 shows an example of this using the structural architecture for entity TWO_CONSECUTIVE presented in Chapter 1. Note that each configuration specification statement, commented as "pointer," is of the form:

```
for INSTANTIATED_COMPONENT use LIBRARY_COMPONENT
```

```
    entity TWO_CONSECUTIVE is
      port(CLK,R,X: in BIT; Z: out BIT);
    end TWO_CONSECUTIVE;
    use work.all;
    architecture STRUCTURAL of TWO_CONSECUTIVE is
      signal Y0,Y1,A0,A1: BIT := '0';
      signal NY0,NX: BIT :='1';
      signal ONE: BIT :='1';
      component EDGE_TRIGGERED_D
        port(CLK,D,NCLR: in BIT;
             Q,QN: out BIT);
      end component;
      for all: EDGE_TRIGGERED_D
        use entity EDGE_TRIG_D(BEHAVIOR); --model pointer
      component INVG
        port(I: in BIT;O: out BIT);
      end component;
      for all: INVG
        use entity INV(BEHAVIOR); --model pointer
      component AND3G
        port(I1,I2,I3: in BIT;O: out BIT);
      end component;
      for all: AND3G
        use entity AND3(BEHAVIOR); --model pointer
      component OR2G
        port(I1,I2: in BIT;O: out BIT);
      end component;
      for all: OR2G
        use entity OR2(BEHAVIOR); --model pointer
    begin
      C1: EDGE_TRIGGERED_D
        port map(CLK,X,R,Y0,NY0);
      C2: EDGE_TRIGGERED_D
        port map(CLK,ONE,R,Y1,open);
      C3: INVG
        port map(X,NX);
      C4: AND3G
        port map(X,Y0,Y1,A0);
      C5: AND3G
        port map(NY0,Y1,NX,A1);
      C6: OR2G
        port map(A0,A1,Z);
    end STRUCTURAL;
```

Figure 3.34   Configuration specification for entity TWO_CONSECUTIVE.

For the case of the D flip-flops the INSTANTIATED_COMPONENTS are "all: EDGE_TRIGGERED_D". The term *all* implies that all instantiated EDGE_TRIGGERED_D components are mapped to the same model. One can map each flip-flop individually, such as:

```
for C1: EDGE_TRIGGERED_D
  use entity EDGE_TRIG_D_A(BEHAVIOR);
for C2: EDGE_TRIGGERED_D
  use entity EDGE_TRIG_D_B(BEHAVIOR);
```

which would map each flip-flop to a different library model.

```
configuration PARTS of TWO_CONSECUTIVE is
  for STRUCTURAL
    for all: EDGE_TRIGGERED_D
        use entity work.EDGE_TRIG_D(BEHAVIOR);
    end for;
    for all: INVG
        use entity work.INV(BEHAVIOR);
    end for;
    for all: AND3G
        use entity work.AND3(BEHAVIOR);
    end for;
    for all: OR2G
        use entity work.OR2(BEHAVIOR);
    end for;
  end for;
end PARTS;
```

Figure 3.35   Configuration declaration for entity TWO_CONSECUTIVE.

In this case, library WORK contains the models. As discussed, the library name WORK is always visible. The statement "use work.all" makes the entity names of the models visible. Alternatively, one can dispense with the "use work.all" statement entirely and use the selected name of the component model. For example, the component specification for the D flip-flops can read:

```
for all: EDGE_TRIGGERED_D
   use entity work.EDGE_TRIG_D(BEHAVIOR); --model pointer
```

Again, this works because the name of library WORK is always visible.

Another possible approach to binding components is through the use of configuration declarations. In this approach, the component instances in the structural architecture are left unbound, and the component specification statements are collected in a separate analyzable unit called a configuration declaration. For example, suppose all the configuration specification statements (those commented as "pointer") are removed from the architecture STRUCTURAL of entity TWO_CONSECUTIVE as shown in Figure 3.34. The component specifications can then be collected in a configuration declaration shown in Figure 3.35. Note that the name of the configuration declaration is PARTS, and is a configuration declaration for entity TWO_ CONSECUTIVE. Inside the configuration declaration, the outer *for* statement gives the name of the architecture being configured. The inner *for* statements are the configuration specifications for the instantiated components.

Configuration declarations are primary library units. However, they must be analyzed after the architecture they are configuring. Also, when the model is being simulated, it is the configuration declaration that is bound to the component being tested in the test bench that tests the model. Figure 3.36 illustrates this.

Configuration declarations are useful when one wants to perform multiple simulations of the same structural architecture with the components bound to different library components for each simulation. In this case, one can analyze the structural architecture once. One can then create any number of different configurations for the architecture. The structural architecture itself

```
entity TB is
  end TB;
use WORK.all;
architecture TCTEMPT of TB is
  signal X,R,CLK,Z: BIT;
  component TWIR
    port (CLK,R,X: in BIT;
          Z: out BIT);
    end component;
    for C1: TWIR use configuration work.PARTS;
begin
  C1: TWIR
    port map(CLK,R,X,Z);
  CLK <='0',              '1'after  10 ns,'0'after  20 ns,
        '1'after  30 ns,'0'after  40 ns,'1'after  50 ns,
        '0'after  60 ns,'1'after  70 ns,'0'after  80 ns,
        '1'after  90 ns,'0'after 100 ns,'1'after 110 ns,
        '0'after 120 ns,'1'after 130 ns,'0'after 140 ns;
  X <= '0','1' after  15 ns,'0' after 55 ns;
  R <= '1','0' after 125 ns,'1' after 127 ns;
end TCTEMPT;
```

Figure 3.36   Testing the configuration.

never has to be analyzed again. This reduces errors and analysis time and promotes reuse of models. In this scenario, the unbound structural architecture corresponds to a printed circuit board with all chip sockets empty. Each configuration corresponds to the insertion of a specific set of chips into the empty sockets. In Chapters 7 and 11, we discuss other useful features of configuration declarations.

3.7.6   File I/O

A crucial requirement in VHDL model development is the ability to drive a model with test vectors. In our discussion of test benches, we showed how this could be done to a limited extent through the use of waveform elements and signal assignment statements. This however, is impractical for large test vector sets. A related problem is that one wants to be able to initialize memories at the start of simulation from an external file. Finally, one would like to be able to write human readable, formatted simulation results to an external file. To address these problems, VHDL provides for file I/O capability, i.e., the ability to read from and write to external files during simulation. All files are sequential and can be regarded as data streams. There are two types of files that can be written and read: formatted and text files. Formatted files must be written by a VHDL simulator; the file format is host dependent and is not human readable in the host environment. Text files are human readable and can be created in the host environment using a text editor. Files can be of mode "in" or "out," but not mode "inout." Thus, a file that is written to, during a given simulation, cannot be read during that same simulation.

3.7.6.1 *Formatted I/O*

Two declarations are required for formatted I/O: a file type declaration and a file declaration using the previously declared file type. For example, a file type representing a stream of bit vectors used to test a model might be declared as:

```
type INP_COMB is file of Bit_Vector;
```

The base type of this file type is Bit_Vector. In general, the base type of a file type may be any type except an access type or another file type. Given this declaration, one can then declare an output file as:

```
file OUTVECT: INP_COMB is out "TEST.VECT";
```

In this declaration, OUTVECT is the logical file name, while "TEST.VECT" is the host file name. The host file name is a string of Characters and, therefore, must be enclosed within double quote marks. Whenever a file of mode "out" is declared, a WRITE procedure is implicitly defined. For example, for the above file, the corresponding WRITE procedure is:

```
WRITE(OUTVECT: out INP_COMB; V: in Bit_Vector);
```

where V is a variable declared in the declaration region in which the WRITE procedure is invoked. Whenever procedure WRITE is executed, the value of V is appended to the file OUTVECT.

Similarly, one can declare an input file:

```
file INVECT: INP_COMB is in "TEST.VECT";
```

A READ procedure and ENDFILE function are implicitly defined for all input files. For the file declared above, the declarations are:

```
READ(INVECT,V,LENGTH)
ENDFILE(INVECT)
```

Each invocation of the READ procedure reads a bit vector from the head of the file INVECT and assigns it to the variable *V.* The next bit vector in the sequential file then moves to the head of the file. For each read, the natural number LENGTH returns the length of the bit vector. This procedure parameter is present only when the base type of the file type is an unconstrained array. The ENDFILE function returns a value of type Boolean. The returned value is TRUE when end of file is reached; otherwise, the value FALSE is returned.

To illustrate the use of formatted I/O, consider Figure 3.37. The VHDL code consists of two entities OBVS and IBVS. Entity OBVS writes a sequence of bit vectors to a host file called TEST.VECT. The test vectors are generated by the entity PULSE_GEN, which we discussed in the section on functions. Recall that for a given value of *N,* it generates a sequence consisting of all $2*N$ possible bit vectors of length *N*. As it outputs each new vector, it toggles the output SYNC. Each time SYNC changes, the process WRITE_VECT executes writing the bit vector to a file.

Entity IVBS "plays" the test vectors back. When input PLAY goes to '1' the first *wait* statement in process READ_VECT is satisfied, and the loop is entered. Each pass through the loop reads one vector and assigns it to the variable *V. V* is then assigned to BVOUT. Next, there is a wait for period PER before doing the next file read. Thus, BVOUT changes every PER time

```
entity OBVS is
  generic(N:INTEGER;PER: TIME);
    port(GEN: in BIT);
end OBVS;
use work.all;
architecture FIO of OBVS is
  type INP_COMB is file of BIT_VECTOR;
  file OUTVECT: INP_COMB is out "TEST.VECT";
  signal VECTORS: BIT_VECTOR(N-1 downto 0);
  signal SYNC: BIT;
  component PG
    generic(N: INTEGER; PER: TIME);
    port(START: in BIT;
         PGOUT: out BIT_VECTOR(N-1 downto 0);
         SYNC: inout BIT);
  end component;
  for C1: PG  use entity work.PULSE_GEN(ALG);
begin
  C1: PG
    generic map(N => N, PER => PER)
    port map(START => GEN, PGOUT => VECTORS, SYNC => SYNC);
  WRITE_VECT: process(SYNC)
    variable V: BIT_VECTOR(N-1 downto 0);
  begin
    V := VECTORS;
    WRITE(OUTVECT,V);
  end process WRITE_VECT;
end FIO;
entity IBVS is
  generic(N:INTEGER;PER: TIME);
  port(PLAY: in BIT; BVOUT: out BIT_VECTOR(N-1 downto 0));
end IBVS;
architecture FIO of IBVS is
  type INP_COMB is file of BIT_VECTOR;
  file INVECT: INP_COMB is in "TEST.VECT";
begin
  READ_VECT: process
    variable LENGTH: NATURAL := N;
    variable V: BIT_VECTOR(LENGTH-1 downto 0);
  begin
    wait on PLAY until PLAY = '1';
    loop
      exit when ENDFILE(INVECT);
      READ(INVECT,V,LENGTH);
      BVOUT <= V;
      wait for PER;
    end loop;
  end process READ_VECT;
end FIO;
```

Figure 3.37 Writing to and reading from a formatted file.

units. The loop runs until end of file is reached. The complete test vector sequence is thus replayed.

Entity OBVS writes to file TEST.VECT, and entity IBVS reads from it. For this to work properly, two conditions must be satisfied:

1. The output file (OUTVECT) and the input file (INVECT), which are in fact the same host file, "TEST.VECT", must be declared in different declarative regions.
2. During a given simulation, one must not attempt to write and then read from the same host file. Thus, in our preceding example, entities OBVS and IBVS cannot be executed during the same simulation.

These conditions ensure that a file is not used in mode *inout* during a given simulation.

3.7.6.2   *Text I/O*

As discussed, formatted files are not human readable. Moreover, they cannot be easily created outside of VHDL simulation in the host operating system environment. In many situations, however, one would like to be able to read files during simulation that were created by a text editor or that are perhaps the object code output of an assembler. VHDL provides text I/O for these applications.

The basic declarations for text I/O are contained within a package TEXTIO, which is contained in library STD. As discussed above, the name of library STD is always visible; however, the name or contents of package TEXTIO must be made visible through *use* clauses. The first two declarations in package TEXTIO are:

```
type LINE is access STRING;
type TEXT is file of STRING;
```

Type LINE is declared to be an access type STRING. *Access types* are used for variables that point to memory locations. These pointers can be created and released during simulation. Thus, storage is not permanently allocated for these pointers. (Details of access types can be found in more advanced books on VHDL.) Type TEXT is a file type whose base type is the type STRING. The type STRING is used to represent a wide variety of input data in the host file environment.

Given these definitions, one can then declare a text file. For example in the case of reading bit vectors from an externally developed file, one can declare:

```
file INVECT: TEXT is "TVECT.TEXT";
```

Thus, the text file with a logical name INVECT is mapped to the host file name "TVECT.TEXT".

The data in VHDL text files is organized in lines, with a variable number of elements per line. To read data from a text file, one first executes a READLINE command to read an entire line, followed by a number of READ commands to strip off individual elements within the line. These functions are also declared in package TEXTIO. For an example of their use, consider the case of reading bit vectors from the text file declared above. In this case, one can invoke the reading process with:

```
READLINE(INVECT,VLINE);
```

```
entity TBVS is
  generic(N:INTEGER;PER: TIME);
  port(PLAY: in BIT;
        BVOUT: out BIT_VECTOR(N-1 downto 0));
end TBVS;
use STD.TEXTIO.all;
architecture TIO of TBVS is
begin
  process
    variable VLINE: LINE;
    variable V:BIT_VECTOR(N-1 downto 0);
    file INVECT: TEXT is "TVECT.TEXT";
  begin
    wait on PLAY until PLAY = '1';
    while not(ENDFILE(INVECT)) loop
      READLINE(INVECT,VLINE);
      READ(VLINE,V);
      BVOUT <= V;
      wait for PER;
    end loop;
  end process;
end TIO;
```

Figure 3.38 VHDL code that reads an input file of type TEXT.

where VLINE is a variable declared to be of type LINE. This is then followed by a number of invocations of:

```
READ(VLINE,V);
```

where V is a variable of type Bit_Vector. Each invocation of this command strips one bit vector off VLINE and copies it into the variable V.

To illustrate the total process, we demonstrate the reading of a text file "TVECT.TEXT" that was created using a text editor and contains the following:

```
000
001
010
011
100
101
110
111
000
```

Figure 3.38 shows VHDL code, which can read this file. For our example, the generic N is set to 3. Then during simulation, when play goes to a '1', the while loop is entered. For each pass through the loop, a line is read, and the single bit vector on that line is stripped and outputted to BVOUT. In this case, READ is only invoked once because only one bit vector appears on each line. If one wants to have multiple bit vectors per line, it is invoked multiple times. When end of file is reached, the loop is exited. The process then returns to the *wait* statement at the top of the process and suspends operation until another event occurs on signal PLAY.

```
package RFP is
  type POWER is range 0 to 1E9
    units pW;
      nw = 1000 pW;
      uW = 1000 nw;
    end units;
  type  FREQUENCY is range 0 to 1E9
    units Hz;
      KHz = 1000 Hz;
      MHz = 1000 KHz;
    end units;
  type RF_SIGNAL is
  record
    STRENGTH: POWER;
    FREQ: FREQUENCY;
  end record;
end RFP;
```

Figure 3.39   Package containing declarations for types.

Writing to text files is performed by assembling lines using a WRITE command, then writing entire lines to a file using the WRITELINE command. We do not illustrate this process here but leave it as an exercise at the end of the chapter.

Note: In this example, the READ function performs a type conversion from type TEXT to type Bit_Vector. These "target" types are restricted to the following:

```
Boolean        Integer
Bit            Real
Bit_Vector     String
Character      Time
```

Although fairly comprehensive, note that physical types other than type Time are not included. User-defined enumeration types are also not included. To illustrate how this type limitation can be overcome, consider the situation where one might want to model radar signals in VHDL. The VHDL code is shown in Figure 3.39. In package RFP, types POWER and FREQUENCY are first declared. Next, a record type RF_SIGNAL is declared with fields STRENGTH and FREQ, which have types POWER and FREQUENCY, respectively. Suppose we create a text file with the name SIGNAL.IN which contains:

```
10 26
30 25
27 24
14 27
```

Each line in this file contains two integer values separated by a space that represents a record value of type RF_SIGNAL. The first integer represents the signal power in picowatts; the second integer represents the signal frequency in megahertz. Entity TB in Figure 3.40 reads this file when the signal RSIG goes to a '1'. For each pass through the loop, the READLINE command reads a line from the file, which consists of the two integers. Next, each integer is stripped off the line using the READ command, and multiplied by the appropriate physical type base unit (pW or MHz) to convert the integers into the appropriate type. The resultant values are copied into the

```
entity TB is
end TB;
use work.RFP.all;
use STD.TEXTIO.all;
architecture TFILE5 of TB is
  signal RFSIG: RF_SIGNAL;
  signal RSIG: BIT;
begin
  process
    variable TEMP_STRENGTH: INTEGER:= 0;
    variable TEMP_FREQ: INTEGER:= 0;
    variable L: LINE;
    file INFILE: TEXT is in "SIGNAL.IN";
  begin
    wait on RSIG until RSIG = '1';
    while not(ENDFILE(INFILE)) loop
      READLINE(INFILE,L);
      READ(L,TEMP_STRENGTH);
      RFSIG.STRENGTH <= TEMP_STRENGTH*1 pW;
      READ(L,TEMP_FREQ);
      RFSIG.FREQ   <= TEMP_FREQ*1 MHz;
      wait for 10 ns;
    end loop;
  end process;
  RSIG <= '1';
end TFILE5;
```

Figure 3.40   Reading radar signals with TEXTIO.

```
/
  0 NS
    SMON1:    ACTIVE /TB/RSIG (value = '1')
    SMON:     ACTIVE /TB/RFSIG (value = (STRENGTH => 10 PW,
              FREQ => 26000000 HZ))
 10 NS
    SMON:     ACTIVE /TB/RFSIG (value = (STRENGTH => 30 PW,
              FREQ => 25000000 HZ))
 20 NS
    SMON:     ACTIVE /TB/RFSIG (value = (STRENGTH => 27 PW,
              FREQ => 24000000 HZ))
 30 NS
    SMON:     ACTIVE /TB/RFSIG (value = (STRENGTH => 14 PW,
              FREQ => 27000000 HZ))
```

Figure 3.41   Simulation results for reading radar signals.

appropriate field of the signal RFSIG. Figure 3.41 shows the results of a simulation of entity TB obtained with the Synopsys simulator.

```
process_statement ::=
   [process_label:]
      process[(sensitivity_list)]
         process_declarative_part
      begin
         process_statement_part
      end process[process_label];

process_statement_part ::=
   [sequential_statement]

sequential_statement ::=
      wait_statement
    | assertion_statement
    | signal_assignment_statement
    | variable_assignment_statement
    | procedure_call_statement
    | if_statement
    | case_statement
    | loop_statement
    | next_statement
    | exit_statement
    | return_statement
    | null_statement

process_declarative_part ::==
   [process_declarative_item]

process_declarative_item :=
      subprogram_declaration
    | subprogram_body
    | type_declaration
    | subtype_declaration
    | constant_declaration
    | variable_declaration
    | file_declaration
    | alias_declaration
    | attribute_declaration
    | attribute_specification
    | use_clause
```

Figure 3.42   Process statement production rules.

3.8   THE FORMAL NATURE OF VHDL

In this chapter we have illustrated the main features of the VHDL language by means of a series of selected examples. A formal definition of the language is given in *IEEE Standard VHDL Language Reference Manual*, which presents the *syntax* and *semantics* of the language. These two terms are defined as:

Syntax.   The set of rules governing the formulation of language statements.

Semantics.   The meaning of language constructs.

The VHDL language syntax is given in production rules. For example, the production rule for the process statement is shown in Figure 3.42. The first rule gives the general structure of a process statement in terms of key words (begin, end) and other constituent parts, such as the optional label, declarative, and statement regions. The rule also shows the manner in which these parts can be interrelated. The two subsequent rules show what language elements can be present in the statement part and the declarative region. These language elements have their own production rules. This syntactic definition proceeds from the specific to the general until primitive language elements are reached that have no production rules.

The *Language Reference Manual* also contains the semantic definition of the language in English. For example, for the *process* statement one of the semantic rules is:

> The execution of a process consists of the repetitive execution of its sequence of statements. After the last statement in the sequence of statements of a process statement is executed, execution will immediately continue with the first statement in the sequence of statements.

This rule explicitly states that process execution is repetitive and that processes loop. Note: The syntax rule in no way implies this.

As the reader becomes more involved with the VHDL language, he will find it necessary to consult the *Language Reference Manual* to resolve questions concerning the details of the language. However, the document is abstract, requiring close attention and effort to understand.

3.9   VHDL 93

Since its original standardization in 1987, the IEEE has received and reviewed input on suggested changes to the language. In 1993, the IEEE released a new standard 1076-1993 for the language incorporating those changes. The changes were not major, and for the most part one can model effectively without them. Nevertheless, we summarize them here, so the reader can use the full power of the language. The reader is warned, however, that as of 1999, not many VHDL vendors have incorporated all the changes in the 1993 standard. We give examples of the most important changes here, but the reader is referred to the 1993 Language Reference Manual for details. The reader should also consult the recent VHDL references given at the end of the book. It should be noted that almost all VHDL 1987 code is still legal in the 1993 standard.

3.9.1   Lexical Character Set

VHDL 87 used the ASCII character set, which is the first 128 characters of the 256 character ISO character set. VHDL 93 uses the full ISO character set. This includes letters with diacritical marks that are used in languages other than English, e.g., the German umlaut.

3.9.2   Syntax Changes

The rather rigid rules on identifiers in the 1987 standard have been relaxed, e.g., the identifier:

```
\7404INV\
```

is now legal. These 1993 standard identifiers (called extended identifiers) must be delimited with back slashes. They can begin with numbers, and are case sensitive. Thus, \7404inv\ is different from the above extended identifier. Extended identifiers can contain reserved words, e.g., \entity\ is a valid identifier. An extended identifier is always different from any short identifier, e.g., \XYZ\ is different from XYZ.

In VHDL 87 the opening and closing syntax for entity and component declarations differed:

```
entity X is
  --------
  --------
end X;

component X
  --------
  --------
end component;
```

Now the component may be written identically:

```
component X is
  --------
  --------
end X;
```

and the VHDL 93 syntax for a component declaration is:

```
component identifier [is]
    [local_generic_clause]
    [local_port_clause]
end component [component_simple_name];
```

Consistency rules also apply to entities, architectures, package declaration, package body, configurations, components, blocks, processes, records, *if* and *case* statements, and subprograms. This allows the user to write code that appears more consistent in syntactic form.

3.9.3 Process and Signal Timing and New Signal Attributes

Postponed process.   In Figure 4.7 and the surrounding discussion in Chapter 4, we discuss the situation, where a number of delta delay-based simulation cycles occur before simulation time advances. Until these delta cycles elapse, the state of the model may not settle out, and other models monitoring it may react improperly from the modeler's view point. VHDL 93 introduces the concept of the postponed process, which is not activated until the last delta cycle of the time point. Consider the following code :

```
A  <= '0', '1' after 1 ns;
B  <= not A;
assert (B = not A)
  report "B is equal to A!"
```

Under steady state conditions, an error should not be detected, but when A switches, B does not assume the inverse of A until delta later. Thus, for one delta cycle, A equals B, and the assertion detects an error. However, suppose the assertion, which is equivalent to a process, is coded as:

```
postponed assert (B = not A)
  report "B is equal to A!"
```

With this construct, the assertion is not executed until after B has changed and no error is reported. The postponed designation can be used on process statements, concurrent signal assignment statements, procedure calls, and assertion statements.

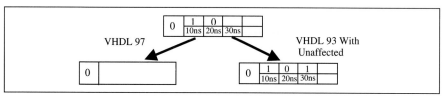

Figure 3.43   Preserving transactions in VHDL 93.

Inertial delay control.   Inertial delay can now be directly controlled independent of propagation delay. Consider this code:

```
A <= reject 2 ns inertial B after 6 ns;
A <= reject 2 ns B after 6 ns;
```

Each of the above statements implies an inertial delay of 2 ns and a propagation delay of 6 ns. Inertial delay is still the default; thus, the use of the key word inertial is optional. In VHDL 87, one would have to write the following code to achieve the same effect:

```
C <= B after 2 ns;-- inertial delay and prop delay of 2ns.
A <= transport C after 4 ns; --- pure propagation delay of
                             --- 4 ns.
```

Unaffected keyword.   Consider the following concurrent signal assignment statement:

```
A <= '1' after 10 ns when CON = '1' else A;
```

A natural interpretation of this statement is that when CON = '1', A takes on the value of '1' after 10 ns, and when CON = '0' there is no change in A. This is not strictly true. The current value of A does not change, but: 1) a transaction will be performed on A. Thus A'*transaction* will toggle and A'*quiet* will be FALSE 2) as a result of the waveform updating rules, all other pending transactions on A will be deleted as shown in the left branch in Figure 3.43. With VHDL 93, the conditional assignment statement can be written as:

```
A <= '1' after 10 ns when CON = '1' else unaffected;
```

By using the *unaffected* keyword, the old transactions on A are preserved when Con='0' as shown in the right branch of Figure 3.43. Also, A'*transaction* does not toggle and A'*quiet* is TRUE.

Another change to the conditional signal assignment statement is that the else branch is optional. Thus:

```
A <= '1' after 10 ns when CON = '1';
```

is equivalent to the statement that used the keyword *unaffected* in the *else* clause. The keyword *unaffected* also applies to selected signal assignment statements.

Signal attributes '*driving* and '*driving_value*.   In VHDL 87, an interface signal of mode *out* cannot be read. Mode *inout* can be used, but many modelers prefer to reserve this for truly bidirectional data. Consider the VHDL 93 code shown in Figure 3.44.

The value of driver of SOUT from this entity is read back using the '*driving_value* attribute. Another attribute of type Boolean, '*driving*, is TRUE if the driver is connected; otherwise, it is FALSE.

```
library IEEE;
use IEEE.STD_LOGIC_1164.all;
entity DRIVING is
  port(SOUT: out STD_LOGIC);
end DRIVING;
architecture ALG of DRIVING is
  signal INT_SOUT: STD_LOGIC;
begin
  SOUT <= '0','1' after 10 ns,'0' after 20 ns,
          '1' after 30 ns;
  INT_SOUT <= SOUT'driving_value;
end ALG;
```

Figure 3.44   'DRIVING_VALUE attribute.

3.9.4   New Operators

Shift and rotate operators.   The operators are of the form:

```
array_object  operation  N
```

where the array object is a one dimensional array whose elements are of type Bit or Boolean. N is an integer specifying the number of times the operation is performed. The operations are: a) shift left logical (sll) - the left input is 'LEFT of the input array element type; b) shift right logical (srl) - the right input is 'LEFT of the input array element type; c) shift left arithmetic (sla) - the right-most bit retains its value and that value is shifted into the next position (sign preservation); d) shift right arithmetic (sra) - the left-most bit retains its value and that value is shifted into the next position (sign preservation); e) rotate left logical (rol), and f) rotate right logical (ror). The packages STD_LOGIC_SIGNED and STD_LOGIC_UNSIGNED contain shift operators for type STD_LOGIC_VECTOR. The package STD_LOGIC_ARITH contains shift operators for type SIGNED and UNSIGNED and for one dimensional arrays of STD_LOGIC.

xnor operator.   This built-in operator applies to types Bit and Boolean and one dimensional vectors of these types. It is of equal precedence to all other logical operators. Like *nand* and *nor*, the *xnor* operator is not associative. It is overloaded for types STD_LOGIC and STD_LOGIC_VECTOR in package STD_LOGIC_1164.

3.9.5   Improvements to Structural Models

Three major improvements have been made to the process of structural modeling:

1. Direct instantiation: The VHDL 87 process of instantiating components involves component declaration, component instantiation, and configuration specification. In VHDL 93 one can perform direct instantiation as illustrated in Figure 3.45, thus, vastly simplifying the code.
2. Expressions on inputs: In VHDL 93, expressions on input ports are now allowed. Note the application of the value '1' to an input port in Figure 3.45.

```
      library IEEE;
      use IEEE.STD_LOGIC_1164.all;
      entity NAND2 is
        generic(DELAY: TIME);
        port(A,B: in STD_LOGIC; C: out STD_LOGIC);
      end NAND2;
      architecture DF of NAND2 is
      begin
        C <= not(A and B) after DELAY;
      end DF;

      library IEEE;
      use IEEE.STD_LOGIC_1164.all;
      entity INVERTER is
        port(I: in STD_LOGIC; O: out STD_LOGIC);
      end INVERTER;
      architecture STRUCTURE of INVERTER is
      begin
        C1: entity work.NAND2(DF)    --- direct instantiation
          generic map(DELAY => 1.0 ns)  -- expressions
          port map(A => I, B => '1',C =>O); --on input ports
      end STRUCTURE;

      library IEEE;
      use IEEE.STD_LOGIC_1164.all;
      configuration NEW_TIMING of INVERTER is
        for STRUCTURE
          for C1: entity work.NAND2(DF)--incremental binding
            generic map(DELAY => 1.23 ns);
          end for;
        end for;
      end NEW_TIMING;
```

Figure 3.45   New structural modeling features in VHDL 93.

3. Incremental binding: VHDL 93 allows a configuration body to override an earlier configuration specification. In Figure 3.45 this feature is used to back annotate a more accurate delay value to the NAND gate.

3.9.6   Shared Variables

Since the early days of the VHDL standardization process, some users have wanted shared variables included in the language. These are very useful in performing system-level simulation. However, the objection to them is that their hardware correspondence is weak. Another issue is: What access control mechanism should be included in the language to support them? VHDL 93 includes shared variables, but there is no built-in access control mechanism built into the language. (The current sentiment seems to favor the use of monitors.) Thus, the user must insure that there is no attempt at simultaneous access of the shared variable by two or more agents. Figure 3.46 shows a situation, where two processes P1 and P2 share access to a shared variable

```
entity SHARED_VAR is
  port(START1, START2: in BIT);
end SHARED_VAR;

architecture ALG of SHARED_VAR is
  type PROC_NUM is (PROC1,PROC2);
  shared variable LATEST: PROC_NUM;
begin
  P1: process
  begin
    wait until START1 = '1';
    LATEST := PROC1;
  end process;
  P2: process
  begin
    wait until START2 = '1';
    LATEST := PROC2;
  end process;
end ALG;
```

Figure 3.46   Shared variable.

LATEST. LATEST contains the value PROC1 or PROC2 depending on which process has accessed it most recently. Note that if START1 and START2 make positive transitions simultaneously, one cannot predict what value will end up in LATEST. Based on the way processes are handled in simulators on uni-processors, one gets the value PROC1 or PROC2. However, for parallel processor execution of the model, if one processor disturbs another during variable access, one can obtain a result that is neither of these values. Thus, the desire for an access control mechanism arose.

3.9.7   Improved Reporting Capability

The reporting mechanism in VHDL 87 is contrived. Using the assert construct one writes:

```
assert FALSE
report "Executing Processor 1"
severity note;
```

Another problem is that the report clause only accepts a string argument. If one wishes to include the state of an integer variable I in the message, the modeler must write a type conversion function to convert I to its string equivalent. In VHDL 93, one writes:

```
report   "Executing processor"&Integer'image(I);
```

to include the value of integer I in the output message. Keyword *report* is the new reporting command. The '*image* attribute performs a type conversion function.

3.9.8   General Programming Features

Impure Function.   A function with side effects. Impure functions do not communicate solely through the call interface and can read or modify external data. With this type of function, one can code such things as a random number generator that uses an external seed or the C

```
entity IMPURE_F is
end IMPURE_F;

architecture ALG of IMPURE_F is
  signal INT,INT_INC: INTEGER := 0;
  impure function INC_AND_UPDATE
    return INTEGER is
  begin
    INT <= INT + 1;
    return INT;
  end INC_AND_UPDATE;
begin
  INT_INC <= INC_AND_UPDATE;
end ALG;
```

Figure 3.47   Use of an impure function.

GET_CHAR routine, which can read a character from an external file and move the character pointer to the next character to be read. Figure 3.47 shows how an impure function accesses and increments an external signal. This is not allowed in VHDL 87.

Foreign Interfaces.   In VHDL 93 Package Standard, the attribute FOREIGN is defined as:

```
attribute FOREIGN is STRING;
```

Suppose you wanted to invoke an input user interface for controlling models during simulation. You might have a function declaration:

```
function USER_INP(INP: in  COMMAND) return Integer;
```

One can associate this with a system program with the declaration:

```
attribute FOREIGN of USER_INP:function is "VHDL_USER_IN";
```

where VHDL_USER_IN is local system software written in C and X windows. Other objects can be FOREIGN, also. A propriety implementation of an entity written in another programming language or HDL can use the same mechanism.

3.9.9   File I/O

In VHDL 93, a file is defined as a fourth object class that can be passed to a procedure as a parameter. Files are mode-less, you can read from and write to a file. In the new Package Standard, the type for opening a file is defined as:

```
type FILE_OPEN_KIND is (READ_MODE, WRITE_MODE, APPEND_MODE);
```

Thus, a text file for reading is declared as:

```
file IN_MEM: TEXT open READ_OPEN is "MEM_CONTENTS"
```

And later, another declaration is possible that allows writing to the same file:

```
file OUT_MEM: TEXT open WRITE_OPEN is "MEM_CONTENTS"
```

APPEND_MODE is a write mode, where the information is appended to the end of the existing file. FILE_OPEN and FILE_CLOSE functions are implicitly declared. Files INPUT (keyboard) and OUTPUT (monitor screen) are now declared open.

3.9.10 Groups

In VHDL 87 one can use attributes to annotate one's description with information that can be extracted by tools, e.g., synthesis. However, this annotation can only be applied to single items. VHDL 93 introduces the concept of a group where many, related items may be annotated with a single value. A common application is the annotation of time constraints on signal paths. Suppose BEGIN_PT and END_PT are two signal objects:

```
group BEGIN_TO_END: PATH(BEGIN_PT,END_PT);
attribute PROPAGATION_DELAY: TIME;
attribute PROPAGATION_DELAY of BEGIN_TO_END group is 100 ns;
```

This VHDL 93 code establishes a constraint delay between signal BEGIN_PT and END_PT. (Note this has no effect on simulation.)

3.9.11 Extension of Bit String Literals

If one is representing the decimal number 35, it can be represented in the three bit string literal forms:

B"100011"

O"43"

X"23"

In VHDL 87 these can only be assigned to objects of type Bit_Vector. In VHDL 93 they can be used with any value system as long as the values used are compatible with the binary, octal, or hex number systems. For example, a STD_LOGIC_VECTOR(7 downto 0) could be initialized to O"43" but not O"4Z".

3.9.12 Additions and Changes to Package Standard

In VHDL 93 the following changes were made to the VHDL 87 package STANDARD.

1. Additions
   a. Types: FILE_OPEN_KIND and FILE_OPEN_STATUS.
   b. Subtype: DELAY_LENGTH.
   c. Interface definitions for all the standard language operators, e.g. AND, *.
   d. Attribute FOREIGN.
2. Changes
   a. ASCII character set was replaced by the full ISO character set.

3.10 SUMMARY

Our purpose in this chapter has been to give the reader a basic understanding of VHDL, particularly those features that are utilized in the models in later chapters. The presentation here is obvi-

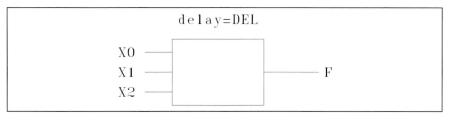

Figure 3.48 Block diagram for circuit for problem 3.3.

ously not complete. Other language features will be illustrated as part of the modeling process. The models will also give complete examples, where all the necessary features of the language are pulled together. In understanding this chapter, the reader should have learned enough VHDL to appreciate the usefulness of the language and to understand model structures.

PROBLEMS

3.1   Develop a VHDL model for a 3-input AND gate. Simulate the AND gate using the following test bench.

```
entity TB is
end TB;
architecture AND3T of  TB is
  signal X,Y,Z,O: Bit;
  component AND3
    port(I1,I2,I3: in Bit; O: out Bit);
  end component;
  for C1: AND3 use entity work.AND3(DATA_FLOW);
begin
  C1: AND3
    port map(X,Y,Z,O);
  X <= '1' after 1 ns, '0' after 2 ns, '1' after 3 ns,
       '0' after 4 ns, '1' after 5 ns, '0' after 6 ns,
       '1' after 7 ns;
  Y <= '0' after 1 ns, '1' after 2 ns, '1' after 3 ns,
       '0' after 4 ns, '0' after 5 ns, '1' after 6 ns,
       '1' after 7 ns;
  Z <= '0' after 1 ns, '0' after 2 ns, '0' after 3 ns,
       '1' after 4 ns, '1' after 5 ns, '1' after 6 ns,
       '1' after 7 ns;
end AND3T;
```

3.2   Develop a VHDL description for the OPAR3 entity used in the ones counter example covered in this chapter. Make the architectural body of OPAR3 purely structural. You can use the 3-input AND gate developed in Problem 3.1. Develop a test bench for the OPAR3 entity and simulate.

3.3   Consider the following truth table for the combinational logic circuit whose block diagram is shown in Figure 3.48.

```
X2 X1 X0        F
---------      ---
 0  0  0        0
 0  0  1        1
 0  1  0        0
 0  1  1        1
 1  0  0        1
 1  0  1        1
 1  1  0        1
 1  1  1        0
```

The circuit has delay DEL.

    a. Develop an algorithmic VHDL description for the circuit. Your code should be efficient.

    b. Develop a data flow VHDL description for the circuit. Your code should be efficient.

3.4   Consider the following VHDL code.

```
entity X is
  port(A: in Bit_Vector(7 downto 0);
       Z: out Bit_Vector(7 downto 0));
end X;
architecture ALG of X is
begin
  process(A)
    variable ZV: Bit_Vector(7 downto 0);
  begin
    for I in 7 downto 0 loop
      ZV(7-I) := A(I);
    end loop;
    Z <= ZV;
  end process;
end ALG;
```

    a. If A changes from all 0's to 10101111, what will the output Z be?

    b. Describe in one sentence the function of entity X.

3.5   Write an entity and algorithmic architectural body for a bit flipper, i.e., it reverses the order of bits in an 8-bit word, e.g., 10101111 will be flipped to 11110101.

3.6   The Fibonacci series is the following series of numbers 0,1,1,2,3,5,8,13,21, etc. This sequence can be defined as follows:

```
F0 = 0
F1 = 1
Fi = F(i-1) + F(i-2)   i > 1
```

Except for the first two members of the series, the ith member of the series is the sum of the (i-1) and (i-2) members of the series. Assume a device is activated with a START command, computes the first 20 members of the series (F0 through F19), and outputs all 20 members of the series 60 ns later. Develop a concise VHDL algorithmic description for the device.

3.7   Design a two-out-of-five code detector. The device receives as input a 5-bit parallel word. The detector output is a logic 1 for any code word that has exactly two 1's in it and is a logic 0, otherwise. Perform the following steps:

    a. Develop a VHDL entity declaration for the detector.

    b. Develop an algorithmic behavioral architectural body for the detector.

    c. Develop a data flow behavioral architectural body for the detector.

    d. Develop a structural architectural body for the detector.

     e. Simulate all three bodies using a VHDL simulation system to verify the correctness of each model.

     f. Prepare a report summarizing your results.

3.8   Design a parallel even-parity detector. The input to the detector is a 4-bit code. The detector output is a logic 1 for any code word that has even parity, i.e., for any code word that has an even number of 1's. Model your circuit by performing the following steps.

     a. Develop a VHDL entity declaration for the checker.

     b. Develop an algorithmic behavioral architectural body for the checker.

     c. Develop a data flow behavioral architectural body for the checker.

     d. Develop a structural architectural body for the checker.

     e. Simulate all three bodies to verify the correctness of each model.

   Prepare a report which contains:

     a. A description of how each model was developed.

     b. VHDL source listings.

     c. Well annotated simulation results.

3.9   A designer wishes to model a local area network protocol without specifying the design at a detailed level. What kind of type definition would he use to model the signal interface? Explain your answer.

3.10 Answer the following questions about lexical elements:

     a. Describe the difference between a separator and a delimiter.

     b. Make a chart that shows the relationship between the two classes and the two types of abstract literals. Give examples of each combination of class and type.

     c. Explain the difference between 'A' and "A". Give examples of type declarations for which each value would be appropriate.

     d. Explain the relationship between identifiers and reserved words.

3.11 Which of the following bit string literals are valid? For each valid bit string literal, determine the length of the literal and the value of the literal in the VHDL language sense.

     a. B"1011_0101_1001"

     b. B"0001_1011"

     c. B"1001-0111"

     d. B"0101 0111"

     e. X"B5_CD"

     f. X"3HA4"

     g. O"237"

     h. O"233_814"

     i. O"0011_7322"

     j. "1011"

     k. "1011_0001"

3.12 Which of the following are valid user-defined identifiers? For each invalid identifier, state the reason it is invalid.

     a. Help

     b. 2nd_item

     c. Case

     d. small_device

     e. This_label_might_be_too_long_what_do_you_think_about_the_possibility:

     f. BODY

     g. register

     h. REPORT

    i. branch

    j. _name_

3.13 The VHDL language requires that all user-defined identifiers be distinct. Which of the following pairs of identifiers are distinct?

    a. MY_NAME, MYNAME

    b. cat, CAT

    c. Dog, dOg

    d. cow, cows

    e. two, too

3.14 Which of the following are valid abstract literals? For each valid abstract literal, give its value as a decimal number without exponents and state whether the literal is decimal, integer, real, or based.

    a. 2e5

    b. 5#224_33#

    c. 2,534,215

    d. 2_534_215

    e. 2.5e2

    f. 12#A_B#

    g. 16e-2

    h. 16.0e-0.5

    i. .5

    j. 2#1011_0001#e10

3.15 Compare and contrast the built-in data types Integer, Natural, and Positive.

3.16 Specify type declarations for the following data types.

    a. A four-valued logic system, MVL4, with values '0','1','X',and 'Z'. Values '0' and '1' have the usual logic meaning. Value 'X' means the logic value is unknown (may be either 0 or 1). Value 'Z' means high impedance state. Any uninitialized data value of this type should have value 'X'.

    b. A DAY_OF_WEEK enumeration data type.

    c. A data type ATOMIC_NUMBER that can have integer values in the range from 1 to 120.

    d. A data type COST that can have real values between $0.00 and $1,000.00.

    e. A descending range data type DEC_16 with integer values from 15 to 0.

    f. A 16-bit descending-index register composite data type, REGISTER_16_BIT_DESCENDING, with index values from the type DEC_16 declared above, and component values of type Bit.

    g. A two-dimensional table, TRISTATE_TABLE_2D, with index values and table entries, all of type TRISTATE (see Enumeration Types in text).

    h. A record data type, PERSONNEL, with fields for last name (LAST) (up to twenty characters); first name (FIRST) (up to twenty characters); middle initial (MID); and social security number (SOC_SEC).

3.17 Refer to the data types defined in Problem 3.16. Write declarations for the following constants.

    a. A constant DONT_CARE that has value 'X' of type MVL4.

    b. A constant MID_WEEK that has value WEDNESDAY of type DAY_OF_WEEK.

    c. A constant HYDROGEN that has value 1 of type ATOMIC_NUMBER.

    d. A constant MINUS_1 that consists of the 16-bit representation for (-1) in twos complement notation of type REGISTER_16_BIT_DESCENDING.

     e. A constant TRISTATE_AND of type TRISTATE_TABLE_2D that defines the logical "and" operation for two data objects of type TRISTATE. Assume TTL technology, i.e., high impedance inputs to a gate are treated as logic 1.

     f. A constant MY_PERSONNEL_RECORD that defines your own data, consistent with type PERSONNEL.

3.18 Refer to Problem 3.16. Write declarations for the following variables and signals.

     a. A signal RESET of type MVL4.

     b. A variable CURRENT_DAY that has a value equal to the current day of the week.

     c. A variable CHICKEN_PRICE that has a value equal to the current price of chicken.

     d. Signals REGISTER_1, REGISTER_2, REGISTER_3, and REGISTER_4 that hold 16-bit data values with descending index ranges.

3.19 Refer to the variables and signals declared in Problem 3.18. Also refer to the constants declared in Problem 3.17. Write the following variable and signal assignment statements.

     a. A variable assignment statement that sets the variable CURRENT_DAY to the value MID_WEEK.

     b. A variable assignment statement that inflates the CHICKEN_PRICE by one cent each time it is executed.

     c. A signal assignment statement that uses the bits of REGISTER_4 to selectively complement the bits of REGISTER_2. Each bit in REGISTER_2 that corresponds to a position in REGISTER_4 that is logic 1 should be complemented. The other bits of REGISTER_2 should remain unchanged. An example calculation is shown:

```
REGISTER_2 initial value:    1011 0101 0001 1100
REGISTER_4 value:            1111 0000 1101 0101
                             -------------------
REGISTER_2 final value:      0100 0101 1100 1001
```

3.20 Refer to the constant TRISTATE_AND declared in Problem 3.17. Assume that V1, V2, and V3 are signals of type TRISTATE. Let V1 and V2 be inputs to an AND gate and let V3 be the gate output. Let the gate delay be 10 ns.

     a. Write a process that will update the output of the gate whenever either input changes.

     b. Write a subprogram that will update the output of an arbitrary 2-input AND gate in which the input and output signals are of type TRISTATE. Write the call statement that will implement the given gate involving signals V1, V2, and V3.

     c. Referring to the models created in parts a and b, answer the following questions.

        1. Compare and contrast the models.

        2. Discuss advantages and disadvantages of each model.

        3. Discuss applications of each model.

3.21 Repeat Problem 3.20 with the following modifications. Add a third input, ENABLE, to the tristate gate. When ENABLE is '0', the output should be 'Z'. When ENABLE is '1', the output should be the "and" of the data inputs. Consistent with TTL technology, a value of 'Z' on the ENABLE input will act like a logic 1.

3.22 Consider the predefined data types defined in Figures 3.16 and 3.17. Find the value of each of the following expressions.

     a. Character'pos('A')

     b. SEVERITY_LEVEL'pos(WARNING)

     c. Character'pos(STX) > SEVERITY_LEVEL'pos(ERROR)

     d. Character'val(65)

     e. SEVERITY_LEVEL'val(2)

    f. Character'('A') > Character'('a')

    g. SEVERITY_LEVEL'(NOTE) < SEVERITY_LEVEL'(FAILURE)

    h. Boolean'high

    i. Natural'low

    j. Positive'left

    k. Bit'right

    l. Character'pred('A')

    m. Character'pred('+') = Character'succ(')')

3.23 Consider the following type and signal declarations.

```
type ANIMAL is (LION, DOG, CAT, MOUSE, HORSE, FOX, COW);
type ANIMAL_VECTOR is array (ANIMAL range <>) of Natural;
signal A1: ANIMAL := HORSE;
signal A2: ANIMAL;
signal BV1: ANIMAL_VECTOR (FOX downto CAT):=(1, 10, 2, 6);
```

Indicate which of the following expressions, if any, are valid. For each valid expression, determine the value at time $t=0$.

    a. A2

    b. BV1(MOUSE)

    c. MOUSE < LION

    d. ANIMAL'val(BV1(FOX))

    e. ANIMAL'pos(CAT)

3.24 Consider the following type and signal declarations in standard VHDL.

```
type COURSE is (MATH, SCIENCE, SOCIAL_STUDY, LANGUAGE);
type COURSE_VECTOR is array (Natural range <>) of COURSE;
  signal A1: COURSE := SCIENCE;
  signal A2: COURSE;
  signal BV1: COURSE_VECTOR(10 downto 8):=(MATH, LANGUAGE,
    MATH);
```

Indicate which of the following expressions, if any, are valid. For each valid expression, determine the value at time $t=0$.

    a. A2

    b. BV1(10)

    c. SCIENCE > SOCIAL_STUDY

    d. COURSE'val(1)

    e. COURSE'pos(BV1(8)) < SCIENCE

3.25 Show how one declares a type representing an array of arbitrary length, where each element of the array is a five-valued logic type whose values are U,0,1,X,Z.

3.26 Given the following declaration:

```
Signal X: Bit_Vector(0 to 3);
```

Find:

    a. X'RANGE

    b. X'LEFT

    c. X'RIGHT

    d. X'HIGH

    e. X'LOW

3.27 Figure 3.49 shows a three-dimensional cube of characters. Give the declarations necessary to declare and initialize a variable to this cube of values.

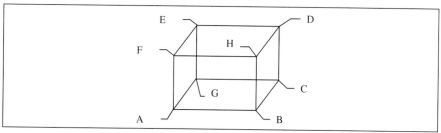

Figure 3.49 3D Character cube for Problem 3.27.

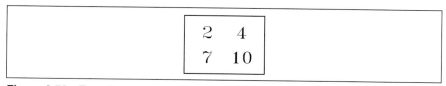

Figure 3.50 Two-dimensional array for Problem 3.28.

3.28 The two-dimensional array in Figure 3.50 contains elements of type Integer. Give the declarations necessary to declare and initialize a variable to this array of values.

3.29 If you don't explicitly initialize a VHDL signal or variable, what will its default initialization value be?

3.30 Answer the following questions about data types Bit and Boolean.
   a. Explain the difference between types Bit and Boolean.
   b. Which of these types can be used with the logical operators?
   c. The result of a relational operator is of which type?
   d. Which type of expressions are tested by *if* statements?

3.31 Consider the following VHDL description of a 3-input NOR gate.

```
entity NOR3 is
  generic (DEL: time);
  port (X1, X2, X3: in Bit; Y: out Bit);
end NOR3;
--
architecture GATE of NOR3 is
begin
  -- Assignment statements
end GATE;
```

Which of the following assignment statements, if used in the preceding model, would correctly implement a 3-input NOR gate? More than one statement may be valid.
   a. Y <= X1 nor X2 nor X3
   b. Y <= (X1 nor X2) nor X3
   c. Y <= (X1 or X2) nor X3
   d. Y <= not (X1 or X2 or X3)
   e. Y <= not (X1 nor X2) nor X3

3.32 Write a VHDL procedure that counts the number of 1's in a bit vector. The procedure must be general, i.e., it must accept an input parameter of arbitrary length.

3.33 Given that X is of type Bit_Vector(0 to 3), write an increment function for X.

3.34 The following questions deal with processes and *wait* statements:

    a. For the following process, explain the effect of the *wait* statement during simulation:

```
PROC1: process
    X <= X + 1;   --- X is type integer
    wait on R,Q until (U = '1') for 200 ns;
    X <= X - 1;
end process PROC1;
```

    b. For the following process, explain what happens to the process as simulation starts and then continues.

```
process
begin
    X   <= X + 1;
    wait on Y;
end process;
```

    c. For the following process, explain what happens during simulation:

```
process
begin
    X <= X + 1;
end process;
```

3.35 Write a process statement that has the same effect as the following concurrent signal assignment statement.

```
with Z select
  A <= transport
    X and Y when 0,
    X or Y   when 1,
    not X    when others;
```

3.36 In the following VHDL description, indicate which signal assignment statements are illegal by drawing a circle around the illegal ones.

```
entity EX1 is
end EX1;
architecture P7 of EX1 is
  signal A1: Bit;
begin
  B1: block
    signal X: Bit;
  begin
    X <= B2.Y;
    B2: block
      signal Y: Bit;
    begin
      Y <= B1.X;
      A1 <= B3.Z;
    end block B2;
  end block B1;
  B3: block
    signal Z: Bit;
  begin
    Z <= A1;
```

```
       Z <= B1.B2.Y;
       A1 <= Z or A1;
     end block B3;
   end P7;
```

3.37 Consider the following VHDL code:

```
package D is
  type D1 is array(0 to 2) of Integer;
  type D3 is array(0 to 1,0 to 1,0 to 2) of Integer;
end D;
use work.D.all;

entity IT is
  port(Q: in Bit; XOUT,YOUT: out Integer);
end IT;

architecture DO of IT is
begin
  process(Q)
    variable A: D3 :=
      (((16,8,7),(3,11,5)),((15,6,9),(17,9,4)));
    variable X: Integer;
    variable Y: Integer;
  begin
    X := 0;
    Y := 0;
    for k in 0 to 2 loop
      X := X + A(0,0,K) + A(1,1,K);
      Y := Y + A(0,1,K) + A(1,1,K);
    end loop;
    XOUT <= X;
    YOUT <= Y;
  end process;
end DO;
```

Answer the following questions about the preceding VHDL code.
   a. Draw a sketch of the data structure represented by variable A.
   b. Assume that the process executes once.
       1. Explain through words and diagrams what computation is carried out by the pro-
           cess.
       2. Give the resultant values of XOUT and YOUT.
3.38 Given the following truth table:

```
X2 X1 X0    f
--------    --
 0  0  0    0
 0  0  1    1
 0  1  0    1
 0  1  1    0
 1  0  0    1
 1  0  1    1
 1  1  0    1
 1  1  1    0
```

Develop a VHDL description that implies a ROM implementation.

3.39 An engineer wishes to create a VHDL model of a Motorola 68000 microprocessor system. She wants to simulate the execution of actual microprocessor code, but wants to write the 68000 program in assembly language. The program is expected to change frequently. What VHDL language features would be used to handle this modeling situation?

3.40 For the following code, explain what the result of model generation will be.

```
arch X of Y is
begin
  for I in 1 to N generate
    S(I) <= V(I) nand W(I) after DEL;
  end generate;
end X;
```

3.41 Give the physical type definition for types voltage, current, power, and resistance. Also declare a record type that can be used to represent voltage sources, i.e., one field represents the voltage output of the voltage source, the other is its source resistance. Use these type definitions to solve for the current and power transferred to an external resistor that is placed across the terminals of a voltage source.

3.42 Implement and apply the following package:
   a. Give the enumeration type definition for a three-valued system, 0, 1, Z, where Z is the high impedance state.
   b. Use overloading to define AND, OR, and NOT operations for this system. Give the circuit rationale for your operators.
   c. Place the results of parts a and b in a package.
   d. Simulate the operation of the following logic equation using the package developed in part c:

$$E = (AB' + C'D)'$$

   where ' denotes inversion.

3.43 Develop models for AND, OR, and NOT gates and place them in a library called GATES. Employ these models in a structural architecture, which implements the logic equation in Problem 3.42. Simulate the architecture.

3.44 Develop an unbound structural architecture that implements the logic equation in Problem 3.42. Bind the architecture to the models in library GATES (Problem 3.43) with a configuration declaration. Simulate the system.

3.45 Consider the following VHDL code:

```
entity BLOCKS is
end BLOCKS;
architecture TEST of BLOCKS is
begin
  L1: block
    signal A: Integer := 10;
    signal B: Integer := 4;
  begin
    L2: block
      signal B: Integer := 5;
    begin
      A <= B after 5 ns;
      B <= L1.B after 10 ns;
    end block;
    B <= A after 15 ns;
```

```
    end block;
  end TEST;
```

    a. Explain the visibility situation in this set of blocks.

    b. If this code were simulated, what would the value of all signals be, as a function of time?

3.46 Consider the following lines from "Elegy In A Country Church Yard," the famous poem by Thomas Gray,

```
The curfew tolls the knell of parting day,
The lowing herd winds slowly o're the leigh,
The plowman homeward plods his weary way,
And leaves the world to darkness and to me.
```

The object of this problem is to use file and text I/O.

    a. Write a VHDL model that constructs each line from individual words, writes each line to a file using file I/O, reads it back with file I/O, and does a word-by-word comparison to ensure that the data was written and read correctly. Use an assertion to output a message signaling the successful completion of the test. Also use the signal and variable monitoring capability of the simulator to prove that the words are read back correctly.

    b. Using a text editor, edit the poem into a file and read the file into a VHDL model using text I/O. Use the signal and variable monitoring capability of the simulator to prove that the words are read correctly.

## CHAPTER 4

Basic VHDL Modeling Techniques

In this chapter we present basic VHDL modeling techniques. We begin by discussing how the VHDL language supports the modeling of propagation delay and concurrency. Next, we discuss the VHDL timing model in detail. Following that we discuss how to model combinational and sequential logic with VHDL. Next, we present models of common primitive elements used in digital logic design. Finally, we explain how to verify models using test benches.

4.1 MODELING DELAY IN VHDL

Delay is an important aspect of digital logic. Therefore, any good hardware description language will have one or more delay models. In this section, we describe the delay features of the VHDL language.

4.1.1 Propagation Delay

Electronic signals must obey the basic laws of physics. The logic values '0' and '1' are usually implemented in actual circuits using two different voltage levels. For example, in TTL technology, it is customary to let +5 volts represent logic '1' and to let 0 volts represent logic '0'. Due to circuit capacitance, voltage levels at a node in the circuit cannot change instantaneously. As a result, in an actual circuit, there is always a finite delay between the time that a gate input changes value and the time that the gate output changes. The VHDL language uses the notation <= to indicate a signal change that will occur after a propagation delay.

In high level models, the standard concept of variable assignment that is familiar to C and C++ programmers is also needed in order to describe algorithms in the traditional way. In this context, assignment of new values to variables is instantaneous. The VHDL language uses the

	Initial	t1	t1+2	t1+4	t1+6
X	1	4	5	5	3
Y	2	2	2	3	2
AS	2	2	8	10	15
Z	0	3	2	2	2
BS	2	2	5	10	12

	Initial	t1	t1+2	t1+4	t1+6
X	1	4	5	5	3
Y	2	2	2	3	2
AV	2	8	10	15	6
Z	0	3	2	2	2
BV	2	11	12	17	8

Figure 4.1 Comparison of simulation results for instantaneous variable assignment and delayed signal assignment statements.

notation := to indicate instantaneous variable assignment. The following examples illustrate the two concepts:

```
(1)    AS <= X*Y after 2 ns;---- delayed signal assignment
(2)    BS <= AS+Z after 2 ns;--- delayed signal assignment
```

These two statements are termed *signal assignment* statements. The propagation delay for each of these signal assignments is specified to be 2 ns. *Instantaneous variable assignment* statements are illustrated as:

```
(3)    AV := X*Y;     ---- instantaneous variable assignment
(4)    BV := AV+Z;    ---- instantaneous variable assignment
```

Although these are straightforward concepts, propagation delay can produce unanticipated results. To illustrate the difference between instantaneous variable assignment and delayed signal assignment, suppose that statements (1) to (4) are sequential statements within processes that are executed in the order written (see Figure 4.2 for details). Further, suppose that the processes that contain the statements are executed at time t=t1 and at 2 ns intervals after time t1. Note: The delay associated with the signal assignment statements is also 2 ns. Also, assume that the initial conditions for both situations are identical as shown in the left-most column of Figure 4.1. Let the waveforms for integer inputs X, Y, and Z be:

Figure 4.1 illustrates the results for both delayed signal assignment statements with propagation delay and instantaneous variable assignment statements. At t=t1, the variables AV and BV take on their new values of 8 and 11, respectively. Since the new values are updated the instant the statement is executed, AV is first updated to its new value of 8 when statement (3) is executed. Statement (4) then uses the new value of 8 for AV when it computes the new value of BV to be 11. BV assumes its new value immediately at t=t1. Similarly, at t=t1, statement (1) computes a new value of 8 for signal AS. However, signal AS will not assume the new value until time t1+2 because of the propagation delay. As a result, when statement (2) is executed at t=t1, the value of AS is still 2. Therefore, the new value for BS will be 5. Again, BS will not assume its new value until t1+2.

At t=t1+2 ns, signals AV and BV are updated to their new values of 8 and 5, respectively; then, all statements are again executed in the order written. The variables again assume their new values immediately, and statement (4) uses the new value of AV computed in statement (3) to compute the new value for BV. Statement (1) computes a new value of 10 for AS. However, since AS does not assume the new value until time t1+4, statement (2) computes a new value of 10 for BS using the current value of 8 for AS. Again, BS will not assume the new value until time t1+4. The reader should verify the values for AS, BS, AV, and BV at times t1+4 and t1+6.

Notice that the change in X at time t1 does not have an effect on signal BS until time t1+4, due to the propagation delays. However, the change in X at time t1 has an immediate effect on variable BV at time t1 because variables are updated instantaneously. More importantly, notice that the final effect on BV and BS are significantly different for the two cases. Clearly, BS is not just a delayed version of BV. The propagation delay has changed the actual values of BS relative to BV, not just delayed the changes. The reader should study this example carefully because propagation delay effects are critical to the understanding of all models in the remainder of this book.

Since variable assignment statements are intended to allow algorithm development and since they do not allow for propagation delay, they are restricted by language syntax for use in processes, functions, and procedures. Therefore, all declarations for variables must be within processes or subprograms. Variable declarations are not allowed in the declarative regions for architectures or blocks.

Since all entity ports correspond to actual circuit signals and since architectures are intended to represent the behavior of entities, all declarations for ports and architecture objects must be signals. Internal signals may be declared in the declarative regions of architectures and blocks, but not within processes or subprograms. However, signal assignment statements may occur anywhere. Note: The model shown in Figure 4.2 satisfies all the restrictions discussed above.

As discussed in Chapter 3, the statements in a process are always executed in the order written. Statements that are always executed in the order written are called *sequential* statements. The signal assignment statements in process PROP_DELAY and the variable assignment statements in process INSTANTANEOUS are, therefore, classified as sequential statements.

In process INSTANTANEOUS, the order of the statements is critical. If statement (4) executes before statement (3), the results will be different. Contrast this situation to that for the signal assignment statements in process PROP_DELAY. Since both signal assignment statements use current values for all signals on the right-hand-side, and since the new values computed for the left-hand-side signals do not take effect until 2 ns later, statements (1) and (2) could be executed in either order without affecting the result computed. In effect, we could execute the two statements simultaneously on different processors and schedule both signal changes to occur

```
entity STATEMENTS is
  port(X,Y,Z: in INTEGER; -- Note: Entity ports are
       B: out INTEGER);   -- always signals.
end STATEMENTS;

architecture  PROP_DELAY of STATEMENTS is
  signal AS: INTEGER;
begin
 process (X,Y,Z)
 begin
   AS <= X*Y after 2 ns;   --Statement (1)
   B  <= AS+Z after 2 ns;  --Statement (2)
 end process;
end PROP_DEL;

architecture INSTANTANEOUS of STATEMENTS is
begin
 process(X,Y,Z)
   variable AV,BV: INTEGER;
 begin
   AV := X*Y;  --Statement (3)
   BV := AV+Z; --Statement (4)
   B <= BV;
 end process;
end INSTANTANEOUS;
```

Figure 4.2 Complete code to illustrate instantaneous and delayed assignment statements.

after a 2 ns delay. The reader should simulate the model with statements in either order to verify these claims.

Statements that can execute simultaneously are called *concurrent* statements. Therefore, statements (1) and (2) could appear as concurrent statements within an architecture body. They do not need to be placed within a process. The simulation result would be the same in either case. This possibility exists because of the delay encountered when updating new values for the signals. In the next section, we show how the VHDL simulator uses delay to implement concurrency.

4.1.2 Delay and Concurrency

Since logic signals flow in parallel, VHDL models of logic circuits must include provision for concurrency of execution. Figure 4.3 illustrates this concept, showing three logic blocks. If one assumes that input set 1 and input set 2 are activated simultaneously, logic blocks 1 and 2 will be activated in parallel. Logic block 3 will be activated as soon as either of the outputs from logic block 1 (Z1) or from logic block 2 (Z2) changes. While signals are propagating through logic block 3, new input signal changes can be propagating their way through blocks 1 and 2. Thus, signal flow can take place through all three blocks simultaneously.

The hardware description language must have a mechanism for modeling this simultaneity. In VHDL this requirement is handled by the process construct. Each process represents a

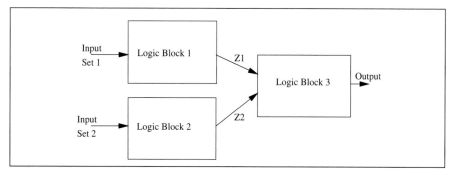

Figure 4.3 Mapping logic blocks to processes.

```
LOGIC_BLOCK1: process(X1,X2,X3)
    variable YINT: BIT;
begin
    YINT := X1 and X2;
    Z1 <= YINT or X3 after 30ns;
end process LOGIC_BLOCK1;
```

Figure 4.4 Process example.

block of logic, and all processes execute in parallel. (Of course, if the simulator is running on a single processor, they actually execute in some sequence, but the observed effect from a simulation point of view is that they execute in parallel.) In VHDL, a process is activated when a signal in its sensitivity list changes. Using Figure 4.3, assuming we know the functionality of blocks, we can create a VHDL process for each logic block. The sensitivity list for a process contains, in general, the input signal set for the logic block. To be specific, logic block 1 might be represented by the process shown in Figure 4.4.

Note: In the figure the process body first contains a declaration section, where the variable YINT, which is "local" to the process, is declared. The executable section between the *begin* and *end* keywords consists of a variable assignment statement, which computes an intermediate value (YINT) followed by a signal assignment statement, which completes the evaluation of the function and incorporates the signal propagation delay across the block.

The idea of a process can also be incorporated into signal assignment statements. Consider the following two signal assignment statements:

```
AS <= X*Y;    ---- Statement S1
BS <= AS+Z;   ---- Statement S2
```

The statements are printed in textual order, so our natural inclination is to say that they execute in the sequence printed. This is referred to as *sequential execution*.

However, these two statements can be interpreted another way. S1 and S2 can be treated as concurrent signal assignment statements. As such, they act just like processes. The sensitivity list for S1 contains X and Y and the list for S2 contains AS and Z. Under this interpretation, statement S1 will be executed any time X or Y changes value. Statement S2 will be executed any time that AS or Z changes value. For example, the *concurrent signal assignment* statement:

	Initial	t1	t1+delta	t1+2*delta
X	1	4	4	4
Y	2	2	2	2
AS	2	2	8	8
Z	0	3	3	3
BS	2	2	5	11

Figure 4.5   Concurrent signal assignment.

```
AS <= X * Y;   -- Concurrent signal assignment statement
```

is equivalent to the following *process* statement:

```
process (X,Y)   -- Equivalent process statement
begin
  AS <= X * Y;
end process;
```

Since there is not an explicit propagation time specified, the default propagation time is *delta delay*. This is a very small time greater than zero, but smaller than any explicitly specifiable time.

Let us now consider the execution of S1 and S2, where they are regarded as concurrent processes. Assume that X and Z both change at time t1. Then both S1 and S2 are activated. However, notice that, in this case, the value of AS used by S2 is the old value of AS at t1 and not the new value for AS computed by statement S1 because the new value of AS calculated at time t1 by statement S1 will not be available until time t1+delta. If this new value reflects a change in AS, S2 will be activated again at time t1+delta to compute the new value for BS, based on the new value for AS.

In Figure 4.5 we illustrate this concept using the example from the previous section. Recall that X and Z both change value at t=t1. Since X changes value at t1, statement S1 executes at t=t1. It computes a new value of 8 for signal AS, but AS does not assume the new value until t1+delta. Similarly, since Z changes at t=t1, statement S2 will execute at t=t1. It computes a new value of 5 for BS since AS still has the value of 2, and Z has value 3. BS does not assume the new value until time t1+delta.

At time t1+delta, AS and BS both assume their new values (8 and 5, respectively) computed at time t1. Since AS changes at this time, statement S2 executes again at t=t1+delta. It computes a new value of 11 for signal BS that will not take effect until time t1+2*delta. The change in X at time t1 has its effect on AS at time t1+delta and on BS at time t1+2*delta. Thus, concurrent signal assignments behave in a manner similar to that caused by finite propagation delays because of the built-in delta delay.

4.1.3   Sequential and Concurrent Statements in VHDL

Since one cannot, in general, infer from the text whether sequential or concurrent execution is implied, one must use either specialized notation or the semantics of the language to specify the mode of execution. In VHDL, language semantics is employed. Concurrent execution is implied if the statements are directly under an architecture declaration. Sequential execution is implied if the statements occur within a process or subprogram.

```
entity STATEMENTS is
  port(X,Y,Z: in INTEGER; -- Note: entity ports are
       BS: out INTEGER);  -- always signals.
end STATEMENTS;

architecture CONCURRENT of STATEMENTS is
  signal AS: INTEGER;
begin
  AS <= X*Y;    --Statement S1
  BS <= AS+Z;   --Statement S2
end PROP_DEL;

architecture SEQUENTIAL of STATEMENTS is
begin
 process(X,Y,Z)
   variable AV,BV: INTEGER;
 begin
   AV := X*Y;   --Statement S3
   BV := AV+Z;  --Statement S4
   BS <= BV;
 end process;
end INSTANTANEOUS;
```

Figure 4.6 VHDL code to illustrate concurrent and sequential statements.

Figure 4.6 illustrates these concepts. The inputs to entity STATEMENTS are X,Y, and Z. With architecture CONCURRENT, AS and BS are declared as signals; AS is an internal signal and BS is a port signal. The two signal assignment statements are inserted in the code section of the architectural body. Therefore, concurrent execution is implied. They can be written in either order, and the effect is the same.

In architecture SEQUENTIAL, the variable assignment statements S3 and S4 are inserted inside a process where two local variables are declared. Since they occur within a process, they will be executed in the exact sequence in which they appear. Here the order of S3 and S4 is important. The result of the computation, BV, is assigned to a signal, so its value can be observed at an external port. Note that signal assignment statements can be either sequential or concurrent statements.

The concepts of sequential and concurrent execution apply to many statement types in addition to signal assignment statements. For example, the process statement is a concurrent statement since it occurs immediately within an architecture. If there were other process statements in the same architecture, they would not be executed in the order written, but instead would be executed only when a signal in the sensitivity list experiences an event. However, unlike signal assignment statements, process statements can only be concurrent. A complete list of sequential statements and concurrent statements can be found in Figure 3.18.

4.1.4 Implementation of Time Delay in the VHDL Simulator

In VHDL, time can be specified in two ways. Consider the following two signal assignment statements:

```
(1)      Y <= X;                    --delta delay
(2)      Y <= X after 10 ns;        --standard time unit delay.
```

In statement (1), signal Y receives the current value of signal X after delta delay. Delta delay is a period of time greater than 0 but less than any standard time unit. The value of Y is not updated immediately. The change is delayed by a very small amount of time represented by delta delay. Other statements that use the value of Y at the current time, will use the old value of Y, even if the statement is executed after statement (1). This is a very important concept, one that is central to the effective use of VHDL, but which is a common source of misunderstanding and error for new VHDL users.

In statement (2), a standard time unit delay is used, i.e., Y takes on the value of X after 10 ns. During simulation, the elapsed time in standard time units is called *simulation time*. Note that because of the definition of delta delay, no number of delta delays added together can cause simulation time to advance.

These time delay mechanisms are implemented in the VHDL simulator so it is helpful to review how the simulation cycle works. In Chapter 2 we introduced a simplified simulation cycle that accounted for time in standard time units only. Here, we present a simulation cycle which allows for both delta delay and standard time unit delay. The steps in the cycle are:

1. If there are no entries on the time queue, then stop; otherwise, advance the simulation time to the time of the next entry on the time queue, and go to step 2.
2. Start a new simulation cycle without advancing simulation time. Remove all entries on the time queue scheduled for the current simulation cycle and update all signal values affected by these entries. Activate all processes triggered by events on the updated signals.
3. Execute the activated processes (no order of execution is implied) and (possibly) schedule new time queue entries. Some of the new time queue entries may involve delta delays.
4. If there are new transactions on signals, due to signal assignments with delta delays scheduled during step 3, go to step 2; otherwise go to step 1.

Notice that there may be many simulation cycles associated with the same simulation time. The simulation cycles executed at the same simulation time are separated by delta time. Thus, an accurate definition of delta delay is that one has merely moved to the next simulation cycle without advancing simulation time. For example, when statement (1) is executed, the value of Y will not change until the beginning of the next simulation cycle. The new value assigned to Y will be available during the next simulation cycle. However, if there are other statements executed during the current simulation cycle that access the value of Y, the old value of Y will be used even if the other statements are executed after statement (1).

In statement (2), Y will take on the new value during a simulation cycle that is executed 10 nanoseconds from the current simulation time. In this case, it is obvious that other statements that access Y before 10 ns have passed will use the old value. What is not so obvious is that the same principle holds for statement (1). The only difference is in the amount of time that passes before the new value of Y is used. For statement (1), the new value of Y is available delta time later; for statement (2), the new value of Y is available 10 ns later.

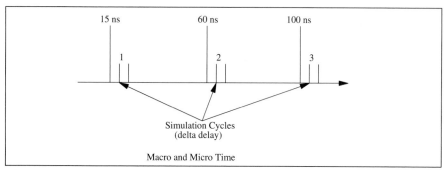

Figure 4.7   Macro and micro time.

Another view to take is that delta cycles represent *micro* time, while standard time units represent *macro* time, and that any point in macro time can contain as many increments, as desired, of micro time. Figure 4.7 illustrates this situation in a possible simulation scenario. At point 1, the simulation time has been advanced to 15 ns. At that value of macro time, two simulation cycles are performed, which require delta delay but do not advance simulation time. The next simulation cycle advances macro time to 60 ns. This situation is repeated at point 2 (60 ns) and point 3 (100 ns).

The two methods for measuring time give rise to three model types:

1. Delta delay only: These are employed for purely functional verification of models.
2. Standard time unit delay only: Used to validate system timing.
3. Mixed: These models contain both types of delay. The delta delay mechanism is used to force an ordering of events that do not incur a significant amount of delay. Standard time unit delay is then used to represent significant delay, perhaps across the modeled logic block. Mixed models are also used to study system timing.

To illustrate the difference between the two different timing models, consider the four architectures for entity BUFF shown in Figure 4.8. Entity BUFF is merely a buffer whose output Z follows the input X after a time delay. In architecture ONE, input X is copied into variable Y1 after zero delay. Z is then scheduled to take on the value of X after 1 ns. The total delay through architecture ONE is 1 ns. In architecture TWO, signal Y2 takes on the value of X after delta delay. The output Z then receives the value of Y2 after another delta delay. The total delay through architecture TWO is 2 delta. In architecture THREE, Y3 follows X after delta delay; Y3 is then copied to Z with a 1 ns delay. Whether the total delay through the architecture is 1 ns or 1 ns plus delta is a philosophical question hinging on whether you consider the delta to be absorbed into the 1 ns time period. In architecture FOUR the total delay is 2 ns.

Figure 4.9 shows the response of each architecture to the following pulse on X:

The outputs for architectures ONE, TWO, THREE, and FOUR are Z1, Z2, Z3, and Z4, respectively. Note: A delay of 1 delta is denoted as +1; a delay of 2 delta is denoted as +2, and time

```
entity BUFF is
  port(X: in BIT; Z: out BIT);
end BUFF;

architecture  ONE of BUFF is
begin
  process(X)
    variable Y1: BIT;
  begin
    Y1 := X;
    Z <= Y1 after 1 ns;
  end process;
end ONE;

architecture TWO of BUFF is
  signal Y2: BIT;
begin
  Y2 <= X ;
  Z <= Y2 ;
end TWO;

architecture THREE of BUFF is
  signal Y3: BIT;
begin
  Y3 <= X;
  Z <= Y3 after 1 ns;
end THREE;

architecture FOUR of BUFF is
  signal Y4: BIT;
begin
  Y4 <= X after 1 ns;
  Z <= Y4 after 1 ns;
end FOUR;
```

Figure 4.8 Four different delay situations.

advance and delta cycle advance are both displayed in the same printout. The notation is interpreted as:

```
(ns)
N        end of simulation cycle at time =  N ns.
+1       end of simulation cycle at time =  N ns.
         plus 1 delta delay
+2       end of simulation cycle at time =  N ns.
         plus 2 delta delays
N+1      end of simulation cycle at time = N+1 ns.
```

When signal assignment statements are imbedded in processes, care must be taken to understand the meaning of delta delay. For example, consider architecture FIVE of entity BUFF shown in Figure 4.10. If X changes at time t1, its value is copied into signal Y5 at time t1+delta. The second statement copies the value of Y5 at time t1, that is, the old value of Y5 into signal Z at time

```
     TIME |----------------------------SIGNAL NAMES----------
     (NS) |X      Z1     Y2     Z2     Y3     Z3     Y4     Z4
        0 |'0'    '0'    '0'    '0'    '0'    '0'    '0'    '0'
       +1 |---    ---    '0'    '0'    '0'    ---    ---    ---
        1 |'1'    '0'    ---    ---    ---    '0'    '0'    '0'
       +1 |---    ---    '1'    ---    '1'    ---    ---    ---
       +2 |---    ---    ---    '1'    ---    ---    ---    ---
        2 |---    '1'    ---    ---    ---    '1'    '1'    ---
        3 |---    ---    ---    ---    ---    ---    ---    '1'
        4 |'0'    ---    ---    ---    ---    ---    ---    ---
       +1 |---    ---    '0'    ---    '0'    ---    ---    ---
       +2 |---    ---    ---    '0'    ---    ---    ---    ---
        5 |---    '0'    ---    ---    ---    '0'    '0'    ---
        6 |---    ---    ---    ---    ---    ---    ---    '0'
```

Figure 4.9 Response of four architectures.

```
   architecture FIVE of BUFF is
     signal Y5: BIT;
   begin
     process(X)
     begin
       Y5 <= X;
       Z  <= Y5;
     end process;
   end FIVE;
   architecture FIVE_A of BUFF is
     signal Y5: BIT;
   begin
     process(X,Y5)
     begin
       Y5 <= X;
       Z  <= Y5;
     end process;
   end FIVE_A;
```

Figure 4.10 Effect of delta delay.

t1+delta. Since all statements in a process are executed at the same time, both signal assignment statements will execute at time t1. The new value of Y5 will not be copied into Z until there is another event on X. Signals Y5 and Z5 in Figure 4.11 illustrate the response of architecture FIVE to a pulse on X.

If the intent of the modeler is for the value of Y5 to be scheduled to be copied onto Z after another delta delay, then Y5 can be added to the sensitivity list as shown in architecture FIVE_A. Now, Y5 will change after delta delay, which generates a new event on Y5 at time t1+delta. This new event will trigger the process again at time t1+delta, and Z will be updated to the new value of Y5 at t1+2*delta. In effect, after two delta delays, Z will have the value of X. Signals Y5A and Z5A in Figure 4.11 illustrate the response of architecture FIVE_A to a pulse on X.

```
          Time|-------signal names------|

          (ns)| X     Y5    Z5    Y5A   Z5A

             0| '0'   '0'   '0'   '0'   '0'
            +1| ---   '0'   '0'   '0'   '0'
             1| '1'   ---   ---   ---   ---
            +1| ---   '1'   '0'   '1'   '0'
            +2| ---   ---   ---   '1'   '1'
             2| ---   ---   ---   ---   ---
             3| ---   ---   ---   ---   ---
             4| '0'   ---   ---   ---   ---
            +1| ---   '0'   '1'   '0'   '1'
            +2| ---   ---   ---   '0'   '0'
             5| ---   ---   ---   ---   ---
             6| ---   ---   ---   ---   ---
```

Figure 4.11   Response of two architectures.

A similar effect could be achieved by eliminating the process altogether and making the statements concurrent statements as shown in architecture FIVE_B.

```
architecture FIVE_B of BUFF is
   signal Y5: BIT;
begin
   Y5 <= X;
   Z  <= Y5;
end FIVE_B;
```

In this architecture, when X changes, the first statement is executed, which causes Y5 to take on the value of X after delta delay. When Y5 changes, the second statement is executed, causing Z to take on the value of X after two delta delays.

4.1.5   Inertial and Transport Delay in Signal Propagation

In VHDL there are two types of delay in signal assignment statements, *inertial delay* and *transport delay*. Here are examples of each:

```
Z  <=  I  after 10 ns;          ----inertial delay
Z  <=  transport I after 10 ns; ----transport delay
```

The first assignment statement implies inertial delay; that is, the signal propagation will take place if and only if input I persists at a given level for 10 ns—the amount of time specified in the *after* clause. Thus, changes in I will affect Z only if the new value of I stays at the new level for 10 ns or more. In the second case, where transport delay is specified, all changes on I will propagate to Z, regardless of how long the value of I stays at the new level.

The inertial delay mechanism filters out inputs that change rapidly. Inertial delay models the effect of capacitance on signal changes. At the circuit level, node voltages represent logic signal values. Because of capacitance, node voltages cannot change instantaneously. The voltage changes gradually as charge builds up, or drains from, the node. If a transistor is being driven by the node, it will not change state until a specific threshold voltage is reached. In other words, the circuit experiences an inertial effect. The difference in time between the initial change in node voltage and the change of state of the transistor is the inertial delay. In VHDL, inertial delay is implied by a signal assignment statement unless the key word *transport* is used. The use of inertial delay is quite natural when modeling real hardware. Since transport delay does not accurately represent real circuits, it is usually employed at higher levels of abstraction, for example, in defining model inputs for testing purposes.

4.2 THE VHDL SCHEDULING ALGORITHM

In this section we explain how transport and inertial delay are implemented. In order to discuss this implementation we need to discuss some preliminary concepts. The first two concepts are those of *transactions* and *waveforms*. Figure 4.12 illustrates these concepts. The signal Z is driven from two processes, Process A and Process B. For each process that drives Z, a driver is created, i.e., DaZ for process A and DbZ for process B. The resolution function F resolves the values in DaZ and DbZ to produce the value of Z. Each time an assignment is made to a driver, a transaction takes place, which is defined in the *VHDL Language Reference Manual* (LRM) as:

Transaction.   A pair consisting of a value and time. The value part represents a future value of the driver; the time part represents the time at which the value part becomes the current value of the driver.

Given this definition, the LRM then defines a waveform as follows:

Waveform.   A series of transactions. Each transaction represents a future value of the driver of the signal. The transactions in a waveform are ordered with respect to time, so one transaction appears before another if the first represents a value that will occur sooner than the value represented by the other.

Figure 4.12 illustrates a series of transactions, i.e., a waveform for driver DaZ of signal Z. Finally, we need one further definition, also from the LRM:

Current value of the driver
1. There exists one transaction whose time component is not greater than the current simulation time.
2. The current value of the driver is the value component of this transaction.
3. If, as the result of simulation time advance, the current simulation time becomes equal to the time component of the next transaction, the first transaction is deleted from the projected output waveform, and the next transaction becomes the current value of the driver.

In Figure 4.12, the current value of the driver DaZ is marked as CV. Figure 4.13 illustrates the effect of simulation time advance on the current value of the driver. When simtime = 15, the left-most transaction shown (10,0) has the current value of the driver since its time component is 10, while the next transaction (22,1) has a time component of 22. When simtime advances to 22, the transaction (22,1) determines the current value of the driver (1), and the transaction (10,0) is deleted from the waveform.

4.2.1 Waveform Updating

When a signal assignment statement is executed, the waveform driver for that signal is immediately updated. The value of the signal itself does not change when the signal assignment state-

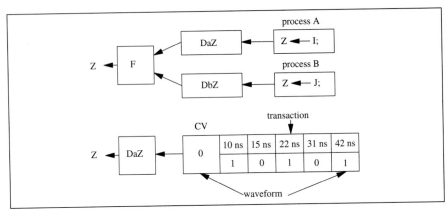

Figure 4.12   Transactions and waveforms.

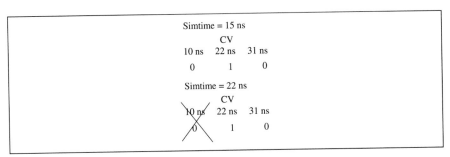

Figure 4.13   Current value of the driver.

ment executes. Only the driver for the signal changes. The *VHDL Language Reference Manual* gives the following steps for waveform updating. This updating algorithm executes every time a signal assignment statement is executed. The updating algorithm may execute many times on the same signal driver during a simulation cycle if more than one assignment is made to that signal in the same process during the simulation cycle.

Waveform Updating Algorithm

1. All old transactions that are projected to occur at or after the time at which the earliest new transaction is projected to occur are deleted from the projected output waveform.
2. The new transactions are then appended to the projected output waveform in the order of their projected occurrence.

   If the reserved word transport does not appear in the corresponding signal assignment, the first delay in the waveform is considered to be inertial delay, and the projected output waveform is further modified as follows:

3. All new transactions are marked.
4. An old transaction is marked if it immediately precedes a marked transaction and its value component is the same as that of the marked transaction.

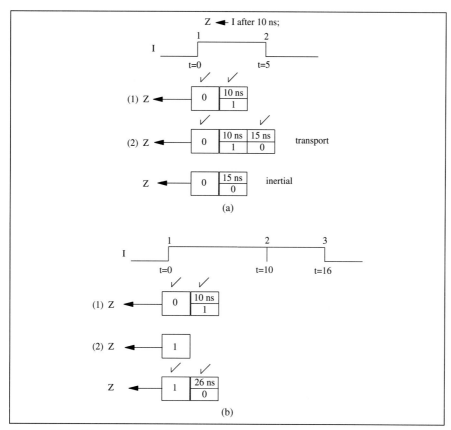

Figure 4.14 Waveform updating.

5. The transaction that determines the current value of the driver is marked.
6. All unmarked transactions (all of which are old transactions) are deleted from the projected output waveform.

In Figure 4.14(a), the delay specified for the signal assignment statement is 10 ns. A pulse of duration 5 ns is applied to I. Assume the current value of I is '0'. At event 1, which occurs at t = 0 ns, I makes a '0' to '1' transition. This causes a (10 ns, 1) transaction to be appended to the waveform for Z as shown in line (1). Both the current value of the driver and the new transaction are marked and, thus, retained. At event 2, which occurs at t = 5 ns, I switches low, and the transaction (15 ns, 0) is appended to the waveform as shown in line (2). If transport is specified, all transactions are retained as shown in line (2). However, if transport is not specified, the marking process removes the middle transaction (10 ns,1) as shown in the line below line (2). The resulting waveform schedules only the value '0' onto Z, and thus, the input pulse on I is suppressed.

In Figure 4.14(b), the situation is changed in that the pulse on I lasts for 16 ns. Again at event 1, which occurs at t=0 ns, I switches high, and a transaction (10 ns, 1) is appended to the waveform as shown in line (1). Both the current value of the driver and the new transaction are marked and thus, retained. At event 2, which occurs at t=10 ns, simulation time is advanced, and '1' becomes the current value of the driver as shown in line (2). At event 3 which occurs at t = 16 ns, I switches low, and a transaction (26 ns, 0) is appended to the waveform as shown in the line below line (2). Both transactions are marked and thus, retained. Because simulation time advanced and a new current value of the driver was formed before the new transaction was added, the input pulse on I was passed to the output Z.

4.2.2 Side Effects

Although the inertial delay mechanism is effective for ordinary logic modeling, it sometimes has unanticipated side effects. For example, one might like to use the following process to schedule two signal changes at the start of simulation:

```
process
begin
   Z <= '1' after 50 ns;
   Z <= '0' after 100 ns;
   wait;
end process;
```

The process, which executes just once at the start of simulation, executes sequentially. Assume that the initial value of Z is '0', then the inertial delay rule eliminates the (50 ns, 1) transaction scheduled by the first signal assignment statement. To fix this problem, one must add the key word *transport* to the statements:

```
process
begin
   Z <= transport '1' after 50 ns;
   Z <= transport '0' after 100 ns;
   wait;
end process;
```

Or the following form:

```
process
begin
   Z <=  '1' after 50 ns,
         '0' after 100 ns;
   wait;
end process;
```

The semantics of the last statement is that the first value assigned is done so with inertial delay, while the second value is assigned with transport delay. Thus, both signal changes would be scheduled.

4.3 MODELING COMBINATIONAL AND SEQUENTIAL LOGIC

VHDL has the necessary constructs to model combinational and sequential logic. Figure 4.15 illustrates the two situations. In Figure 4.15(a), a combinational logic circuit is illustrated where the output Z is strictly a function of the input X and responds to changes in X after delay DEL. If its circuit is modeled as a network of interconnected gates, its network graph is *acyclic*, i.e.,

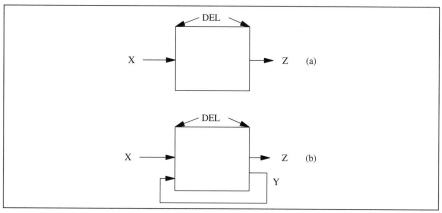

Figure 4.15 Basic combinational and sequential logic.

feedback free. The combinational circuit can be modeled in VHDL using the process construct as follows:

```
COMBINATIONAL: process(X)
   --- declare process variables
   variable ZVAR:BIT;
begin
   ---   represent circuit functionality
   ---   compute ZVAR
   Z <= ZVAR after DEL; ---model circuit delay
end process;
```

Note that the process models the combinational logic circuit as follows:

1. When the signal X in the process sensitivity list changes, the process executes, i.e., the sensitivity list is equivalent to the circuit inputs.
2. The process represents the functionality of the circuit by computing the value of the variable ZVAR.
3. The circuit delay is modeled by assigning ZVAR to the signal Z after delay DEL.

Thus, the process output Z is a function of its input X after delay DEL, effectively modeling the combinational circuit.

Sequential circuits have feedback, e.g., in Figure 4.15(b), the circuit is again driven by the input X, but in this case two sets of outputs are produced: (1) Z, the data outputs of the circuit and (2) Y, the circuit *state variables*, which are fed back as additional inputs to the circuit. This feedback mechanism causes the circuit to exhibit sequential behavior. Process SEQUENTIAL, shown below, models sequential behavior. The state variable is stored in the signal Y, which is driven by the process. Y is also in the sensitivity list of the process, and thus, the process will be triggered when Y changes, modeling the feedback in the sequential circuit. This feedback causes sequential circuits to exhibit memory. Depending on the level of abstraction used, however, the feedback mechanism may be "buried" in the model. For example, in modeling a register behaviorally, as opposed to a structural model of an interconnection of gates, the feedback will not be

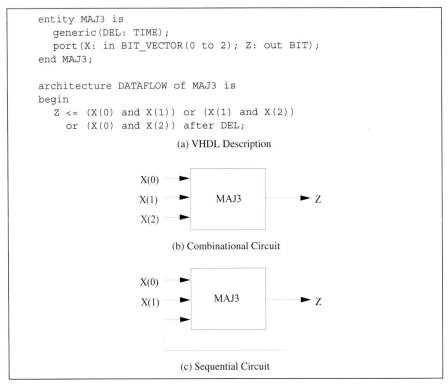

```
        entity MAJ3 is
          generic(DEL: TIME);
          port(X: in BIT_VECTOR(0 to 2); Z: out BIT);
        end MAJ3;

        architecture DATAFLOW of MAJ3 is
        begin
          Z <= (X(0) and X(1)) or (X(1) and X(2))
             or (X(0) and X(2)) after DEL;
```

(a) VHDL Description

(b) Combinational Circuit

(c) Sequential Circuit

Figure 4.16   A majority/consensus element.

evident in the model. We will see that other language constructs can be used to model sequential behavior at these higher levels of abstraction.

```
SEQUENTIAL: process(X,Y)
   --- declare process variables
   variable YVAR,ZVAR:BIT;
begin
   ---  represent circuit functionality
   ---  compute YVAR and ZVAR
   Y <= YVAR after DEL; ---state variable delay
   Z <= ZVAR after DEL; ---output delay
end process;
```

We conclude this discussion with an example. Figure 4.16(a) gives the VHDL description for the majority function of three inputs. Note that the behavior of the circuit is implemented in a single signal assignment statement, which has an equivalent process. In Figure 4.16(b) the model is used open loop and acts as a combinational logic circuit. In Figure 4.16(c) the output Z is connected back to input X(2). This feedback connection converts the model to a sequential circuit, which one could call the "consensus" element, in that the output Z is equal to the last value for which X(0) and X(1) agreed.

Combinational Primitives	Sequential Primitives
Gates	Flip-flops
Buffers	Registers
Adders	Latches
Multiplexers	Counters
Decoders	RAMs
Encoders	
Comparators	
Shifters	
Arithmetic logic units	
Population counters	
ROMs	
PLAs	

Figure 4.17   Design primitives.

4.4 LOGIC PRIMITIVES

In Table 1.1 we listed the common structural domain primitives that are used at different abstraction levels. In Chapter 9, we will see that these are typical functions found in an ASIC library. In this section we present VHDL models for some of those primitives. In doing so we will illustrate additional techniques for modeling combinational and sequential logic. Figure 4.17 shows a collection of primitives that are classified as being combinational or sequential logic. We give models for most of them in this chapter. Some are described elsewhere in the text or are left as homework problems. We give rather simple forms of each of the primitive models, using generics to represent time delay, but the timing models for the devices are simple. Also, we restrict ourselves to the use of types Bit and Bit_Vector for representing signals and data. In Chapter 5 we use multiple valued logic. In Chapter 7 we incorporate more complicated timing into models.

In this chapter, we use two basic approaches to modeling, algorithmic and data flow, as indicated by the architecture names ALG and DF.

4.4.1 Combinational Logic Primitives

We now present models for the combinational logic primitives.

4.4.1.1 Gate Primitive

Gate primitives are the most basic of the combinational logic primitives. Figure 4.18 shows the model for a 2-input AND gate. The model computes the logical AND of the two inputs and copies that result onto an output O after generic delay DEL. Primitive models for other basic gates are left as exercises for the reader.

4.4.1.2 Buffer Primitive

Buffer primitives copy their input I to the output O provided that their enable input E is '1'. If their E input is a '0', the buffer output O will go to a high impedance condition. Figure 4.19 gives a model for a buffer element. In this case, the high impedance condition on the output is mod-

```
entity AND2 is
  generic(DEL: TIME);
  port(I1,I2: in BIT; O: out BIT);
end AND2;
architecture DF of AND2 is
begin
  O <= I1 and I2 after DEL;
end DF;
```

Figure 4.18   AND gate primitive.

```
entity BUF is
  generic(DATA_DEL,Z_DEL: TIME);
  port(I,EN: in BIT; O: out BIT);
end BUF;
architecture ALG of BUF is
begin
  process(I,EN)
  begin
    if EN = '1' then
      O <= I after DATA_DEL;
    else
      O <= '1' after Z_DEL;
    end if;
  end process;
end ALG;
```

Figure 4.19   Buffer primitive.

eled by a logic '1'. Thus, the output of this buffer can be directly connected to the output of another buffer using a wired AND bus. For many situations, this is a sufficiently accurate model. Note: We do incorporate different propagation delays for enabling and disabling the buffer output. In Chapter 5 we show how to use multivalued logic to model busses more accurately.

4.4.1.3   Adder Primitive

Adders are important combinational logic circuits. The basic primitive in adders is the full adder, which adds two data bits and an input carry to generate sum and carry outputs. Figure 4.20 gives a data flow model for a full adder. Note: The sum output is the odd parity function of the three inputs. The carry output is the majority function of the three inputs. Different delays are modeled for the two functions.

Multibit adders are formed by cascading full adders together. Figure 4.21 shows a four-bit adder implemented in this fashion.

4.4.1.4   Multiplexer Primitive

It is frequently necessary to select from a number of data sources. Generally the number of data sources is 2^N, and we speak of a 2^N to 1 multiplexer. Also, we multiplex vectors that are B bits wide where B ranges from 1 to W, with W being the data word size in the system being imple-

```
entity FULL_ADDER is
  generic(SUM_DEL,CARRY_DEL:TIME);
  port(A,B,CI: in BIT; SUM,COUT: out BIT);
end FULL_ADDER;

architecture DF of FULL_ADDER is
begin
  SUM <= A xor B xor CI after SUM_DEL;
  COUT <= (A and B) or (A and CI) or (B and CI)
        after CARRY_DEL;
end DF;
```

Figure 4.20 Full adder primitive.

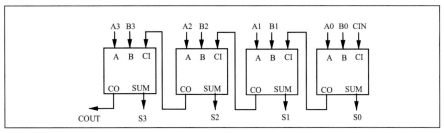

Figure 4.21 Four-bit adder.

```
entity FOUR_TO_1_MUX is
  generic(DEL: TIME);
  port(IN0,IN1,IN2,IN3: in BIT_VECTOR(3 downto 0);
       SEL: in BIT_VECTOR(1 downto 0);
       O: out BIT_VECTOR(3 downto 0));
end FOUR_TO_1_MUX;

architecture DF of FOUR_TO_1_MUX is
begin
  O <= IN0 after DEL when SEL = "00" else
       IN1 after DEL when SEL = "01" else
       IN2 after DEL when SEL = "10" else
       IN3 after DEL;
end DF;
```

Figure 4.22 Multiplexer primitive.

mented. Figure 4.22 shows a model for a 4 to 1, 4-bit wide multiplexer. Note that this is modeled compactly using a conditional signal assignment statement.

```
entity TWO_TO_4_DEC is
  generic(DEL: TIME);
  port(I: in  BIT_VECTOR(1 downto 0);
       O: out BIT_VECTOR(3 downto 0));
end TWO_TO_4_DEC;

architecture ALG of TWO_TO_4_DEC is
begin
  process(I)
  begin
    case I is
      when "00" => O<= "0001" after DEL;
      when "01" => O<= "0010" after DEL;
      when "10" => O<= "0100" after DEL;
      when "11" => O<= "1000" after DEL;
    end case;
  end process;
end ALG;
```

Figure 4.23   Decoder primitive.

4.4.1.5   Decoder Primitive

Decoders accept an N-bit input and activate one of 2^N outputs. The position number, i, of the output that is activated is equal to the decimal equivalent of the code present on the N-bit input. In that sense, the decoder functions as a binary to decimal converter. Decoders are used to convert bit patterns that correspond to addresses into single-line select signals, which can be used to enable the outputs of a device. They are also used to interpret bit patterns in the opcode field of computer instructions in order to select which instruction to execute. Figure 4.23 gives a model for a 2 to 4 decoder.

4.4.1.6   Encoder Primitive

Encoders perform the inverse function of decoders, i.e., they convert 2^N inputs to an N-bit code, where the N-bit code represents the highest priority input that is active. Thus, a priority ranking is necessary for the inputs, e.g., input 0 could be the highest priority and input $2^N - 1$ could be the lowest priority. Figure 4.24 gives a VHDL model for a 4 to 2 encoder. Again, a conditional signal assignment statement concisely implements the priority mechanism in the model. Note that the output of the decoder will be "11" if I(3) is the only active input and also in the case where no inputs are active.

4.4.1.7   Shifter Primitive

Shifting is an important data operation on bit vectors because under certain conditions, shifting right implements a division by two; shifting left implements a multiplication by two. In a purely logical sense, shifting corresponds to a permutation of a bit pattern. Figure 4.25 gives a VHDL model for a shifter which shifts the input right one position when SR ='1' and SL ='0', and left one position when SR ='0' and SL='1'. For the other two combinations of SR and SL, the input is transferred to the output unshifted. IL and IR are the values shifted in for left and right shifts respectively. Note that concatenation is used in modeling the shift operations.

```
entity FOUR_TO_2_ENC is
  generic(DEL: TIME);
  port(I: in  BIT_VECTOR(3 downto 0);
       O: out BIT_VECTOR(1 downto 0));
  end FOUR_TO_2_ENC;

architecture DF of FOUR_TO_2_ENC is
begin
  O <= "00" after DEL when I(0) = '1' else
       "01" after DEL when I(1) = '1' else
       "10" after DEL when I(2) = '1' else
       "11" after DEL;
end DF;
```

Figure 4.24 Encoder primitive.

```
entity SHIFTER is
  generic(DEL: TIME);
  port(DATA_IN: in BIT_VECTOR(3 downto 0);
       SR,SL: in BIT; IL,IR: in BIT;
       DATA_OUT: out BIT_VECTOR(3 downto 0));
  end SHIFTER;

architecture  ALG of SHIFTER is
begin
  process(SR,SL,DATA_IN,IL,IR)
    variable CON: BIT_VECTOR(0 to 1);
  begin
    CON := SR&SL;
    case CON is
      when "00" => DATA_OUT <= DATA_IN after DEL;
      when "01" => DATA_OUT <= DATA_IN(2 downto 0) & IL
                     after DEL;
      when "10" => DATA_OUT <= IR & DATA_IN(3 downto 1)
                     after DEL;
      when "11" => DATA_OUT <= DATA_IN after DEL;
    end case;
  end process;
end ALG;
```

Figure 4.25 Shifter primitive.

4.4.1.8 Data Operations Package

To implement some of the remaining primitive models, it is useful to define some basic functions and procedures. Figure 4.26 gives a package which contains these functions. All the functions operate on bit vectors of unconstrained length. In every case it is assumed that these bit vectors represent a binary number, and that the least significant bit is on the right. Note: In each case an internal variable is created, which has a descending range, and the parameter value is assigned to it. Thus, the calling parameters can either have ascending or descending ranges as

```vhdl
package PRIMS is
  procedure ADD(A,B: in BIT_VECTOR; CIN: in BIT;
               SUM: out BIT_VECTOR; COUT: out BIT);
  function INC(X : BIT_VECTOR) return BIT_VECTOR;
  function DEC(X : BIT_VECTOR) return BIT_VECTOR;
  function INTVAL(VAL : BIT_VECTOR) return INTEGER;
end PRIMS;
package body PRIMS is
  procedure ADD(A,B: in BIT_VECTOR; CIN: in BIT;
               SUM: out BIT_VECTOR; COUT: out BIT) is
    variable SUMV,AV,BV: BIT_VECTOR(A'LENGTH-1 downto 0);
    variable CARRY: BIT;
  begin
    AV := A;
    BV := B;
    CARRY := CIN;
    for I in 0 to SUMV'HIGH loop
      SUMV(I) := AV(I) xor BV(I) xor CARRY;
      CARRY := (AV(I) and BV(I)) or (AV(I) and CARRY)
               or (BV(i) and CARRY);
    end loop;
    COUT := CARRY;
    SUM := SUMV;
  end ADD;

  function INC(X : BIT_VECTOR) return BIT_VECTOR is
    variable XV: BIT_VECTOR(X'LENGTH-1 downto 0);
  begin
    XV := X;
    for I in 0 to XV'HIGH loop
      if XV(I) = '0' then
        XV(I) := '1';
        exit;
      else XV(I) := '0';
      end if;
    end loop;
    return XV;
  end INC;
function DEC(X : BIT_VECTOR) return BIT_VECTOR is
    variable XV: BIT_VECTOR(X'LENGTH-1 downto 0);
  begin
    XV := X;
    for I in 0 to XV'HIGH loop
      if XV(I) = '1' then
        XV(I) := '0';
        exit;
      else XV(I) := '1';
      end if;
    end loop;
    return XV;
  end DEC;

  function INTVAL ( VAL: BIT_VECTOR) return INTEGER is
    variable VALV: BIT_VECTOR(VAL'LENGTH - 1 downto 0);
    variable SUM: INTEGER := 0;
  begin
    VALV := VAL;
    for N in VALV'LOW to VALV'HIGH loop
      if VALV(N) = '1' then
        SUM := SUM + (2**N);
      end if;
    end loop;
    return SUM;
  end INTVAL;
end PRIMS;
```

Figure 4.26 Data operations package.

```
      use work.PRIMS.all;
      entity ALU is
        generic(DEL: TIME);
        port(A,B: in BIT_VECTOR(3 downto 0); CI: in BIT;
              FSEL: in BIT_VECTOR(1 downto 0);
              F: out BIT_VECTOR(3 downto 0); COUT: out BIT);
      end ALU;
      architecture ALG of ALU is
      begin
        process(A,B,CI,FSEL)
          variable  FV: BIT_VECTOR(3 downto 0);
          variable COUTV: BIT;
        begin
          case FSEL is
            when "00" => F <= A after DEL;
            when "01" => F <= not(A) after DEL;
            when "10" => ADD(A,B,CI,FV,COUTV);
                          F <= FV after DEL;
                          COUT <= COUTV after DEL;
            when "11" => F <= A and B after DEL;
          end case;
        end process;
      end ALG;
```

Figure 4.27 ALU primitive.

long as the least significant bit is on the right. Further comments on the specific functions or procedures are:

1. ADD: A *for* loop implements the full adder logic equations. The number of loop iterations is equal to the length of the arguments. It is assumed that the arguments are the same length.
2. INC and DEC: Again using a *for* loop, the INC(DEC) function begins in the low-order position and inverts all 1's (0's) until the first 0 (1)is reached. The first 0 (1) is also inverted, and all subsequent bits are copied.
3. INTVAL: This function uses a *for* loop to sum appropriate powers of two to convert a bit vector to an integer.

4.4.1.9 ALU Primitive

Arithmetic logic units (ALUs) are combinational circuits used to perform the basic arithmetic and logic operations in computer systems. Figure 4.27 gives a model of a simple ALU. It has three data inputs: A and B are bit vectors, and CI is an input carry. There are two data outputs: F is a bit vector output and COUT is an output carry. The function select input, FSEL, controls the function performed by the ALU. In this case FSEL is a 2-bit vector, and the ALU performs four operations:

1. F = A.
2. F = not(A).

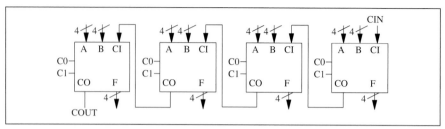

Figure 4.28   16-bit ALU.

3. $F = A + B$, where $+$ denotes twos complement addition. The input carry (CI) is included
    in the sum and an output carry (COUT) is generated.
4. $F = A$ and B.

This simple ALU provides for basic logic and arithmetic capability. More complicated operations can be built by repeated passage of data through the ALU and the storage of intermediate results in registers. Several interesting questions arise in ALU design: (1) Is the set of operations functionally complete, i.e., can an arbitrary logic or arithmetic function be implemented by repeated application of the ALU? and (2) What are the trade-offs between the functional capability of the ALU and speed of execution and the number of other primitives required, e.g., registers and multiplexers? These problems are addressed in the homework problems at the end of the chapter. Finally, simple ALUs are frequently cascaded to form ALUs which can operate on longer words. Figure 4.28 shows a 16-bit ALU formed from four 4-bit ALUs. Note: Control bits C0 and C1 are applied to the function select inputs.

4.4.1.10   PLA Primitive

In current VLSI technology, combinational logic is frequently implemented with a programmable logic array (PLA). Figure 4.29 shows a PLA that implements four functions (Z1, Z2, Z3, Z4) of three variables (X1, X2, X3). Although the array can realize a variety of logic functions of the four input variables, the total number of product terms over all four output functions is limited to four. Compare this capacity to that of a ROM with three address inputs and 4-bit words. The ROM can implement any four functions of the same three input variables. In general, PLAs cost less than ROMS with the same number of inputs and outputs. However, PLAs are less versatile.

The PLA in Figure 4.29 is divided into two planes, the AND and OR planes. The AND plane generates product terms by ANDing literals together. The OR plane sums (ORs) product terms together to form the output function. The basic electronic circuit of the PLA implements NOR-NOR logic. This particular version is modified for AND-OR logic by adding inverters at the outputs to produce NOR-OR logic. The NOR gates can be converted into AND gates by using DeMorgan's law:

$$\overline{A} \text{ and } B \text{ and } \overline{C} = \text{not}(A \text{ or } \overline{B} \text{ or } C)$$

To produce a term like $\overline{A}B\overline{C}$, we must connect the A, $\overline{B}$, and C inputs to the AND plane input. Each horizontal line, e.g., R1, then corresponds to a product term. The outputs of the AND plane are connected directly to the OR plane inputs to produce the logic OR of a subset of the product terms. The desired connections are made by placing transistor switches between appropriate

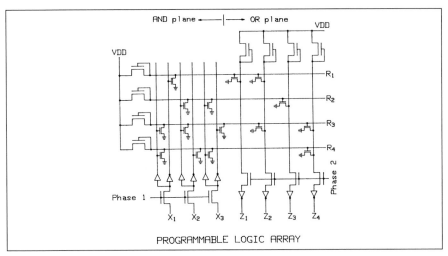

AND plane ◄────|────► OR plane

Figure 4.29 Programmable logic array.

wires. This particular PLA has been "programmed" by placing switches that implement the following logic equations.

$$R1 = X1$$
$$R2 = \overline{X2}\,\overline{X3}$$
$$R3 = \overline{X1}\,\overline{X2}X3$$
$$R4 = \overline{X1}X2\overline{X3}$$

$$Z1 = R1 = X1$$
$$Z2 = R1 + R3 = X1 + (\overline{X1}\,\overline{X2}X3)$$
$$Z3 = R2 = \overline{X2}\,\overline{X3}$$
$$Z4 = R3 + R4 = (\overline{X1}\,\overline{X2}X3) + (\overline{X1}X2\overline{X3})$$

Line R2 corresponds to the product term $\overline{X2}\,\overline{X3}$ because X2 and X3 are provided as inputs to the first level NOR logic. Summarizing, if variable X is needed in a product term, connect $\overline{X}$ to a switch; if $\overline{X}$ is needed, connect X to a switch. The reader should verify the other product terms. In the PLA in Figure 4.29, the AND plane has 24 potential connection points, the OR plane 16.

One could model the PLA at the switch level. However, multivalued logic is required to do this, which we discuss in Chapter 7. Also, this level of detail may be unnecessary for many applications. Thus, we create a behavioral model which scans two connection matrices and the AND and OR plane input values to compute the PLA output. A connection matrix is an array of 1's and 0's in which a 1 indicates that a switch is connected, and a 0 indicates that it is not. Figure 4.30 gives the behavioral model for the PLA. The model is divided into two processes: AND_PLANE and OR_PLANE. A signal (R) connects the AND plane output with the OR plane input.

In process AND_PLANE, the connection matrix is stored in a three-dimensional array. The first dimension input selects the AND plane function being computed (Ri), the second dimension input selects the input literal (Xj), and the third dimension input indicates whether the TRUE or FALSE literal (but not both) is used as a switch input.

```
entity PLA is
  generic(AND_DEL,OR_DEL: TIME);
  port(X: in BIT_VECTOR(1 to 3);
       Z: out BIT_VECTOR(1 to 4));
end PLA;
architecture CONNECTION_MATRIX of PLA is
  signal R: BIT_VECTOR(1 to 4);
begin
 AND_PLANE: process(X)
   variable RV: BIT_VECTOR(1 to 4);
   type AND_ARRAY is array( 1 to 4, 1 to 3, 1 to 2 ) of BIT;
   variable AND_PL: AND_ARRAY :=
     ((('0','1'),('0','0'),('0','0')),
      (('0','0'),('1','0'),('1','0')),
      (('1','0'),('1','0'),('0','1')),
      (('1','0'),('0','1'),('1','0')));
   begin
     for I in 1 to 4 loop
       RV(I) := '0';
       for J in 1 to 3 loop
       assert not(AND_PL(I,J,1) = '1' and AND_PL(I,J,2) = '1')
         report "Error in AND plane wiring";
         if AND_PL(I,J,1) = '1' then
           RV(I) := RV(I) or X(J);
         end if;
         if AND_PL(I,J,2) = '1' then RV(I) := RV(I) or not X(J);
         end if;
       end loop;
       R(I) <= not RV(I) after AND_DEL;
     end loop;
   end process AND_PLANE;
 OR_PLANE: process(R)
   variable ZV: BIT_VECTOR(1 to 4);
   type OR_ARRAY is array(1 to 4,1 to 4) of BIT;
   variable OR_PLANE: OR_ARRAY :=
     (('1','0','0','0'),('1','0','1','0'),
      ('0','1','0','0'),('0','0','1','1'));
   begin
     for I in 1 to 4 loop
       ZV(I) := '0';
       for J in 1 to 4 loop
         if OR_PLANE(I,J) = '1' then ZV(I) := ZV(I) or R(J);
         end if;
       end loop;
       Z(I) <= ZV(I) after OR_DEL;
     end loop;
   end process OR_PLANE;
 end CONNECTION_MATRIX;
```

Figure 4.30 PLA model.

In the process, code two nested loops are used to scan the matrix. The outer loop index scans variable *i*; the inner loop index scans variable *j*. If AND_PL(I,J,1) is '1', the true literal for variable J is included in the Ith product term. This should be done if $\overline{J}$ is needed in the logic equation for the Ith product term. If AND_PL(I,J,2) is '1', the false literal for variable J is included in the Ith product term. This should be done if J is needed in the logic equation for the Ith product term. Note: The true and false literal should not both be included; an assertion checks for this. After the loop is completed, the result is inverted to form the desired NOR function. Note: The NOR operator can not be used directly in a loop because it is not associative.

In the OR_PLANE process, the connection matrix is a two-dimensional array. OR_PLANE(I, J) = '1' if and only if product term J is included in output function I. The OR function is again implemented with two nested loops. The delay across the AND plane and OR plane is modeled with the two generic delays AND_DEL and OR_DEL.

There are two remaining primitives in the combinational logic column of Figure 4.17, the population counter and the comparator. The 1's counter example in Chapter 3 illustrated how to implement a population counter. Modeling of a comparator is given as a homework problem at the end of this chapter.

4.4.2 SEQUENTIAL LOGIC

We now present models for sequential logic primitives.

4.4.2.1 *Flip-Flop Primitive*

Flip-flops are the most basic of sequential logic primitives. Figure 4.31 gives an algorithmic model of a JK flip-flop. Note: Direct setting or resetting overrides the clocked action of the flip-flop. If R=S=1, nothing happens. Since this is an illegal condition, the output is, in general, unspecified. Therefore, doing nothing is consistent with the specification. In Chapter 7 we discuss other ways to react to unusual conditions.

Models for other types of basic flip-flops are easily defined.

4.4.2.2 *Register Primitive*

To store data words, one uses registers. Although structural models of registers can be built from flip-flops and gates, we define a behavioral model here for a register. Figure 4.32 gives a register model. Note: The register is entirely synchronous, i.e., both reset and load functions are synchronous, with reset having priority. Other models might make the reset asynchronous with respect to the clock. Using registers that are totally synchronous results in very structured timing that can avoid timing errors. Note: The register is a controlled register, i.e., it will only be loaded when LOAD = '1' and the clock rises. Note: A guarded block is employed (discussed in the next section). Having a control signal (LOAD) makes the register more easily controllable with signals from a control unit. Control units are discussed in detail in Chapters 6 and 8.

4.4.2.3 *Latch Primitive*

The clocked register element modeled above samples its inputs and updates its outputs at the instant the clock makes a positive transition. Obviously, clocked register elements that are triggered by the fall of the clock can also be defined. Another form of memory element, which is similar to the register, is a latch, in which the output follows the input when the clock is asserted,

```
entity JKFF is
  generic(SRDEL,CLKDEL: TIME);
  port(S,R,J,K,CLK: in BIT; Q,QN: inout BIT);
end JKFF;

architecture ALG of JKFF is
begin
  process(CLK,S,R)
  begin
    if S = '1' and R = '0' then
      Q <= '1' after SRDEL;
      QN <= '0' after SRDEL;
    elsif S = '0' and R = '1' then
      Q  <= '0' after SRDEL;
      QN <= '1' after SRDEL;
    elsif CLK'EVENT and CLK = '1' and S='0' and R='0' then
      if J = '1' and K = '0' then
        Q <= '1' after CLKDEL;
        QN <= '0' after CLKDEL;
      elsif J = '0' and K ='1'  then
        Q <= '0' after CLKDEL;
        QN <= '1' after CLKDEL;
      elsif J= '1' and K= '1'  then
        Q <= not Q after CLKDEL;
        QN <= not QN after CLKDEL;
      end if;
    end if;
  end process;
end ALG;
```

Figure 4.31   JK flip-flop model.

```
entity REG is
  generic(DEL: TIME);
  port(RESET,LOAD,CLK: in BIT;
       DATA_IN: in BIT_VECTOR(3 downto 0);
       Q: inout BIT_VECTOR(3 downto 0));
end REG;

architecture DF of REG is
begin
  REG: block(not CLK'STABLE and CLK ='1')
  begin
    Q <= guarded "0000" after DEL when RESET ='1' else
         DATA_IN after DEL when LOAD ='1' else
         Q;
  end block REG;
end DF;
```

Figure 4.32   Register model.

```
LATCH is
   generic(LATCH_DEL:TIME);
   port(D: in BIT_VECTOR(7 downto 0);
        CLK: in BIT; LOUT: out BIT_VECTOR(7 downto 0));
end LATCH;

architecture DFLOW of LATCH is
begin
   LATCH: block(CLK = '1')
   begin
      LOUT <= guarded D after LATCH_DEL;
   end block LATCH;
end DFLOW;
```

Figure 4.33 Latch primitive

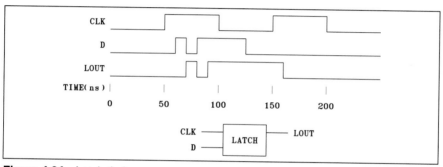

Figure 4.34 Latch timing.

and stores the input value when the clock transitions from asserted to deasserted. Thus, when clock is asserted, a latch acts like a logic buffer, while at the subsequent clock transition, it acts as a memory element. Latches have an advantage over edge triggered memory elements in that new output values can occur earlier and thus, reduce propagation delay.

Figure 4.33 gives a VHDL description for a latch element. Note that when the guard (CLK) for block LATCH goes to a '1', the value of D is transferred to LOUT after LATCH_DEL, and any subsequent changes on D will be propagated to LOUT as long as CLK is '1'. When CLK falls, the guarded statement is disabled and the signal LOUT maintains the last value propagated to it during the period when CLK was '1'. Figure 4.34 shows the timing of the latch element for LATCH_DEL = 10 ns, when the input D changes from all 0's to all 1's.

A word about terminology: Writers of some books and specifications use the terms "register" and "latch" interchangeably. In our view, however, the two terms imply different behaviors and we, therefore, use them as defined here.

4.4.2.4 Shift Register Primitive

Shift registers are important elements for data operations. Figure 4.35 gives the model for the primitive. Note that the element can be loaded (LOAD = '1'), shifted right (SR = '1' and SL = '0'),

```
entity SHIFTREG is
  generic(DEL: TIME);
  port(DATA_IN: in BIT_VECTOR(3 downto 0);
       CLK,LOAD,SR,SL: in BIT; IL,IR: in BIT;
       Q: inout BIT_VECTOR(3 downto 0));
end SHIFTREG;

architecture  DF of SHIFTREG is
begin
  SH:block(not CLK'STABLE and CLK ='1')
  begin
    Q <= guarded DATA_IN after DEL when LOAD= '1' else
    Q(2 downto 0) & IL after DEL when SL='1' and SR='0' else
    IR & Q(3 downto 1) after DEL when SL='0' and SR='1' else
    Q;
  end block SH;
end DF;
```

Figure 4.35   Shift register primitive.

and shifted left (SR = '0' and SR = '1'). IR and IL are the serial inputs for right and left shifting, respectively. Loading has priority over shifting.

4.4.2.5   Counter Primitive

Counting is a fundamental operation in digital systems. Figure 4.36 gives a counter model that can be reset (RESET = '1'), loaded (LOAD = '1'), and can count up or down as controlled by the signal UP when the clock rises. Note that the priority order is (1) reset, (2) load, and (3) count. Counting uses the INC and DEC functions in PACKAGE PRIMS.

4.4.2.6   Ram Primitive

An important sequential primitive is the RAM primitive. It models an array of registers that are selected by a decoding function. Thus, it can be built structurally from previously defined primitives. However, designing a behavioral model for RAM is a much more efficient approach.

Figure 4.37 gives the primitive model. The model is implemented by a single process, which is activated by a change in NCS (the low-active chip select line), RD (the read line), WR (the write line), DATA (the data input), and ADDRESS (the address input). Internal to the process, a test for NCS low is made. If NCS is low, tests are made for RD = '1' and WRITE = '1', and if one of them is high, the operation indicated is performed. If NCS is high, the high-impedance condition "1111" is transferred to the bus DATA. Note that a type conversion is employed in memory accessing; that is, signal ADDRESS, an array of bits of type BIT_VECTOR, is converted to type INTEGER to index the array representing the RAM. Note that we model the RAM array as a variable inside a process. It can also be modeled as a signal internal to the architecture.

This simple model of a RAM is adequate for many modeling applications. In Chapter 5 we illustrate other approaches.

```
        entity COUNTER is
          generic(DEL:TIME);
          port(RESET,LOAD,COUNT,UP,CLK: in BIT;
               DATA_IN: in BIT_VECTOR(3 downto 0);
               CNT: inout BIT_VECTOR(3 downto 0));
        end COUNTER;
        use work.PRIMS.all;
        architecture ALG of COUNTER is
        begin
          process(CLK)
          begin
            if CLK = '1' then
              if RESET = '1' then
                CNT <= "0000" after DEL;
              elsif LOAD ='1' then
                CNT <= DATA_IN after DEL;
              elsif COUNT ='1' then
                if UP = '1' then
                  CNT <= INC(CNT) after DEL;
                else
                  CNT <= DEC(CNT) after DEL;
                end if;
              end if;
            end if;
          end process;
        end ALG;
```

Figure 4.36   Counter primitive.

4.4.2.7   *Oscillator Primitive*

In any clocked sequential system, an oscillator is necessary to produce the system clock. Figure 4.38 gives a clock generator model in which the oscillator action is implemented by a feedback delay mechanism (the output of the process (CLOCK) is one of the signals in the process's sensitivity list). When RUN rises, the first clock period is scheduled, and CLKE is set to '1'. After that, for each rise of CLOCK, another clock period will be scheduled. When RUN falls, CLKE is reset. The current oscillator period will finish, and the clock will then stop. Note that the process output CLOCK is buffered from the port signal CLK because CLOCK is an input to a process, and thus, can not be a port signal of mode out.

Oscillators can also be modeled using *wait* statements. Figure 4.39 gives such an example. As long as RUN = '1', CLOCK will be '1' for HI_TIME and '0' for LO_TIME. The sum of HI_TIME and LO_TIME is the oscillator period; their relative magnitude determines the oscillator "duty cycle." Note that this model must be used in a system where RUN is initialized to '0'. If RUN is initialized to '1', the oscillator will not start because it needs an event on RUN to start.

It might seem that the first statement in the process, "wait until RUN ='1';" is redundant. However, without this statement, if simulation is started with RUN = '0', an infinite loop will result and simulation time will not advance.

```
use work.PRIMS.all;
entity RAM is
  generic (RDEL,DISDEL: TIME);
  port (DATA: inout BIT_VECTOR(3 downto 0);
        ADDRESS: in BIT_VECTOR(4 downto 0);
        RD,WR,NCS: in BIT);
end RAM;

architecture SIMPLE of RAM is
  type MEMORY is array(0 to 31) of BIT_VECTOR(3 downto 0);
begin
  process(RD,WR,NCS,ADDRESS,DATA)
    variable MEM: MEMORY;
  begin
    if NCS='0' then
      if RD='1' then
        DATA <= MEM(INTVAL(ADDRESS)) after RDEL;
      elsif WR='1'then
        MEM(INTVAL(ADDRESS)) := DATA;
      end if;
    else
      DATA <= "1111" after DISDEL;
    end if;
  end process;
end SIMPLE;
```

Figure 4.37   RAM primitive.

4.4.3 Testing Models: Test Bench Development

Any VHDL model of consequence (longer than two lines) must be thoroughly simulated before one can conclude that it is correct. Models are tested in a *test bench*. Examples of simple test benches were given in Chapter 3, but here, we elaborate on their development.

Figure 4.40 shows a model test bench. The Model Under Test (MUT) receives test vectors from a Stimulus Generator, which also provides an expected (GOLD) response. The Comparator compares the MUT response with the expected response and issues GO/NO GO signals. A test bench can have two types of feedback. First, the model state can be fed back to the Stimulus Generator to model interaction between these two elements. Second, there can be feedback from the Comparator to the Stimulus Generator that allows adaptive testing, i.e., the results of one test dictate what the next test will be. The function of the Stimulus Generator is to model the effect of the environment surrounding the system that the model represents.

The design of the Stimulus Generator is the major effort in test bench development. There are five major approaches to implementing a Stimulus Generator:

1. Use signal assignment statements.   The ONES_CNT test bench shown in Figure 3.9 and the TWOS_CONSECUTIVE test bench in Figure 3.36 illustrate the use of signal assignment statements. In the ONES_CNT test bench, the signal assignment statement is quite simple, but in the TWOS_CONSECUTIVE test bench, the signal assignments are very complex.

```
entity CLOCK_GENERATOR
  generic(PER: TIME);
  port(RUN: in BIT; CLK: out BIT);
end CLOCK_GENERATOR;

architecture ALG of CLOCK_GENERATOR is
  signal CLOCK: BIT;
begin
  process (RUN,CLOCK)
    variable CLKE: BIT := '0';
  begin
    if RUN='1' and not RUN'STABLE then
      CLKE := '1';
      CLOCK <= transport '0' after PER/2;
      CLOCK <= transport '1' after PER;
    end if;
    if RUN='0' and not RUN'STABLE then
      CLKE := '0';
    end if;
    if CLOCK='1' and not CLOCK'STABLE  and CLKE = '1'then
      CLOCK <= transport '0' after PER/2;
      CLOCK <= transport '1' after PER;
    end if;
    CLK <= CLOCK;
  end process;
end ALG;
```

Figure 4.38 Feedback oscillator.

```
entity COSC is
  generic(HI_TIME,LO_TIME: TIME);
  port(RUN: in BIT; CLOCK: out BIT := '0');
end COSC;
architecture ALG of COSC is
begin
  process
  begin
    wait until RUN ='1';
    while RUN = '1'  loop
      CLOCK <= '1';
      wait for HI_TIME;
      CLOCK <= '0';
      wait for LO_TIME;
    end loop;
  end process;
end ALG;
```

Figure 4.39 *Wait* statement oscillator.

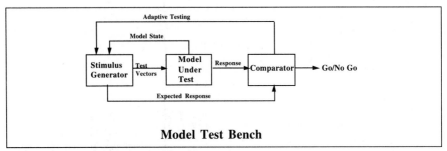

Figure 4.40   Organization of a test bench.

```
entity TEST_BENCH
is end TEST_BENCH;

use STD.TEXTIO.all;
use WORK.all;
architecture ONES_CNT1 of TEST_BENCH is
  signal PLAY: BIT;
  signal A: BIT_VECTOR(2 downto 0);
  signal C: BIT_VECTOR(1 downto 0);
  component ONES_CNTA
    port (PLAY: in BIT; A: in BIT_VECTOR(2 downto 0);
          C: out BIT_VECTOR(1 downto 0));
  end component;
  for L1: ONES_CNTA use entity ONES_CNT(ALGORITHMIC);
begin
  L1: ONES_CNTA
    port map(PLAY, A, C);
  process
    variable VLINE: LINE;
    variable V1: BIT_VECTOR(2 downto 0);
    variable V2: BIT_VECTOR(1 downto 0);
    file INVECT: TEXT is "TVECT.TEXT";
  begin
    PLAY<= '0', '1' after 3 ns;
    wait on PLAY until PLAY = '1';
    while not(ENDFILE(INVECT)) loop
      READLINE(INVECT, VLINE);
      READ(VLINE, V1);
      READ(VLINE, V2);
      A<= V1;
      wait for 1 ns;
      assert (V2 = C)
        report "WARNING: C is NOT equal to (V2)"
          severity WARNING;
    end loop;
  end process;
end ONES_CNT1;
```

Figure 4.41   Text I/O test bench.

```
     S <= '1' after 10 ns,  '0' after 220 ns,  '1' after 920 ns,
           '0' after 1120 ns;
     R <= '1' after 230 ns,  '0' after 420 ns,  '1' after 1320 ns,
           '0'after 1420 ns;
     J <= '1' after 500 ns,  '0' after 600 ns,  '1' after 700 ns;
     K <= '1' after 600 ns;
     CLK <='1' after 50 ns,   '0' after 100 ns,  '1' after 150 ns,
           '0' after 200 ns,  '1' after 250 ns,  '0' after 300 ns,
           '1' after 350 ns,  '0' after 400 ns,  '1' after 450 ns,
           '0' after 500 ns,  '1' after 550 ns,  '0' after 600 ns,
           '1' after 650 ns,  '0' after 700 ns,  '1' after 750 ns,
           '0' after 800 ns,  '1' after 850 ns,  '0' after 900 ns,
           '1' after 950 ns,  '0' after 1000 ns,'1' after 1050 ns,
           '0' after 1100 ns,'1' after 1150 ns,'0' after 1200 ns,
           '1' after 1250 ns,'0' after 1300 ns,'1' after 1350 ns,
           '0' after 1400 ns,'1' after 1450 ns,'0' after 1500 ns,
           '1' after 1550 ns,'0' after 1600 ns,'1' after 1650 ns,
           '0' after 1700 ns;
```

Figure 4.42   Generating test vectors for a JK flip-flop using signal assignment statements.

The next three forms are meant to alleviate the necessity to create complex signal assignment statements.

2. Use text I/O.   In many cases text I/O is the most efficient way to produce model inputs and provide the correct outputs. For example, the following text file could be used to test the 1's counter:

$$000 \; 00$$
$$001 \; 01$$
$$010 \; 01$$
$$011 \; 10$$
$$100 \; 01$$
$$101 \; 10$$
$$110 \; 10$$
$$111 \; 11$$

The left-most 3-bit vector is the input A; the right-most 2-bit vector is the correct output C, i.e., the Gold value. Figure 4.41 shows a test bench that reads this text I/O file. When the signal PLAY makes a 0 to 1 transition, reading of the file begins. A line is read every nanosecond. The first vector from the line (V1) is applied to the input A. The second vector from the line (V2) is compared with the output C in the *assert* statement.

3. Use input generator primitives.   In this approach, a structural model is created in which the components are primitives that generate sequences of input vectors.

Figure 4.42 shows a set of signal assignment statements that generate test vectors to exercise the JK flip-flop model. This model illustrates two common requirements for testing sequential models (1) generation of clocks (signal CLK) and (2) generation of input combinations for a circuit (signals S, R, J, and K). Instead of using the complex signal assignment statement for

```
entity PULSE_GEN is
  generic(N: INTEGER; PER: TIME);
  port(START: in BIT; PGOUT: out BIT_VECTOR(N-1 downto 0));
end PULSE_GEN;
architecture ALG of PULSE_GEN is
  function INT_TO_BIN (INPUT : INTEGER;N : POSITIVE)
    return BIT_VECTOR is
    variable FOUT: BIT_VECTOR(0 to N-1);
    variable TEMP_A: INTEGER:= 0;
    variable TEMP_B: INTEGER:= 0;
  begin    -- Beginning of function body.
    TEMP_A := INPUT;
    for I in N-1 downto 0 loop
      TEMP_B := TEMP_A/(2**I);
      TEMP_A := TEMP_A rem (2**I);
      if (TEMP_B = 1) then
        FOUT(N-1-I) := '1'; else
        FOUT(N-1-I) := '0';
      end if;
    end loop;
    return FOUT;
  end INT_TO_BIN;
  begin    -- Beginning of architecture body.
    process(START)
    begin
      for I in 0 to 2**N-1 loop
        PGOUT <= transport INT_TO_BIN(I,N) after I*PER;
      end loop;
    end process;
end ALG;
```

Figure 4.43   Input combination generator.

CLK shown in Figure 4.42, clock generation can easily be handled by using one of the oscillator primitives, e.g., the *wait* statement oscillator shown in Figure 4.39. To generate input combinations, one can use the model shown in Figure 4.43. For a given integer value N, this model generates the integer sequence 0 through $2^N - 1$, and converts each integer to its N-bit binary equivalent. The binary output changes with period PER. This pulse generator is particularly effective for testing combinational logic primitives and was, in fact, used to check out the primitives defined in this chapter.

Figure 4.44 shows a test bench for the JK flip-flop using input generator primitives. The JK clock input (CLK) is driven by the oscillator output (OSC). The S, R, J, and K inputs are driven by the four-bit PG output. The oscillator and the input combination generators have the same period, but the IGC is started 25 ns later, so its output transitions occur in the middle of the clock period.

4. Graphics generation.   There are commercial tools one can use to prepare inputs by creating a waveform graphically and letting the tool automatically generate the VHDL code for a test bench. Figure 4.45 shows waveforms created with TestBencher Pro from Synapticad. The VHDL generated for each signal is of the following form:

```
        entity TEST_BENCH is
        end TEST_BENCH;

        use work.all;
        architecture JKFF_TEST of TEST_BENCH is
          signal RUN, CLOCK, START, S, R, J, K, CLK, Q, QN: in BIT;
          signal PGOUT: BIT_VECTOR(3 downto 0);
        ------------ OSCILLATOR-------------------------
          component OSCILLATOR
            generic(HI_TIME, LO_TIME: TIME);
            port(RUN: in BIT; CLOCK: out BIT:= '0');
          end component;
        ------------ INPUT COMBINATION GENERATOR --------
          component ICG
            generic(N: INTEGER; PER: TIME);
            port(START: IN BIT; PGOUT: out BIT_VECTOR(N-1 downto 0));
          end component;
        ------------ JKFF -----------------------------
          component JK
            generic(SRDEL, CLKDEL: TIME);
            port(S, R, J, K: in BIT; CLK: in BIT; Q, QN: inout BIT);
          end component;
          for T1: OSCILLATOR use entity COSC(ALG);
          for T2: ICG use entity PULSE_GEN(ALG);
          for T3: JK use entity JKFF(ALG);
          signal PG: BIT_VECTOR(3 downto 0);
          signal OSC: BIT;
        begin
          T1: OSCILLATOR
            generic map(25 ns, 25 ns)
            port map(RUN, OSC);
          T2: ICG
            generic map(4, 50 ns)
            port map(START, PG);
          T3: JK
            generic map( 12 ns, 15 ns)
            port map(PG(3), PG(0), PG(2), PG(1), OSC, Q, QN);
          process
          begin
            RUN<='0', '1' after 50 ns, '0' after 9000 ns;
            START<= '1', '0' after 25 ns;
            wait;
          end process;
        end JKFF_TEST;
```

Figure 4.44 Test bench using input generator primitives.

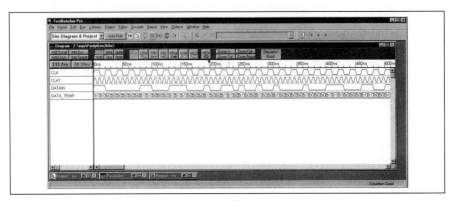

Figure 4.45 Graphics-based input generation.

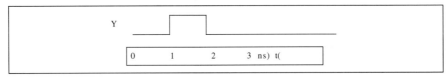

Figure 4.46 Waveform for signal Y.

```
process
   begin
      CLK <= '0';
      wait for 0 ns;
      while true loop
         CLK <= '1';
         wait for 10 ns;
         CLK <= '0';
         wait for 10 ns;
      end loop;
   end process;
```

5. Sensor Modeling. In many cases the stimulus generator is a model of a sensor that is an input to a physical system. We discuss this case in Chapters 9 and 11.

PROBLEMS

4.1 Develop complete VHDL descriptions that implement the code in Figure 4.2.
   a. Simulate the descriptions with 2 ns between input changes, and validate the results shown in Figure 4.1.
   b. Reverse the order of statements (1) and (2) and verify that the results do not change.
   c. Reverse the order of statements (3) and (4) to verify that the results do change.

4.2 Explain the concept of delta time and give an example to illustrate this concept.

4.3 Given the two statements:
```
X1 <= Y after 2 ns;
X2 <= transport Y after 2 ns.
```
Plot the response of X1 and X2 for the Y waveform shown in Figure 4.46.

Figure 4.47 Waveform for signal X.

4.4 Consider these two signal assignment statements:

```
Y <= transport X after 2 ns;
Z <= X after 2 ns;
```

If X has the waveform shown in Figure 4.47, sketch the waveforms for Y and Z.

4.5 Assume the following VHDL program is simulated.

```
--
entity EX2 is
end EX2;
--
architecture TEST of EX2 is
    signal A,B,C,D: BIT_VECTOR (7 downto 0) := "11001100";
    procedure X(signal Y: in BIT_VECTOR (7 downto 0);
                signal Z: out BIT_VECTOR (7 downto 0)) is
    variable TEM: BIT_VECTOR (7 downto 0);
    begin
       for I in 0 to 7 loop
          TEM(I) := Y(7-I);
       end loop;
        Z <= TEM after 1 ns;
    end X;
begin
    S1: X(C,D);
    S2: C <= A and B after 2 ns;
    S3: A <= "10110110" after 5 ns, "10001100" after 10 ns;
    S4: B <= "11110000" after 5 ns, "00001111" after 10 ns;
end TEST;
```

Answer these questions:
a. For each label (S1, S2, S3, and S4), list all the times (in ns) that the associated statement will be executed. Assume time starts at t=0 when simulation begins.
b. At what times will there be events on signal C?
c. Show the waveform for signal C by listing its value during all time steps.
d. What is the final value of signal D, and at what time does signal D change to its final value? Explain how you obtained the answer.
e. State in words what procedure X does.

4.6 Assume the following VHDL program is simulated.

```
--
entity FINAL is
end FINAL;
--
architecture TEST of FINAL is
    signal A,B,C,D: BIT_VECTOR (7 downto 0) := "11001100";
    procedure X(signal Y: in BIT_VECTOR (7 downto 0);
                signal Z: out BIT_VECTOR (7 downto 0)) is
```

```
        variable TEM: BIT_VECTOR (7 downto 0);
      begin
        for I in 6 downto 0 loop
          TEM(I) := Y(I+1);
        end loop;
        TEM(7) := Y(7);
        Z <= TEM after 2 ns;
      end X;
    begin
      S1: C <= A xor B after 3 ns;
      S2: X(C,D);
      S3: A <= "10110110" after 7 ns, "10001100" after 15 ns;
      S4: B <= "10110110" after 7 ns, "00001111" after 15 ns;
    end TEST;
```

Answer these questions.

    a. For each label (S1, S2, S3, and S4), list the time that the associated statement will be executed. Assume time starts at t=0 when simulation begins.

    b. At what times will there be events on signal C?

    c. Show the waveform for signal C by listing its value during all time steps.

    d. What is the final value of signal D, and at what time does signal D change to its final value? Explain how you obtained the answer.

    e. State in words what procedure X does.

4.7   Shown is an entity QUESTION and its test bench TB. Assume they are analyzed and the test bench is simulated. Sketch a plot of CON, INP, Q, and QOUT. Clearly label the time scale.

```
entity QUESTION is
    port(INP: in BIT; CON: in BIT; QOUT: out BIT);
end QUESTION;

architecture ANSWER of QUESTION is
    signal Q: BIT;
begin
    B1: block(CON='1')
    begin
       Q <= guarded INP after 10ns;
       QOUT <= Q after 5 ns;
    end block;
end ANSWER;

entity TB is
end TB;

use WORK.all;
architecture QT of TB is
    signal INP,CON,QOUT: BIT;
    component QUEST
       port(INP: in BIT; CON: in BIT; QOUT: out BIT);
    end component;
    for C1: QUEST use entity QUESTION(ANSWER);
begin
    C1: QUEST
       port map(INP,CON,QOUT);
```

```
        process
        begin
          INP <= '1' after 5 ns;
          CON <= '1' after 10ns, '0' after 30 ns;
          wait;
        end process;
     end QT;
```

4.8 This is the VHDL description of a black box entity (BLKBOX).

```
     entity BLKBOX is
        generic(T: TIME);
        port(I: in BIT; Z: out BIT);
     end BLKBOX;

     architecture INSIDE of BLKBOX is
        signal Y1,Y2: BIT;
     begin
        Y1 <=  I after T;
        Y2 <=  transport I after T;
        Z  <=  Y1 xor Y2;
     end INSIDE;

     entity TB is
     end TB;

     use work.all;
     architecture BLKBOXT of TB is
        signal I,Z: BIT;
        component BLKBOX
           generic(T: TIME);
           port(I: in BIT; Z: out BIT);
        end component;
        for C1: BLKBOX use entity work.BLKBOX(INSIDE);
     begin
        C1: BLKBOX
           generic map(2 ns)
           port map(I,Z);
        I <= '1' after 3 ns,'0' after 4 ns,  '1' after 6 ns,
             '0' after 7 ns,'1' after 11 ns, '0' after 14 ns;
     end BLKBOXT;
```

Perform the following tasks:
   a. For the test bench shown (TB), plot I, Y1, Y2, and Z.
   b. Describe in general terms what entity BLKBOX does.
4.9 Given the signal assignment statement:

```
     C <= A or B after 2 ns;
```

and the waveforms in Figure 4.48, sketch the transactions that illustrate how the waveform updating algorithm would work for this input.
4.10 Both loops in architecture LOOPs (of entity INT_SEQ) are seemingly capable of outputting integers 0 through 9. However, one of the loops does not work. Which one? Why? Verify your answer by simulating the description.

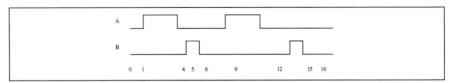

Figure 4.48   Waveforms for signals A and B.

```
entity INT_SEQ is
   port(OUT1: out INTEGER; OUT2: out INTEGER);
end INT_SEQ;
architecture LOOPS of INT_SEQ is
   signal START: BIT;
begin
   START <= '1';
   process(START)
   begin
      for I in 0 to 9 loop
         OUT1 <= I after I*(1 ns);
      end loop;
   end process;
   process(START)
   begin
      for I in 0 to 9 loop
         OUT2 <= transport I after I*(1 ns);
      end loop;
   end process;
end LOOPS;
```

4.11 Simulate the Majority/Consensus element and verify that feeding back the output to one of the inputs does produce consensus action.

4.12 Develop VHDL descriptions for two approaches to the design of 4-bit binary comparators:
   a. As a cascade of four identical comparator cells.
   b. A parallel implementation.

4.13 Design a 2x1 vector multiplexer (MUX) with an enable input. Each data input and output should be an 8-bit vector. The unit should be combinational in nature. If the enable input is logic 0, the output vector should be "11111111". If the enable input is logic 1, then the output should be determined by the select input and the data inputs. Include these three generic delays:
   a. DATA_DELAY is the delay in output change after a change on one of the data inputs.
   b. SELECT_DELAY is the delay in output change after a change on the select input.
   c. ENABLE_DELAY is the delay in output change after a change on the enable input.
   Include an entity description and an algorithmic-level, behavioral domain architecture description.

4.14 Perform the following modifications to the PLA model shown in Figure 4.30.
   a. Modify the PLA model, so it can implement a different logic function.
   b. Make the PLA model general, so it can implement logic functions with an arbitrary number of inputs and outputs.

4.15 Develop behavioral models for 7400 series MSI parts. Represent time delays with generics. Develop models for one or more of the following:

   a. BCD to 7 segment decoder (SN 7446)
   b. 4-bit binary full adder with fast carry (SN 7483)
   c. Synchronous decade rate multiplier (SN 74167)
   d. 16-bit multiple-port register file with tristate outputs (SN 74172)
   e. Arithmetic logic unit (SN 74181)
   f. Carry look ahead generator (SN 74182)
   g. First-in, first-out memory (SN 74S225)

4.16 Shown is the state table of a T (toggle) flip-flop. Note: It has an overriding reset input.

R	CLOCK	T	Q(t)	Q(t+1)
0	⋮	1	1	0
0	⋮	1	0	1
0	⋮	0	1	1
0	⋮	0	0	0
1	X	X	X	0

Give an interface description (entity) and behavioral architectural body for this flip-flop. Assume the model has delta delay timing.

4.17 Shown is the state table for a J K flip-flop. Write a VHDL behavioral description for the flip-flop. Represent any delays with generics.

J	K	CLOCK	Q(t)	Q(t+1)
X	X	0	X	Q(t)
X	X	1	X	Q(t)
0	0	⋮	X	Q(t)
1	0	⋮	X	1
0	1	⋮	X	0
1	1	⋮	X	not Q(t)

Note: X denotes a "don't care"; the value can be either 1 or 0.

4.18 Given the design trees in Figure 4.49, develop a design of a three stage, binary counter circuit using VHDL. The design will be done in two phases. Phase 1 will use (a) Design Tree 1 to construct the circuit. Phase 2 will use (b) Design Tree 2. In both phases an oscillator will be used. It should have the following entity declaration:

```
entity OSC is
   generic(HITIME,LOWTIME:TIME);
   port(RUN: in BIT; CLK: out BIT);
end OSC;
```

HITIME and LOWTIME are the durations of the high time and low time of the oscillator period. When RUN = '1', the oscillator runs; when RUN = '0', it stops.

Design Phase 1

Use (a) Design Tree 1 in this phase. Develop an algorithmic behavioral model for the counter. Its entity declaration should be:

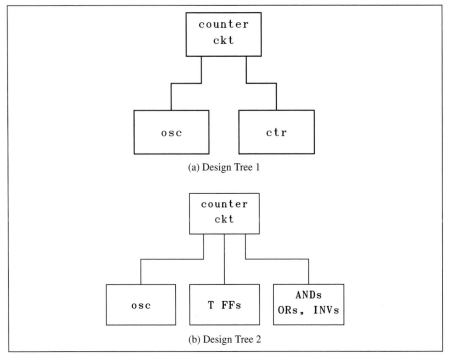

Figure 4.49   Design trees.

```
entity UP_CNT is
   generic((R_DEL,CLK_DEL: TIME);
   port(RESET,COUNT,CLK: in BIT;
        CNT: inout BIT_VECTOR(2 downto 0));
end UP_CNT;
```

R_DEL and CLK_DEL are the counter reset and clock delays. RESET = '1' holds the counter reset. The counter counts positive clock transitions when COUNT = '1' and RESET='0'. In writing the model, develop and use an increment function. Develop a test bench for the model, simulate it, and observe its response.

Design Phase 2

Use (b) Design Tree 2 for this phase in which the counter employs T flip-flops and any necessary AND, and OR gates, and inverters. The Phase 2 design must meet the following requirements:

a. The counter shall be a 3-stage, synchronous counter (not a ripple carry counter).
b. You must use the hierarchy as shown; thus the oscillator, the T flip-flops, and the miscellaneous gates must be behavioral models.
c. The oscillator outputs a 25% duty cycle, 10 MHz square wave.
d. The T flip-flop, shown in Figure 4.50, toggles when T = '1' and the clock makes a positive transition. If T = '0' the flip-flop retains its previous state. R is an overriding clear. If R = '1', the flip-flop resets, independent of the clock. The timing specifications for the T flip-flop are reset delay = 10 ns and clock delay = 15 ns.

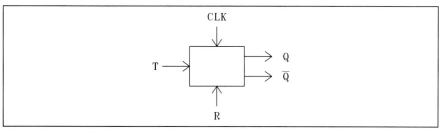

Figure 4.50   T flip-flop.

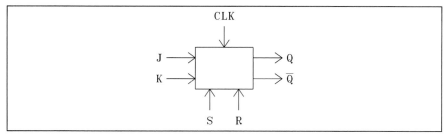

Figure 4.51   JK flip-flop.

    e. All gates have a delay of 5 ns.

    f. For each model in the counter, use generics to make the models general, i.e., entity dec-
        larations for the T flip-flop and a 2-input AND gate would look like:

```
entity TFF is
   generic(RDEL,CLOCKDEL: TIME);
   port(R,T,CLK: in BIT; Q: out BIT);
end TFF;
entity AND2 is
   generic(DEL: TIME);
   port(I1,I2: in BIT; O: out BIT);
and AND2;
```

    g. Develop a structural architecture of the counter system, including the oscillator. Plug this
        architecture into the test bench you developed for Phase 1 and simulate.

    Prepare a report which contains:

    1. A description of how each model was developed.

    2. VHDL model and test bench listings.

    3. Well-annotated simulation results.

4.19 Repeat the structural design of the counter described in Problem 4.18. Use the JK flip-flop
     with a block diagram as shown in Figure 4.51 as the basic flip-flop type.

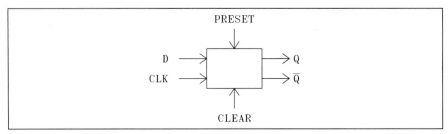

Figure 4.52   D flip-flop.

The timing specifications for the JK flip-flop are:

Parameter	From Input	To Output	Specifications
TPLH	S	Q	16 ns
TPHL	S	$\overline{Q}$	25 ns
TPLH	R	$\overline{Q}$	16 ns
TPHL	R	Q	25 ns
TPLH	CLK	Q or $\overline{Q}$	16 ns
TPHL	CLK	Q or $\overline{Q}$	25 ns

The timing specifications for logic gates are:
```
TPLH    5 ns
TPHL    3 ns
```
The timing specification for inverters are:
```
TPLH    3 ns
TPHL    2 ns
```
Explain any special timing anomalies you observe in simulating the counter built from these flip-flops.

4.20 Repeat Problem 4.19. Use a D flip-flop with the block diagram shown in Figure 4.52 as the basic flip-flop type.

The timing specifications for the D flip-flop are:

Parameter	From Input	To Output	Specificatio
TPLH	PRESET	Q	16 ns
TPHL	PRESET	Q	25 ns
TPLH	CLEAR	Q	16 ns
TPHL	CLEAR	Q	25 ns
TPLH	CLK	Q or $\overline{Q}$	16 ns
TPHL	CLK	Q or $\overline{Q}$	25 ns

Assume the flip-flop to be negative edge triggered.

4.21 Use the primitives developed in this chapter to implement the system shown in Figure 4.53. Your system model must meet the following requirements:

a. The Data Unit may use only one, four-function ALU, but as many of the other primitives as necessary. The Data Unit must be capable of performing these two calculations:

```
Z = X xor Y --- calculation 1
Z = 2*(X-Y) --- calculation 2
```

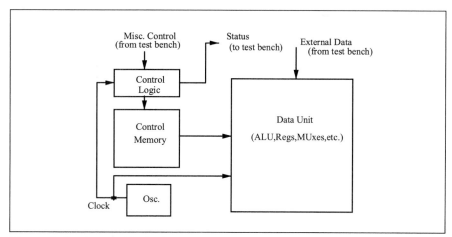

Figure 4.53 System block diagram.

The values X and Y come from the test bench via the external data port. The result Z can be left in a Data Unit register. Twos complement arithmetic is assumed.

b. For each calculation a sequence of control words will be stored in a Control Memory (use the RAM primitive). The Control Logic shall cause the appropriate control sequence to be fetched, causing calculation 1 or 2 to be performed.

c. The Control Logic may interact with the test bench to control the starting, stopping, and loading of data.

d. The clock shall be developed by an oscillator primitive.

e. Do your design such that the two calculations can be done in minimum time. The same design must be used for both calculations.

f. Simulate your design for both calculations for a suitable number of test cases.

g. Prepare a report summarizing your results.

4.22 Design a more powerful ALU than the one presented in this chapter. Repeat Problem 4.21. Discuss the effect the new ALU has on execution time and the number of required multiplexers and registers.

Algorithmic Level Design

In this chapter we begin our discussion of the structured design process. Figure 5.1 illustrates the top-level, structured design blocks. The specification for the design is written in a *natural language*, such as English. This natural language description may be accompanied by a block or timing diagram. The first step in the process is to translate the English description into one of three forms (1) a truth table, (2) a state diagram, or (3) an algorithmic model. All three forms represent the behavior of the system. The truth table and the state diagram are primarily pictorial, and the algorithmic model is textual. In this book we write the algorithmic description in VHDL. Any one of these forms can be developed initially, and in this chapter we explain the process for the algorithmic model. The truth table and state diagram are discussed in Chapter 8. To develop a truth table or state diagram, one must have made a judgment as to the nature of the system function implied by the word description, i.e., does it imply a combinational or sequential circuit. To develop an algorithmic model, one does not need to have made that determination. This model can be formed by mapping groups of sentences to constructs in the VHDL language.

Figure 5.1 shows a bidirectional connection between truth tables, state diagrams, and algorithmic models. Even though the truth table or the state diagram might be created first, one might wish to validate the table or state diagram by then developing and simulating a behavioral model. Alternatively, one might develop the algorithmic model first, but then want to express it later in terms of truth tables or state diagrams. Thus, it is important to develop transformations that allow us to move between the different forms.

Figure 5.1 is, in a sense, just "the beginning of the story," as other steps must be carried out before a design is synthesized. In Chapters 8, 9, 10, and 11, we discuss other aspects of the process.

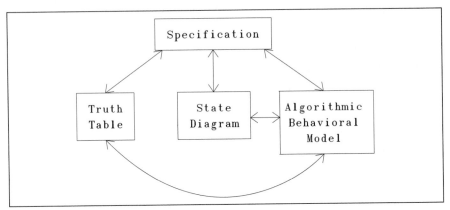

Figure 5.1 Top-level, structured design blocks.

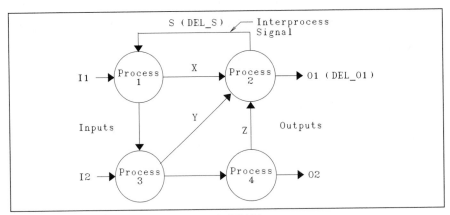

Figure 5.2 Typical process model graph (PMG).

5.1 GENERAL ALGORITHMIC MODEL DEVELOPMENT IN THE BEHAVIORAL DOMAIN

In this section we describe how to develop an algorithmic description from an English specification. To do this, we perform the following steps:

1. Map sentence groups in the requirements to VHDL processes.
2. Assign an activity list to each process.
3. Develop the VDHL code that implements each activity.

As an aid to step 1, we introduce a pictorial representation in Figure 5.2 for VHDL behavioral models called the process model graph (PMG).

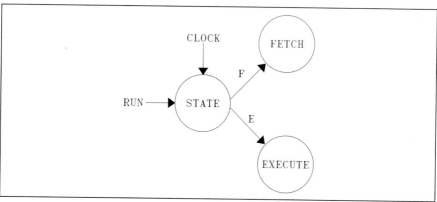

Figure 5.3 Processor PMG.

5.1.1 Process Model Graph

In VHDL an architectural body is used to define the behavior of a device. Each architectural body is a set of concurrently running processes. The processes are either process blocks or various forms of the signal assignment statement. One can give a pictorial representation to a behavioral architectural body by means of a *process model graph* (PMG). Figure 5.2 illustrates this situation for a typical behavioral model. The nodes of the graph are the set of processes in the model. The graph arcs denote signal passage between the process nodes. Each arc is labeled with a designator of the form S (DEL_S) where S is the signal name and DEL_S is the delay of the transmitting process. For example, the behavioral description of Node 2 in the graph might have the following form:

```
PROC_2: process(X,Y,Z)
   variable SVL,OVL: BIT;
begin
   --------  Function of the process: compute SVL and OVL
   S <= SVL after DEL_S;
   O1 <= OVL after DEL_O1;
end PROC_2;
```

DEL_S is the time delay associated with the signal assignment statement driving the signal S, and DEL_O1 is the time delay associated with the signal assignment statement driving signal O1.

The process model graph represents a partitioning of the model. This partitioning can either be *physical* or *functional*. With physical partitioning, the delay values on the arcs represent the propagation delay across specific blocks of real logic gates and flip-flops. In some cases, particularly in an early stage of the design, a PMG will represent a purely functional partitioning of the model. Such a case is illustrated in Figure 5.3, which shows the process model graph for a processor. Here the nodes are fetch, execute, etc., which represent functions that do not correspond to a particular block of hardware. In fact, two functional nodes may share a block of logic in some (future) physical implementation. For example, both the fetch and execute nodes could access the same "real" memory.

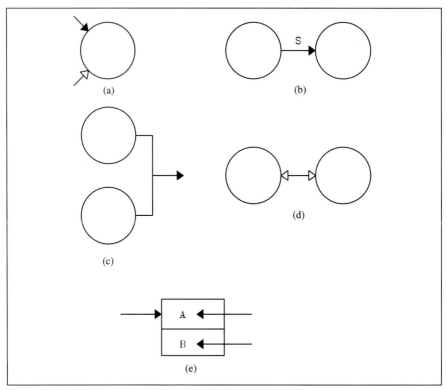

Figure 5.4   PMG constructs.

When developing process model graphs, a number of situations must be handled. Figure 5.4 illustrates some of them.

1. Figure 5.4(a) shows two signals arriving at a node. The signal with the solid arrowhead is in the sensitivity list for the process. When there is a transaction on that signal, the process is activated. The signal with the open arrowhead is sampled by the process but does not activate it.

2. In Figure 5.4(b) the arc has a signal name (S), but no delay value is given. This implies that the driving signal assignment has delta delay.

3. Figure 5.4(c) shows unidirectional signal flow, where the signal is driven by two processes, and a bus resolution function is implied.

4. Figure 5.4(d) illustrates bidirectional signal flow between two processes. A bus resolution function is implied.

5. Some models use guarded blocks as a major model "part." However, a guarded block is a collection of assignment statements with a special trigger (the guard signal). These assignment statements are merely a collection of processes. Figure 5.4(e) illustrates a notation that can be used for blocks. The signal touching the outer edge of the block is the guard. The other arcs point to the internal processes of the block and are inputs to them.

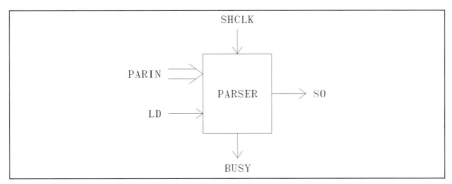

Figure 5.5 Block diagram for parallel to serial converter.

In summary, one can say that the process model graph clearly illustrates the structure of an algorithmic-level model in the behavioral domain. When the graph represents a physical partitioning of the model, both propagation delay and system structure are clearly illustrated. All but simple models are made more understandable by the use of process model graphs. And as we shall see, they are a good starting point in the development of the algorithmic model.

5.1.2 Algorithmic Model of a Parallel to Serial Converter

In this example, we develop the algorithmic model of a parallel to serial converter (PARSER). Figure 5.5 gives a block diagram of the converter. The English description for the system is as follows:

> The 8-bit parallel word (PARIN) is loaded into the converter when the control signal LD makes a zero to one transition. At this time the status signal BUSY is set high. The data is shifted out serially at a rate controlled by the input shift clock SHCLK. Shifting occurs at the rise of the clock. BUSY remains high until shifting is complete. While BUSY is high, no further loads will be accepted.

As already indicated, the algorithmic model development process involves mapping English sentences to VHDL processes. In general, there is more than one way to do this. In this case a straightforward choice is to use two processes: LOAD and SHIFT. The mapping of English requirements to these processes is:

1. Process LOAD: (a) "The 8-bit parallel word (PARIN) is loaded into the converter when the control signal LD makes a zero to one transition. At this time the status signal BUSY is set high." (b) "BUSY remains high until shifting is complete. While BUSY is high, no further loads will be accepted."
2. Process SHIFT: (a) "The data is shifted out serially at a rate controlled by the input shift clock SHCLK. Shifting occurs at the rise of the clock." (b) "BUSY remains high until shifting is complete."

Note that the requirement that "BUSY remains high until shifting is complete" is shared by the two processes. This is because process SHIFT will detect that shifting is complete and signal that condition to process LOAD, which will then reset BUSY.

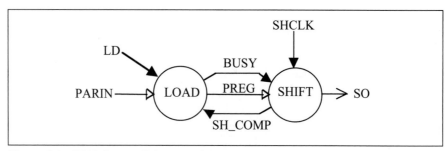

Figure 5.6 PMG for parallel to serial converter.

Figure 5.6 shows the process model graph for the algorithmic model. Process LOAD is triggered by signal LD to load the value of input vector PARIN into an internal register (PREG) and to set BUSY. BUSY going to a 1 triggers process SHIFT. Process SHIFT copies PREG into a serial register, and shifts the register contents out on the serial line SO. When the shift is complete, process SHIFT sets signal SH_COMP to a 1, which triggers process LOAD to reset BUSY. At this point one has specified the shell of the model, that is:

```
entity PAR_TO_SER is
   port(LD,SHCLK: in BIT; PARIN: in BIT_VECTOR(0 to 7);
        BUSY: inout BIT; SO: out BIT);
end PAR_TO_SER;

architecture TWO_PROC of PAR_TO_SER is
   signal SH_COMP: BIT :='0';
   signal PREG: BIT_VECTOR(0 to 7);
begin
   LOAD:process(LD,SH_COMP)
   begin
      ---- Activities:
         ----1)Register Load
         ----2)Busy Set
         ----3)Busy Reset
   end process LOAD;

   SHIFT:process(BUSY,SHCLK)
   begin
      ----Activities:
         ----1)Shift Initialize
         ----2)Shift
         ----3)Shift Complete
   end process SHIFT;
end TWO_PROC;
```

The process model graph implies the interface description and the *shell* of each process. It also implies the signals between processes. The process shell shows the signals in the sensitivity list for each process. Based on the mapping between the English specification and the processes, we have annotated the process shell with activity lists that describe the function of the process. To develop the model further, we will develop VHDL code for each activity. Following is the result of doing this. Note that the code for each activity has two parts (1) a control statement, in this case an *if* statement, which checks to see whether an event has occurred or a condition is true and (2) the data operations that are carried out if the event or condition is detected.

```
entity PAR_TO_SER is
   port(LD,SHCLK: in BIT; PARIN: in BIT_VECTOR(0 to 7);
        BUSY: inout BIT := '0'; SO: out BIT);
end PAR_TO_SER;

architecture TWO_PROC of PAR_TO_SER is
   signal SH_COMP: BIT :='0';
   signal PREG: BIT_VECTOR(0 to 7);
begin
   LOAD:process(LD,SH_COMP)
   begin
      ---- Activities:
      if LD'EVENT and LD='1'and BUSY='0' then
         ----1)Register Load
         PREG <=  PARIN;
         ----2)Busy Set
         BUSY <= '1';
      end if;
         if SH_COMP'EVENT and SH_COMP='1' then
            ----3)Busy Reset
            BUSY <= '0';
         end if;
   end process LOAD;
   SHIFT:process(BUSY,SHCLK)
      variable COUNT: INTEGER;
      variable OREG: BIT_VECTOR(0 to 7);
   begin
      ----Activities:
      if BUSY'EVENT and BUSY = '1' then
         ----1)Shift Initialize
         COUNT := 7;
            OREG := PREG;
         SH_COMP <= '0';
      end if;
      if SHCLK'EVENT and SHCLK= '1'and BUSY='1' then
         ----2)Shift
         SO<=OREG(COUNT);
         COUNT := COUNT - 1;
         ----3)Shift Complete
         if COUNT < 0 then
            SH_COMP <= '1';
         end if;
      end if;
   end process SHIFT;
end TWO_PROC;
```

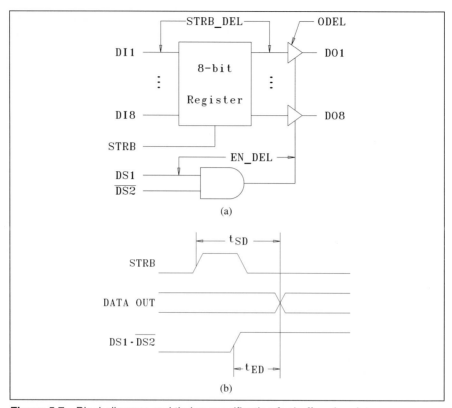

Figure 5.7 Block diagram and timing specification for buffered register.

5.1.3 Algorithmic Models with Timing

In the previous example, no timing was modeled. In this section we present techniques that illustrate how to model the input/output timing of the device. As a simple example of this, we model a buffered 8-bit register. Figure 5.7 shows a block diagram for the buffered register and its timing specification. The English description for the register is:

> The register is loaded on the rise of the strobe (STRB), and assuming that the output buffers are enabled, the output of the buffers will change t_{SD} nanoseconds later. The enable condition for the register buffer is the AND of the DS1 and $\overline{DS2}$ inputs. Any change in the enable condition will cause the outputs to change t_{ED} nanoseconds later.

In analyzing this specification, we note that there are two signal paths from inputs to the output: (1) the data path through the register to the outputs and (2) the control path from the enable inputs to the output. Each of these paths must be represented by a separate signal. In fact, one can state the following rule:

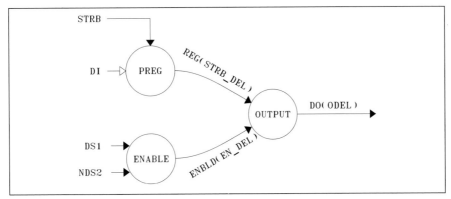

Figure 5.8 PMG for buffered register.

Rule 1. In an algorithmic model incorporating timing, separate input/output signal paths must be represented by separate VHDL signals. The signals in parallel paths are activated concurrently.

Following this rule ensures that possible simultaneous signal activity on each path is modeled.

To model the two paths in the buffered register model, we assume t_{SD} = STRB_DEL + ODEL and t_{ED} = EN_DEL + ODEL. STRB_DEL is the delay from the rise of the strobe to the arrival of the data at the register output. EN_DEL is the delay from the enable inputs to the output buffers: ODEL is the delay through the output buffers.

We will use three processes, which are mapped to the English specification as:

1. PREG: "The register is loaded on the rise of the strobe (STRB)."
2. ENABLE: "The enable condition for the register buffer is the AND of the DS1 and $\overline{DS2}$ inputs."
3. OUTPUT: (a) "Assuming that the output buffers are enabled, the output of the buffers will change t_{SD} nanoseconds later" and (b) "Any change in the enable condition will cause the outputs to change t_{ED} nanoseconds later."

Figure 5.8 shows a process model graph for the register. Process PREG receives the strobe and data inputs, and outputs a delayed value of the data (REG). Process ENABLE similarly produces a delayed value of the enable signal (ENBLD). Process OUTPUT receives REG and ENBLD, and produces the output DO. The delays for the three processes (STRB_DEL, EN_DEL, ODEL) are also indicated on the graph. Thus, with regard to Rule 1, the two separate data paths are modeled by the two signals REG and ENBLD.

The full VHDL description for this three process model is shown in Figure 5.9. In the interface description, the three delays in the model have been specified as "generics." This allows the model to be general. When the model is instantiated in an interconnected system description, the value of these generic parameters will have to be specified. In light of the discussion above, the VHDL in Figure 5.9 is self-explanatory. However, note that in the OUTPUT process, we represent the high-impedance condition for the output as a vector of all 1's because we have not used multivalued logic.

```
entity BUFF_REG is
    generic(STRB_DEL,EN_DEL,ODEL: TIME);
    port(DI:  in BIT_VECTOR(1 to 8);
     STRB: in BIT;DS1:  in BIT;
     NDS2: in BIT;
     DO: out BIT_VECTOR(1 to 8));
end BUFF_REG;
--
architecture THREE_PROC of BUFF_REG is
    signal REG: BIT_VECTOR(1 to 8);
    signal ENBLD: BIT;
begin
PREG: process(STRB)
    begin
        if (STRB = '1') then
            REG <=DI after STRB_DEL;
        end if;
end process PREG;
--
ENABLE: process(DS1,NDS2)
    begin
        ENBLD <= DS1 and  not NDS2 after  EN_DEL;
end process ENABLE;
--
OUTPUT: process(REG,ENBLD)
    begin
        if (ENBLD = '1')
        then
            DO <= REG after ODEL;
        else
            DO <= "11111111" after ODEL;
        end if;
end process OUTPUT;
end THREE_PROC;
```

Figure 5.9 Algorithmic VHDL description for buffered register.

Other model configurations are possible; for example, processes PREG and ENABLE can be combined into a single process, which performs the combined function of these two processes:

```
FRONT_END: process(STRB,DS1,NDS2)
begin
   if STRB'EVENT and STRB = '1' then
     REG <=DI after STRB_DEL; --activity 1
   end if;
   if DS1'EVENT or NDS2'EVENT then
     ENBLD <= DS1 and  not NDS2 after EN_DEL;   --activity 2
   end if;
end process FRONT_END;
```

Thus, we have replaced two single activity processes with one dual activity process. Note however, that the signals REG and ENBLD are controlled in the same fashion in each approach. A basic question which might be asked is, what is the trade-off between using more or fewer processes in an algorithmic description? The answer to this question is based on the following factors:

1. Number of signals: Using more processes may result in more signals in the description. More signals will require more queue processing which could adversely effect simulation efficiency.

2. Individual process complexity: If individual processes are too complex, the process code may not account for all possible combinations of input changes. In this case the model is restricted in application. Using simpler processes can avoid this problem.

3. Ease of mapping: The use of more processes may allow for a very natural mapping between specification requirements and processes; for example, using PREG, ENABLE, and OUTPUT processes is a very natural mapping to the specification requirement.

As stated in Chapter 3, individual signal assignment statements are processes; therefore, the same model structure can be expressed more compactly. An alternative architecture of the register model, which uses this approach, follows:

```
architecture DATA_FLOW of BUFF_REG is
begin
   B:block(STRB = '1'and not STRB'STABLE)
      signal REG: BIT_VECTOR(1 to 8);
      signal ENBLD: BIT;
   begin
      REG <= guarded DI after STRB_DEL;        ---process REG
      ENBLD <= DS1 and not NDS2 after EN_DEL; ---proc ENABLE
      DO <=REG after ODEL when ENBLD = '1'    ---proc OUTPUT
            else "11111111" after ODEL;
   end block B;
end DATA_FLOW;
```

This approach uses a block that is guarded by a 0 to 1 transition on the strobe. Each of the three processes PREG, ENABLE, and OUTPUT is implemented by a single statement. The statement corresponding to process REG is guarded and will be executed only when the guard condition is TRUE. This form of the process model is more concise, but the structure of the model is not as evident. For more complicated modeling situations, implementing a complicated I/O formula within a process block can be more effective. As indicated by the name of the architecture, this concise form of the model is a data flow description. In the next chapter, we examine these descriptions in greater detail.

5.1.4 Checking Timing

As discussed in Chapter 4, the built-in inertial delay feature of the signal assignment statement filters out a pulse on a signal shorter than the minimum duration specified by the delay value in the "after" clause. Thus, in the case of the register model, any new values assigned to the signals REG, ENBLD, and DO must persist, respectively, for at least an amount of time equal to delay

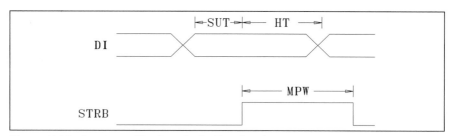

Figure 5.10   Input specifications for buffered register.

STRB_DEL, ENBLD, or ODEL, for the new values to have an effect. However, there is no indication to the modeler that a short pulse has been deleted, other than the fact that the signal does not change.

It is useful to be able to print a message when such a short pulse is deleted to alert the user because this situation frequently indicates a possible modeling error. In addition, complex timing constraints are often specified that require checking. For example, let us again consider the buffered register just modeled. It is a common requirement for a register that the data input be stable for a duration of time prior to the clock transition that strobes data into the register. This requirement is known as the *setup time* specification on the data relative to the clock. A similar requirement states that the data should remain stable a minimum amount of time after the clock makes its transition (i.e., a *hold-time* specification). VHDL allows us to detect violations of timing constraints by using assertions that have the following form:

```
assert     Boolean Expression
   report     Error Message;
```

When the assertion is executed, it checks the Boolean expression. A TRUE value indicates a normal condition. A FALSE value indicates an error condition, and the error message is printed. Assertions can be used to check practically anything, but in this application, we are using them to check the timing of signals. For timing assertions, the components of the Boolean expressions are signal attributes. Two particularly useful signal attributes for this purpose are X'stable(T) and X'delayed(T). As explained in Chapter 3, X'stable(T) is TRUE if and only if signal X has been stable for T time units. X'delayed(T) is the value X had T time units earlier.

We illustrate the use of assertions in timing modeling with the following example. Figure 5.10 shows typical input specifications for the register in Figure 5.7. The specification says:

1. DI should be stable for SUT nanoseconds before STRB rises (setup time),
2. DI should be stable for HT nanoseconds after STRB rises (hold time), and
3. STRB should have a minimum duration at the one level of MPW nanoseconds (minimum pulse width).

The following assertion checks the setup time specification:

```
assert not (not STRB'stable and (STRB = '1')
            and not DI'stable(SUT))
   report   "Setup Time Failure";
```

Thus, if the strobe has just made a positive transition, and the data input has not been stable for the previous SUT ns, a setup time failure will be reported.

Using DeMorgan's theorem, one could convert the assert statement to the simpler form:

```
assert STRB'stable or (STRB = '0') or DI'stable(SUT)
    report  "Setup Time Failure";
```

Again, using DeMorgan's theorem, one could also write the simplified form for the hold-time test:

```
assert  STRB'delayed(HT)'stable or (STRB'delayed(HT) = '0')
        or DI'stable(HT)
    report "Hold Time Failure";
```

Similarly, for the minimum pulse width check, the assertion is:

```
assert  STRB'stable or (STRB = '1')
        or STRB'delayed'stable(MPW)
    report "Minimum pulse width failure";
```

In the last term of this expression, we use STRB'delayed'stable instead of STRB'stable because we wish to check STRB for stability one delta cycle before STRB has changed. It makes no sense to check for stability right after it changes.

Timing assertions can also be developed using such attributes as S'last_event and the function NOW from Package Standard. We leave it as an exercise for the reader to do so.

Assertions can be inserted in the interface description or architectural body of a design entity. If placed in an architectural body, the assertions are specific to a particular implementation of the design entity. If placed in the interface description, they can be used to check the timing of signals in and out of any architectural body of the design entity. Following is the interface description for the buffered register model with the assertions inserted in its declarative region:

```
entity BUFF_REG is
   generic(STRB_DEL,EN_DEL,ODEL,SUT,HT,MPW: TIME);
            port(DI:  in BIT_VECTOR(1 to 8);
                 STRB: in BIT;DS1:  in BIT;
                 NDS2: in BIT;
                 DO: out BIT_VECTOR(1 to 8));
begin
   assert STRB'stable or (STRB = '0') or DI'stable(SUT)
     report  "Setup Time Failure";
   assert STRB'delayed(HT)'stable
          or (STRB'delayed(HT) = '0')
          or DI'stable(HT)
   report "Hold Time Failure";
   assert STRB'stable or (STRB = '1')
          or STRB'delayed'stable(MPW)
     report "Minimum pulse width failure";
end BUFF_REG;
```

We said at the outset that we're trying to map English requirements to VHDL processes. This is also true for timing requirements represented by assertions, as each assertion is represented during simulation by a "passive" process, i.e., a process that can monitor signals but not change them.

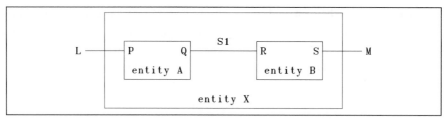

Figure 5.11   System interconnection model.

5.2   REPRESENTATION OF SYSTEM INTERCONNECTIONS

In VHDL the interconnection of system components is implemented in a structural architectural body representing the system entity. Consider the simple system X shown in Figure 5.11.

System X contains two components, entity A and entity B, which have the following interface descriptions:

```
entity A is
   port(P: in BIT; Q: out BIT);
end A;

entity B is
   port(R: in BIT; S: out BIT);
end B;
```

The interconnection structure for system X is defined in the following entity declaration and accompanying architectural body:

```
entity X  is
   port(L: in BIT; M: out BIT);
end X;

architecture STRUCTURAL of X is
   signal S1: BIT; -- Internal interconnection signal declared
   component A      -- Component declared
     port(P: in BIT; Q out BIT);
   end component;
   component B      -- Component declared
     port(R: in BIT; S out BIT);
   end component;
begin
   A1: A port map(L,S1);   ----Component instantiation
   B1: B port map(S1,M);   ----Component instantiation
end STRUCTURAL;
```

The interconnection of system components is defined by the port maps in the component instantiation statements. The port maps define the system signals that are connected to each port of each component. In the simple case shown above, the signal S1 is defined as an interconnection between components A and B because it is specified to be an output of component A and an input to component B.

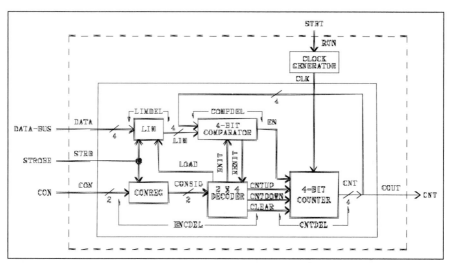

Figure 5.12   Block diagram for a two module system.

Note in the preceding example, when the components are instantiated, the interconnect is specified by position matching of port map entries in the instantiation statements with port entries in the component declarations. This can also be specified by named association. For example, one could write the instantiation statements as:

```
A1: A port map(Q => S1, P => L);
B1: B port map(S => M , R => S1);
```

Note that position is unimportant, and the named association identifies the correspondence.

5.2.1   Comprehensive Algorithmic Modeling Example

In our concluding example of algorithmic modeling, we develop a model of a multimodule system that further illustrates specification interpretation and mapping to processes.

Figure 5.12 shows the block diagram of a two-module system consisting of a clock generator and a 4-bit controlled counter. The clock generator produces a 50% duty cycle clock with a period of PER ns. The counter consists of five logic blocks (1) 2-bit control register (CONREG), (2) 2-to-4 decoder, (3) 4-bit limit register (LIM), (4) 4-bit comparator, and (5) 4-bit counter. The counter module receives a 2-bit control input (CON), which is stored in CONREG, then decoded to perform the following four commands:

1. (CON = 00). Clear the counter
2. (CON = 01). Load the LIM (limit) register with the value on the DATA input lines.
3. (CON = 10). Count up until the COUNTER equals the value in the LIM register.
4. (CON = 11). Count down until the COUNTER equals the value in the LIM register.

The counter, when commanded to count and enabled (EN = 1) counts positive clock transitions, in an up or down mode, until the limit is reached. When a counting command is decoded, the ENIT signal is sent to the comparator by the decoder logic. The comparator uses the rising edge of ENIT to set EN=1 initially. After that, EN is controlled by the comparator.

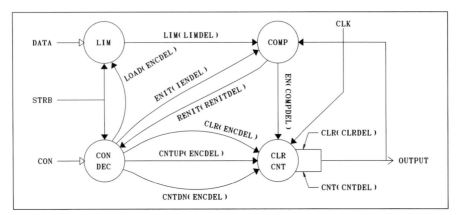

Figure 5.13   Process model graph for the controlled counter.

The diagram shows four propagation delays (1) ENCDEL—the combined delay across the control register (CONREG) and the decoder, (2) LIMDEL—the delay through the limit register, (3) COMPDEL—the delay through the comparator, and (4) CNTDEL—the delay through the counter. The interface timing functions as follows. CONREG is loaded on the rise of STRB, and LIM is loaded on the fall of STRB. There are setup time requirements on both CON and DATA, relative to their respective STRB transitions of SUT nanoseconds. There is a minimum pulse width requirement of MPW on the STRB signal itself.

Figure 5.13 shows the Process Model Graph for the controlled counter. Note that the functions of the control register (CON) and decoder are represented by a single node. This is because the timing specification gives the total delay across the two blocks, but not individual delays. Thus, this node will be modeled with a single VHDL process.

Figure 5.14 and Figure 5.15 give the VHDL for the system just described. Note that there are two entities—one for the clock generator (Figure 5.14) and one for the 4-bit controlled counter (Figure 5.15).

In the clock generator model, the oscillator action is implemented in a single process by a feedback delay mechanism. The signal CLOCK, which is generated by the process, is one of the signals in the process sensitivity list. The output of the entity is CLK, which equals CLOCK after one delta delay. It is necessary to use this buffering mechanism since VHDL does not allow a signal of mode out to be referenced within the entity. Note also that Clock generation will occur only after the RUN input has made a 0 to 1 transition and will be inhibited after RUN falls.

The entity description for the controlled counter (Figure 5.15) first specifies a set of generics for the time delays in the model. Between the *begin* and *end* keywords, the timing assertions are given. The architectural body contains four processes: DECODE, LOAD_LIMIT, CTR, and LIMIT_CHK. Process DECODE decodes the 2-bit input CON into the 4-bit signal CONSIG. It also generates the signal ENIT, which is used initially to enable the comparator. Process LOAD_LIMIT loads the LIM register when STRB falls and the appropriate decoder output is a 1. Process CTR will respond to its clock input (CLK) by counting only if the clock enable signal (CLKE) is set. CLKE is set by the rise of EN, the comparator enable signal, and reset by its fall. Process LIMIT_CHK sets the enable signal EN to 1 when ENIT is received. It responds with

```
entity CLOCK_GENERATOR is
   generic (PER: TIME);
   port(RUN: in BIT;CLK: out BIT);
end CLOCK_GENERATOR;
--
architecture FEEDBACK of CLOCK_GENERATOR is
   signal CLOCK: BIT;
begin
  process (RUN,CLOCK)
    variable CLKE: BIT := '0';
  begin
    if RUN'EVENT then
      if RUN = '1' then
        CLKE := '1';
        CLOCK <= transport '0' after PER/2;
        CLOCK <= transport '1' after PER;
      else
        CLKE := '0';
      end if;
    end if;
    if(CLOCK'EVENT and CLOCK = '1' and CLKE = '1' ) then
      CLOCK <= transport '0' after PER/2;
      CLOCK <= transport '1' after PER;
    end if;
    CLK <= CLOCK;
  end process;
end FEEDBACK;
```

Figure 5.14   Algorithmic description of clock generator.

RENIT, which is used by process DECODE to reset ENIT. Process LIMIT_CHK uses the fall of ENIT to reset RENIT. The enable signal EN is reset when the count limit is reached.

The descriptions given in Figures 5.14 and 5.15 specify the clock generator and the controlled counter as generic design entities. To implement the complete system shown in Figure 5.12, one has to specify the connection between the two design entities. This is done in Figure 5.16. First, an entity is defined for the total system. In Figure 5.12 this system entity is shown inside the dashed box. The port statement for this entity defines the signals that cross the system boundary. Note that the names for signals can be different at the system boundaries and internal to the system. The architectural body is structural. First, the two components and an internal signal (CLK) are declared. Between the *begin* and *end* keywords, the two components are instantiated; that is, generic maps and port maps are used to give values to the generic parameters and connect signals to the correct inputs.

Thus, we have a full system model. In the next section, we discuss some of the problems of system modeling in detail.

```
entity CONTROLLED_CTR is
  generic(SUT,MPW,ENCDEL,ENITDEL,RENITDEL,CLRDEL,
          CNTDEL,LIMDEL,COMPDEL: TIME);
  port(CLK,STRB: in BIT;
       CON:  in BIT_VECTOR(0 to 1);
       DATA:  in BIT_VECTOR(0 to 3);
       COUT: out BIT_VECTOR(0 to 3));
begin
  assert STRB'STABLE or STRB = '1' or DATA'STABLE(SUT)
    report "Set up time failure on DATA input"
    severity NOTE;
  assert STRB'STABLE or STRB = '0' or CON'STABLE(SUT)
    report "Set up time failure on CON input"
    severity NOTE;
 assert STRB'STABLE or STRB = '1' or STRB'DELAYED'STABLE(MPW)
    report "Pulse width failure on STRB"
    severity NOTE;
end CONTROLLED_CTR;
use work.counter_pac.all;
architecture PROCESS_IMPL of CONTROLLED_CTR is
  signal ENIT,RENIT: BIT;
  signal EN: BIT;
  signal CONSIG,LIM: BIT_VECTOR(0 to 3);
  signal CNT: BIT_VECTOR(0 to 3);
begin
  --
DECODE:process(STRB,RENIT)
  variable CONREG: BIT_VECTOR(0 to 1) := "00";
 begin
  if (STRB = '1') then
    CONREG := CON;
    case CONREG is
    -- Signal CLEAR is CONSIG(0).
      when "00" => CONSIG <= "1000" after ENCDEL;
    -- Signal LOAD is CONSIG(1).
      when "01" => CONSIG <= "0100" after ENCDEL;
    -- Signal CNTUP is CONSIG(2).
      when "10" => CONSIG <= "0010" after ENCDEL;
                   ENIT <= '1' after ENITDEL;
    -- Signal CNTDOWN is CONSIG(3).
      when "11" => CONSIG <= "0001" after ENCDEL;
                   ENIT <= '1' after ENITDEL;
    end case;
  end if;
  if RENIT'EVENT and RENIT = '1' then
    ENIT <= '0' after ENITDEL;
  end if;
 end process DECODE;
```

Figure 5.15 Algorithmic description of controlled counter in behavioral domain.

```
      LOAD_LIMIT:process(STRB)
       begin
        if(CONSIG(1)='1' and STRB'EVENT and STRB ='0') then
          LIM <= DATA after LIMDEL;
        end if;
       end process LOAD_LIMIT;
      --
      CTR: process(CONSIG(0),EN,CLK)
         variable CNTE: BIT:= '0';
       begin
        if (CONSIG(0)='1' and CONSIG(0)'EVENT) then
          CNT <= "0000" after CLRDEL;
        end if;
        if EN'EVENT then
          if EN = '1' then
            CNTE := '1';
          else
            CNTE := '0';
          end if;
        end if;
        if (CLK'EVENT and CLK =  '1' and CNTE = '1') then
          if (CONSIG(2)='1') then
            CNT <= INC(CNT);
          elsif (CONSIG(3)='1') then
            CNT <= DEC(CNT);
          end if;
        end if;
       end process CTR;
      --
      LIMIT_CHK: process(CNT,ENIT)
       begin
        if ENIT'EVENT then
          if ENIT = '1' then
            EN <= '1' after COMPDEL;
            RENIT <= '1' after RENITDEL;
          else
            RENIT <= '0' after RENITDEL;
          end if;
        elsif ((EN = '1') and (CNT = LIM)) then
          EN <= '0' after COMPDEL;
        end if;
       end process LIMIT_CHK;
      COUT <= CNT;
      end PROCESS_IMPL;
```

Figure 5.15 Algorithmic description of controlled counter in behavioral domain (continued).

```
use work.all;
entity COUNT_SYS is
  port(STRT,STROBE: in BIT;
       CON: in BIT_VECTOR(0 to 1);
       DATA_BUS: in BIT_VECTOR(0 to 3);
       CNT: out BIT_VECTOR(0 to 3));
end COUNT_SYS;
--
architecture TWO_COMPONENT of COUNT_SYS is
  signal CLK: BIT;
  component CLOCK_GEN
    generic(PER: TIME);
    port (RUN: in BIT; CLK: out BIT);
  end component;
  component CON_CTR
    generic(SUT,MPW,ENCDEL,ENITDEL,RENITDEL,
            CLRDEL,CNTDEL,LIMDEL,COMPDEL:TIME);
    port (CLK,STRB: in BIT;
          CON: in BIT_VECTOR(0 to 1);
          DATA: in BIT_VECTOR(0 to 3);
          COUT: out BIT_VECTOR(0 to 3));
  end component;
  for CLKGEN: CLOCK_GEN use entity CLOCK_GENERATOR(FEEDBACK);
  for CTR: CON_CTR use entity CONTROLLED_CTR (PROCESS_IMPL);
begin
  CLKGEN: CLOCK_GEN
    generic map(100 ns)
    port map(STRT,CLK);
  CTR: CON_CTR
    generic map(20 ns, 30 ns, 25 ns, 11 ns, 11 ns, 10 ns,
                15 ns,12 ns,10 ns)
    port map(CLK,STROBE,CON,DATA_BUS,CNT);
end TWO_COMPONENT;
```

Figure 5.16 Structural architecture body of counter system.

5.3 ALGORITHMIC MODELING OF SYSTEMS

In this section we discuss the algorithmic modeling of systems. In doing so, we address the following issues:

1. Multivalued logic systems
2. Multiplexing
3. Intermodule communication protocols

5.3.1 Multivalued Logic Systems

Fundamentally, our main purpose is to model binary logic. However, real logic gates exhibit behavior, which cannot be adequately represented by two logic levels. This situation exhibits itself also when modeling systems. For example, consider the system bus shown in Figure 5.17. Two bus drivers are shown, DR1 and DR2. For a given bus driver, its data value, D_i will be con-

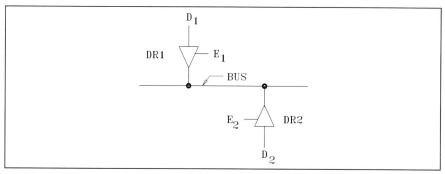

Figure 5.17 System bus.

nected to the bus when E_i = '1'. First, suppose all bus drivers are off, i.e., E_i = '0' for all i. In this case, each driver is in a high impedance state; in fact, the whole bus "floats" at this high impedance state. We will designate this high impedance state as 'Z'. Second, suppose both DR1 and DR2 are enabled (E1 = E2 ='1'), and D1 = '0' and D2 = '1'. This is probably not a "normal" condition, but since one of the main applications of models is the detection of faults, situations of this type must be considered. What value should we assign to the bus? Depending on the technology and circuit design used in the drivers, one might be able to say that the '0' would win out over the '1' or vice versa. However, in general, one might have to say that the output is unknown. We designate this unknown value as 'X'.

Thus, to begin algorithmic modeling of systems, we use a four-valued logic system consisting of the values: '0', '1', 'X', 'Z'. These four values are adequate for many modeling situations at this level of abstraction. Later, we will extend the set to seven, and then nine values to more accurately model detailed circuit effects.

5.3.1.1 Logic Package SYSTEM_4

We now present a package for a four-valued logic system. In this section, we present and discuss the important features of the package. The complete SYSTEM_4 package is given on the CD-ROM. Initially, one defines the basic scalar and vector types:

```
type MVL4 is ('X','0','1','Z');
type MVL4_VECTOR is array(NATURAL range <>) of MVL4;
```

Note that in defining MVL4, we have made 'X' the left-most value of the type, so all signals and variables of this type will be initialized by default to the unknown state.

Logic and resolution functions in the four-valued system are stored in tables. This makes it easier to understand the functions, and more importantly, results in an efficient simulation run-time implementation of the functions. To implement these tables, we require the following type definitions:

```
    -- one dimensional array
type MVL4_TAB1D is array (MVL4) of MVL4;
    -- two dimensional array
type MVL4_TABLE is array (MVL4, MVL4) of MVL4;
    --three dimensional array
type MVL4_TAB3D is array (MVL4, MVL4, MVL4) of MVL4;
```

Note that the index values for arrays of these types are of type MVL4.

5.3.1.2   Resolution Functions

As pointed out in Chapter 3, whenever a signal is driven by more then one process, a resolution function is required. These functions are one of the primary means of multiplexing in VHDL. A resolution function receives as input an array consisting of the drivers of the signal and produces a scalar output whose value is the resolved value. The resolution functions described here are implemented as two-dimensional tables, whose arguments are a pair of drivers. The array of drivers is scanned and each adjacent pair of drivers in the array is compared by the table function.

We now consider the truth tables for six alternative resolution functions:

1. The WiredAnd Function: '0's dominate and two 'Z's resolve to a 'Z', modeling a floating bus.

```
-- truth table for "WiredAnd" function
constant table_WIREDAND: MVL4_TABLE :=
--  --------------------------------
--  |  X    0    1    Z  |
--  --------------------------------
   (('X', '0', 'X', 'X'),   --  |  X
    ('0', '0', '0', '0'),   --  |  0
    ('X', '0', '1', '1'),   --  |  1
    ('X', '0', '1', 'Z'));  --  |  Z
```

2. The WiredOr Function: '1's dominate and two 'Z's resolve to a 'Z', modeling a floating bus.

```
-- truth table for "WiredOr" function
constant table_WIREDOR: MVL4_TABLE :=
--  --------------------------------
--  |  X    0    1    Z  |
--  --------------------------------
   (('X', 'X', '1', 'X'),   --  |  X
    ('X', '0', '1', '0'),   --  |  0
    ('1', '1', '1', '1'),   --  |  1
    ('X', '0', '1', 'Z'));  --  |  Z
```

3. The WiredPullup Function: '0's dominate and two 'Z's resolve to a '1', modeling a bus with a pull-up resistor.

```
-- truth table for "WiredPullUp" function
constant table_WIREDPULLUP: MVL4_TABLE :=
--  --------------------------------
--  |  X    0    1    Z  |
--  --------------------------------
   (('X', '0', 'X', 'X'),   --  |  X
    ('0', '0', '0', '0'),   --  |  0
    ('X', '0', '1', '1'),   --  |  1
    ('X', '0', '1', '1'));  --  |  Z
```

4. The WiredPulldown Function: '1's dominate and two 'Z's resolve to a '0', modeling a bus with a pull-down resistor.

```
-- truth table for "WiredPullDown" function
constant table_WIREDPULLDOWN: MVL4_TABLE :=
--  --------------------------------
--  |  X    0    1    Z  |
--  --------------------------------
   (('X', 'X', '1', 'X'),   --  |  X
    ('X', '0', '1', '0'),   --  |  0
    ('1', '1', '1', '1'),   --  |  1
    ('X', '0', '1', '0'));  --  |  Z
```

5. The WiredX Function: Detection of a '0' and a '1' results in an X. Two Z's resolve to a 'Z', modeling a floating bus.

```
-- truth table for "WiredX" function
constant table_WIREDX: MVL4_TABLE :=
--  ------------------------------
--  |  X    0    1    Z  |
--  ------------------------------
     (('X', 'X', 'X', 'X'),   -- | X |
      ('X', '0', 'X', '0'),   -- | 0 |
      ('X', 'X', '1', '1'),   -- | 1 |
      ('X', '0', '1', 'Z'));  -- | Z |
```

6. The WiredOne Function: Assumes only one driver is active, i.e., equal to '0' or '1'. Two 'Z's resolve to a 'Z', modeling a floating bus.

```
-- truth table for "WiredOne" function
-- this truth table allows only one driver to be active
constant table_WIREDONE: MVL4_TABLE :=
--  ------------------------------
--  |  X    0    1    Z  |
--  ------------------------------
     (('X', 'X', 'X', 'X'),   -- | X |
      ('X', 'X', 'X', '0'),   -- | 0 |
      ('X', 'X', 'X', '1'),   -- | 1 |
      ('X', '0', '1', 'Z'));  -- | Z |
```

Each of these tables is accessed within the body of a function, e.g., the WiredX function can be implemented as follows:

```
function WiredX (V: MVL4_VECTOR) return MVL4 is
   variable result: MVL4 := 'Z';
begin
   for i in V'range loop
      result := table_WIREDX(result, V(i));
      exit when result = 'X';
   end loop;
   return result;
end WiredX;
```

Note that we have no knowledge of the order in which drivers are inspected, and we assume that comparing drivers two at a time is equivalent to comparing all n drivers at the same time. The first part of the previous statement implies that table_WIREDX defines a commutative function; the second part implies that the table function must be an associative function.

The implementations for the other functions are similar and are given on the CD-ROM. One exception to this is the WiredOne function. Although it could be implemented by accessing the table, the table is really not required, and the function is implemented as follows:

```
function WiredOne (V: MVL4_VECTOR) return MVL4 is
   variable result: MVL4 := 'Z';
   variable got_one: BOOLEAN := FALSE;
begin
   for i in V'range loop
      next when V(i) = 'Z';
      if got_one then
         assert false
         report "Multiple contributors to WiredSingle node."
```

```
            severity WARNING;
            result := 'X';
            return result;
         end if;
         got_one := TRUE;
         result := V(i);
      end loop;
      return result;
   end WiredOne;
```

Given the definition of the following functions:

```
function WiredPullUp (V: MVL4_VECTOR) return MVL4;
function WiredPullDown (V: MVL4_VECTOR) return MVL4;
function WiredX (V: MVL4_VECTOR) return MVL4;
function WiredOne (V: MVL4_VECTOR) return MVL4;
function WiredAnd (V: MVL4_VECTOR) return MVL4;
function WiredOr (V: MVL4_VECTOR) return MVL4;
```

One can define subtypes for "wired" scalar signals, that is:

```
subtype DotPU is WiredPullUp MVL4;
subtype DotPD is WiredPullDown MVL4;
subtype DotX is WiredX MVL4;
subtype Dot1 is WiredOne MVL4;
subtype DotAnd is WiredAnd MVL4;
subtype DotOr is WiredOr MVL4;
```

Signals that are declared to be of these types will automatically inherit the attached bus resolution function. Finally, one defines the resolved bus types, which again have their associated bus resolution functions:

```
type BusPU is array (Natural range <>) of DotPU;
type BusPD is array (Natural range <>) of DotPD;
type BusX is array (Natural range <>) of DotX;
type Bus1 is array (Natural range <>) of Dot1;
type BusAnd is array (Natural range <>) of DotAnd;
type BusOr is array (Natural range <>) of DotOr;
```

5.3.1.3   Sense and Drive Functions

The wired signal types just defined are used to represent bussed signals on the interconnect between devices. However, because logic operations must be redefined for the four-valued system, it is more efficient to represent internal signals using types MVL4 and MVL4_VECTOR. Then, the redefinition will only have to occur once. Package SYSTEM_4 contains type conversion functions to convert from the bus types to MVL4_VECTOR and vice versa. For example, a *Sense* function, which converts from type BusX to type MVL4_VECTOR, is:

```
function Sense (V: BusX; vZ: MVL4) return MVL4_VECTOR is
   alias Value: BusX (V'length-1 downto 0) is V;
   variable Result: MVL4_VECTOR (V'length-1 downto 0);
begin
   for i in Value'range loop
      if ( Value(i) = 'Z' ) then
         Result(i) := vZ;
      else
```

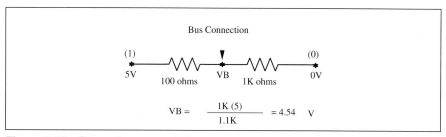

Figure 5.18 Strong 1 vs. a weak 0.

```
        Result(i) := Value(i);
      end if;
    end loop;
    return Result;
end Sense;
```

The vZ input for the function allows the user to select the value of the function when the input is 'Z'. This value is technology dependent. For example, for TTL circuitry, one might want to interpret a 'Z' as a logic '1'.

A *Drive* function is used to convert from type MVL4_VECTOR to type BusX.

```
function Drive (V: MVL4_VECTOR) return BusX is
begin
  return BusX(V);
end Drive;
```

This function uses the concept of a closely related type to make the conversion. Note that it is not necessary to convert between types MVL4 and DotX because DotX is a subtype of MVL4. However, BusX is not a subtype of MVL4_VECTOR; thus, conversion between these two types is necessary.

The drive and sense functions correspond in a hardware sense to driver and receiver circuits. Package SYSTEM_4 contains sense and drive functions for the various wired signal types presented above.

5.3.1.4   *Logic Systems with More Values*

In the four-valued logic system, MVL4, the motivation for the value 'X' was to represent bus contention. Referring again to Figure 5.18, we said that if bus driver 1 had an output of '0' and if bus driver 2 had an output of '1', then, the resultant bus value would be unknown, i.e., 'X'. We used this value because we had no knowledge of the *strength* of the two outputs, that is, would the '0' win out over the '1' or vice versa. The use of an 'X' to represent this situation is for many cases pessimistic. In many technologies, the source impedances for logic zeros and ones differ, and if the two values are tied together, one will "win" over the other. Consider the situation shown in Figure 5.18 where a strong '1' has source impedance of 100 ohms, while a weak '0' has source impedance of 1000 ohms. If tied together, the resultant bus voltage, VB, is 4.54 volts, which would clearly be interpreted by receiving circuits as a logic 1. Since this bus wiring problem is very common in logic systems, it is important to be able to represent signal strength. This produces more accurate simulations by eliminating overly pessimistic unknown ('X') signal values.

A basic question is, how many strength values should there be? When modeling at the circuit level using SPICE, an essentially infinite number of strengths are possible. For digital modeling, however, this is obviously not required. For our purposes, we begin by concentrating on a system that has two strengths: strong and weak. These strengths are utilized in a seven-valued system called MVL7, which was developed by ZYCAD (now Synopsys) and which eventually became a candidate for an IEEE standard.

5.3.1.5   System MVL7

System MVL7 is a logic system of seven *values* that is obtained by combining three *states* with two *strengths*. The three states are 0, 1, and X, and are considered to represent information internal to the devices. Values combine state with strength to represent information at device interfaces. MVL7 uses two strengths, strong and weak. The basic type definition for system MVL7 is:

```
type MVL7 is ('X', -- strong X (strong unknown)
              '0', -- strong 0 (strong low)
              '1', -- strong 1 (strong high)
              'Z', -- tri-state X (high impedance)
              'W', -- weak X (weak unknown)
              'L', -- weak 0 (weak low)
              'H');-- weak 1 (weak high)
```

Three states and two strengths yield six values. However, MVL7 also contains a third strength Z. The source impedance of Z is much higher than the weak strength. Using the strict concepts of strength and state, this should give rise to three values, e.g., Z0, Z1, and ZX. However, the impedance is generally so high that the state is not considered to be significant. Therefore, Z0 = Z1 = ZX = Z, and we have a seven-valued system.

5.3.1.6   IEEE 9 Valued Logic System

IEEE added two additional values, 'U' and '-', to MVL7. Value 'U' is interpreted as uninitialized and is the value used to represent the power-on state of a device. Value '-' is the traditional don't care value and is used to aid in the minimization of combinational logic during synthesis. This nine-valued logic system became the basis for IEEE standard 1164, which is encapsulated in package STD_LOGIC_1164.

The basic type definitions are:

```
type STD_ULOGIC is ('U','X','0','1','Z','W','L','H','-');
type STD_ULOGIC_VECTOR is array(NATURAL range <>)
     of STD_ULOGIC;
type STDLOGIC_TABLE is array (STD_ULOGIC, STD_ULOGIC)
     of STD_ULOGIC;
```

The "U" in ULOGIC means unresolved. The resolution function is defined in the following constant:

```
constant RESOLUTION_TABLE: STDLOGIC_TABLE := (
   --------------------------------------------------------------
   --|  U     X     0     1     Z     W     L     H     -         |   |
   --------------------------------------------------------------
   ( 'U',  'U',  'U',  'U',  'U',  'U',  'U',  'U',  'U' ),  --  |   U
   ( 'U',  'X',  'X',  'X',  'X',  'X',  'X',  'X',  'X' ),  --  |   X
   ( 'U',  'X',  '0',  'X',  '0',  '0',  '0',  '0',  'X' ),  --  |   0
   ( 'U',  'X',  'X',  '1',  '1',  '1',  '1',  '1',  'X' ),  --  |   1
   ( 'U',  'X',  '0',  '1',  'Z',  'W',  'L',  'H',  'X' ),  --  |   Z
   ( 'U',  'X',  '0',  '1',  'W',  'W',  'W',  'W',  'X' ),  --  |   W
   ( 'U',  'X',  '0',  '1',  'L',  'W',  'L',  'W',  'X' ),  --  |   L
   ( 'U',  'X',  '0',  '1',  'H',  'W',  'W',  'H',  'X' ),  --  |   H
   ( 'U',  'X',  'X',  'X',  'X',  'X',  'X',  'X',  'X' )   --  |   -
                                                              );
```

This table encapsulates the relative strength of values, i.e., strong dominates weak. If R is the function implemented by the table, then:

$$0 \ R \ H = H \ R \ 0 = 0 \ R \ L = L \ R \ 0 = 0$$
$$1 \ R \ L = L \ R \ 1 = 1 \ R \ H = H \ R \ 1 = 1$$

Also note that X means the value is 0 or 1; W means the value is L or H. Thus, X and W are unknown values that arise from bus contention:

$$0 \ R \ 1 = 1 \ R \ 0 = X$$
$$L \ R \ H = H \ R \ L = W$$

X's can also arise from error condition, e.g. a flip-flop transits into an unknown state. Strength is applied among values and unknowns:

$$L \ R \ X = X, \quad 1 \ R \ W = 1.$$

Z represents the high impedance state. U, the uninitialized state, is the left-most value of the type and is the default initialization value. One can test to see if logic starts up properly, as all Us should change to 0's or 1's as simulation is carried forward. Value '-' means don't care for synthesis, but for simulation it is treated as an X.

The resolution function is:

```
function RESOLVED (S: STD_ULOGIC_VECTOR)
         return STD_ULOGIC is
   variable RESULT : STD_ULOGIC := 'Z'
begin
   if (S'LENGTH = 1) then
      return S(S'LOW);
   else
      for I in S'RANGE loop
         RESULT := RESOLUTION_TABLE(RESULT, S(I));
      end loop;
   end if;
   return RESULT;
end RESOLVED;
```

Given the resolution function RESOLVED, one can then make the following declarations:

```
subtype STD_LOGIC is RESOLVED STD_ULOGIC;
type STD_LOGIC_VECTOR is array(NATURAL range <>)
      of STD_LOGIC;
```

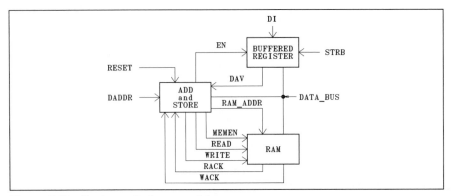

Figure 5.19   System with three modules.

This subtype and type are universally used by most organizations, especially when modeling for synthesis. However the Boolean operators are overloaded for both the unresolved and resolved types, and there are SENSE and DRIVE functions in package STD_LOGIC_MISC for converting between types STD_ULOGIC_VECTOR and STD_LOGIC_VECTOR—in the unlikely event this would be necessary.

5.3.2   Comprehensive System Example

We now present a comprehensive example using SYSTEM_4, which illustrates the use of multi-valued logic in system models.

Figure 5.19 shows a three module system consisting of a buffered register, RAM, and ADD and STORE units. The system performs the following sequence of actions:

1. When STRB goes high, data (DI) is strobed into the buffered register, and data available (DAV) goes to '1'.
2. ADD and STORE responds to DAV = '1' by setting EN = '1'. This gates the data from the buffered register onto the DATA_BUS and resets DAV.
3. ADD and STORE reads the data from DATA_BUS and stores the data in an internal holding register.
4. ADD and STORE initiates a RAM read operation from address RAM_ADDR (which has previously been set equal to DADDR) by asserting the READ signal. RACK = '1' signals the completion of the read process and results in deasserting signal READ. Signals READ and RACK are commonly referred to as handshaking signals.
5. ADD and STORE copies the data off the bus from the memory read, and adds it to the contents of the internal holding register, placing the result back into the holding register.
6. ADD and STORE initiates a RAM write operation, which writes the contents of the internal holding register into the RAM location specified by RAM_ADDR by asserting signal WRITE. WACK = '1' signals the completion of the write process and results in deasserting signal WRITE. WRITE and WACK are handshaking signals.
7. ADD and STORE returns to its initial state waiting for other data available (DAV) from the buffered register.

```
use work.SYSTEM_4.all;
entity BUFF_REG is
  generic(STRB_DEL,DAV_DEL,ODEL: TIME);
  port(DI:  in MVL4_VECTOR(7 downto 0);
       STRB,EN: in MVL4;
       DAV: out MVL4;
       DO: out BUS1(7 downto 0):="ZZZZZZZZ");
end BUFF_REG;
--
architecture TWO_PROC of BUFF_REG is
  signal REG: MVL4_VECTOR(7 downto 0);
begin
  FRONT_END: process(STRB,EN)
  begin
    if STRB'EVENT and STRB = '1' then
      REG <=DI after STRB_DEL;
      DAV <= '1' after DAV_DEL;
    end if;
    if EN'EVENT and EN='1' then
      DAV <= '0' after DAV_DEL;
    end if;
  end process FRONT_END;
  --
  OUTPUT: process(REG,EN)
  begin
    if (EN = '1') then
      DO <= DRIVE(REG) after ODEL; else
      DO <= "ZZZZZZZZ" after ODEL;
    end if;
  end process OUTPUT;
end TWO_PROC;
```

Figure 5.20 Buffered register model.

Since DATA_BUS is driven by all three modules, a resolution function is required. We will use the WiredOne bus function, which requires bus type BUS1.

Figures 5.20, 5.21, and 5.22 give the algorithmic models for the buffered register, RAM and ADD_STORE respectively. The register model is self-explanatory; however, note that the DRIVE function is used to convert the internal type MVL4_VECTOR to BUS1. The RAM model is also straightforward. It reacts to changes in CS, RD, or WRITE to perform the indicated READ or WRITE operations. For WRITE, function SENSE converts type BUS1 to MVL4_VECTOR. An incorrect, all 'Z's word will be converted to all '1's. READ employs the DRIVE function for the type conversion in the opposite direction. Acknowledge signals RACK and WACK are high for duration ACK_PW. Note that an aggregate is used to initialize the memory to all '0's.

ADD_STORE is a unit, which controls a sequence of activities. It goes through a sequence of control steps, and in each step, a particular set of activities is carried out. Figure 5.22 gives the algorithmic model for ADD & STORE. The control steps are labeled in comments as CS0

```
    use work.SYSTEM_4.all;
    --
    entity RAM is
      generic(RDEL,DISDEL,ACK_DEL,ACK_PW: TIME);
       port(DATA: inout BUS1(7 downto 0):="ZZZZZZZZ";
            ADDR: in MVL4_VECTOR(4 downto 0);
            RD,WRITE,CS: in MVL4;
            RACK,WACK: out MVL4);
    end RAM;
    --
    architecture SIMPLE of RAM    is
    begin
     MEM: process (CS,RD,WRITE)
      type MEMORY is array(0 to 31) of MVL4_VECTOR(7 downto 0);
      variable MEM: MEMORY:= (others => (others  => '0'));
     begin
       if CS = '1' then
         if RD = '1' then
           DATA <= DRIVE(MEM(INTVAL(ADDR))) after RDEL;
           RACK <= '1' after ACK_DEL,
                   '0' after ACK_DEL + ACK_PW;
         elsif WRITE = '1' then
           MEM (INTVAL(ADDR)):= SENSE(DATA,'1');
           WACK <= '1' after ACK_DEL, '0' after ACK_DEL+ACK_PW;
         end if;
       else
         DATA <= "ZZZZZZZZ"   after DISDEL;
       end if;
     end process MEM;
    end SIMPLE;
```

Figure 5.21   Simple RAM model.

through CS6. Each step consists of a sequence of VHDL statements ending with a *wait* statement. The *wait* statements have the following two forms:

```
wait on signal until signal = value;
wait for time;
```

Wait statements cause a process to suspend (see Chapter 3). For the first form, the process resumes when the signal in the *wait* statement changes to the proper value. This form can be used to model the detection of handshaking signals, which are used to advance the control unit to the next state. Note that there are three cases of this in the model of ADD & STORE, and they check for the occurrence of the three signals: DAV, RACK, and WACK. The second form is used in the model to delay by the fixed amount of time, CLK_PER, before proceeding to the next step. As the name implies, this time period corresponds to the period of the clock used to advance the control unit. This delay is sufficiently long for the devices triggered by the control step to respond before advancement to the next step takes place. This second form models an open-ended communication mechanism.

```
    use work.SYSTEM_4.all;
    entity ADD_STORE is
      generic(CON_DEL, DO_DEL, MA_DEL, DIS_DEL, CLK_PER: TIME);
      port(RESET,DAV,RACK,WACK: in MVL4;
           MEMEN: out MVL4;
           EN,READ,WRITE: inout MVL4;
           DATA: inout BUS1(7 downto 0):="ZZZZZZZZ";
           DADDR: in MVL4_VECTOR(4 downto 0);
           MADDR: out MVL4_VECTOR(4 downto 0));
    end ADD_STORE;
    architecture ALG of ADD_STORE is
    begin
     CON: process
       variable DATA_REG: MVL4_VECTOR(7 downto 0);
     begin
       if RESET = '1' then
         DATA <= "ZZZZZZZZ"after DIS_DEL; --CS0
       end if;
       wait on DAV until DAV = '1';
    ------------------------------------------
         EN <= '1' after CON_DEL;        --CS1
       wait for CLK_PER;
    ------------------------------------------
         EN <= '0' after CON_DEL;
         DATA_REG := SENSE(DATA,'1');  --CS2
       wait for CLK_PER;
    ------------------------------------------
         MADDR <=  DADDR after MA_DEL;
         MEMEN <= '1' after CON_DEL;    --CS3
         READ <= '1' after CON_DEL;
       wait on RACK until RACK ='1';
    ------------------------------------------
         DATA_REG := ADD8(SENSE(DATA,'1'),DATA_REG);
         READ <= '0'after CON_DEL;
         MEMEN <= '0'after CON_DEL;      --CS4
       wait for CLK_PER;
    ------------------------------------------
         DATA <= DRIVE(DATA_REG) after DO_DEL;
         WRITE <= '1'after CON_DEL;
         MEMEN <= '1'after CON_DEL;     --CS5
       wait on WACK until WACK ='1';
    ------------------------------------------
         WRITE <= '0'after CON_DEL;
         MEMEN <= '0'after CON_DEL;     --CS6
         DATA <= "ZZZZZZZZ" after DIS_DEL;
       wait for CLK_PER;
     end process CON;
    end ALG;
```

Figure 5.22 Add & Store model.

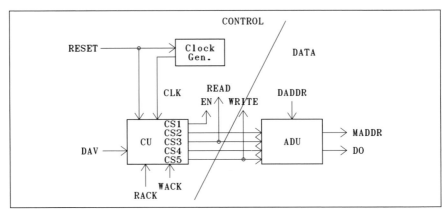

Figure 5.23   Control/data partition for ADD & STORE.

A point to note about all three models is that their outputs, which are connected via the signal DATA_BUS, are all initialized to "ZZZZZZZZ". Recall that the default initialization value for signals of type MVL4 is 'X'; thus, all drivers of the signal DATA_BUS will receive "XXXXXXXX" as an initial value unless otherwise specified. In order for bus resolution to work properly, any signal driver considered to be inactive should contain the value 'Z'. Thus, they should be initialized to this value. In addition, whenever a particular output is disconnected from the bussed signal, its driver should also be returned to the all 'Z' state. Each of the three models has provisions for doing this.

5.3.2.1   *Hardware Interpretation of the Control Steps*

In the algorithmic model for ADD & STORE, we used a control step sequence employing *wait* statements. It is our intent to only use modeling styles for algorithmic descriptions, which have a correspondence to real hardware. Thus, it is appropriate to ask what the hardware interpretation of the control steps is. To do this, we must first perform a *control unit/data unit partition* of ADD_STORE. Figure 5.23 shows such a partition. The control unit section consists of a control unit (CU), and a clock generator. The data unit section consists of ADU, the add unit.

First, let us look at algorithmic models for the clock generator and ADU to understand how CU interacts with them (see Figures 5.24 and 5.25, respectively). The clock generator model is started by the fall of RESET. This sets the variable CLKE to '1' and schedules the first clock period. After that, the clock generator process triggers itself and continues to schedule clocks until RESET goes high, at which time CLKE is set to '0'. The current clock cycle then finishes and the clock generator remains high until RESET goes low again.

The add unit (ADU), shown in Figure 5.25, reacts to positive going changes in CS2, CS3, CS4, and CS5 to (1) load the data register, (2) perform an addition, (3) output a memory address, and (4) output a data value to the data bus.

Figure 5.26 shows a schematic of a structural model for the control unit. The CU receives inputs RESET, CLK, DAV, RACK, and WACK. Its outputs are the five control states CS1 through CS5 and memory enable (MEMEN). Note that CS1, CS3, and CS5 are connected, respectively, to the EN, READ, and WRITE interface control signals. Each control state (CSi)

```
use work.SYSTEM_4.all;
entity CLOCK_GENERATOR is
  generic(PER: TIME);
  port(RESET: in MVL4; CLK: out MVL4);
end CLOCK_GENERATOR;

architecture IMPL_1 of CLOCK_GENERATOR is
  signal CLOCK: MVL4;
begin
  process (RESET,CLOCK)
    variable CLKE: BIT := '0';
  begin
    if RESET='0' and not RESET'STABLE then
      CLKE := '1';
      CLOCK <= transport '0' after PER/2;
      CLOCK <= transport '1' after PER;
    end if;
    if RESET='1' and not RESET'STABLE then
      CLKE := '0';
    end if;
    if CLOCK='1' and not CLOCK'STABLE and CLKE = '1'then
      CLOCK <= transport '0' after PER/2;
      CLOCK <= transport '1' after PER;
    end if;
      CLK <= CLOCK;
  end process;
end IMPL_1;
```

Figure 5.24   Clock generator model.

maps to a flip-flop. All flip-flops are clocked by the signal CLK, so the control unit is *synchronous*. The control unit acts like a shift register in that a '1' propagates from CS0 to CS6. However, there are three modes of *advance* from one state to another:

1. Automatic advance: Control state CSi always advances to state (CSi+1). For example, CS1 is the D input for flip-flop C2. CS1 will last for one clock period only. It is assumed that external activities timed by the control state can be completed in one clock period. In this case, CS1= '1' causes EN = '1', which gates the data onto the data bus that is connected to the inputs of the internal register in the ADD & STORE unit. At the rise of CS2, this data is clocked into the register. This data transfer operation must be completed in one clock period. One can always select the clock frequency to ensure that enough time is available.
2. Handshaked advance: The advance from CS0 to CS1 is an example of this. The control unit waits in CS0 until DAV = '1'. At the next clock transition, it advances to CS1. Note that this places strict timing requirements on the signal DAV. It must not be released from its one level until the control unit advances to the next state (CS1). This can be done by setting up a *handshaking protocol* between the control unit and the device being controlled. In this case, the device being controlled is the buffered register. The protocol proceeds as follows: The buffered register sets DAV = '1' and holds it there

```
use work.SYSTEM_4.all;
entity ADU is
  generic(DO_DEL,MA_DEL: TIME);
  port(CS2,CS3,CS4,CS5: MVL4;
       DATA: inout BUS1(7 downto 0) := "ZZZZZZZZ";
       DADDR: in MVL4_VECTOR(4 downto 0);
       MADDR: out MVL4_VECTOR(4 downto 0));
end ADU;
architecture BEHAVIOR of ADU is
begin
 DU: process(CS2,CS3,CS4,CS5)
   variable DATA_REG : MVL4_VECTOR(7 downto 0);
 begin
   if CS2'EVENT or CS4'EVENT then
     if CS2 = '1' then
       DATA_REG:= SENSE(DATA,'1');
     end if;
     if CS4 = '1' then
       DATA_REG:= ADD8(SENSE(DATA,'1'),DATA_REG);
     end if;
   end if;
   if CS3'EVENT and CS3 = '1' then
     MADDR <= DADDR after MA_DEL;
   end if;
   if CS5'EVENT then
     if CS5 = '1' then
       DATA <= DRIVE(DATA_REG) after DO_DEL;
     else
       DATA <= "ZZZZZZZZ";
     end if;
   end if;
  end process DU;
end BEHAVIOR;
```

Figure 5.25 ADD unit model.

until the control unit advances to CS1 and sets EN = '1'. When EN = '1', the buffered register releases DAV to the '0' level.

3. Asynchronous advance: The advance from CS3 to CS4 is an example of this. CS3 going to a '1' initiates a memory read. RAM signals the completion of memory read with the signal RACK (read acknowledge). However, RACK is pulsed high for a duration that is typically less than a clock period. Therefore, the 0 to 1 transition on RACK sets the RK flip-flop to the '1' state. On the next rising edge of CLK, CS4 is set, which resets flip-flop C10 using its asynchronous R input. Thus, this mechanism provides a way for the control unit to advance when an asynchronous response is anticipated.

5.3.2.2 *Accuracy of Algorithmic Models Using* Wait *Statements*

If one examines the algorithmic model for ADD & STORE, one notes two apparent deficiencies in the approach:

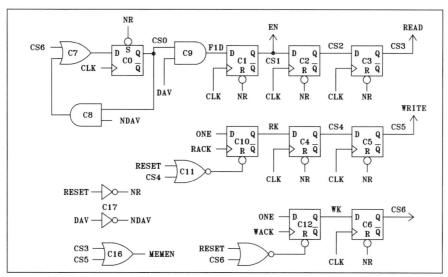

Figure 5.26 Control unit structural model.

1. The algorithmic model (Figure 5.22) only samples reset in control step CS0, although in the structural model of the control unit (Figure 5.26), the unit will respond to a reset at any time.
2. The control unit sequence for ADD & STORE is a simple loop. It is not clear how this approach might be used if the control sequence involves conditional branching among the states.

5.3.2.3   Resetting

One approach to the reset problem is to say that it is not really a problem. A case can be made for the position that the *wait* statement algorithmic model is a high-level model, while incorporation of detailed reset logic is a lower-level design activity, which is best incorporated at the gate level. However, this means that the algorithmic and gate level models differ in their responses, which is undesirable.

Figure 5.27 and Figure 5.28 show a better approach. In architecture ONE of entity WAIT_STEPS, a control unit is modeled that allows resetting from any state. This is accomplished by inserting the control sequence in a loop. At each step in the control sequence, the *wait* statement and the *next* statement are used to check for the occurrence of reset. When reset occurs, the program branches to the loop label, i.e., to the start of the loop. Since the loop condition is now FALSE (because R='1'), the loop will terminate. Since the loop statement is the last statement in the process, execution moves to the top of the process, i.e., to the *wait* statement immediately after the *begin* keyword. The program pauses at the top of the process until the reset goes low. Thus, in architecture ONE, at the start of simulation, the model waits for RUN = '1'. After RUN = '1', the unit loops through the states 0,1, and 2 until R = '1', at which time it exits the loop and waits for R to become '0' again. This initial wait statement is necessary to prevent the occurrence of an infinite looping process when R = '1', which would prevent simulation time from advancing.

```
package INTRES is
   type INTARRAY is array(NATURAL range <>) of INTEGER;
   function INTBUS(S: INTARRAY) return INTEGER;
   subtype RINTEGER is INTBUS INTEGER;
end INTRES;

package body INTRES is
   function INTBUS(S: INTARRAY) return INTEGER is
      begin
         for I in S'RANGE loop
         return S(I);
      end loop;
   end INTBUS;
end INTRES;

use work.INTRES.all;
use work.SYSTEM_4.all;
entity  WAIT_STEPS is
   generic(CLK_DEL,DIS_DEL: TIME);
   port(RUN,X,R: in DOT1 := '0'; S: out RINTEGER);
end WAIT_STEPS;

architecture ONE of WAIT_STEPS is
begin
   process
   begin
      wait until R = '0' and RUN = '1';
LOOP_START: while R = '0'and RUN = '1' loop
         S <= 0;
         next LOOP_START when R = '1';      ---Step 0
      wait until R = '1' for CLK_DEL;
      ----------------------------
         S <= 1;
         next LOOP_START when R = '1';      ---Step 1
      wait until R = '1' for CLK_DEL;
      ----------------------------
         S <= 2;
         next LOOP_START when R = '1';      ---Step 2
      wait until R = '1' for CLK_DEL;
      end loop;
   end process;
end ONE;
```

Figure 5.27   Resetting.

5.3.2.4   Control Sequence Branching

It would seem that control sequencing could be achieved by making each state a simple loop with a label, then using the *next* statement to test conditions and branch to the appropriate label. However, the semantics of the VHDL *next* statement do not allow this. The *next* statement can only force a branch to the enclosing loop label. Branching to arbitrary loop labels is not allowed. This restriction is necessary to enforce structured programming.

```
architecture TWO of WAIT_STEPS is
 signal TRIGGERB,TRIGGERBA,TRIGGERC,TRIGGERCA: DOT1 := '0';
 signal SINT: RINTEGER register;
begin
A: process
  begin
    SINT <= null;
    wait on RUN,TRIGGERBA,TRIGGERCA until RUN = '1';
    SINT <= 0;                          ---Step 0
    wait for CLK_DEL;
    SINT <= 1;                          ---Step 1
    wait for CLK_DEL;
    SINT <= null;
    if X = '1' then
      TRIGGERB <= not(TRIGGERB);
    else
      TRIGGERC <= not(TRIGGERC);
    end if;
  end process A;
B: process
  begin
    SINT <= null;
    wait on TRIGGERB;
    SINT <= 2;                          ---Step 2
    wait for CLK_DEL;
    SINT <= 3;                          ---Step 3
    wait for CLK_DEL;
    SINT <= null;
    TRIGGERBA <= not(TRIGGERBA);
  end process B;
C: process
  begin
    SINT <= null;
    wait on TRIGGERC;
    SINT <= 4;                          ---Step 4
    wait for CLK_DEL;
    SINT <= 5;                          ---Step 5
    wait for CLK_DEL;
    SINT <= null;
    TRIGGERCA <= not(TRIGGERCA);
  end process C;
  S <= SINT;
end TWO;
```

Figure 5.28   Branching.

Control sequence branching can be achieved by placing each simple (non-conditional) sequence within a process, then at the end of that process using *if* and *case* statements to decide which other simple sequence to activate. Figure 5.27 and Figure 5.28 illustrate this. In architecture TWO of entity WAIT_STEPS, when RUN goes to 1, the simple state sequence 0,1 is executed. Then, if X = '1', sequence 2,3 is executed; if X = '0' sequence 4,5 is executed. In either

```
use work.SYSTEM_4.all;
entity TIME_MUX is
   generic(DEL1,DEL2: TIME);
   port(PHASE_ONE,PHASE_TWO: in MVL4;
        Z: out DOTX := '0');
end TIME_MUX;
architecture  GUARDED_BLOCK0 of TIME_MUX is
begin
  PH_ONE: block(PHASE_ONE = '1')
  begin
    Z <= guarded '0' after DEL1;
  end block PH_ONE;
  PH_TWO: block(PHASE_TWO = '1')
  begin
    Z <= guarded '1' after DEL2;
  end block PH_TWO;
end GUARDED_BLOCK0;
```

Figure 5.29   Time multiplexing with two guarded blocks.

case, control returns to sequence 0,1. Transitions on trigger signals activate the processes. Since signal SINT is driven by three processes, a resolution function is required. Package INTRES provides this. Finally, at various places in the three processes, the statement SINT <= null; is encountered. This is done to disconnect that process's driver from the bus resolution function in order to time multiplex the driving of signal SINT. This situation is discussed in detail in the next section.

In summary, we can say that the algorithmic style with *wait* statements is useful, in a synthesis sense, when one takes an architectural view of a system in which the control sequences are mostly simple non-branching sequences. For situations where more complex branching is involved, the state machine style discussed in Chapter 8 is more appropriate.

5.3.3   Time Multiplexing

In an earlier section, we used a bus resolution function to implement a time shared bus. This approach to *time multiplexing* was implemented by insuring that all drivers of the bus signal were maintained in the 'Z' state when they were inactive. We now consider other time multiplexing possibilities. Figure 5.29 shows a model of a time multiplexing situation implemented with two guarded blocks.

Block PH_ONE assigns the value '0' to signal Z when PHASE_ONE = '1'. Block PH_TWO0 assigns the value '1' to Z when PHASE_TWO = '1'. PHASE_ONE and PHASE_TWO are clock pulses that are never at the '1' level at the same time (see Figure 5.30).

Signal Z is declared to be of type DOTX and thus, inherits the wired X resolution function. Assume the intent of the modeler is to (1) have block PH_ONE control Z when PHASE_ONE is high, and have block PH_TWO control Z when PHASE_TWO is high and (2) have Z retain the last value assigned to it when both PHASE_ONE and PHASE_TWO are low. However, with the model as written, Figure 5.30 shows what actually happens. If Z begins in the state 'X', its value will switch to '0' when PHASE_ONE = '1'. When PHASE_TWO = '1', one might expect Z to

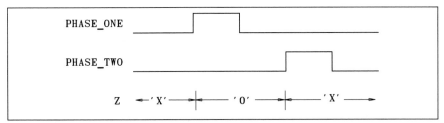

Figure 5.30   Timing for two guarded blocks.

```
architecture  GUARDED_BLOCK1 of TIME_MUX is
  signal ZINT: DOTX register;
begin
  PH_ONE: block(PHASE_ONE = '1')
  begin
    ZINT <= guarded '0' after DEL1;
  end block PH_ONE;
  PH_TWO: block(PHASE_TWO = '1')
  begin
    ZINT <= guarded '1' after DEL2;
  end block PH_TWO;
  Z <= ZINT;
end GUARDED_BLOCK1;
```

Figure 5.31   ZINT as a register.

take on the value '1'. Instead, Z returns to the value 'X' because the driver for block PH_ONE contains a '0', and the driver for block PH_TWO contains a '1'. The bus resolution function WiredX will resolve this to the value 'X'. A bus resolution function is a static function of its drivers, regardless of how long ago the values were placed in the drivers. In this case, we want the value of the driver that most recently changed to control the value of the signal Z. We could apparently fix this problem by modifying the code of architecture GUARDED_BLOCK0 so that when the guard signal of a particular block goes to zero, the value 'Z' is assigned to its driver:

```
if guard'event and guard = '0' then
  Z <= 'Z';
```

This allows the non-'Z' driver value to come through. However, there is one problem that this does not solve: What happens when both guards are false? Then, assigning 'Z' to both guards places that value in each driver for the signal Z, and the resulting signal value is 'Z'. Thus, this approach does not cause Z to retain the last value assigned to it. In fact, it is not possible to write a resolution function that incorporates this notion of "last." Resolution functions can only represent combinational logic circuits.

To handle this problem of bus resolution combined with guarded signal assignment statements, VHDL provides special mechanisms for signals used in guarded signal assignments. If the modeler wants to imply time multiplexing, the signals are designated as being a "register" or "bus." For example, signal ZINT is designated as a register in Figure 5.31 and a bus in Figure 5.32. If a guarded signal is designated as being a register or a bus, and if the guarded statement

```
architecture  GUARDED_BLOCK2 of TIME_MUX is
  signal ZINT: DOTX bus;
begin
  PH_ONE: block(PHASE_ONE = '1')
  begin
    ZINT <= guarded '0' after DEL1;
  end block PH_ONE;
  PH_TWO: block(PHASE_TWO = '1')
  begin
    ZINT <= guarded '1' after DEL2;
  end block PH_TWO;
  Z <= ZINT;
end GUARDED_BLOCK2;
```

Figure 5.32 ZINT as a bus.

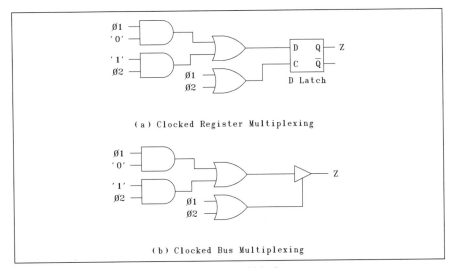

Figure 5.33 Equivalent logic for two guarded blocks.

driving it has its block guard become FALSE, the statement's driver is assumed to be disconnected and is ignored by the bus resolution function. The block whose guard is TRUE will have its value come through the bus resolution function. In the case where all the block guards are off, if ZINT is a "register," it will retain its last value. If ZINT is a "bus," it will be assigned the default value provided by the bus resolution function. For the bus resolution functions in package SYSTEM_4, the value of 'Z' is returned as a default value if the function has no active inputs. Note that the signal ZINT assumes the desired output, 'Z', after one delta delay.

In a hardware sense, when ZINT is a register, clocked register multiplexing is implied. If ZINT is a bus, clocked bussing is implied. The equivalent logic for these two situations is shown in Figure 5.33.

```
architecture PROC of TIME_MUX is
begin
  PH_ONE: process(PHASE_ONE)
  begin
    if PHASE_ONE = '1' then
      Z <= '0' after DEL1;
    end if;
  end process PH_ONE;
  PH_TWO: process(PHASE_TWO)
  begin
    if PHASE_TWO = '1' then
      Z <= '1' after DEL2;
    end if;
  end process PH_TWO;
end PROC;
```

Figure 5.34   Two processes driving Z.

The same situation can exist in processes that are not guarded blocks. Figure 5.34 shows a similar effect for two processes that both drive signal Z. Z is a static function of both drivers no matter how "old" the value in a particular driver is.

The solution to this problem is similar to the one for guarded blocks. Figure 5.35 illustrates the solution. Again internal signal ZINT is designated as a register. In each process, when PHASE_ONE or PHASE_TWO turns off, a value "null" is assigned to ZINT, which has the effect of disconnecting the driver from the resolution function. Again, if PHASE_ONE and PHASE_TWO do not equal '1' at the same time, only the value of the active driver (non null) will get through. Because ZINT has been designated as a register, if both drivers are null, ZINT will retain the most recent non-null value. If ZINT is designated as a bus, ZINT will take on the value 'Z' when both drivers are null.

It should be noted that this problem of time multiplexing with the resolution function exists because we wish to impart certain semantics to signal resolution at the algorithmic level of abstraction. At lower levels of abstraction, when one desires time multiplexing, one can just design a real circuit that does it, then develop a VHDL model of the real-time multiplexing circuit. However, it is very convenient to have the time-multiplexing capability at the algorithmic level that does not require an explicit hardware multiplexer. On the other hand, if we plan to synthesize algorithmic descriptions that have implied time multiplexing, this must be recognized by the synthesis process.

The use of an IF statement, which assigns a value or null to a driver, tends to make the descriptions cluttered and verbose. Figure 5.36 shows another approach that dispenses with use of the resolution function entirely. Instead, it uses separate signals for each process, then multiplexes them using (not SIGNAL'quiet). The assumption is that the two signals are not active at the same time.

It might seem to the reader that the time-multiplexing problem rarely occurs in VHDL modeling, but in fact, this is not so. Consider the outline of an algorithmic model of a processor shown in Figure 5.37.

```
architecture PROC_NULL of TIME_MUX is
  signal ZINT: DOTX register;
begin
  PH_ONE: process(PHASE_ONE)
  begin
    if PHASE_ONE = '1' then
      ZINT <= '0' after DEL1;
    else
      ZINT <= null;
    end if;
  end process PH_ONE;
  PH_TWO: process(PHASE_TWO)
  begin
    if PHASE_TWO = '1' then
      ZINT <= '1' after DEL2;
    else
      ZINT <= null;
    end if;
  end process PH_TWO;
  Z <= ZINT;
end PROC_NULL;
```

Figure 5.35 Solution for two processes driving a single signal.

```
entity TIME_MUX is
  generic(DEL1,DEL2: TIME);
  port(PHASE_ONE,PHASE_TWO: in MVL4;
       Z: buffer MVL4);
end TIME_MUX;

architecture QUIET_MUX of TIME_MUX is
  signal  PH1,PH2,Z1,Z2: MVL4;
begin
  PH_ONE: process(PHASE_ONE)
  begin
    if PHASE_ONE = '1' then
      Z1 <= '0' after DEL1;
    end if;
  end process PH_ONE;
  PH_TWO: process(PHASE_TWO)
  begin
    if PHASE_TWO = '1' then
      Z2 <= '1' after DEL2;
    end if;
  end process PH_TWO;
  Z <= Z1 when not Z1'quiet else
       Z2 when not Z2'quiet else
       Z;
end QUIET_MUX;
```

Figure 5.36 'Quiet multiplexer.

```
-----Computer Model

FETCH: process
   wait until (F = '1');
   -----
   -----
   ----- Fetch  instruction
   -----
   PC <= INC(PC); -- Increment the Program counter
end process FETCH;
EXECUTE: process
 wait until (E = '1');
 ------
 ------
 case IR(15 downto 12) is
 -----
 -----
 ---- Jump instruction
 when "0011" => PC < IR(11 downto 0);
 ----
 ----
 end process EXECUTE;
```

Figure 5.37 Processor model outline.

Again, assume that E and F are disjoint pulses, and PC is a resolved signal. There is no way to write a resolution function that can cause PC to retain the last value assigned to it; null drivers or 'quiet muliplexers must be employed.

Synthesis tools cannot handle the high-level multiplexer, as they cannot generate a circuit to determine when a signal is "quiet." However, it is very convenient at the algorithmic level to have a method for time multiplexing that does not require an explicit hardware multiplexer.

PROBLEMS

5.1   List different ways that algorithmic-level VHDL models can be used in the design process. Below the list, describe each way in two or three sentences.

5.2   Figure 5.38 shows the block diagram of a two module digital system. The basic function of the system implements one-half of a UART, i.e., it receives a 4-bit word in parallel and shifts it out serially. The shift rate is under control of an oscillator (CLKGEN) external to the parallel to serial converter. Whenever RUN=1, CLKGEN puts out a clock (CLK) with a 50 ns period. When RUN=0, the CLK output is at a logic 0 level. The parallel to serial converter functions like this: Whenever the shift out signal SHOUT makes a 0 to 1 transition, the parallel data is loaded into the internal register SR and BUSY is set. Data is shifted out on the line SO in conjunction with the first 0 to 1 transition on CLK after BUSY goes high and continues for three more clock pulse transitions. After the four bits have been shifted out, BUSY is reset. It can be assumed that loading of parallel data will not be attempted when BUSY is a 1. Develop a complete system model using VHDL .

5.3   Write a segment of VHDL code that can check to see if the time between a rise on a signal X and a fall on a signal Y is less than time T.

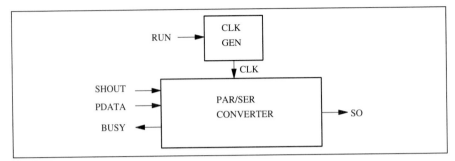

Figure 5.38   Parallel to serial converter.

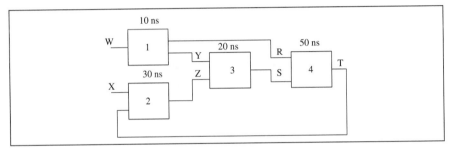

Figure 5.39   Interconnected blocks.

5.4   Figure 5.39 shows a logic structure with a number of interconnected blocks. For each block a propagation delay value is shown, but the function described by each block is not described. Give an outline of a VHDL description for the model. The description should consist of a single entity. Assume W, X, and T are entity interface signals, while Q, Y, Z, R, and S are internal signals. Assume that a change on any block input will cause all block outputs to change after the indicated propagation delay time. Include as much detail as possible using the information given to you. Note: The term "block" used in this question refers to a physical block of logic, and does not necessarily mean a VHDL block.

5.5   Design a 2x4 decoder with an enable input. Use the four-valued logic system MVL4 defined in package SYSTEM_4. You may let the inputs and outputs be either vectors or scalars. When the enable input is '0', all outputs should be 'Z'. When the enable input is '1', exactly one output should normally be '1' with the other outputs '0'. When either the data inputs or enable inputs are 'Z' or 'X', the outputs should be selected appropriately. Please indicate how you decided on the output values in these cases. The unit should be combinational in nature. Include these two generic delays:

   1. DATA_DELAY: The delay in output change after a change on one of the data inputs when the output is already enabled.
   2. ENABLE_DELAY: The delay in output change after a change on the enable input.

   Include an entity description and an algorithmic level, behavioral domain architecture description.

5.6   Design a 2x1 vector multiplexer (MUX) with an enable input. Each data input and output should be an 8-bit vector. Use the four-valued logic system in package SYSTEM_4. The unit should be combinational in nature. If the enable input is logic 0, the output vector should be

"ZZZZZZZZ". If the enable input is logic 1, the output should be determined by the select input and the data inputs. Include these three generic delays:

    1. DATA_DELAY: The delay in output change after a change on one of the data inputs when the output is already enabled.

    2. SELECT_DELAY: The delay in output change after a change on the select input when the output is already enabled.

    3. ENABLE_DELAY: The delay in output change after a change on the enable input.

Include an entity description and an algorithmic level, behavioral domain architecture description.

5.7   Design a 4-bit up/down counter. The counter has four control inputs, UP, DOWN, LOAD, and ENABLE. All counter state changes are synchronized on the rising edge of input CLK. Asynchronous input CLEAR causes all bits in the counter to assume the ZERO state independent of the CLK signal. There is a 4-bit input, DATA_IN, and a 4-bit output, DATA_OUT. When ENABLE=1, DATA_OUT is equal to the 4-bit value stored in the counter. When ENABLE=0, DATA_OUT is 'ZZZZ'. ENABLE works independent of the clock. If UP=1, the counter counts up each time there is a rising edge on CLK. If DOWN=1, the counter counts down each time there is a rising edge on CLK. If LOAD=1, the counter executes a parallel load operation from DATA_IN to the counter on the rising edge of CLK. There is an output C that is logic '1' only when the counter is counting up and contains '1111' or when the counter is counting down and contains '0000'. The counter counts modulo 16. If more than one of the signals UP, DOWN, and LOAD are active at the same time, the result is unspecified. You may assume that a function INC4 and a function DEC4 exist that perform a 4-bit increment and a 4-bit decrement operation, respectively. See package PRIMS in Figure 4.26(a) and Figure 4.26(b).

    a. Draw a process model graph for the UP/DOWN COUNTER. Maximum credit will be given for having several simple processes as opposed to a few complex processes.

    b. Write an algorithmic level VHDL description of the counter that is based on the process model graph constructed in part a.

5.8   The object of this assignment is to create a model of the 8214 priority interrupt control unit. Do this modeling in the following manner:

    a. The model should be a single architecture, chip level, behavioral model.

    b. Present an analysis showing how you developed the path delays for the model.

    c. Draw a process model graph for the model.

    d. Include checks for setup and hold time violations.

    e. Test your model thoroughly.

    f. Write a well-prepared report that illustrates your approach to modeling and interprets your results.

5.9   In this assignment you will create a model of the SN74ALS617 16-bit parallel error detection and correction circuit. Do this modeling in the following manner:

    a. The model should be a single architecture, chip level, behavioral model.

    b. Present an analysis showing how you developed the path delays for the model. These should be developed from the I/O delay data given in the data sheet.

    c. Draw a process model graph for the model.

    d. Include checks for setup and hold time violations. Use the setup and hold time specifications given in the handout (right-hand column).

    e. Test your model thoroughly. Do this as follows:

        1. Test the EDAC in a structural model, which also includes a RAM and Micro_shell. For the RAM you can start with a simple RAM model. Modify the model, so it will induce: single-bit errors, double-bit errors, triple-bit errors, all 1's errors, and all 0's

errors. Give the RAM some control inputs, so these errors can be induced upon command. Micro_shell is a simple model that contains just enough logic to control the EDAC and the RAM.

    2. Simulate the writing of words into the memory using the EDAC function "generate check word" (Table 1 in the specification). Under various conditions of error and no error, read the the words back using "read,flag, and correct function" (Table 4 in the specification).

  f. Write a well-prepared report illustrating your approach to modeling and interpreting your results. In interpreting your results, annotate your printouts, so it is obvious how you modeled the timing.

5.10   Derive the code for a wired AND bus resolution function, which can be used for signals of type MVL ('0','1','Z').

5.11   Assuming signals of type MVL ('0','1',Z), write a wired OR bus resolution function.

5.12   Given is a resolution function for a signal of type BIT. Explain how the function does its job. What physical circuit property does the resolution function implement?

```
function RES_FUNC(signal X: BIT_VECTOR) return BIT is
begin
    for i in X'range loop
        if X(i) = '0' then
            return '0';
        end if;
    end loop;
end RES_FUNC;
```

5.13   Develop a two argument truth table for a wired_X resolution function with pull up for the four-valued logic system MVL4.

5.14   Here is the truth table and function declaration for the wired pull-down bus function:

```
constant table_WIREDPULLDOWN: MVL4_TABLE :=
--  -----------------------------
--  |  X    0    1    Z  |
--  -----------------------------
    (('X', 'X', '1', 'X'),   --  | X |
     ('X', '0', '1', '0'),   --  | 0 |
     ('1', '1', '1', '1'),   --  | 1 |
     ('X', '0', '1', '0'));  --  | Z |
function WiredPullDown (V: MVL4_VECTOR) return MVL4 is
    variable result: MVL4 := 'Z';
begin
    for i in V'range loop
        result := table_WIREDPULLDOWN(result, V(i));
        exit when result = '1';
    end loop;
    return result;
end WiredPullDown;
```

What circuit property does this function implement? Include a circuit diagram of the bus.

5.15   Shown below is a partial VHDL description of a model with two processes. Process ONE and process TWO are activated by two, mutually exclusive pulses P1 and P2. Both processes load register X. It is desired that time multiplexing be performed, i.e. process ONE should control X when P1 goes to the '1' level and process TWO should control X when P2 goes to the '1' level.

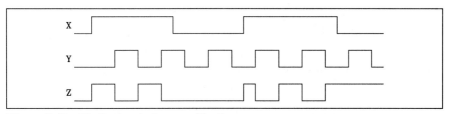

Figure 5.40   Timing for device specification.

    a. Does the given code work correctly? If not, why not?

    b. If you answered negatively to question a, modify the code, so it will work.

```
type MVL is ('0','1',Z);
signal X: F MVL;   ----- F a previously declared bus
                   ----- resolution function.

ONE: process(P1)
begin
  --------
  --------
  X <= Y after 100 ns;
end process ONE;

TWO: process(P2)
begin
  -------
  -------
  X <= Z after 100 ns;
end process TWO;
```

5.16 Design a device with two binary inputs (X and Y) and one binary output (Z). Whenever there
is a '0' to '1' transition on input X, the output Z must assume the value '1' and remain at that
value until there is a change on input Y. Any change on input Y while X='1' will cause the output to change. If there are simultaneous changes on inputs X and Y, the output is unspecified.
Figure 5.40 illustrates the timing for the device.

When presented with the previous design specification, a student wrote the following algorithmic level description:

```
entity DEVICE is
   port (X,Y: in VLBIT; Z: inout VLBIT);
end DEVICE;

architecture ALG of DEVICE is
begin
 Px: process (X)
 begin
  if X = '1' then
    Z <= '1';
  end if;
 end process;
 Py: process (Y)
```

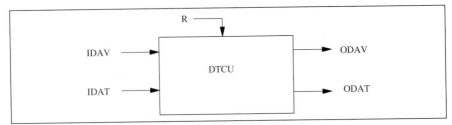

Figure 5.41   Data transfer control unit.

```
begin
  if X = '1' then
    Z <= not Z;
  end if;
 end process;
end ALG;
```

    a. The code given above contains a fundamental error. Describe the error.

    b. Design an architecture that will work correctly. Perform the functions shown in the two processes above in separate processes. You may need to add more processes and additional signals or variables.

5.17 A signal R in a VHDL model represents a hardware register. It is desirable to have R driven by two different processes. It is also desirable that R always contains the last value assigned to it. Outline how you can do this in VHDL.

5.18 Write an algorithmic description of the Data Transfer Control Unit shown in Figure 5.41. The unit does nothing when R = '1'. When R falls, the unit enters a state, where it waits for the rise of input data available (IDAV). When IDAV rises, the unit enters a state, where it reads the input data (IDAT) and transfers it to the output ODAT. It also places a '1' on output data available (ODAV). The output data and ODAV are held stable for a period of time equal to CLK_PER, after which the unit returns to the state where it waits for another rise on IDAV. Your algorithmic description should use WAIT statements to control asynchronous and synchronous advance.

5.19 Design a variable delay unit. In pipelined computer systems, data marches in streams. Since the streams can come from various sources, it is sometimes necessary to delay one stream, so it arrives at the pipeline input at the proper time. Analog approaches to the problem tend to be unreliable, and the use of shift registers is only practical for small delays. Figure 5.42 shows a better approach, where a two port memory is used to delay a parallel data stream. In a two port memory, one can simultaneously write to and read from the memory at the same time, provided the same memory location is not involved. Also shown is a read/write controller for the memory. The systems works as follows: A RESET initializes the device. When START is asserted for one clock period, the write address (WADDR) is initialized to 0, and the read address (RADDR) is initialized to DELAY (the number of clock cycles that one wants the data delayed). During the next clock period, the first data word is presented to the memory input. For every clock pulse, data is written, and WADDR is incremented. However, for the first DELAY clock pulses, RADDR is decremented and no reading takes place. When RADDR reaches 0, reads begin, and RADDR is incremented for each clock pulse after that point. Thus, after the start up delay, the DATA_OUT will equal DATA_IN, delayed by DELAY clock cycles. When STOP goes to 1, the system finishes the transmission of the remaining data stored in memory. Note that this is a totally synchronous system, i.e., both the control unit and

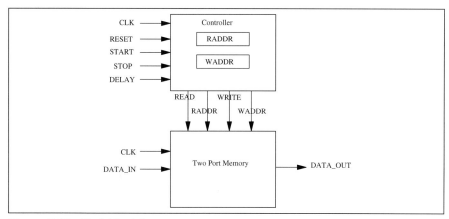

Figure 5.42 Variable delay.

the two port memory are triggered off the same clock. The two port memory can be viewed as being a clocked register array. It can be assumed that the source providing the data is also triggered off the same clock and that START and STOP also change with the clock.

    a. Develop an algorithmic description for the controller and the two port memory. The controller description should use a modeling style employing WHILE loops and WAIT statements. The controller should be able to delay the input data from 0 to 15 clock pulses. The two port memory should contain the minimum number of locations required to support this amount of delay. Use four-bit data words.

    b. Simulate the algorithmic description with a command file, which presents a data stream consisting of the binary sequence 0 to 15 repeated several times. Simulate for different delay values.

    c. Synthesize a gate-level design of the control register using primitive gates and flip-flops. For the part of the control unit that manipulates and/or tests WADDR and RADDR, create an algorithmic VHDL model, which interfaces to the gate-level control register. Use the same model of the two port memory that you developed in part a.

    d. Simulate this mixed-level design using the same approach as in part b.

    e. Submit a well written report that describes and evaluates your design.

5.20 The object of this problem is to understand how to model multi-module systems that communicate over a data bus. Figure 5.43 shows a three module system that consists of a Buffered Register, RAM, and an ADD and STORE unit. The system functions as follows:

    a. When STRB goes high, data (DI) is strobed into the Buffered Register, and data available (DAV) goes to '1'.

    b. ADD and STORE responds to DAV = '1' by setting EN = '1'. This gates the data from the Buffered Register onto the DATA_BUS and resets DAV.

    c. ADD and STORE reads the data from DATA_BUS, adds it to the input value NUM, and stores the data in an internal holding register.

    d. ADD and STORE initiates a RAM write operation, which writes the the contents of the internal holding register into the RAM location specified by the input DADDR.

    e. ADD and STORE returns to its initial state, waiting for another data available (DAV) from the Buffered Register.

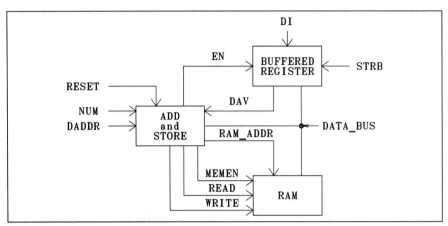

Figure 5.43   Add and Store.

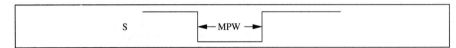

Figure 5.44   Zero active input pulse.

You may use the Buffered Register model from the text. The RAM model is quite similar to that given in the text. Develop the architectural body for the ADD and STORE unit. Simulate the whole system to verify that it works properly.

5.21 Model a DMA controller interface with VHDL. The interface is discussed in Chapter 11 of Morris Mano's book *Computer System Architecture*, second edition (Prentice Hall, 1982, pages 428 – 434). The entire discussion is useful for the assignment, but specifically, you will model the system shown in Fig. 11–18 to:

a. Develop complete models for the DMA controller and RAM.

b. Develop partial models for the microprocessor and the peripheral device. The partial microprocessor model only needs to contain the logic to model the interface between the microprocessor and the bus and the DMA controller. The peripheral device model merely interacts with the DMA controller and sends and receives blocks of data from it.

c. Your models are to include realistic timing. Exactly what timing is included is up to you, with one exception. The RAM memory model unit must include a check on the set-up time on the address and data, vs. the chip select and read/write controls. Whatever timing that is modeled should be well documented.

d. The bus shall employ four state logic (X,0,1,Z). Use package SYSTEM_4 for this purpose.

e. Validate your models through simulation. In addition to validating normal operations, simulate situations, where: (1) One of the devices fails, and as a result, bus contention occurs, and (2) a failure results in incorrect timing on RAM inputs.

f. Prepare a report summarizing your results.

5.22 Figure 5.44 shows a zero active input pulse. A specification on it states that in order to activate the device that it drives, it must have a minimum low pulse width of MPW ns. Any zero-level pulse narrower than this should be reported. Give the VHDL code for detecting and reporting the error.

5.23 Explain in English what the following VHDL code does.

```
entity DAVCON is
  generic(CLK_PER:TIME);
  port(R,IDAV: in BIT;
       IDAT: in BIT_VECTOR(7 downto 0);
       ODAV: out BIT;
       ODAT: out BIT_VECTOR(7 downto 0));
end DAVCON;

architecture WAIT_LOOP of DAVCON is
begin
  process
  begin
    WLOOP: while R = '0' loop
      ODAV <= '0';
      wait until  IDAV = '1' and IDAV'EVENT;
      ODAT <= IDAT; ODAV <= '1';
      wait for CLK_PER;
    end loop;
  end process;
end WAIT_LOOP;
```

5.24 Convert the system model described in section 5.3.2 to IEEE nine-valued logic. Simulate and check for correct results.

CHAPTER 6

Register Level Design

In this chapter we discuss design at the register level of abstraction. At this level, behavior will be defined by VHDL data flow descriptions.

6.1 TRANSITION FROM ALGORITHMIC TO DATA FLOW DESCRIPTIONS

In the previous chapter we developed algorithmic VHDL models suitable for system level modeling. Two basic questions are (1) how do data flow descriptions differ from algorithmic descriptions, and (2) what design activities can be carried out at the register level that cannot be done at the chip level? Let us first address question (1). Although algorithmic descriptions imply a register transfer process, data flow descriptions explicitly represent it. Specifically, in data flow descriptions:

1. Signals are declared, which represent the data movement and connectivity of the real circuit.
2. There is a direct mapping between the statements of the data flow description and a register structural model.
3. There is an implied placement because register level elements are identified and the connectivity between them specified.
4. Multiplexers and buses are identified.
5. The clocking mechanism for registers is identified.
6. Abstract data types are usually transformed to types, such as BIT, BIT_VECTOR, MVL4, MVL4_VECTOR, STD_LOGIC and STD_LOGIC_VECTOR, i.e., state assignment is performed. This allows identification of registers and specifies their lengths.

```
use work.funcs.all;
entity REG_SYS is
  port(C: in BIT; COM: in BIT_VECTOR(0 to 1);
       INP: in BIT_VECTOR(0 to 7));
end REG_SYS;
architecture ALG of REG_SYS is
  signal R1,R2: BIT_VECTOR(0 to 7);
begin
  process(C)
  begin
    if C='1' then
      case COM is
        when "00" => R1 <= INP;
        when "01" => R2 <= INP;
        when "10" => R1 <= ADD8(R1,R2);
        when "11" => R1 <= ADD8(R1,INC8(not(R2)));
      end case;
    end if;
  end process;
end ALG;
```

Figure 6.1   Algorithmic model of register system.

Regarding the second question, because the register interconnect is explicitly represented, we can use the data flow description to study:

1. Timing relationships between register-level elements.
2. Resource allocation.
3. Scheduling.
4. Microcoded control unit design.
5. Bus design.

In this chapter we illustrate items 1 and 4. Item 5 is a homework problem and items 2 and 3 are covered in Chapter 12.

6.1.1   Transformation Example

Consider the following example of transformation from an algorithmic to a data flow description:

A system consists of two 8-bit registers R1 and R2 and an adder. A 2-bit command input COM is used to specify one of four commands:

1. COM = "00". Load R1.
2. COM = "01". Load R2.
3. COM = "10". Add R2 to R1.
4. COM = "11". Subtract R2 from R1.

Figure 6.1 shows an algorithmic description for the register system. Even though it clearly indicates *what* happens, it does not explain *how*.

```
architecture DF1 of REG_SYS is
   signal MUX_R1,R1,R2,R2C,R2TC,MUX_ADD,SUM:
          BIT_VECTOR(0 to 7);
   signal D00,D01,D10,D11,R1E: BIT;
begin
  D00 <= not COM(0) and not COM(1);
  D01 <= not COM(0) and COM(1);       ---Command Decoder
  D10 <= COM(0) and not COM(1);
  D11 <= COM(0) and COM(1);
  MUX_R1  <= SUM when D00 = '0' else  --Register 1 Mux INP;
     R1E <= D00 or D10 or D11;        --Register 1
R1_REG: block(R1E = '1' and C='1' and not C'STABLE)
   begin
     R1 <= guarded MUX_R1;
   end block R1_REG;
R2_REG: block(D01 = '1' and C='1' and not C'STABLE)
   begin
     R2 <= guarded INP;               ---Register 2
   end block R2_REG;
   R2C <= not R2;                      ---Complement
   R2TC <= INC8(R2C);                  ---Increment
   MUX_ADD <= R2TC when D11 = '1' else
            R2;                        ---Adder Mux
     SUM <= ADD8(R1,MUX_ADD);          ---Adder
end DF1;
```

Figure 6.2   Detailed data flow description.

Figure 6.2 shows a detailed data flow description of the register system. Note that (1) signals have been declared illustrating the connectivity; (2) registers, multiplexers, decoders, and data operations—i.e., add, increment, and invert—have been strictly identified; and (3) the signal assignments illustrate how the data moves. It's natural at this point to draw a schematic of the system.

Other data flow descriptions are possible. Figure 6.3 shows a concise data flow description implemented with just two guarded signal assignment statements. Although this representation is economical, its representation of connectivity is poor and would not allow some of the design activities we discuss below to be carried out.

One enhancement that could be made to the detailed data flow description in Figure 6.2 is to do the decoding locally. (This was actually done in the concise description in Figure 6.3.) With local decoding, there is no central decoder; instead, decoding is done at the register or multiplexer, where the control is required. In our fairly simple register system, local decoding is probably superior. Figure 6.4 shows a detailed data flow description with local decoding. Figure 6.5 gives a schematic for local decoding.

In general, centralized decoding can offer certain advantages. First, it provides a more natural form of functional decomposition, which may be easier to synthesize automatically. Second, the existence of a central decoder accommodates growth easier if new functions are added to the specification. In specific, completely defined designs, however, local decoding uses less logic.

```
architecture DF2 of REG_SYS is
  signal R1,R2: BIT_VECTOR(0 to 7);
begin
R1_REG: block((COM(0) or not COM(1))='1' and C='1'
              and not C'STABLE)
  begin
    R1 <= guarded
        ADD8(R1,R2) when (COM(0) and not COM(1)) = '1' else
    ADD8(R1,INC8(not(R2))) when (COM(0) and COM(1)) ='1' else
        INP;
  end block R1_REG;
R2_REG: block(( not COM(0) and COM(1)) = '1' and C='1'
              and not C'STABLE)
  begin
    R2 <= guarded INP;
  end block R2_REG;
end DF2;
```

Figure 6.3   Concise data flow description of register system.


```
architecture DF3 of REG_SYS is
  signal MUX_R1,R1,R2,R2C,R2TC,MUX_ADD,SUM:
        BIT_VECTOR(0 to 7);
  signal R1E,R2E: BIT;
begin
  MUX_R1 <= SUM when COM(0) = '1' else   ---Register 1 Mux
          INP;
  R1E  <= COM(0) or not COM(1);          ---Register 1
R1_REG: block(R1E = '1' and C='1' and not C'STABLE)
  begin
    R1 <= guarded MUX_R1;
  end block R1_REG;
  R2E <= not R1E;
R2_REG: block(R2E = '1' and C='1' and not C'STABLE)
  begin
    R2 <= guarded INP;                   ---Register 2
  end block R2_REG;
  R2C <= not R2;                         ---Complement
  R2TC <= INC8(R2C);                     ---Increment
  MUX_ADD <= R2TC when COM(1) = '1' else
            R2;                          ---Adder Mux
  SUM <= ADD8(R1,MUX_ADD);               ---Adder
end DF3;
```

Figure 6.4   Detailed data flow description with local decoding.

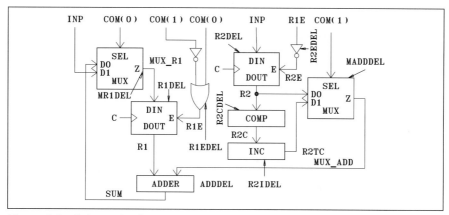

Figure 6.5   Schematic of detailed data flow description with local decoding.

6.2 TIMING ANALYSIS

We now show how to use the data flow description of the register-level model to perform timing analysis. To do this, we will use architecture DF3 and incorporate timing into it. Figure 6.6 shows how timing can be incorporated into the model. Note: Generics have been used to represent delays. The generic delay for each register level element is listed next to the element in Figure 6.5.

Analysis of the register-level model with timing illustrates two potential timing problems with the register system: (1) setup time requirement on the command input COM vs. the rise of clock C and (2) a minimum clock period requirement for clock C. Figure 6.7 illustrates these two requirements.

We now consider each requirement in detail:

1. Setup time on COM vs. the rise of C (COMSUTC): When the command input COM changes value, enough time must elapse so that inputs to the registers will stabilize before the rise of C. We can list these requirements as follows:

   a. Register 1   Enable (R1E) stable: COMSUTC > R1EDEL.

   b. Register 2   Enable (R2E) stable: COMSUTC > R1EDEL + R2EDEL.

   c. Register 1   Data Input (MUX_R1) stable: COMSUTC > MADDDEL + ADDDEL + MR1DEL.

Requirement c is the most stringent requirement, and therefore, we use it as the set up specification for COM vs. the rise of C.

2. Minimum clock period (CPER) requirement on the clock C: If one clocks data into Register 2 (COM = "00"), and follows that with a subtract (COM = "11") during the next clock period, then the clock period must meet the following requirement:

CPER > R2DEL + R2CDEL + R2IDEL + MADDDEL + ADDDEL + MR1DEL

Even though we are dealing with a specific circuit here, both of these requirements are typical of what is required for systems in general.

```
use work.funcs.all;
entity REG_SYS_T is
  generic(MR1DEL,R1EDEL,R1DEL,R2EDEL,R2DEL,
           R2CDEL,R2IDEL,MADDDEL,ADDDEL: TIME);
  port(C: in BIT; COM: BIT_VECTOR(0 to 1);
       INP: in BIT_VECTOR(0 to 7));
begin
  process(C)
    variable COLDT,CNEWT: TIME:= 0 ns;
  begin
    assert not(C'EVENT and C = '1'
      and not COM'STABLE(MADDDEL+ADDDEL+MR1DEL))
      report "COM Set Up Time Failure" severity WARNING;
    if C'EVENT and C = '1' then
    CNEWT := NOW;-- NOW returns the current simulation time.
      assert (CNEWT - COLDT) > (R2DEL + R2CDEL + R2IDEL
                                + MADDDEL + ADDDEL + MR1DEL)
        report "Clock Period Too Short" severity WARNING;
    end if;
    COLDT := CNEWT;
  end process;
end REG_SYS_T;
architecture DF3T of REG_SYS_T is
  signal MUX_R1,R1,R2,R2C,R2TC,MUX_ADD,SUM:
         BIT_VECTOR(0 to 7);
  signal R1E,R2E: BIT;
begin
  MUX_R1 <= SUM after MR1DEL when COM(0) = '1' else
            INP after MR1DEL;
  R1E <= COM(0) or not COM(1) after R1EDEL; ---Register 1
R1_REG: block(R1E = '1' and C='1' and not C'STABLE)
  begin
    R1 <= guarded MUX_R1 after R1DEL;
  end block R1_REG;
  R2E <= not R1E after R2EDEL;
R2_REG: block(R2E = '1' and C='1' and not C'STABLE)
  begin
    R2 <= guarded INP after R2DEL;            ---Register 2
  end block R2_REG;
  R2C <= not R2 after R2CDEL;                 ---Complement
  R2TC <= INC8(R2C) after R2IDEL;             ---Increment
  MUX_ADD <= R2TC after MADDDEL when COM(1) = '1' else
             R2 after MADDDEL;                ---Adder Mux
  SUM <= ADD8(R1,MUX_ADD) after ADDDEL;       ---Adder
end DF3T;
```

Figure 6.6 Register system with timing.

In using the model of the register system with timing for modeling activities, it is important that these specifications on COM and C be met when employing the model. To ensure this, assertions have been added to the interface description. The assertions are contained in a process

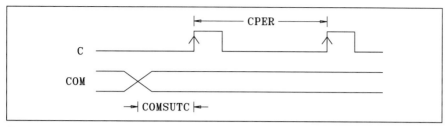

Figure 6.7 Register system timing specification.

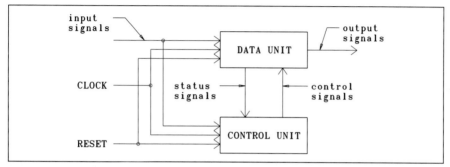

Figure 6.8 Partition of register level device into basic units.

that is triggered by a change in clock C. The assertion check on COM is straightforward and requires little discussion. The minimum clock period check measures the time between successive positive clock period transitions (CNEWT-COLDT) and compares it to the specification. It uses the function NOW from Package Standard, which returns the present value of simulation time when invoked.

6.3 CONTROL UNIT DESIGN

One of the major design activities at the register level is the design of control units because, at this level, all control signals are explicitly represented. Devices are partitioned into two major components as shown in Figure 6.8. The *data unit* contains data registers that hold operands and results and combinational logic units that manipulate and process data values. The *control unit* generates a sequence of *control signals* that determines when data transfers are to occur in the data unit. The control unit needs *status* information from the data unit to control conditional branch operations in the control unit. The data unit also generates the device output signals using data, status, and control signal information.

6.3.1 Types of Control Units

Control units can either be *microcoded* or implemented with standard gates (*hardwired*). Hardwired control units must be custom designed for each device and can take on many forms. Figure 6.9 shows the Huffman model for a finite state machine controller. Block MEM contains all

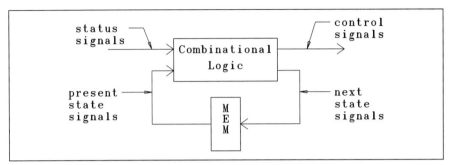

Figure 6.9 Huffman model of hardwired controller.

the control flip-flops, and block CL (combinational logic) contains all the logic gates in the controller. The current values of control signals are dependent in general on status information from the data unit and on the present states of control flip-flops (Mealy type) or just on the present states of the control flip-flops (Moore type). The next states of the control flip-flops are determined by the current states and the status information from the data unit.

The controller design described in Chapter 8, the one-hot state assignment approach, is a special case of the Huffman model in which a separate flip-flop is dedicated to each state in the controller. This special case has the advantage of being easily automated. Although it uses more flip-flops than necessary, the CL is often much less complex, possibly resulting in an overall reduction in complexity.

Hardwired controllers are inherently faster than microcoded controllers. The primary disadvantages are greater complexity, greater cost, and the need for custom designs.

The microcoded control unit simply reads the value of all control signals from ROM at each time step as shown in Figure 6.10. At each time step, the present values of all control signals are read from ROM and stored in the memory instruction register (MIR). The level control signals are available at the outputs of the MIR until the next values are read from ROM. The correct sequence of control signals is obtained by generating the proper sequence of addresses at the ROM inputs. This type of controller works best if the sequencing of addresses is relatively simple. Usually, the address generator (AG) circuit is a counter with the option to parallel load a new address when one wishes to jump to a new point in the control sequence. If the controller usually goes to the next higher sequential address for the next set of control signal values with just an occasional need to parallel load a new address, the address generator (AG) circuit can be a low complexity circuit. Additional trade-offs in address generation are discussed in a later section of this chapter.

The primary advantages of a microcoded controller are:

1. A standard design can be developed for a series of devices. The only difference is in the values of control signals that are stored in the ROM. Therefore, by changing the ROM, a different controller is obtained. A standard board can be mass-produced. When an order is received, the specific ROM for that controller is installed on the board and shipped to the customer.

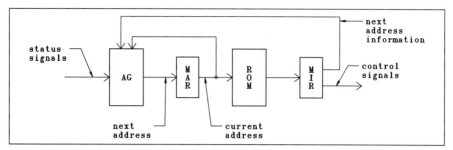

Figure 6.10 Block diagram for a microcoded control unit.

2. Design changes are easier to accommodate. It is only necessary to change the ROM program. The controller hardware does not have to be changed. As a result, design changes are less costly for microcoded controllers than for hardwired controllers.

3. Design time and design cost is greatly reduced because the hardware is already designed and debugged. The controller design is reduced to programming the ROM, which can make effective use of existing software tools, such as compilers, assemblers, simulators, and debuggers. Normally, fewer errors will occur in the design process, and those that do occur can be more easily corrected by changing the program, rather than by changing the hardware.

The primary disadvantages are:

1. Slower operation when compared to hardwired design because the read time for the ROM must be accommodated.

2. The microcoded design will also be more costly for small devices because it includes a ROM and two registers minimally. However, for very large designs, the microcoded controller will be less costly because of the very large number of gates and flip-flops needed in the hardwired designs.

3. Using a standard controller limits the designer to the use of features built into the controller. More features implies higher cost. Everyone must pay the higher cost, even if they do not use all features of the controller. Every designer must observe the constraints imposed by the controller. These constraints include such things as the number of control signals that can be generated, number of status signals that can be handled, and limits imposed by the address generation logic. The controller design is a trade-off between flexibility and cost.

We will illustrate these basic concepts by discussing the design of a small RISC machine.

6.4 ULTIMATE RISC MACHINE

The term *RISC* stands for reduced instruction set computer, implying that the instruction set is very simple, as opposed to CISC machines which have complex instruction sets. RISC machines have very efficient VLSI implementations, which allow them to execute their instruction set at high speed, thereby offsetting the disadvantage of a smaller instruction set. Having decided to design a machine with a small instruction set, we might as well consider the ultimate RISC architecture (URISC) developed by Mavaddat and Parham. URISC is the Ultimate RISC machine because it has only one instruction; however, Mavaddat and Parham claim that URISC is a universal machine because any complex operation can be programmed using only this one instruction. Its simplicity allows many design issues to be explored with a minimum of detail.

6.4.1   Single URISC Instruction

The URISC instruction is a SUBTRACT AND BRANCH ON NEGATIVE instruction. Since there is only one instruction, there is no need for an operation code to define the instruction to be executed. The instruction format only needs to identify the operands and the branch address if the result is negative. The single instruction in the URISC machine has the following format:

1st Operand Address	2nd Operand Address	Branch Address

The instruction executes as follows:

1. The first operand is subtracted from the second operand, and the result is stored in the second operand's location.
2. If the result of the subtraction is negative, a branch to the target address is executed; otherwise, the next instruction in sequence is executed.
3. A branch to location 0 will stop the machine.

Thus, all instructions are 3-byte instructions, and all instructions require three memory cycles. Therefore, URISC will only be fast when it can access high speed memory. More will be said about the required memory speed later.

We will use an assembly language notation for an instruction of the form:

$$L: F,S,T$$

where:

L is a label representing the address where the instruction is to be stored.
F is the symbolic name of the first operand address.
S is the symbolic name of the second operand address.
T is the symbolic name of the branch address.

We also assume the existence of an assembly language pseudo-operation of the form:

$$L: WORD\ C$$

which causes the constant value C to be loaded into symbolic location L.

As an example of the use of URISC, consider the program shown in Figure 6.11, which performs the operation $Z = (X+Y)/2$. The program was developed by Salahuddin Almajdoub, a graduate student at Virginia Tech. Note: Division of X+Y by 2 is accomplished by the successive subtraction method. When the program is finished, a jump to STOP (location 0) occurs, which causes the machine to halt. How this is accomplished is described later.

6.4.2   URISC Architecture

In this section we describe the architecture of the URISC Machine. Figure 6.12 shows the data unit for URISC. The program counter (PC) will hold the address of the next word of the instruction being executed. The memory data register (MDR), and memory address register (MAR) are

Memory	Symbolic	Instruction	Comments
0	STOP:	WORD 0	;Halt
1	READ_Y:	Y,TEMP1,NEXT1	;TEMP1<=(-Y)
2	NEXT1:	TEMP1,X,NEXT2	;X<=Y+X
3	NEXT2:	Z,Z,TEST	;Z<=0
4	TEST:	X,TEMP2,POSITIVE	;TEMP2<=-(X+Y)
5	NEGATIVE:	TWO,TEMP2,STOP	;TEMP2<=-(X+Y)-2
6	COUNT_NEG:	ONE,Z,NEXT3	:Z<=Z-1
7	NEXT3:	TWO,TEMP3,NEGATIVE	;Go to NEGATIVE
8	POSITIVE:	TWO,X,STOP	;X<=(X+Y)-2
9	COUNT_POS:	MONE,Z,NEXT4	;Z<=Z+1
10	NEXT4:	TWO,TEMP4,POSITIVE	;Go to POSITIVE
	TEMP1:	WORD 0	
	TEMP2:	WORD 0	
	TEMP3:	WORD 0	
	TEMP4:	WORD 0	
	X:	WORD 0	
	Z:	WORD 0	
	ONE:	WORD 1	
	MONE:	WORD-1	
	TWO:	WORD 2	

Figure 6.11 URISC program to compute (X+Y)/2.

used to interface with the program memory. Register R holds the first operand prior to the subtract operation. The subtract operation is performed by a twos complement adder (ADDER). One input to the adder is the data value on BUS_A. The other input is either the bit-by-bit complement of the R register (if COMP is '1') or a constant "00000000" (if COMP is '0'). The output of ADDER is always connected to BUS_B. With COMP='1' and Cin='1', the adder performs a twos complement subtraction of the contents of the R register from the data on BUS_A. With COMP='0' and Cin='1', the ADDER will add a constant 1 to BUS_A. This capability is used to increment the PC. With COMP='0' and Cin='0', the data on BUS_A is transferred directly to BUS_B since a constant zero is added to BUS_A. This configuration is used to transfer data from the PC or MDR to the MAR. The adder contains two status outputs Z and N, which indicate when the result was zero or negative, respectively. These status values are loaded into the respective status flip-flops when Zin and Nin are asserted.

Shown on the diagram are labeled *control points* which are indicated by an X. Each control point is labeled with a signal name. Control points with suffix "out" (e.g., PCout) act as enable signals to gate data onto BUS_A. When the signal is asserted, the indicated data register contents is gated onto BUS_A. For example, when PCout is asserted, the contents of the PC register is gated onto BUS_A.

Control points with suffix "in" control the loading of data into the indicated register. For example, when MARin is asserted, the data on BUS_B is loaded into the MAR. When Zin is asserted, the Z status output from the ADDER is loaded into the Z status flip-flop.

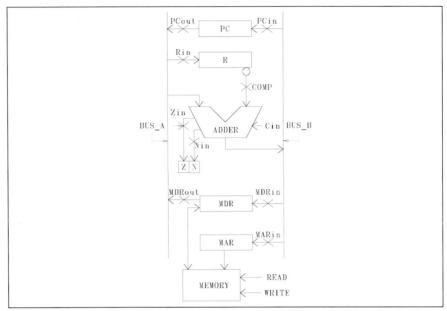

Figure 6.12   Data unit for URISC machine.

The program counter is incremented by asserting control signals PCout, Cin, and PCin. PCout connects the PC to BUS_A. Since COMP is not asserted, the second input to the adder is "00000000". With Cin=1, PC+1 will be placed on BUS_B. With PCin active, PC+1 will then be loaded into PC at the end of the cycle.

The first operand is subtracted from the second operand by loading the first operand into the R register and the second operand into the MDR register. Then, with MDRout, Cin, COMP, and MDRin asserted, the desired twos complement subtraction operation will be performed with the result going to the MDR register.

The address in the PC can be moved to the MAR by asserting only PCout and MARin. A constant zero is added to the address in PC, which does not change it.

6.4.3   URISC Control

To control the URISC, various control points in the data unit must be activated at the proper time. This service is provided by the control unit. Figure 6.13 shows a block diagram of the URISC data and control units plus the memory unit. The URISC control unit receives as input a system clock that it uses to time the changes in control signals. The outputs of the control unit are the control signals shown in Figure 6.12 (PCout, COMP, MDRin, etc.). In addition, the control unit generates two special control signals ZEND and NNEND, which are only used internally. Their purposes will be clarified in the next section.

6.4.3.1   URISC Instruction Cycle Control Sequence

At each clock period, the control unit must output the state of each control signal. Figure 6.14 shows a control sequence, provided by Mavaddat and Parham, that will cause one instruction to

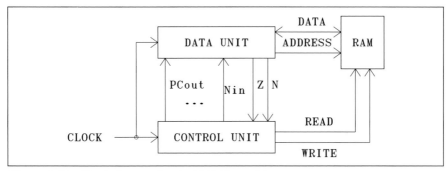

Figure 6.13 Block diagram for the URISC machine.

```
0.  PCout,Zin,MARin,READ,ZEND
1.  MDRout,MARin,READ
2.  MDRout,Rin
3.  PCout,Cin,PCin,MARin,READ
4.  MDRout,MARin,READ
5.  MDRout,COMP,Cin,Nin,MDRin,WRITE
6.  PCout,Cin,PCin,MARin,READ
7.  PCout,Cin,PCin,NNEND
8.  MDRout,PCin
```

Figure 6.14 Control sequence for URISC instruction execution.

be fetched from memory and executed. At each time step, the control signals that must be asserted are listed. Those signals not listed are not asserted. It is assumed that the PC register holds the address of the first word of the next instruction to be executed when the sequence is entered. It is also assumed that each instruction occupies three consecutive memory locations to hold, respectively, the address of the first operand, address of the second operand, and the branch address to take if the result of the subtraction of the first operand from the second operand is negative. At the end of the sequence, the address of the first word of the next instruction to be executed will be left in the PC register. By repeating the sequence, a program can be executed.

These steps can be explained as follows:

1. Steps 0 through 2: Read the first operand into the R register. Step 0 loads the address of the instruction in the MAR register by asserting PCout and MARin. Since COMP and Cin are not asserted, the adder will add zero to the PC address. Also, since Zin is asserted, the Z flip-flop will be set if the address in the PC register is address 0. After this transfer, since READ is asserted, the address of the first operand will be transferred to the MDR. If the PC address was address 0, then since ZEND is asserted, the control sequence will return to step 0. In effect, step 0 in the control sequence will execute indefinitely, resulting in a dynamic halt of the URISC machine. Assuming that address 0 was not present, at step 1, the address of the first operand is moved from the MDR register to the MAR register by asserting only MDRout and MARin. At the end of the control step, since READ is asserted, the first operand value will be loaded into the

MDR register. During control step 2, the first operand value will be moved from MDR to R.

2. Steps 3 through 5: Read the second operand, subtract it from the first, and place the result in MDR. Set the N flip-flop if the result is negative. At control step 3, the address in PC is incremented to point to the second word of the current instruction, which holds the address of the second operand. The incremented address is put into PC and MAR. This is accomplished by asserting PCout, Cin, MARin, and PCin. PC+1 then goes to PC and MAR. At the end of the control step, the address of the second operand is read from memory into the MDR. Control step 4 then moves the address of the second operand from MDR to MAR. At the end of control step 4, the value of the second operand is read from memory into the MDR register. Control step 5 then causes the first operand to be subtracted from the second operand and the result put in the MDR. This is accomplished by asserting MDRout, Cin, COMP, and MDRin. At the same time, flip-flop N is set if the result is negative since Nin is asserted. After all this is done, at the end of control step 5, the result is written into the second operand's memory location since MAR was not changed.

3. Step 6: Reads the branch address into the MDR. With PCout, Cin, and PCin asserted, the PC is incremented to hold the address of the third word of the instruction. At the same time the address of the third word is loaded into MAR. At the end of control step 6, the branch address is read from memory into MDR.

4. Step 7: Increments the program counter to point to the first word of the next sequential instruction in memory. Since NNEND is asserted, the control sequence returns to step 0 if N was NOT set in step 5. This occurs only if the result was NOT negative. Therefore, a non-negative result causes the next instruction in sequence to be executed; whereas, a negative result causes control to go to step 8.

5. Step 8: Copies the branch address into the program counter. This occurs only when the result of the subtraction is negative. The next instruction is fetched from the branch address, instead of from the next sequential memory location. Control always returns to step 0 after step 8 is executed.

6.4.3.2   URISC Timing

By examining the control sequence for the URISC instruction execution, we notice that two consecutive actions must be taken during most control steps. For example, during step 0, the contents of the PC register must be moved to the MAR register, then a memory read must be performed. Further, we notice that in each case, there is an action inside the URISC data unit followed by a memory operation. Therefore, a two-phase clocking scheme in which the first phase accomplishes a data transfer within the URISC data unit, and the second phase accomplishes a memory operation, seems appropriate. Figure 6.15 defines the proposed timing for the URISC machine. At the beginning of the control step, the falling edge of PH1 updates the control signals which are then held constant for the duration of the control step. The falling edge of PH2 causes a data transfer within the URISC data unit. The falling edge of PH1 at the end of the control step causes a memory operation and updates the control signals for the next control step. Figure 6.16 shows the timing of operations for the first three control steps.

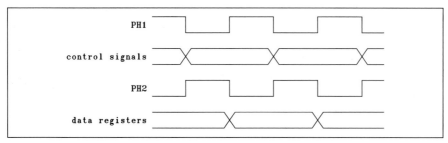

Figure 6.15   Two phase clock for URISC processor.

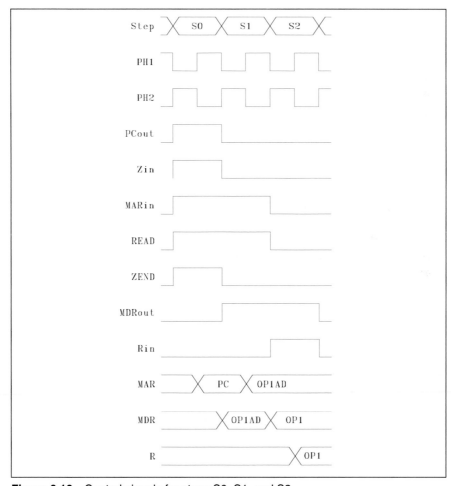

Figure 6.16   Control signals for steps S0, S1, and S2.

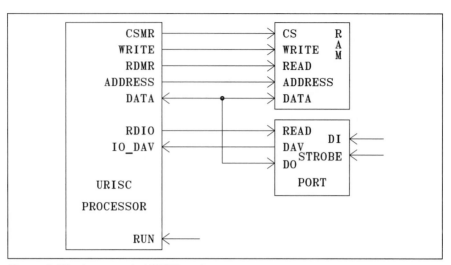

Figure 6.17   An example URISC system.

6.4.4   URISC System

Figure 6.17 shows a block diagram of a computer system using the URISC processor. It includes a RAM memory module and an input PORT module. A RUN input has been added to start and stop the processor.

Signals DATA and ADDRESS are the external data bus and address bus, respectively. Signal RDMR is asserted for a data transfer from RAM to the processor, and signal WRITE is asserted for a data transfer from the processor to RAM. Signal DAV, also added to the basic URISC, is asserted by the input PORT when it has received a new data item to alert the processor to that fact. The port gates data onto the DATA bus only when it detects an RDIO signal.

6.4.5   Design of the URISC at the Register Level

Figure 6.18 shows the outline for a register level VHDL description of the URISC. As discussed in Chapter 5, package SYSTEM_4 from the CD ROM contains declarations and functions for a four-valued logic system MVL4 ('X', '0', '1', 'Z') in which 'X' is interpreted as unknown, and 'Z' is interpreted as the high impedance state for a tristate signal line.

Most of the generic declarations in the entity declaration are timing delays for functions in the URISC that are self-explanatory. ENABLE_DELAY is the delay associated with enabling data onto a data bus, for example. Parameter PER is the clock period. DATA is the external data bus that connects to the RAM and input PORT. ADDRESS is the external address bus that connects to the RAM. Type WORD is an eight-position vector with base type MVL4. The other port declarations are consistent with the system diagram shown in Figure 6.17.

The architecture declaration contains a set of signals that are used internally to the URISC processor. Some of them, such as COMP and ZEND, are already familiar to the reader. The others are defined later. The architecture contains a block for each register, a block for each combinational logic unit within the URISC processor, and a block for the control unit. Some of the

```
      -- The entity declaration of the URISC processor
      use work.all;use work.SYSTEM_4.all;
      entity URISC is
        generic(ENABLE_DEL,DISBL_DEL,REG_DEL,ADD_DEL,PER,
               COUNT_DEL, ROM_DEL,OR_DEL,AND_DEL,INV_DEL,
               MUX_DEL: TIME);
        port(DATA: inout WORD:="ZZZZZZZZ";
             ADDRESS: inout WORD:="00000001";
             RUN,IO_DAV: in BIT; RDIO,WRITE: inout BIT;
             RDMR,CSMR: out BIT);
      end URISC;
      -- The architecture of the URISC ------------------------
      use work.all;use work.SYSTEM_4.all;
      architecture BEHAVIORAL of URISC is
        signal PC1,R,R_NOT,BUS_A,BUS_B,MDR1,MDR: WORD;
        signal PC: WORD:="00000001";
        signal Z,ZERO,ZIN,N,PH1,PH2,R_IN,MDR_IN,N_IN,MAR_IN,
               C_IN,CLK,CLEAR: BIT;
        signal COMP,MDR_OUT,PC_IN,PC_OUT,ZEND,NNEND,READ: BIT;
        signal C: BIT_VECTOR(3 downto 0);
      begin
        PC_REG : block (PCIN and PH2='0'and not PH2'stable)
      ---Insert Code for Program Counter Register ----------
        end block;
        R_REG : block (R_IN='1'and PH2='0'and not PH2'STABLE)
      ---Insert Code for Register R -----------------------
        end block;
        -------- The Adder -------------------------------
        BUS_B <=  ADD(BUS_A,R_NOT,C_IN) after ADD_DEL;
        N_REG : block(N_IN= '1' and PH2='0'and not PH2'STABLE)
      ---Insert Code for Register N-----------------------
        end block;
        Z_REG : block(Z_IN= '1' and PH2='0'and not PH2'STABLE)
      ---Insert Code for Z Register -----------------------
        end block;
        MDR_REG : block((MDRIN and PH2='0'and not PH2'STABLE)
                   or (READ and PH1='0' and not PH1'STABLE))
      ---Insert Code for Memory Data Register --------------
        end block;
        MAR : block (MAR_IN='1' and PH2='0' and not PH2'STABLE)
      ---Insert Code for Memory Address Register MAR -------
        end block;
        process
      ---Insert Code for Internal Two Phase Clocks ---------
        end process;
        -------- The Control Unit -----------------------
      ---Insert Code for the Control Unit
      end BEHAVIORAL;
```

Figure 6.18   Outline for register level description of the URISC processor.

```
      ---------The Program Counter Register -------------
      -- PC is a negative edge triggered register--------
      PC_REG : block (PCIN and PH2='0'and not PH2'stable)
      begin
        PC1 <= guarded BUS_B after REG_DEL;
        BUS_A <= PC1 after ENABLE_DEL when PC_OUT='1' else
                 "ZZZZZZZZ" after DISBL_DEL;
      end block PC_REG;
```

Figure 6.19   Design of PC register.

blocks are defined later. The adder block contains a reference to a package function that computes the twos complement addition of two vectors with a carry-in.

Each of the registers will be designed as a negative edge triggered register with negative edge triggered flip-flops. For example, Figure 6.19 shows the complete design of the PC register. It is described as a guarded block with the guard condition being activated when PCin is asserted, and the PH2 clock changes from 1 to 0. The contents of BUS_B is loaded into the PC when the guard is true. The contents of PC is gated onto BUS_A when PCout is asserted independent of the guard condition. If PCout is not asserted, the PC register output is put in the high impedance state. The other registers have similar specifications. Their design is left as an exercise to the reader (see Problem 6.13). Not shown is the internal oscillator, which generates PH1 and PH2. This oscillator is started when RUN goes high.

6.4.6   Microcoded Controller for the URISC Processor

Since the control sequence for the URISC machine instruction execution has only two branch points, and since the branch points are both to step 0, the address generation logic for the microcoded controller can be very simple indeed. We shall use a modulo 9 counter with the option to parallel load all zeros into the counter if either of the branch conditions is true. The counter will, therefore, cycle from 0 to 8 and back to 0 again unless the branch conditions are true. If a branch condition is true, the counter will be loaded with all zeros to affect a branch to step 0. Figure 6.20 shows the VHDL description of the controller. This description can be directly transformed into a hardware circuit as shown in Figure 6.10 in a straightforward manner.

The COUNTER is loaded in response to a transition in PH1 clock from 1 to 0. Signal CLEAR defines the branch conditions that require a branch to step 0. The conditions Z and ZEND indicate that address zero was detected in the PC at the end of step 0. Therefore, the COUNTER will continue to reset to ZERO indefinitely realizing a dynamic halt of the processor. Condition (not N and NNEND) indicate that the result of the subtraction was NOT negative and the controller is at the end of step 7. The desired action is to return to step 0 and execute the next sequential instruction in memory. If N were TRUE, step 8 would be executed to load the branch address into the PC, causing a branch to a new address in program memory. The COUNTER is also cleared after step 8 because we want to start execution of the next instruction. For all other conditions, the COUNTER is incremented to cause execution of the next control step.

```
---------The COUNTER -----------------------------
-- The Counter has a synchronous CLEAR ------------
COUNTER: block (PH1= '0' and not PH1'STABLE )
begin
  C <= guarded "0000" after COUNT_DEL when CLEAR='1' else
               INC_COUNTER(C) after COUNT_DEL;
end block COUNTER;
-------- The microinstructions ROM  ----------------------
ROM: process(C)
  type SQ_ARRAY is array(0 to 8,0 to 8) of BIT;
  constant MEM : SQ_ARRAY:=
-- 0       1    2    3    4    5    6    7    8   COLUMN
MDR_OUT,MAR_IN,N_IN,R_IN,PC_IN,ZEND,C_IN,WRITE,NNEND,micins
(('0',    '1', '0', '0', '0', '1', '0', '0', '0'), --0
 ('1',    '1', '0', '0', '0', '0', '0', '0', '0'), --1
 ('1',    '0', '0', '1', '0', '0', '0', '0', '0'), --2
 ('0',    '1', '0', '0', '1', '0', '1', '0', '0'), --3
 ('1',    '1', '0', '0', '0', '0', '0', '0', '0'), --4
 ('1',    '0', '1', '0', '0', '0', '1', '1', '0'), --5
 ('0',    '1', '0', '0', '1', '0', '1', '0', '0'), --6
 ('0',    '0', '0', '0', '1', '0', '1', '0', '1'), --7
 ('1',    '0', '0', '0', '1', '0', '0', '0', '0'));--8
begin
  MDR_OUT  <=   MEM(INTVAL(C),0) after ROM_DEL;
  MAR_IN   <=   MEM(INTVAL(C),1) after ROM_DEL;
  N_IN     <=   MEM(INTVAL(C),2) after ROM_DEL;
  R_IN     <=   MEM(INTVAL(C),3) after ROM_DEL;
  PC_IN    <=   MEM(INTVAL(C),4) after ROM_DEL;
  ZEND     <=   MEM(INTVAL(C),5) after ROM_DEL;
  C_IN     <=   MEM(INTVAL(C),6) after ROM_DEL;
  WRITE    <=   MEM(INTVAL(C),7) after ROM_DEL;
  NNEND    <=   MEM(INTVAL(C),8) after ROM_DEL;
end process ROM;
LOGIC: block
begin
  ZIN     <=  ZEND;
  ZERO    <=  NOR_BITS(BUS_B) after OR_DEL;
  CLEAR <= (Z and ZEND) or (not N and NNEND) or
          (C = "1000") after AND_DEL+OR_DEL;
  PC_OUT  <=  not MDR_OUT after INV_DEL;
  READ    <= MAR_IN;
  COMP    <= N_IN;
  MDR_IN  <= N_IN;
  RDMR <= READ and not MVL4toBIT(ADDRESS(7)) after AND_DEL;
  RDIO <= READ and MVL4toBIT(ADDRESS(7)) after AND_DEL;
  CSMR    <= not MVL4toBIT(ADDRESS(7)) after INV_DEL;
end block LOGIC;
```

Figure 6.20   Microcoded control unit for URISC.

The COUNTER output is used as an address input to ROM. Whenever the COUNTER value changes, a new set of control signals is read out of ROM. In this design, the control signals are used directly from the ROM outputs instead of being loaded into an MIR register. This would only be possible if the driving circuits of the ROM were sufficient to handle the loading of the control signals. In most cases, an MIR or buffer would be necessary (see Problem 6.17). The design is left as an exercise for the reader.

To reduce the number of bits in the ROM word, it was noted that the control signals were not completely independent. For example, by examining the control sequence, one notices that Zin and ZEND are always active at the same time. Therefore, instead of using separate bits in the ROM word for these two control signals, Zin was made equal to ZEND. Other dependencies are indicated by statements in the LOGIC portion of the control unit.

6.4.7   Hardwired Controller for the URISC Processor

The design of a hardwired unit is less structured than the design of a microcoded unit since each hardwired unit requires a custom design. Figure 6.21 shows a VHDL register-level description of the control hardware. A more structured approach would use one flip-flop per control step (see Problem 6.18). For a systematic design approach, see Chapter 8.

PROBLEMS

6.1   Study the scheduling and allocation problems discussed in Chapter 12. Which steps in sched-uling and allocation must use a register-level description, as opposed to an algorithmic repre-sentation?

6.2   Assume that a VHDL data flow description contains the following two concurrent statements in a guarded block:
```
B:block(not CLK'STABLE and CLK = '1')
begin
   X <= guarded ADD(Y,Z);
   R <= guarded ADD(Q,S);
end block B;
```
In the VHDL model, these two statements execute concurrently. However, in real hardware, one could execute them concurrently or in sequence. If done in sequence, they could share the same adder. Assume that signals Y, Z, Q, and S are data inputs and signals X and R are clocked registers. Draw hardware block diagrams for the two approaches, showing muxes, adders, and registers. Compare the concurrent and sequential approach in terms of required hardware and clock cycle. What register-level design activities are we carrying out?

6.3   A data flow VHDL description contains the following concurrent statements:
```
C <= A;
D <= B;
```
Assume A, B, C, and D are all registers and that the transfer between them is via a data bus or buses. A concurrent operation requires two buses; a sequential operation requires only one bus. Find chips that can implement an 8-bit wide bus in a TTL data book. Compare required hard-ware and transfer time for the two cases. What register-level design activities are we carrying out?

6.4   This problem considers some of the essential elements of bus timing. Figure 6.22 shows a block diagram of a bus and its associated control logic. The select signal, SEL, gates one of the

```
-------- The COUNTER ----------------------------
-- The Counter has a synchronous CLEAR ------------
COUNTER: block (PH1= '0' and not PH1'STABLE )
  begin
    C <= guarded "0000" after COUNT_DEL
                when (CLEAR='1'or C="1000") else
         INC_COUNTER(C) after COUNT_DEL;
  end block COUNTER;
  --Hard Wired Control Unit
  --Decoder
  --First Stage Decoding
  ST0 <= not C(2) and not C(1) and not C(0) after AND_DEL;
  ST1 <= not C(2) and not C(1) and C(0) after AND_DEL;
  ST2 <= not C(2) and C(1) and not C(0) after AND_DEL;
  ST3 <= not C(2) and C(1) and  C(0) after AND_DEL;
  ST4 <= C(2) and not C(1) and not C(0) after AND_DEL;
  ST5 <= C(2) and not C(1) and C(0) after AND_DEL;
  ST6 <= C(2) and C(1) and not C(0) after AND_DEL;
  ST7 <= C(2) and C(1) and C(0) after AND_DEL;
  --Second Stage Decoding
  ST07 <= ST0 or ST7 after OR_DEL;
  ST25 <= ST2 or ST5 after OR_DEL;
  ST36 <= ST3 or ST6 after OR_DEL;
  ST57 <= ST5 or ST7 after OR_DEL;
  ST78 <= ST7 or C(3) after OR_DEL;
  --Control Signals
  PC_OUT <= (ST07 or ST36) and not C(3)
            after (OR_DEL + AND_DEL);
  C_IN <= ST36 or ST57 after OR_DEL;
  PC_IN <= ST36 or ST78 after OR_DEL;
  MAR_IN <= not(ST25 or ST78) after (OR_DEL + INV_DEL);
  MDR_OUT <=not PC_OUT after INV_DEL;
  READ <= MAR_IN; COMP <= ST5; N_IN <= ST5; MDR_IN <= ST5;
  WRITE <= ST5; R_IN <= ST2; ZIN <= ST0; ZEND <=ST0;
  NNEND <= ST7;
  --Register and Counter Controls
  ZERO <= NOR_BITS(BUS_B) after OR_DEL;
  CLEAR <= (Z and ZEND) or (not N and NNEND)
            after AND_DEL+OR_DEL;
  RDMR <= READ and not MVL4toBIT(ADDRESS(7)) after AND_DEL;
  RDIO <= READ and MVL4toBIT(ADDRESS(7)) after AND_DEL;
  CSMR <= not MVL4toBIT(ADDRESS(7)) after INV_DEL;
```

Figure 6.21 VHDL description of hardwired control unit for URISC.

data items, DA or DB, onto the bus. Provided that R_EN is 1 when CLK rises, the data is copied off the bus. Signals SEL and R_EN are produced by flip-flops (clocked by CLK), which in turn drive combinational logic blocks (CLs). Draw a timing diagram for the bus showing CLK, SEL, R_EN, and BUS. What is the critical timing requirement? Include the setup time on the register receiving the data.

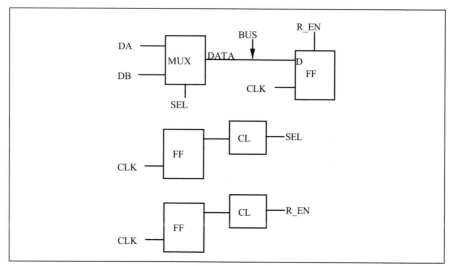

Figure 6.22   Bus timing block diagram.

6.5   Develop a VHDL model for the bus system described in the previous problem. Include assertion checks for bus timing violations. Validate the model through simulation.

6.6   This problem pertains to the register system with timing in Figure 6.6. Using data obtained from device data books for the technology of your choice, e.g., TTL, ECL, or CMOS, determine realistic values for the delay generics in the system. Simulate the system first by using input data that satisfies the timing constraints and second by using input data that violates the timing constraints.

6.7   Redo the design of the register system with timing (Figure 6.6) using the ALU primitive defined in Chapter 4.

6.8   Write a program for the URISC machine that implements unsigned multiplication of two binary numbers.

6.9   It is contended that the URISC machine is universal, i.e., given enough instructions it can implement any computation. Can Boolean logic operations be implemented by the URISC? Support your answer with analysis.

6.10 Develop two complete timing diagrams for step 0 through step 8 for the URISC processor. Figure 6.16 shows a sample diagram for steps 0 through 2. One timing diagram should illustrate a case where the result of subtraction is negative and the other diagram a case where the result is non-negative.

6.11 Determine the maximum allowable clock frequency for the URISC processor.
       a. Develop an expression which relates delay generics in the URISC processor model to the maximum allowable clock frequency for the processor. Do this for both the microprogrammed and the hardwired control unit.
       b. For the technology of your choice, e.g., TTL, ECL, or CMOS, determine realistic values for the delays and calculate the maximum allowable clock frequency for the two cases.
       c. Simulate both models to validate your results.

6.12 Make a list of the pros and cons of microprogrammed vs. hardwired control units.

6.13 Complete the VHDL specifications for the following components in the URISC processor outline of Figure 6.18.

      a. Register R
      d. Register N
      e. Register Z
      f. Memory data register
      g. Memory address register
      h. Two phase clocks.

6.14 Using the results of Problem 6.13, fully test a URISC processor module, such as that outlined in Figure 6.18.

6.15 Design and test the RAM and STROBE modules in the URISC system shown in Figure 6.17.

6.16 Integrate the URISC system components (Figure 6.17) into a fully operational VHDL model of a URISC system. Use the URISC processor module developed in Problem 6.14. Use the RAM and STROBE modules developed in Problem 6.15. Simulate the system using the program shown in Figure 6.11.

6.17 In the microcoded control unit of Figure 6.20, the control signals are provided directly at the ROM outputs. This would only be possible if the ROM outputs could drive the loads of all the control signals. Frequently, the outputs of ROM can drive only one gate load. Therefore, a buffer register is needed between the ROM outputs and the control points in the circuit. Modify the VHDL specification of the microcoded control unit to incorporate a memory instruction register (MIR) to act as a buffer register at the ROM outputs. Discuss the implications on timing and on system clocks.

6.18 Modify the hardwired control unit for the URISC processor (Figure 6.21) to include one flip-flop per control step. Describe how the design process for this form of controller is more structured than the design process for the form shown in Figure 6.21.

C H A P T E R   7

Gate Level and ASIC Library Modeling

In this chapter we discuss how to accurately model designs at the gate level and which design activities are carried out at this level. Accurate gate-level descriptions must model timing accurately, must account for varying circuit output impedance, and must check for error conditions. Such models are very useful for studying hazards and race problems in logic circuits. Simulation of large, gate-level models can be computationally intensive; therefore, it is important that the models be efficient. The configuration management constructs of the VHDL language provide efficient methods for creating and maintaining a design data base related to gate level modeling.

The ASIC design process requires accurate models of the cells in the ASIC library. It is particularly important to accurately model cell delay. In this chapter we discuss various approaches to this problem.

7.1 ACCURATE GATE LEVEL MODELING

In this section we define what constitutes an accurate gate level model. We introduce the required features with a running example that involves a 2-input OR gate. Figure 7.1 shows three architectures of a 2-input OR gate that a person with a rudimentary knowledge of VHDL could write. Architecture DELTA_DEL implements the functionality of the gate, and specifies delta delay. Delta serves as a *unit delay*. Models of this type are useful only for verification of the function of a logic circuit. Architecture FIXED_DEL models the propagation delay of the device, but is not a general model. To change the delay, one must change the delay value in the architecture code. Although this is feasible for a small number of gates, for a large, gate level model, say, thousands of gates, this approach is impractical. Architecture GNR_DEL solves this problem by using a generic delay for the gate. This generic delay is declared in the interface description, and its value is specified when the gate is instantiated.

```
entity OR2 is
  port(I1,I2: in BIT; O: out BIT);
end OR2;

architecture DELTA_DEL of OR2 is
begin
  O <= I1 or I2;
end DELTA_DEL;

architecture FIXED_DEL of OR2 is
begin
  O <= I1 or I2 after 3 ns;
end FIXED_DEL;

entity OR2G is
  generic(DEL: TIME);
  port(I1,I2: in BIT; O: out BIT);
end OR2G;

architecture GNR_DEL of OR2G is
begin
  O <= I1 or I2 after DEL;
end GNR_DEL;
```

Figure 7.1   Three basic OR gate models.

In many instances, designers maintain three sets of delay data for device models: minimum, maximum, and typical. For a purely functional verification, one might also wish to have zero, i.e., delta delay. Thus, it would be desirable to have a delay function which could return one of these four values. In addition, in some modeling situations, one might wish to make this selection globally, i.e., have all device models employ the same delay mode. Figure 7.2 shows how this is done. In a package called TIMING_CONTROL, a type TIMING is declared, which defines the four timing modes. Then, the constant TIMING_SEL is declared with default value of typical (TYP). The function T_CHOICE in the package selects the correct timing mode, based on the value of TIMING_SEL. An OR gate model that employs the function T_CHOICE follows the package code in Figure 7.2. The contents of the package are made visible to all gate models by the statement use work.TIMING_CONTROL.all. To change the global timing option, one must reanalyze the package with a different value for the constant TIMING_SEL. When the value of TIMING_SEL is changed, all models then use the new delay value.

In many modeling situations, one might wish to have different delays for classes of gates or even for individual gates. Later, when we discuss configuration declarations for gate-level modeling, we address this issue.

7.1.1   Asymmetric Timing

The previous delay modeling approaches are useful for basic timing studies; however, examination of a manufacturer's data book reveals that real gate delays are specified in a different fash-

```
package TIMING_CONTROL is
  type TIMING is (MIN,MAX,TYP,DELTA);
  constant TIMING_SEL: TIMING := TYP;
  function T_CHOICE(TIMING_SEL: TIMING; TMIN,TMAX,TTYP: TIME)
    return TIME;
end TIMING_CONTROL;
package body TIMING_CONTROL is
  function T_CHOICE(TIMING_SEL: TIMING; TMIN,TMAX,TTYP: TIME)
    return TIME is
  begin
    case TIMING_SEL is
      when DELTA => return 0 ns;
      when TYP => return TTYP;
      when MAX => return TMAX;
      when MIN => return TMIN;
    end case;
  end T_CHOICE;
end TIMING_CONTROL;
use work.TIMING_CONTROL.all;
entity OR2_TV is
  generic(TMIN,TMAX,TTYP: TIME);
  port(I1,I2: in BIT; O: out BIT);
end OR2_TV;

architecture VAR_T of OR2_TV is
begin
  O <= I1 or I2 after T_CHOICE(TIMING_SEL,TMIN,TMAX,TTYP);
end VAR_T;
```

Figure 7.2   Timing control package.

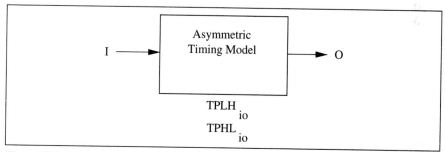

Figure 7.3   Asymmetric timing model.

ion. Figure 7.3 shows a general *asymmetric* timing model. The model has input I and output O, and two time delay parameters, TPLHio and TPHLio, which are defined as:

TPLHio.   The time required for output O to change from low ('0') to high ('1') in response to a change at input I.

TPHLio.   The time required for output O to change from high ('1') to low ('0') in response to a change at input I.

```
entity OR2GV is
  generic(TPLH,TPHL: TIME);
  port(I1,I2: in BIT; O: out BIT);
end OR2GV;

architecture VAR_DEL of OR2GV is
begin
  process(I1,I2)
    variable OR_NEW,OR_OLD:BIT;
  begin
    OR_NEW := I1 or I2;
    if OR_NEW = '1' and OR_OLD = '0' then
       O <= OR_NEW after TPLH;
    elsif OR_NEW = '0' and OR_OLD = '1' then
       O <= OR_NEW after TPHL;
    end if;
    OR_OLD := OR_NEW;
  end process;
end VAR_DEL;
```

Figure 7.4   A variable delay OR gate model.

Note that for TPLHio, the low to high transition occurs on the output O in response to a change on input I. The change on input I could either be from low to high (non-inverting gate) or from high to low (inverting gate). The fact that TPLHio and TPHLio are different for a device is caused primarily by the difference in output source impedance when driving to a high or low value.

Figure 7.4 is a model of a 2-input OR gate, which incorporates the possible difference between TPLH and TPHL. The model uses two internal variables OR_NEW and OR_OLD to store the state of the function for the present and previous process call. These two variables are compared to determine whether to schedule an output value at all (a model efficiency consideration), and if it is scheduled, which delay (TPLH or TPHL) to use.

Delay functions can directly implement delays. Figure 7.5 shows architecture FUNC_DEL of entity OR2GV (originally modeled in Figure 7.4). This architecture invokes the function VAR_DEL from package DELAY to determine whether to use TPLH or TPHL (see Figure 7.7). Architecture FUNC_DEL is, thus, a concise model. However, it is slightly less efficient than architecture VAR_DEL because it will always schedule an output transaction when there is an input change even if the new output is the same as the old output. Architecture VAR_DEL schedules an output transaction only if the old and new output values are different.

7.1.2 Load Sensitive Delay Modeling

In real circuits the delay of the circuit is a strong function of the load on the circuit. A great deal of engineering effort is devoted to measuring this load and adjusting the circuit model delay accordingly. In this section we present a number of approaches to modeling load sensitive delays.

```
use work.DELAY.all;
architecture FUNC_DEL of OR2GV is
begin
  process(I1,I2)
    variable OR_NEW:BIT;
  begin
    OR_NEW := I1 or I2;
    O <= OR_NEW after VAR_DEL(TPLH,TPHL,OR_NEW);
  end process;
end FUNC_DEL;
```

Figure 7.5   Using a function to select delays.

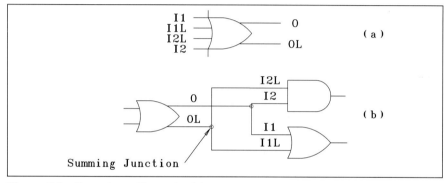

Figure 7.6   Fanout sensitive gate delay.

7.1.2.1   *Fanout Sensitive Delay*

For TTL circuits on printed circuit boards, delay due to input capacitance of the gates is the most significant factor. Usually, delay due to metal runs can be neglected. Later in this chapter, we discuss how the output delay values can be back annotated, based on fanout in a specific circuit. However, here we present a different approach in which the gate model allows each gate instance to measure its own fanout at the start of simulation. This idea was first suggested by Z. Navabi of Northeastern University. Figure 7.6(a) shows the approach. In addition to its normal inputs (I1,I2) and output (O), the model has *load ports* I1L, I2L, and OL, which are used to pass load information between gate models. I1L and I2L are actually outputs; whereas, OL is an input. Figure 7.6(b) shows how the loading information is passed between gates. The OL input of the driving gate is connected to the IL outputs of the driven gates. Each of the driven gates puts out an integer value (e.g., 1) on its IL port. These IL signals are summed at a summing junction and OL reads the value from the summing junction, for use in its delay equation.

To perform the summing, we use a special resolution function that works on integer signal values. Package Delay in Figure 7.7(a) contains this function, S_FANOUT. An integer subtype FINT is defined, which inherits this resolution function. Figure 7.7(b) gives the gate model for a 2-input OR gate. Not shown is a structural model that defines the interconnection of the gates. Note in the interface description that I1L, I2L, and OL are all subtype FINT. Also, two generics are defined: DEL_UNIT and I_LOAD. In the architectural body, the Input Load Process, which

```
package DELAY is
   type INT_VECT is array (NATURAL range <> ) of INTEGER;
   function S_FANOUT(S: INT_VECT) return INTEGER;
   subtype FINT is S_FANOUT INTEGER;
   function VAR_DEL(TPLH,TPHL: TIME; FNEW: BIT) return TIME;
end DELAY;
package body DELAY is
   function S_FANOUT(S: INT_VECT) return INTEGER is
     variable SUM: INTEGER := 0;
   begin
     for I in S'RANGE loop
       SUM := SUM + 1;
     end loop;
     return SUM;
   end S_FANOUT;
   function VAR_DEL(TPLH,TPHL: TIME; FNEW: BIT)
     return TIME is
   begin
     if FNEW = '0' then
       return TPHL;
     else
       return TPLH;
     end if;
   end VAR_DEL;
end DELAY;
use work.DELAY.all;
                                   (a)


entity ORF is
   generic(DEL_UNIT: TIME := 1 ns;I_LOAD: INTEGER := 1);
   port(I1,I2: in BIT; I1L,I2L: out FINT;
        O: out BIT; OL: in FINT);
end ORF;
architecture FANOUT of ORF is
  signal DELAY: TIME:= 0 ns;
begin
   process      ----Input Load Process
   begin
     I1L <= I_LOAD;
     I2L <= I_LOAD;
     DELAY <= DEL_UNIT*OL;
     wait;
   end process;
   O <= I1 or I2 after DELAY;
end FANOUT;
                                   (b)
```

Figure 7.7 Fanout sensitive delay model.

will run only once at the start of simulation, assigns the value of I_LOAD to I1L and I2L and computes the product of DEL_UNIT*OL and assigns it to DELAY, which is the value of gate delay used during simulation. OL is a resolved signal of subtype FINT, and because of the action

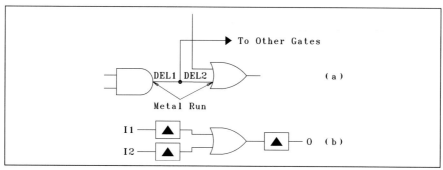

Figure 7.8   Modeling input delay.

of the resolution function, OL will have a value equal to the gate load. Generics DEL_UNIT and I_LOAD have default values of 1 ns and 1, respectively, but of course, when the gate is instantiated, they can be assigned other values.

7.1.2.2   Input Delay Models

Architecture VAR_DEL in Figure 7.4 is a good model of a gate in "isolation." However, it does not consider two factors (1) delay due to interconnect and (2) delay induced by the load driven by the gate being modeled. We consider the interconnect effect first. Figure 7.8(a) shows an OR gate being driven by an AND gate through a metal or polysilicon run. Note that this particular run contributes two delays (1) delay DEL1, which is the delay from the output of the AND gate to the signal fanout point and (2) delay DEL2, which is the delay from the fanout point to the input I2 of the OR gate. A common approach treats the sum (DEL1 + DEL2) as a single input delay for input I2 of the OR gate. If this approach is used for all gates in the system, the effect of interconnect delay for the complete system is correctly modeled. Figure 7.8(b) shows a block diagram of the input delay model for the OR gate. Each input has a delay assigned to it. The OR function itself is considered to have no delay. The propagation delay of the gate is modeled by the output delay.

Figure 7.9 gives VHDL code for the input delay model. Note that a front-end process models the input delay. For each input i, there are generics TPLHi and TPHLi. These generics are given values that define the model input delay when the entities are declared. The values are generated during a back annotation process that measures the length of signal runs in a specific gate layout, and multiplies that length by a pf/unit length factor for the metal or polysilicon material used. The input capacitance of the driven gate is added to this. The resultant value of capacitance is then multiplied by a ns/pf factor peculiar to the circuit being modeled. The values thus calculated for the ith input of a gate become the values assigned to TPLHi and TPHLi in the model.

7.1.2.3   Output Delay Models

Delay models based on output delay are also commonly employed. For example, in the EIA Timing and Environmental Specification a delay equation that uses output loading information is described as:

$$T'_{pd} = (T_{pd} + l_{delay}(C_{load})) * K_v * K_t$$

```
entity OR2GI is
   generic(TPLH1,TPHL1,TPLH2,TPHL2,TPLHO,TPHLO: TIME);
   port(I1,I2: in BIT; O: out BIT);
end OR2GI;

architecture VAR_DEL of OR2GI is
   signal I1_D,I2_D: BIT;
begin
  process(I1,I2)
  begin
    if I1'EVENT then
      if I1 = '0' then
        I1_D <= I1 after TPHL1;
      else
        I1_D <= I1 after TPLH1;
      end if;
    end if;
    if I2'EVENT then
      if I2 = '0' then
        I2_D <= I2 after TPHL2;
      else
        I2_D <= I2 after TPLH2;
      end if;
    end if;
  end process;
  process(I1_D,I2_D)
    variable OR_NEW,OR_OLD:BIT;
  begin
    OR_NEW := I1_D or I2_D;
    if OR_NEW = '1' and OR_OLD = '0' then
      O <= OR_NEW after TPLHO;
    elsif OR_NEW = '0' and OR_OLD = '1' then
      O <= OR_NEW after TPHLO;
    end if;
    OR_OLD := OR_NEW;
  end process;
end VAR_DEL;
```

Figure 7.9   VHDL code for input delay model.

where:

T'_{pd} is the calculated propagation delay in ns used in the model output assignments.

T_{pd} is the manufacturer's propagation delay in ns.

C_{load} is the back annotated load value (calculated capacitive load) in pf.

l_{delay} is the capacitance to time delay function of the technology

and includes C_{ref}—the manufacturer's specified loading for the port type.

K_v is the voltage derating factor.

K_t is the temperature derating factor.

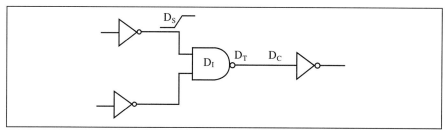

Figure 7.10 Cell delay model.

If one uses a linear model for l_{delay}, then:

$$l_{delay}(C_{load}) = (C_{load} - C_{ref}) * K_c$$

where:

C_{load} is the total load in pf driven by the output.

C_{ref} is the load in pf specified for the data sheet values.

K_c is the straight line capacitive load factor in units of time/pf.

Note that as a result of back annotation, the value of l_{delay} can turn out negative; however, the resultant value for T_{pd} is constrained to be non-negative. The value of C_{load} would be measured in the back annotation process, and would be the sum of effects due to metal or poly runs and the input capacitance for each driven gate.

7.1.3 ASIC Cell Delay Modeling

Figure 7.10 presents a delay model used for ASIC cell modeling. It is a refinement of the output delay model previously discussed that is presented in the *Synopsys Library Compiler Reference Manual*. The calculated delay is the total delay (D_{TOTAL}) between the input pin of a gate and the input pin of the next gate. D_{TOTAL} is given by the following equation:

$$D_{TOTAL} = D_I + D_S + D_T + D_C$$

where:

D_I - Intrinsic delay inherent in the gate and independent of any particular instantiation.

D_S - Slope delay caused by the ramp time of the input signal.

D_T - Transition delay caused by loading of the output pin.

D_C - Connect media delay to an input pin (wire delay).

Intrinsic Delay (D_I). Intrinsic delay of a circuit element (D_I) is the portion of the total delay that is independent of the circuit element's usage. This portion is the fixed (or zero load) delay from the input pin to the output pin of a circuit element. Constant values for the intrinsic delay are stored in the timing group of the load pin as floating-point numbers using these two attributes:

```
intrinsic_rise : float;
intrinsic_fall : float;
```

Slope Delay (D_s). Slope delay is the element delay caused by the slope of the input signal. Normally, library cells are characterized with a nominal ramp time (1.5 ns is common) to determine the delay from input pin to output pin. However, in some technologies, this delay is a strong function of the ramp time. The ramp time is characterized by the slope sensitivity factor. This factor accounts for the time during which the input voltage begins to rise but has not reached the threshold level at which channel conduction begins. The attributes that define it in the timing group of the driving pin are:

```
slope_rise
slope_fall
```

Depending on the direction of input transition, one of these attributes is multiplied by the input ramp time to get the value of D_S.

Transition Delay (D_T).   Transition delay is the time it takes the output pin to change state. It is a function of the source (R_{DRIVER}) resistance of the circuit and the load capacitance. The load capacitance is the sum of the input capacitance of the driven gates (C_{PINS}) and the interconnect capacitance (C_{WIRE}). Thus, the transition delay is given by:

$$D_T = R_{DRIVER}*(C_{WIRE} + C_{PINS})$$

The source resistance is stored in the timing group of the driving pin as floating point numbers using these following two attributes:

```
rise_resistance : float ;
fall_resistance : float ;
```

C_{PINS} can be measured exactly from the known input capacitance of the driven gates.

This model is used by Synopsys for delay calculations directly after synthesis. Thus there is no fabrication information to *back annotate*. Therefore, C_{WIRE} must be estimated. Wire length is computed with the actual number of fanout pins on the net and the fanout_length specifications in the wire_load group. The estimated value is scaled by the capacitance factor, which is defined in the wire_load group as capacitance. If no wire_load group is designated at runtime, the value of C_{WIRE} is 0.

Connect Delay (D_C).   Even after the driving gate output has switched, some time is required for the new value to transit the interconnect and arrive at a gate input. This time is known as connect delay. The interconnect can be modeled as a distributed RC network. Figure 7.11 shows the three generic cases of distributed RC networks used in the Synopsys system.

The *best case* interconnect delay is:

$$D_C(\text{best}) = R_{WIRE} (C_{WIRE} + C_{PIN}) = 0.$$

In this case, R_{WIRE} is considered to be "at the end of the line." Since it is in series with the high input impedance of the gate, it effectively contributes nothing to the RC delay.

The *worst case* interconnect delay is given by:

$$D_C(\text{worst}) = R_{WIRE} (C_{WIRE} + C_{PIN}).$$

Because R_{WIRE} is at "the head of the line," its full effect is felt.

In the *balanced case*, the effect of the wire resistance is on separate equal branches of the interconnect wire.

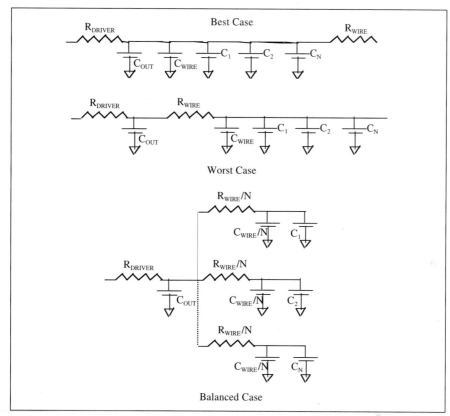

Figure 7.11 Synopsys interconnection delay models.

$$D_C(balance) = R_{WIRE}/N \ (C_{WIRE}/N + C_{PIN})$$

R_{WIRE} is the wire resistance on the net that is determined by the wire_load model. Wire length is computed with a global estimation function whose parameter is the number of fanout pins on the net being estimated. The estimated value is scaled by the resistance factor, which is defined in the wire_load group as resistance.

Example

In an ASIC, an inverter drives a network with a fanout of 4. Each gate input has an input capacitance of 1.5. The wiring capacitance is 4.5. The wire resistance of the network is negligible. The inverter input I is driven by the waveform in Figure 7.12.
The following data is specified for the inverter:

Intrinsic rise: 0.5
Intrinsic fall: 0.4
Rise resistance: 0.15
Fall resistance: 0.07
Slope sensitivity factor: 0.03

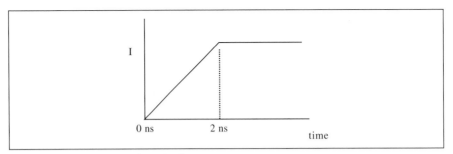

Figure 7.12   Input waveform for inverter.

Compute the total delay from the inverter input to one of the driven gate inputs.

$D_I = 0.4$ns, $D_S = (0.03)(2 \text{ ns}) = 0.06$ ns, $D_T = 0.07(4.5 + 4(1.5))) = 0.735$ ns, $D_C = 0$

Thus, $D_{TOTAL} = 1.195$ ns.

The delay data for library elements is stored in a technology library representation. Figure 7.13 shows such a representation for an inverter from the LSI 10K library. The delay data is given for pin Y (the output) relative to the input pin A. First, the function of the output pin is given (A') followed by the delay data. For this inverter, no slope sensitivity factor is given. A particular factor is assumed and the slope sensitive delay, which is small, is included in the intrinsic delay.

7.1.4 Back Annotation of Delays

When performing post layout simulation for ASIC sell-off (Figure 9.29), it is important to use realistic delays. The ASIC foundry has tools, which can measure wire lengths and thus, compute realistic interconnect delays. These delays must be inserted into the gate-level simulation model for the ASIC. A de facto standard for these delays has evolved called the Standard Delay Format (SDF). It was first developed by Cadence for their tools, but has since been adopted by other vendors. A standardized version was released in 1995 by Open Verilog International (OVI). Version 3.0 of the SDF Specification cites the following features for SDF:

"The Standard Delay Format (SDF) file stores the timing data generated by EDA tools for use at any stage in the design process. The data in the SDF file is represented in a tool-independent way and can include:

- Delays: module path, device, interconnect, and port
- Timing checks: setup, hold, recovery, removal, skew, width, period, and no change
- Timing constraints: path, skew, period, sum, and diff
- Timing environment: intended operating timing environment
- Incremental and absolute delays
- Conditional and unconditional module path delays and timing checks
- Design/instance-specific or type/library-specific data
- Scaling, environmental, and technology parameters

```
   cell(IVDA) {
     area : 1 ;
     pin(A) {
       direction : input ;
       capacitance : 1.3 ;
     }
     pin(Y) {
       direction : output ;
       function : "A'" ;
       timing() {
         intrinsic_rise : 0.49 ;
         intrinsic_fall : 0.37 ;
         rise_resistance : 0.1438 ;
         fall_resistance : 0.0623 ;
         related_pin : "A" ;
       }
       internal_power() {
         related_pin : "A" ;
         rise_power(output_by_cap_and_trans) {
           values("2.5, 3.7, 4.3") ;
         }
         fall_power(output_by_cap_and_trans) {
           values("2.6, 3.8, 4.4") ;
         }
       }
     }
   }
```

Figure 7.13 Technology library representation of LSI 10K inverter.

Throughout a design process, you can use several different SDF files. Some of these files can contain pre-layout timing data. Others can contain path constraints or post-layout timing data."

Figure 7.14 shows the SDF timing for a D flip-flop. First, a "CELL" definition is given. A CELL can be a lower level component, as it is here, or it can be a region of the design. Here, the CELL is the instance top/b/c (hierarchical name) of a D flip-flop. The delays here are ABSO-LUTE values, which would replace the original model values when back annotated. INCRE-MENTAL delays can also be used, in which case, they are added to the original model values when back annotated. The IOPATH specifies a delay from an input (posedge clk) to an output (q). Input posedge clk is a conditional clock input. SDF also contains a more general form, COND(X), in which the specification on a path takes effect when X is true. The delay values are contained in the two triples of minimum, typical, and maximum values. The values are multi-plied by a user-controlled TIMESCALE value, (e.g., 100 ps) to get the actual delays. The first triple is for output rise; the second triple is for output fall. These are pin to pin delays; thus, in terms of the ASIC delay model given above, they would include the effect of intrinsic, slope, and transition delays. The PORT command represents input delays on the CLR input; again, there are separate triples for rise and for fall. This is one of the methods for representing interconnect delay and is used with the model given in Figure 7.9.

```
(CELL
  (CELLTYPE "DFF")
  (INSTANCE top/b/c)
  (DELAY
    (ABSOLUTE
      (IOPATH (posedge clk) q (2:3:4) (5:6:7))
      (PORT clr (2:3:4) (5:6:7))
    )
  )
  (TIMINGCHECK
    (SETUPHOLD d (posedge clk) (3:4:5) (-1:-1:-1))
    (WIDTH clk (4.4:7.5:11.3))
  )
)
```

Figure 7.14   SDF timing for a D flip-flop.

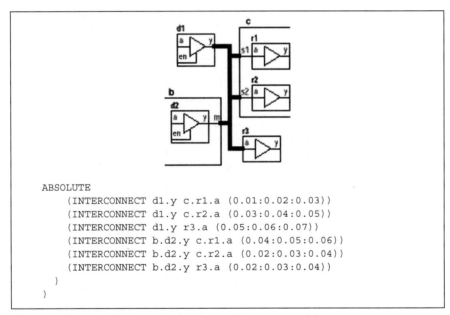

```
ABSOLUTE
    (INTERCONNECT d1.y c.r1.a (0.01:0.02:0.03))
    (INTERCONNECT d1.y c.r2.a (0.03:0.04:0.05))
    (INTERCONNECT d1.y r3.a (0.05:0.06:0.07))
    (INTERCONNECT b.d2.y c.r1.a (0.04:0.05:0.06))
    (INTERCONNECT b.d2.y c.r2.a (0.02:0.03:0.04))
    (INTERCONNECT b.d2.y r3.a (0.02:0.03:0.04))
  )
)
```

Figure 7.15   Specification of point-to-point interconnect delays.

Alternatively, point-to-point interconnect delays can be represented with the INTERCON-NECT construct shown in Figure 7.15. Note that there are three levels of hierarchy involved and the delays are independent of rise or fall. Again the listed delay values are multiplied by the TIMESCALE to get the actual delay values. Delays measured in this fashion correspond to the connect delay in the ASIC delay model. For more information on SDF, the reader is referred to the OVI web site.

7.1.5 VITAL : A Standard for the Generation of VHDL Models of Library Elements

As the Cadence Corporation developed both SDF and Verilog, it was natural they would lead the way in developing Verilog models of ASIC library elements. The implementation of these models in their simulator proved to be very fast; thus, a Verilog description of a synthesized ASIC was an efficient representation. Tool vendors were slow in developing efficient VHDL models of ASIC library elements. Thus, in many cases, one had the interesting situation where the ASIC behavioral model that was synthesized was written in VHDL because of its more powerful modeling features, but the structural model of the synthesized circuit would be generated in Verilog to provide for efficient post-synthesis simulation.

The IEEE responded to this situation by forming the VHDL Initiative Towards ASIC Libraries (VITAL) committee. From the VITAL LRM: "The objectives of the VITAL initiative can be summed up in one sentence: Accelerate the development of sign-off quality ASIC macrocell simulation libraries written in VHDL by leveraging existing methodologies of model development. … VITAL obtains its leverage from (a) the Standard Delay Format, in addition to (b) certain contributed elements of the Std_Timing package of the VHDL Technology Group and specialized timing and behavioral techniques provided by Ryan & Ryan, and (c) the existence of numerous ASIC libraries and tools developed using SDF timing implementations. VITAL used many ideas about the primitives and timing models described in the Verilog language. In particular, Verilog's support for representing truth/state tables and its mechanism for performing pin-to-pin delay selection were found to be highly useful." The result of this effort was a set of standardized VHDL ASIC cell model generation techniques. They are supported by two IEEE packages: VITAL_timing and VITAL_primitives.

In the vendor's system, the generation of these models is called library compilation. We illustrate this with an example from the *Synopsys Library Compiler*. Figures 7.16, 7.17, and 7.18 show the library source code for a 4-input ONAND gate and the resultant compiled VITAL model.

Note that there are two sets of delays in the generic declaration section: propagation delays (e.g., tpd_A_Y) and input delays (e.g., tipd_A). Thus, this model represents interconnect delay using an input delay model. The propagation delays represent intrinsic delays, which include the slope sensitive delay. Default values are given for these, which can be altered during back annotation. The effect of transition delay can be included in the back-annotated input delay.

Internal signals are declared to represent the wires in the input delay model, e.g., A_ipd. The code section of the architecture is divided into three main sections:

1. Input path delays: The function VitalWireDelay computes the proper delay, based on the new signal value, the old signal value, and the value of the wire delay (e.g., tipd_A) and assigns the new value to the wire signal (e.g., A_ipd) after the computed delay. The delay is transport.
2. Functionality section: The function of the ONAND gate is evaluated using the normal Boolean operators for STD_LOGIC. However, for more complicated models, such as, flip-flops, primitives (e.g., state table) from the VITAL_primitive package are used. These primitives are optimized for high performance.

```
Combinational Cell ONAND
/* Library Source Code */
cell(ONAND)
 {area : 9.0000 ;
  pin(Y) {direction : output ;
          capacitance : 0.5000 ;
          max_fanout : 15.0000 ;
          function : "!((A + B + C) D)" ;
          timing() {intrinsic_rise : 0.1234 ;
                    rise_resistance : 0.1500 ;
                    intrinsic_fall : 0.4500 ;
                    fall_resistance : 0.1600 ;
                    related_pin : "A" ;}
          timing() {intrinsic_rise : 0.1234 ;
                    rise_resistance : 0.1500 ;
                    intrinsic_fall : 0.4500 ;
                    fall_resistance : 0.1600 ;
                    related_pin : "B" ;}
          timing() {intrinsic_rise : 0.1234 ;
                    rise_resistance : 0.1500 ;
                    intrinsic_fall : 0.4500 ;
                    fall_resistance : 0.1600 ;
                    related_pin : "C" ;}
          timing() {intrinsic_rise : 0.1800 ;
                    rise_resistance : 0.0500 ;
                    intrinsic_fall : 0.5300 ;
                    fall_resistance : 0.0200 ;
                    related_pin : "D" ;}
          }
  pin(A) {direction : input ;
          capacitance : 1.0000 ;/
          fanout_load : 1.0000 ;}
  pin(B) {direction : input ;
          capacitance : 1.0000 ;
          fanout_load : 1.0000 ;}
  pin(C) {direction : input ;
          capacitance : 1.0000 ;
          fanout_load : 1.0000 ;}
  pin(D) {direction : input ;
          capacitance : 1.0000 ;
          fanout_load : 1.0000 ;}
 }
```

Figure 7.16   VITAL source code for combinational cell ONAND.

3. Path delay section: The output delay is selected, based on the time of input signal changes with adjustments made in the case of glitches.

To fully understand how these three functions are performed, the reader should consult the package VITAL_timing for details.

```
/* VITAL Model Output Code for Cell ONAND */
----- CELL ONAND -----
library IEEE;
use IEEE.STD_LOGIC_1164.all;
library IEEE;
use IEEE.VITAL_Timing.all;
-- entity declaration --
entity ONAND is
  generic(
    TimingChecksOn: Boolean := True;
    InstancePath: STRING := "*";
    Xon: Boolean := False;
    MsgOn: Boolean := True;
    tpd_A_Y    : VitalDelayType01:=(0.123 ns, .450 ns);
    tpd_B_Y    : VitalDelayType01 := (0.123 ns, 0.450 ns);
    tpd_C_Y    : VitalDelayType01 := (0.123 ns, 0.450 ns);
    tpd_D_Y    : VitalDelayType01 := (0.180 ns, 0.530 ns);
    tipd_A     : VitalDelayType01 := (0.000 ns, 0.000 ns);
    tipd_B     : VitalDelayType01 := (0.000 ns, 0.000 ns);
    tipd_C     : VitalDelayType01 := (0.000 ns, 0.000 ns);
    tipd_D     : VitalDelayType01 := (0.000 ns,0.000 ns));
  port(
    Y                          : out   STD_LOGIC;
    A                          : in    STD_LOGIC;
    B                          : in    STD_LOGIC;
    C                          : in    STD_LOGIC;
    D                          : in    STD_LOGIC);
  attribute VITAL_LEVEL0 of ONAND : entity is TRUE;
end ONAND;
```

Figure 7.17   Entity in VITAL model output code for cell ONAND.

7.2 ERROR CHECKING

We wish to model both good and faulty behavior. When modeling faulty behavior, it is useful for the modeler to be alerted when an error occurs. We have previously discussed the use of VHDL assertions for this purpose. Here we discuss how assertions can be used to build error checks into gate-level models. We will focus on checking three different types of error conditions (1) incorrect data values on a given input, (2) incorrect input combinations, and (3) incorrect timing on a signal. Checking these three different types of error conditions is illustrated in Figure 7.19.

1. **Incorrect data values on a given input:** Consider the case where the input signal values are a multivalued logic type, such as MVL4. Then, for certain inputs, the values '0' and '1' are the "normal" values while the occurrence of 'X' and 'Z' are not and indicate a fault condition. In the RS flip-flop model shown in Figure 7.19, occurrence of an 'X' or 'Z' on either the R or S inputs will be detected and an error message will be printed.

2. **Incorrect input combinations:** In some cases, certain input combinations should never occur. If they do, an error condition should be reported. For the RS flip-flop, such a condition is R = S = '1'. The RS flip-flop model in Figure 7.19 implies an RS flip-flop

```
library IEEE;
use IEEE.VITAL_Primitives.all;
library LIBVUOF;
use LIBVUOF.VTABLES.all;
architecture VITAL of ONAND is
  attribute VITAL_LEVEL1 of VITAL : architecture is TRUE;
  SIGNAL A_ipd  : STD_ULOGIC := 'X';
  SIGNAL B_ipd  : STD_ULOGIC := 'X';
  SIGNAL C_ipd  : STD_ULOGIC := 'X';
  SIGNAL D_ipd  : STD_ULOGIC := 'X';
begin
--  INPUT PATH DELAYS
  WireDelay : block
  begin
    VitalWireDelay (A_ipd, A, tipd_A);
    VitalWireDelay (B_ipd, B, tipd_B);
    VitalWireDelay (C_ipd, C, tipd_C);
    VitalWireDelay (D_ipd, D, tipd_D);
  end block;
--  BEHAVIOR SECTION
  VITALBehavior : process (A_ipd, B_ipd, C_ipd, D_ipd)
    -- functionality results
    VARIABLE Results: STD_LOGIC_VECTOR(1 to 1):=(others => 'X');
    ALIAS Y_zd : STD_LOGIC is Results(1);
    -- output glitch detection variables
    VARIABLE Y_GlitchData : VitalGlitchDataType;
  begin
    --  Functionality Section
    Y_zd := (NOT ((D_ipd) AND ((B_ipd) OR (A_ipd) OR C_ipd))));
    --  Path Delay Section
    VitalPathDelay01 (
      OutSignal => Y,
      GlitchData => Y_GlitchData,
      OutSignalName => "Y",
      OutTemp => Y_zd,
      Paths => (0 => (A_ipd'last_event, tpd_A_Y, TRUE),
                1 => (B_ipd'last_event, tpd_B_Y, TRUE),
                2 => (C_ipd'last_event, tpd_C_Y, TRUE),
                3 => (D_ipd'last_event, tpd_D_Y, TRUE)),
      Mode => OnDetect,
      Xon => Xon,
      MsgOn => MsgOn,
      MsgSeverity => WARNING);
  end process;
end VITAL;
configuration CFG_ONAND_VITAL of ONAND is
  for VITAL
  end for;
end CFG_ONAND_VITAL;
```

Figure 7.18 Architecture in VITAL model output code for cell ONAND.

```
   use work.SYSTEM_4.all;
   entity FFRS is
     generic(FFDEL,SPIKE_WIDTH: TIME);
     port(R,S: in MVL4; Q,QN: out MVL4);
   end FFRS;
   architecture BEHAV of FFRS is
   begin
     process(R,S)
       variable R_LAST_EVT, S_LAST_EVT: TIME := 0 ns;
       variable BOTH: BOOLEAN:= FALSE;
     begin
     -----------Check for X's and Z's on inputs
     assert not(R='X') report "X  on R" severity WARNING;
     assert not(R='Z') report "Z  on R" severity WARNING;
     assert not(S='X') report "X  on S" severity WARNING;
     assert not(S='Z') report "Z  on S" severity WARNING;
     -------Spike Detection
     assert (NOW = 0 NS) or ((NOW - R_LAST_EVT) > SPIKE_WIDTH)
       report "Spike On R" severity WARNING;
     R_LAST_EVT := NOW;
     assert (NOW = 0 NS) or ((NOW - S_LAST_EVT) > SPIKE_WIDTH)
       report "Spike On S" severity WARNING;
     S_LAST_EVT := NOW;
 if R = '0' and S ='1' then
       Q <= '1' after FFDEL;
       QN <= '0' after FFDEL;
      end if;
     if R = '1' and S='0' then
       Q <= '0' after FFDEL;
       QN <= '1' after FFDEL;
      end if;
     assert not(R = '1' and S = '1') --Check for R and S
       report "R and S Both 1"       --Both  '1'
       severity WARNING
     if R = '1' and S = '1' then
       Q <= '0' after FFDEL;
       QN <= '0' after FFDEL;
       BOTH := TRUE;
      end if;
     if R /= '1' and S /= '1' and BOTH then  --Previous
       Q <= '0' after FFDEL;        -- inputs were R=S='1'
       QN <= '1' after FFDEL;
       BOTH := FALSE;
      end if;
    end process;
   end BEHAV;
```

Figure 7.19   RS flip-flop model.

implemented with two cross-coupled NOR gates. Both flip-flop outputs are forced to a
'0' when R=S='1', which is actually what happens in the real circuit. If the inputs are

then switched from R = S = '1', to R = S = '0', the final state of the flip-flop will depend on the relative path delays internal to the flip-flop. In writing the model, one must decide what the final state will be if the inputs both switch "simultaneously" from '1' to '0'. In our model, we chose to reset the flip-flop. Note that the Boolean variable BOTH is used to indicate that the previous input conditions had been R = S = '1'. In a real circuit, one input would lag or lead the other, and also, with R = S = '0' and Q = QN = '0', the flip-flop is not in a stable state. Noise and minute differences in gate delays will drive the flip-flop to one state or another. However, in a model of this type, one cannot account for these low-level effects. Other approaches are to detect the occurrence of the R = S = '1', print the error message, and not change the state of the flip-flop at all, or to set the flip-flop state to 'X'. This would result in a simpler model, but it would not be as accurate.

3. **Incorrect timing on a signal:** In the case of the RS flip-flop model, we are interested in checking for spikes on an input. These are caused by hazards in the driving circuits. Detecting their occurrence at the driven circuit is an efficient method. The detection is carried out by storing the "time of last event" for both R and S. Whenever the process modeling the flip-flop is called, the current simulation time, as measured by the function NOW, is subtracted from these times. The resultant time value is then compared with the spike width specification to see if an error has occurred. Note that the flip-flop model itself will react to spikes on R or S no matter how narrow. An alternate approach is to leave the state of the flip-flop undisturbed when spikes occur.

7.3 MULTIVALUED LOGIC FOR GATE LEVEL MODELING

A basic question is, how many strength values should there be? When modeling at the circuit level using SPICE, an essentially infinite number of strengths are possible. For digital modeling, however, this is obviously not required. In Chapter 5 we concentrated on a system that had two strengths: strong and weak. These strengths are utilized in the MVL 7 and the IEEE nine-valued system. As we seek to achieve more accurate modeling of gate and lower level devices, we will see that there may be a need for more values.

7.3.1 Additional Values for MOS Design

The seventh value of system MVL7 (Z) was obtained by collapsing Z0, Z1, and ZX into Z. The rationale was that the impedance was so high that the underlying state was not significant. For modeling at the gate level, this assumption is probably justified. However, increasingly, modeling is being done at the switch level, which is a hybrid level between the circuit and gate levels. This level of modeling has been necessitated by the MOS technology used in VLSI circuit design. When modeling at the switch level, one augments the basic logic of modeling primitives with greater control over signal source impedance. Consider the situation shown in Figure 7.20, which shows an NMOS inverter driven by a transmission gate. When CON = '1', the gate is on, and point I drives point Q. However, if CON goes to '0', point Q remains charged at the 1 level for a relatively long period of time, milliseconds in most cases. This is because the source impedance of the signal Q is very high. If we were using MVL7, we would say that Q has the

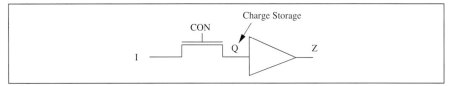

Figure 7.20   Charge storage in MOS circuits.

Z0	Z1	ZX
R0	R1	RX
F0	F1	FX

Figure 7.21   Nine-valued system.

value 'Z'. But what about its state? Charge storage is the major memory storage mechanism in MOS circuits, and we must be able to model it if we want to perform switch-level modeling.

The solution is to have three, high-impedance states: Z0, Z1, and ZX. The result is a nine-valued system built from three strengths (strong, weak, and high impedance) and three states (0,1,X). Figure 7.21 shows a possible value labeling for such a system and Figure 7.22 shows the value lattice for the nine valued system. Signal resolution (R) corresponds to the least upper bound (lub) in the lattice, e.g., (R0)R(R1)=RX. To conform to more or less standard terminology we have used the term forcing (F) for strong, and the term resistive (R) for weak.

7.3.2   Generalized State/Strength Model

As one tries to model more switch-level effects, more strengths may be required. We discuss now a generalization of the state/strength model. It consists of the states X, 0, and 1, and an arbitrary number of strengths, which are represented by a natural number N. 0 is the greatest strength, with strengths decreasing monotonically with N. A value in the system is a 2-tuple of the form (Si,N), where Si is X, 0, or 1, and N is the strength value. The value lattice for this system is shown in Figure 7.23. Note: The greatest element in the lattice is (X,0), which is the least upper bound of (1,0) and (0,0), i.e., the result of combining the strongest 1 with the strongest 0. The least element in the lattice is the value (X,inf), which has the weakest strength, i.e., N = inf.

This structure was applied by Smith and Acosta, who were then at MCC, to model MOS switch networks. Figure 7.24 shows the basic situation. Signals are passed through a series of MOS switches without signal regeneration. The strength of the signal weakens as it passes through each switch, generating the requirement for a large number of signal strengths.

In the MCC system, the number of signal strengths is selectable up to a maximum value of 255. The value selected is dictated by the longest path through the network. The MCC switch model is bi-directional in that signal flow is modeled from source to drain and vice versa. Their network model is a network of switches interconnected at nodes. MCC implemented the switch model and the node model to be discussed later as native system primitives for performance reasons, but they also developed equivalent VHDL code shown in Figures 7.25, 7.26, and 7.27. Figure 7.25 shows package BIT_TYPES, which is employed by the switch and node models. Each switch model receives source and drain signals, which consist of a 2-tuple, whose value consists

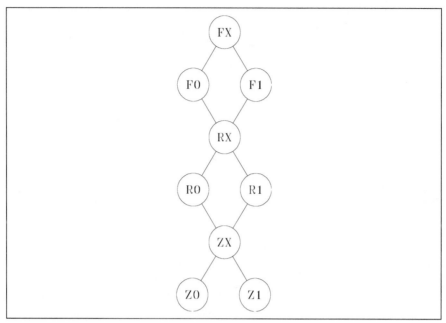

Figure 7.22   Value lattice for nine-valued system.

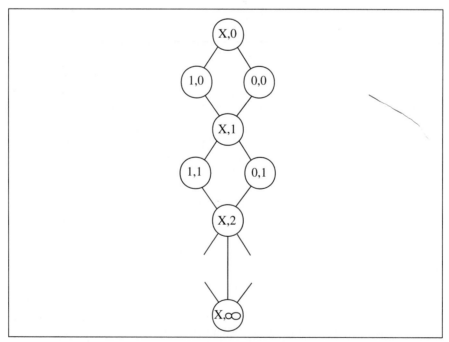

Figure 7.23   Generalized value lattice for a strength/state system.

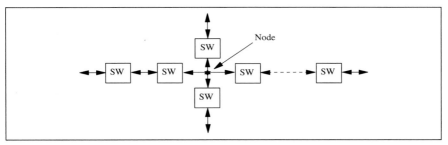

Figure 7.24   MOS switch network.

```
package BIT_TYPES is
  type BIT_CODE is ('0', '1', 'X', 'Z');
  type BIT_CODE_VECTOR is array (Natural range <>)
      of BIT_CODE;
  type WEAKNESS_VALUE is range 0 to 255;
  type WEAKNESS_VALUE_VECTOR is array (Natural range <>)
      of WEAKNESS_VALUE;
  type SWITCH_STATE is (mos_turn_off, mos_turn_on,
                        mos_turn_x);
  procedure LATTICE_LUB(NEW_VALUE: inout BIT_CODE;
                        NEW_WEAKNESS: inout WEAKNESS_VALUE;
                        VALUE: in BIT_CODE;
                        WEAKNESS: in WEAKNESS_VALUE);
  function NEW_NODE_WINS(NEW_VALUE: BIT_CODE;
                        NEW_WEAKNESS: WEAKNESS_VALUE;
                        OLD_VALUE: BIT_CODE;
                        OLD_WEAKNESS: WEAKNESS_VALUE)
      return Boolean;
end BIT_TYPES;
```

Figure 7.25   Switch network package.

of a state of type BIT_CODE and a strength of type WEAKNESS_VALUE. The node models will receive inputs, which are of type BIT_CODE_VECTOR and type WEAKNESS_VALUE_VECTOR. An enumeration type, SWITCH_STATE, is used to represent the state of a switch. The functions LATTICE_LUB and NEW_NODE_WINS are also used by the node models. Only their interface is shown, but if one understands the value lattice, their operation is evident.

Figure 7.26 shows the model NSWITCH. The model reacts to a change in its GATE input to set its state appropriately. Delta time later, when signal STATE changes, the appropriate action is taken:

1. State: MOS_TURN_OFF: Source and drain receive the value 'X', and their weaknesses are set to the greatest weakness (255).
2. State: MOS_TURN_ON: Source and drain receive each other's value, and their weaknesses are increased by the channel degradation of the switch (default value of 1).

```
use work.BIT_TYPES.all;
entity NSWITCH is
  generic (CHANNEL_DEGRADATION: WEAKNESS_VALUE := 1);
  port (GATE: in BIT_CODE;
        SOURCE, DRAIN: inout BIT_CODE;
        SOURCE_WEAKNESS, DRAIN_WEAKNESS: inout WEAKNESS_VALUE);
end NSWITCH;

architecture ARCH of NSWITCH is
  signal STATE: SWITCH_STATE;
begin
  process (GATE, SOURCE, DRAIN, STATE)
  begin
    if (GATE'event) then
      case (GATE) is
        when '0'    => STATE <= mos_turn_off;
        when '1'    => STATE <= mos_turn_on;
        when others => STATE <= mos_turn_x;
      end case;
    end if;
    case (STATE) is
      when mos_turn_off =>  -- infinite weakness
        DRAIN              <= 'X';
        DRAIN_WEAKNESS  <= WEAKNESS_VALUE'high;
        SOURCE             <= 'X';
        SOURCE_WEAKNESS <= WEAKNESS_VALUE'high;
      when mos_turn_on  =>
        DRAIN           <= SOURCE;
    DRAIN_WEAKNESS <= SOURCE_WEAKNESS + CHANNEL_DEGRADATION;
        SOURCE          <= DRAIN;
    SOURCE_WEAKNESS <= DRAIN_WEAKNESS + CHANNEL_DEGRADATION;
      when mos_turn_x   =>
        DRAIN        <= 'X';
    DRAIN_WEAKNESS <= SOURCE_WEAKNESS + CHANNEL_DEGRADATION;
        SOURCE       <= 'X';
    SOURCE_WEAKNESS <= DRAIN_WEAKNESS + CHANNEL_DEGRADATION;
    end case;
  end process;
end ARCH;
```

Figure 7.26 Switch model.

3. State: MOS_TURN_X: Source and drain receive the value 'X', and their weaknesses are increased by the channel degradation of the switch.

Figure 7.27 shows the node model. Nodes are formed when the outputs of switches (sources or drains) are tied together. Entity IDEAL_NODE, therefore, receives vectors of strengths and values representing the outputs of the switches connected to the node as inputs. It then:

1. Determines the least upper bound of all its inputs.
2. Then, if the new value of the node is stronger than the old value of the node, all input values that are weaker than the new node value are set equal to the new value.

```
use work.BIT_TYPES.all;
entity IDEAL_NODE is
  port (STATES: inout BIT_CODE_VECTOR;
        STRENGTHS: inout WEAKNESS_VALUE_VECTOR);
end IDEAL_NODE;

architecture ARCH of IDEAL_NODE is
begin
  process (STATES, STRENGTHS)
    variable NODE_VALUE, RESULT_VALUE: BIT_CODE;
    variable NODE_WEAKNESS, RESULT_WEAKNESS: WEAKNESS_VALUE;
  begin
    RESULT_VALUE := 'X'; -- infinite weakness
    RESULT_WEAKNESS := WEAKNESS_VALUE'high;
    for i in STATES'range loop
      LATTICE_LUB(RESULT_VALUE, RESULT_WEAKNESS, STATES(i),
                  STRENGTHS(i));
    end loop;
    if (NEW_NODE_WINS(RESULT_VALUE, RESULT_WEAKNESS,
        NODE_VALUE, NODE_WEAKNESS)) then
      for i in STATES'range loop
        if (NEW_NODE_WINS(RESULT_VALUE, RESULT_WEAKNESS,
            STATES(i), STRENGTHS(i))) then
          STATES(i)    <= RESULT_VALUE;
          STRENGTHS(i) <= RESULT_WEAKNESS;
        end if;
      end loop;
    end if;
    NODE_VALUE := RESULT_VALUE;
    NODE_WEAKNESS := RESULT_WEAKNESS;
  end process;
end ARCH;
```

Figure 7.27   Node model.

Interval Length	Interval	Symbol
2	1H	P
2	HZ	R
2	ZL	F
2	L0	N
3	1HZ	T
3	HZL	W
3	ZL0	B
4	1HZL	U
4	HZL0	D
5	1HZL0	X

Figure 7.28   HILO logic intervals.

Using this approach, MCC has modeled networks of approximately 1,000 switches. Modeling such a network would be impractical at the circuit level using SPICE.

7.3.3 Interval Logic

Another effective way to develop higher-valued logic systems is based on the concept of *interval logic*. In this approach, one starts with a *basic value set*, ordered in a particular fashion, based on state and strength. Other logic values are then defined as intervals in the basic value set. As an example, consider the value system used in the original HILO system. The basic value set consists of the values {'1','H','Z','L','0'}, where the values have the same meaning as defined for MVL7. The basic value set is divided into a high-value subset and a low-value subset by the value 'Z'. Thus, starting at the leftmost value '1', one decreases in strength through the value 'H' to the value 'Z', which has no state information. When moving to the right in the low subset, the strength increases through the value 'L' to '0'. For this basic value set, one can define the following numbers of intervals: four intervals of length 2, three intervals of length 3, two intervals of length 4, and one interval of length 5. Figure 7.28 gives the definitions of these intervals and their corresponding value symbols.

The type definition for this system would be:

```
type HILO_BIT is(  'X',
                   'U','D'
                  'T','W','B'
                 'P','R','F','N'
                '1','H','Z','L','0');
```

The pyramidal shape of the type declaration implies the way the values are formed.

In summary, we started with a basic value set containing strength and state information, and formed intervals, which represent a range of values. As the length of the range increases, the uncertainty as to the value of the signal holding this value increases. The biggest range, representing the greatest uncertainty, is the value X, i.e., the signal could be equal to any of the basic values. The purpose of establishing a value system of this type is to allow the resolution function to produce resolved values that are less pessimistic than would result from a system with a smaller number of values. Figure 7.29 shows the HILO resolution function.

The concept of interval logic can be generalized. Consider a system with a basic value set {V1,V2, ... ,Vn). One can then generate $n-i+1$ intervals of length of i. Moreover, the total number of logic values in the system is:

$$\sum_{i=1}^{n} i = \frac{n(n+1)}{2}$$

For system HILO, this equation evaluates to 15.

7.3.4 Vantage System

We now consider an interval logic system developed by Vantage Analysis. It uses the following basic value set:

{F0,R0,W0,Z0,D,Z1,R1,F1}

	X	1	0	U	D	P	N	T	B	W	H	L	R	F	Z
X	X	X	X	X	X	X	X	X	X	X	X	X	X	X	X
1	X	1	X	1	X	1	X	1	X	1	1	1	1	1	1
0	X	X	0	X	0	X	0	X	0	0	0	0	0	0	0
U	X	1	X	U	X	U	X	U	X	U	U	U	U	U	U
D	X	X	0	X	D	X	D	X	D	D	D	D	D	D	D
P	X	1	X	U	X	P	X	P	X	U	P	U	P	U	P
N	X	X	0	X	D	X	N	X	N	D	D	N	D	N	N
T	X	1	X	U	X	P	X	T	X	U	P	U	T	U	T
B	X	X	0	X	D	X	N	X	B	D	D	N	D	B	B
W	X	1	0	U	D	U	D	U	D	W	W	W	W	W	W
H	X	1	0	U	D	P	D	P	D	W	H	W	B	W	B
L	X	1	0	U	D	U	N	U	N	W	W	L	W	L	L
R	X	1	0	U	D	P	D	T	D	W	H	W	R	W	R
F	X	1	0	U	D	U	N	U	B	W	W	L	W	F	F
Z	X	1	0	U	D	P	N	T	B	W	H	L	R	F	Z

Figure 7.29 HILO resolution function.

The basic value set has four strengths: F(forcing), R(resistive), W(weak), and Z(high imped-ance). The value D represents a totally disconnected node with no charge storage. From this basic value set, one can generate a 45-valued system. To this Vantage adds the value U, which represents an uninitialized signal. Thus, we have 46 values. Figure 7.30 shows the definition of the values for this system. Value symbols are selected to reflect the end points of the interval. If the 0/1 boundary is crossed, an X is appended. Thus, the value FZX represents the interval from F0(forcing 0) to Z1(high impedance 1).

A basic advantage of systems of this type is that when resolving signals, the results of the resolution can be less pessimistic than with lower valued systems. Consider the circuit shown in Figure 7.31, where two bus drivers have their outputs tied together. Figure 7.31(a) shows a situa-tion where MVL7 is used. The upper driver has its enable tied to X. For normal CMOS or TTL technology (strength parameter = XO1, as explained below), the output will be X. The lower driver's output is 0. X resolved with 0 yields X. However, since both buffers have an input of 0, this is a conservative result: Regardless of whether the upper buffer is enabled, the resolved value should be '0'. Figure 7.31(b) shows the same situation where the 46-valued system is used. Here, the output of the upper buffer will be FZX, which implies that the output must either be a 0 of varying strength, a 1 of high impedance strength, or disconnected. The value 1 of high impedance accounts for the possibility that the previously stored value on the output could have been a Z1. In this case, the resolution function will logically resolve the FZX and the F0 to F0, which is the correct and less pessimistic result. The resolution function for the Vantage system is a 46 × 46 table that is given in David Coelho's book, *The VHDL Handbook*.

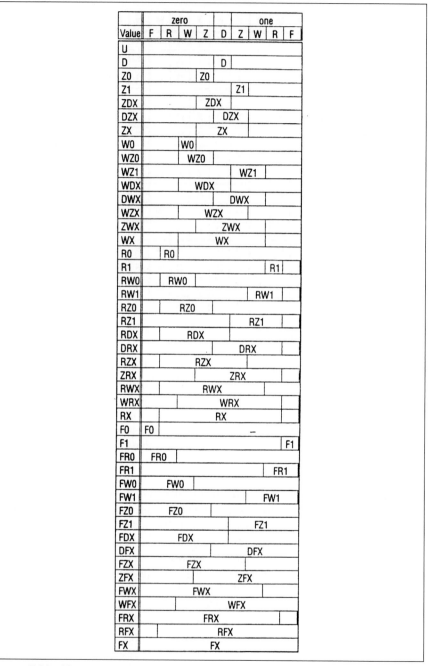

Figure 7.30   Vantage 46-valued system.

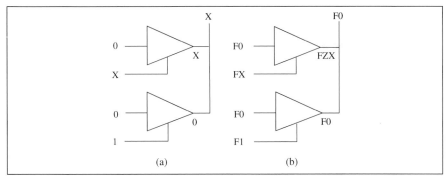

Figure 7.31   Advantage of the Vantage logic system.

```
-- truth table for "or" function
constant OR_TABLE : STDLOGIC_TABLE := (
---------------------------------------------------------
--|  U    X    0    1    Z    W    L    H    -          |  |
---------------------------------------------------------
  ( 'U', 'U', 'U', '1', 'U', 'U', 'U', '1', 'U' ),  -- | U |
  ( 'U', 'X', 'X', '1', 'X', 'X', 'X', '1', 'X' ),  -- | X |
  ( 'U', 'X', '0', '1', 'X', 'X', '0', '1', 'X' ),  -- | 0 |
  ( '1', '1', '1', '1', '1', '1', '1', '1', '1' ),  -- | 1 |
  ( 'U', 'X', 'X', '1', 'X', 'X', 'X', '1', 'X' ),  -- | Z |
  ( 'U', 'X', 'X', '1', 'X', 'X', 'X', '1', 'X' ),  -- | W |
  ( 'U', 'X', '0', '1', 'X', 'X', '0', '1', 'X' ),  -- | L |
  ( '1', '1', '1', '1', '1', '1', '1', '1', '1' ),  -- | H |
  ( 'U', 'X', 'X', '1', 'X', 'X', 'X', '1', 'X' ));  -- | - |
```

Figure 7.32   Truth table for OR function.

7.3.5 Multivalued Gate-Level Models

In many cases designers wish to simulate structural models of gate-level circuits they have designed. In this section we present gate-level models, which use the IEEE standard logic system. We will use the OR gate as an example. Figure 7.32 shows the truth table for the OR function. It is from the STD_LOGIC_1164 package. Note: For this function, A OR 1 = 1 OR A = A OR H = H OR A = 1, for all A of type STD_LOGIC, i.e. 1 and H are equivalent and dominate.

Figure 7.33 gives the VHDL description for the OR Gate entity, from the Synopsys package std_logic_entities. Note that it has four generics (1) N, the number of gate inputs, i.e., it is a general OR model, (2) TLH, (3) THL, and (4) strength, which determines the gate output values.

The strength parameter is defined as follows:

```
type STRENGTH is (strn_X01, strn_X0H, strn_XL1, strn_X0Z,
                  strn_XZ1, strn_WLH, strn_WLZ, strn_WZH,
                  strn_W0H, strn_WL1);
```

This strength parameter is not to be confused with the basic strengths of the value system, i.e., weak and strong. Rather, it is a *generic strength parameter* which is used to determine the

```
----------------------------------------------------------
--Primitive name: ORGATE
--Purpose: An OR gate for multiple value logic STD_LOGIC,
-- N inputs, 1 output.
--(see package IEEE.STD_LOGIC_1164 for truth table)
----------------------------------------------------------
library IEEE;
use IEEE.STD_LOGIC_1164.all;
use IEEE.STD_LOGIC_MISC.all;
entity ORGATE is
  generic (N:    Positive := 2;-- number of inputs
           tLH: Time := 0 ns; -- rise inertial delay
           tHL: Time := 0 ns; -- fall inertial delay
           STRN: STRENGTH := STRN_X01);-- output strength
  port (INPUT: in STD_LOGIC_VECTOR (1 to N);-- inputs
        OUTPUT: out STD_LOGIC);  -- output
end ORGATE;
architecture A of ORGATE is
  signal CURRENTSTATE: STD_LOGIC := 'U';
  subtype TWOBIT is STD_LOGIC_VECTOR (0 to 1);
begin
  P: process
    variable NEXTSTATE: STD_LOGIC;
    variable DELTA: Time;
    variable NEXT_ASSIGN_VAL: STD_LOGIC;
  begin
  -- evaluate logical function
    NEXTSTATE := '0';
    for i in INPUT'range loop
      NEXTSTATE := INPUT(i) or NEXTSTATE;
      exit when NEXTSTATE = '1';
    end loop;
    NEXTSTATE := STRENGTH_MAP(NEXTSTATE, STRN);
    if (NEXTSTATE /= NEXT_ASSIGN_VAL) then
    -- compute delay
      case TWOBIT'(CURRENTSTATE & NEXTSTATE) is
        when "UU"|"UX"|"UZ"|"UW"|"U-"|"XU"|"XX"|"XZ"|"XW"|
             "X-"|"ZU"|"ZX"|"ZZ"|"ZW"|"Z-"|"WU"|"WX"|"WZ"|
             "WW"|"W-"|"-U"|"-X"|"-Z"|"-W"|"--"|"00"|"0L"|
             "LL"|"L0"|"11"|"1H"|"HH"|"H1"
          => DELTA := 0 ns;
        when "U1"|"UH"|"X1"|"XH"|"Z1"|"ZH"|"W1"|"WH"|"-1"|
             "-H"|"0U"|"0X"|"01"|"0Z"|"0W"|"0H"|"0-"|"LU"|
             "LX"|"L1"|"LZ"|"LW"|"LH"|"L-"
          => DELTA := tLH;
        when others => DELTA := tHL;
      end case;
    -- assign new value after internal delay
      CURRENTSTATE <= NEXTSTATE after DELTA;
      OUTPUT <= NEXTSTATE after DELTA;
      NEXT_ASSIGN_VAL := NEXTSTATE;
    end if;
  -- wait for signal changes
    wait on INPUT;
  end process P;
end A;
```

Figure 7.33 OR gate model.

```
    -- truth table for output strength --> STD_ULOGIC lookup
    constant TBL_STRN_STD_ULOGIC: STRN_STD_ULOGIC_TABLE :=
    ----------------------------------------------------------
    --| X01 X0H XL1 X0Z XZ1 WLH WLZ WZH W0H WL1 | strn/output|
    ----------------------------------------------------------
     (('U','U','U','U','U','U','U','U','U','U'), --  | U |
      ('X','X','X','X','X','W','W','W','W','W'), --  | X |
      ('0','0','L','0','Z','L','L','Z','0','L'), --  | 0 |
      ('1','H','1','Z','1','H','Z','H','H','1'), --  | 1 |
      ('X','X','X','X','X','W','W','W','W','W'), --  | Z |
      ('X','X','X','X','X','W','W','W','W','W'), --  | W |
      ('0','0','L','0','Z','L','L','Z','0','L'), --  | L |
      ('1','H','1','Z','1','H','Z','H','H','1'), --  | H |
      ('X','X','X','X','X','W','W','W','W','W'));-- | - |
    function STRENGTH_MAP(INPUT: STD_ULOGIC; STRN: STRENGTH)
      return STD_LOGIC is
    begin
      return TBL_STRN_STD_ULOGIC(INPUT, STRN);
    end STRENGTH_MAP;
```

Figure 7.34 Output strength table and strength function.

strength of a gate output for different technologies. Each strength parameter is a three character label of the form ABC, where:

A: Denotes the output value (X or W) representing unknowns and Z.

B: Denotes the output value (0 or L or Z) representing a low output.

C: Denotes the output value (1 or H or Z) representing a high output.

Thus, a strength parameter of WLZ indicates (1) the output value representing unknowns will be W, (2) the output value representing a low will be L, and (3) the output value representing a high will be Z (a high impedance condition). Table 7.1 shows the correspondence between some strength parameters and technologies as defined by Synopsys. The strength parameters are used with the table in Figure 7.34 to produce the correct output strength for a given internal STD_LOGIC type. Output values developed by the internal model are combined with the output strength parameter to produce the final output. For example, for the strength parameter WLZ (1) internal model values X, Z, and W map to W, (2) internal model values 0 or L map to L, and (3) internal model values 1 and H map to Z. For the OR gate model, the default strength is XO1, which is standard CMOS or TTL.

The architecture begins by evaluating the logic function. This is done with a *for* loop that OR's inputs two at a time, using the table defined in Figure 7.32. This *for* loop approach is necessary to make the model work for arbitrary N. However, it does require that the operation be associative. Next, the STRENGTH_MAP is applied to determine the output value. Then, a check is made to determine if the value of the OR function has changed since the last input change, i.e., NEXTSTATE is compared with NEXT_ASSIGN_VAL. If there is a difference, a new output will be scheduled. The *case* statement is used to determine the value of delta, the delay to be used. To do this, it compares NEXTSTATE with CURRENTSTATE. Any change in strength only is executed in 0 ns, i.e., delta delay. The first *case* clause accounts for this situa-

Table 7.1   Strength parameter technology.

Parameter	Typical Technology
X01	Normal CMOS or TTL (default)
X0H	NMOS
XL1	PMOS
X0Z	Open collector/open drain
XZ1	CMOS open source pullup
WLH	Bus keeper, jam latch
WLZ	Bus keeper weak pulldown
WZH	Alternative bus keeper
WOH	ECL wired AND
WL1	ECL wired OR

tion. The next two clauses account for rise and fall conditions where both strength and state are changing. Next, the values of CURRENTSTATE and output are scheduled to change after delta time units Finally, NEXT_ASSIGN_VAL is set equal to NEXTSTATE in preparation for the next time the gate model is called. The process then waits for an input change.

To understand this model, it is important to note the difference between NEXTSTATE, NEXT_ASSIGN_VAL, and CURRENTSTATE. Variable NEXTSTATE is the value of the OR of the inputs at the present simulation time. Variable NEXT_ASSIGN_VAL is the value of the OR of the inputs the previous time the inputs changed. Signal CURRENTSTATE is the value of the OR function the last time a scheduled output change occurred. In general, NEXT_ASSIGN_VAL and CURRENTSTATE will have different values, based on this delayed state update mechanism.

7.3.6   More Accurate Delay Modeling

Earlier we discussed the definitions of TPHL and TPLH for modeling different delays, depending on whether the output is switching to a high or low. In the OR gate model, signal changes which involved only strength changes, were modeled as zero delay. Other changes employed TPLH and TPHL. In some cases, however, it is important to give a separate set of delays to situations where one is switching to or from a value with a high impedance strength because, for MOS circuits, transitions to and from the high-impedance state can take a very long time compared to normal propagation delays.

7.4   CONFIGURATION DECLARATIONS FOR GATE LEVEL MODELS

In Chapter 3 we discussed the use of configuration specifications and configuration declarations as means of binding components in a structural model to specific behavioral models. As we indicated then, the purpose of these constructs is to bind a component declaration to a model stored in a library. In this section we discuss additional uses of configuration declarations that are particularly useful for gate-level modeling.

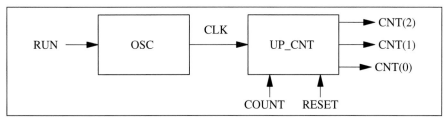

Figure 7.35 Counter system.

Recall that, in a configuration declaration, we specify in a separate analyzable entity all the component binding for a structural architecture. The main reasons for doing this are:

1. The bindings can be postponed or can be altered as a design progresses.
2. If the bindings are changed, the configuration declarations are reanalyzed instead of the structural architecture. The configuration declaration for a structural architecture will contain significantly less code than the architecture itself. This is particularly true of a structural architecture that instantiates thousands of gates. Thus, analysis time is significantly reduced when only the configuration declarations are reanalyzed instead of the entire structural architecture.
3. The configuration declaration is a natural place to perform *name mapping*, which allows the use of various component library models in the same structural architecture.
4. The configuration declaration is also a natural place to collect details on the components in structural models. This information can be altered to perform studies on the nature of the design.

For example, configuration declarations are a good approach to handling the back annotation problem. Delay values collected for a gate array implementation can be inserted into a configuration declaration for the circuit being modeled.

We illustrate the use of a configuration declaration by means of the counter system shown in Figure 7.35. An oscillator component outputs a periodic clock signal (CLK) when RUN = '1'. If inputs COUNT and RESET to component UP_CNT are '1' and '0', respectively, the UP_CNT module counts the clocks modulo eight, i.e., it is a divide by eight counter. If RESET is a '1', the clock action is overridden and the counter is reset. As shown in Figure 7.36, the UP_CNT module can be further decomposed structurally into three T flip-flops and an AND gate.

The design of the counter system can be represented by two different design trees. Figure 7.37(a) shows the design tree for the case where the UP_CNT module is modeled behaviorally. Figure 7.37(b) shows the design tree for the case where the UP_CNT module is modeled structurally. Note that on each diagram, the box representing the leaf component has a box within it. This internal box represents a connector the component can be plugged into. Until a component is plugged into the connector slot, that slot is *unbound*.

Figure 7.38 gives the VHDL structural architecture for the counter system, i.e., the oscillator and the up counter. It corresponds to the schematic in Figure 7.35. Note that even though components OSC and UP_CNT are declared, they are not bound. As already described, these component declarations represent the connectors of the components that will be used. The component instantiations, e.g., C0: OSC port map(RUN => RUN, CLOCK => CLK), specify both

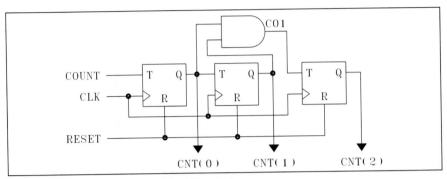

Figure 7.36 Structural model for UP counter.

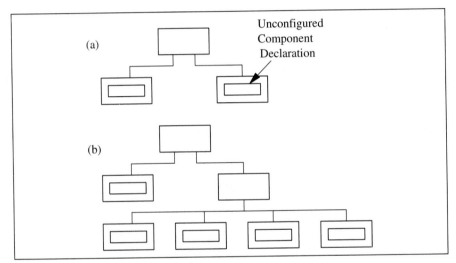

Figure 7.37 Counter system design trees.

the number of components and the interconnection wiring. Figure 7.39 shows an unbound structural architecture for the up counter implemented with T flip-flops and an AND gate. It corresponds to the circuit shown in Figure 7.36.

Figure 7.40 shows configuration ALL_BEHAV of the entity CTR. The configuration body is nested. Inside the *for STRUCTB* structure are configuration statements, which relate to architecture STRUCTB of entity CTR, i.e., there are component specification statements for components C0: OSC and C1: UP_CNT of architecture STRUCTB. Included in these component specifications are generic maps, which specify the actual timing of the two components. Note that configuration ALL_BEHAV configures the design tree shown in Figure 7.37(a).

Figure 7.41 shows configuration PART_STRUCT of entity CTR. In this case a structural architecture is specified for UP_CNT, and is configured in an inner nesting.

```
entity CTR is
  port(RESET,RUN,COUNT: in BIT;
       CNT: inout BIT_VECTOR(2 downto 0));
end CTR;

architecture STRUCTB of CTR is
  signal CLK: BIT;
  component OSC
    port(RUN: in BIT; CLOCK: out BIT := '0');
  end component;
  component UP_CNT
    port(RESET,COUNT,CLK: in BIT;
         CNT: inout BIT_VECTOR(2 downto 0));
  end component;
begin
  C0: OSC
    port map(RUN => RUN,CLOCK => CLK);
  C1: UP_CNT
    port map(RESET => RESET, COUNT=> COUNT,
             CLK => CLK, CNT => CNT);
end STRUCTB;
```

Figure 7.38   Unbound structural architecture of counter system.

```
entity UP_CNT is
  generic(R_DEL,CLK_DEL:TIME);
  port(RESET,COUNT,CLK: in BIT;
       CNT: inout BIT_VECTOR(2 downto 0));
end UP_CNT;
architecture STRUCT of UP_CNT is
  signal C01: BIT;
  component T
    port(R,T,CLK: in BIT; Q: out BIT);
  end component;
  component AND2G
    port(I1,I2: in BIT; O: out BIT);
  end component;
begin
  C1: T
    port map(R => RESET,T => COUNT,CLK => CLK,Q => CNT(0));
  C2: T
    port map(R => RESET,T => CNT(0),CLK => CLK,Q => CNT(1));
  C3: T
    port map(R => RESET,T => C01, CLK => CLK, Q => CNT(2));
  C4: AND2G
    port map(I1 => CNT(0), I2 => CNT(1), O => C01);
end STRUCT;
```

Figure 7.39   Unbound structural architecture of UP counter.

```
configuration ALL_BEHAV of CTR is
  for STRUCTB
    for C0: OSC use entity work.COSC(WAIT_DEL)
      generic map(60 ns, 40 ns);
    end for;
    for C1: UP_CNT use entity work.UP_CNT(ALG)
      generic map(10 ns, 15 ns);
    end for;
  end for;
end ALL_BEHAV;
```

Figure 7.40   Behavioral configuration of counter system.

```
configuration PART_STRUCT of CTR is
  for STRUCTB
    for C0: OSC use entity work.COSC(WAIT_DEL)
      generic map(60 ns, 40 ns);
    end for;
    for C1: UP_CNT use entity work.UP_CNT(STRUCT)
      generic map(10 ns, 15 ns);
        for STRUCT
          for all: T use entity work.TFF(ALG)
            generic map(10 ns, 15 ns);
          end for;
          for C4: AND2G use entity work.AND2(BEH)
            generic map(5 ns);
          end for;
        end for;
    end for;
  end for;
end PART_STRUCT;
```

Figure 7.41   Partly structural configuration of counter system.

7.4.1   Default Configuration

In the default configuration, the system looks for model entity names that match the declared component name. No binding statements are necessary.

Figure 7.42 shows algorithmic and structural implementations of a buffer. Figure 7.43 shows a test bench that tests the buffer. The default binding for the test bench is contained in configuration DEFAULT1. It binds the most recently analyzed version of the entity BUFF to the component BUFF in the test bench. If the structural entity is the most recently analyzed entity, it will also bind the most recently analyzed entity INV to the component INV in the structural architecture. In other words, the default binding extends down through the hierarchy. Thus, the same test bench can be used to test both the behavioral and structural models.

Although the default configuration provides a simple, powerful method for specifying library bindings, subtle problems can arise when one forgets which architecture was last analyzed for the default entity. This is particularly problematic if several individuals share a com-

```
entity BUFF is
  port( I: in BIT; O: out BIT);
end BUFF;
architecture ALG of BUFF is
begin
  O <= I;
end ALG;
                              (a) Behavioral Buffer


entity INV is
  port(I: in BIT; O: out BIT);
end INV;
architecture ALG of INV is
begin
  O <= not I;
end ALG;


entity BUFF is
  port( I: in BIT; O: out BIT);
end BUFF;


architecture STRUCTURE of BUFF is
  component INV
    port(I: in BIT; O: out BIT);
  end component;
  signal S: BIT;
begin
  C1: INV
    port map(I,S);
  C2 : INV
    port map(S,O);
end STRUCTURE;
                              (b) Structural Buffer
```

Figure 7.42 Default configuration for behavioral and structural buffers.

mon library. The "other guy" may have analyzed a different architecture without informing you. Such events can result in test benches that use incorrect default configurations, and therefore, do not produce repeatable results.

7.4.2 Configurations and Component Libraries

Configurations are important in the use of component libraries. Different simulator companies have standard component model libraries. Design groups also develop their own libraries as a design develops. These libraries contain generic gates and register elements. The interface descriptions for these gates and flip-flops are frequently general; thus, the configuration body must map the descriptions used in the structural architecture under study to the library interface descriptions. This is the name mapping function referred to previously.

As an example, consider a structural model of the TWO_CONSECUTIVE circuit shown in Figure 1.1. Figure 7.44 gives a structural model of the circuit. In this example we use compo-

```
entity TB is
end TB;

architecture BUF_TEST of TB is
  signal I,O: BIT;
  component BUFF
    port( I: in BIT; O: out BIT);
  end component;
begin
  C1: BUFF
    port map(I,O);
  I <= '0', '1' after 10 ns, '0' after 20 ns;
end BUF_TEST;
configuration DEFAULT_1 of TB is
  for BUF_TEST
    for C1:BUFF
    end for;
  end for;
end DEFAULT_1;
```

Figure 7.43   Default configuration for test bench.

nents from two libraries, the Synopsys IEEE library, which contains models developed by that company, and the WORK library. In the structural model of TWO_CONSECUTIVE, four components: EDGE_TRIGGERED_D, INVG, AND3G, and OR2G are declared and instantiated, but are not bound. Figure 7.45 shows the interface description for the library models being employed. Descriptions are shown for DFFREG, ANDGATE, and INVGATE elements. The description for the ORGATE model is similar to the ANDGATE description and is not shown. These interface descriptions are taken from the SYNOPSYS IEEE library, package STD_LOGIC_COMPONENTS with the exception of the DFFREG model. This model is a Synopsys model that was modified to produce both noninverted and inverted state outputs. The reset output was also made low active. Each model has generics TLH, THL, and strength. DFFREG, ANDGATE, and ORGATE have a generic N, which specifies the data width in the case of the DFFREG, and the number of inputs in the case of gate models. Note that all generics have default values, which are used if other values are not specified. This is sometimes the case when component declarations in structural models are bound to the models.

Figure 7.46 shows two configurations of architecture STRUCTURAL of entity TWO_CONSECUTIVE. Configuration PARTS1 binds the architecture components to the four library components discussed previously. Component DFFREG comes from the library WORK. The other three components are in the Synopsys IEEE library, package STD_LOGIC_ENTITIES. Configuration PARTS1 uses the default values for timing, which are zero values for TLH and THL. Thus, this configuration would be used to simulate delta delay timing. Configuration PARTS2 specifies a set of TLH and THL values and thus, would be used to model a specific timing situation. Both configurations PARTS1 and PARTS2 use the default value for strength, which implies an output strength of X01, which is appropriate for standard TTL or CMOS.

```
library IEEE;
use IEEE.std_logic_1164.all;
use IEEE.std_logic_entities.all;
use work.all;
entity TWO_CONSECUTIVE is
  port(CLK, R: in STD_LOGIC; X: in STD_LOGIC;
          Z: out STD_LOGIC);
end TWO_CONSECUTIVE;

architecture STRUCTURAL of TWO_CONSECUTIVE is
  signal Y0, Y1, A0, A1: STD_LOGIC:= '0';
  signal NY0, NX: STD_LOGIC:= '1';
  signal ONE: STD_LOGIC:= '1';
  component EDGE_TRIGGERED_D
    port(CLK, D, NCLR: in STD_LOGIC; Q, QN: out STD_LOGIC);
  end component;
--------------- edge triggered register -------------
  component INVG
    port(I: in STD_LOGIC; O: out STD_LOGIC);
  end component;
--------------------- inverter ---------------------
  component AND3G
    port(I1, I2, I3: in STD_LOGIC; O: out STD_LOGIC);
  end component;
--------------------- ANDGATE ----------------------
  component OR2G
    port(I1, I2: in STD_LOGIC; O: out STD_LOGIC);
  end component;
--------------------- ORGATE -----------------------
begin
  C1: EDGE_TRIGGERED_D
    port map(CLK, X, R, Y0, NY0);
  C2: EDGE_TRIGGERED_D
    port map(CLK, ONE, R, Y1, open);
  C3: INVG
    port map(X, NX);
  C4: AND3G
    port map(X, Y0, Y1, A0);
  C5: AND3G
    port map(NY0, Y1, NX, A1);
  C6: OR2G
    port map(A0, A1, Z);
end STRUCTURAL;
```

Figure 7.44   TWO_CONSECUTIVE structural model.

7.5   MODELING RACES AND HAZARDS

In this section we show how configuration declarations are useful for modeling races and hazards. As an example of modeling a race with configuration declarations, let us again consider the

```
entity DFFREG is
   generic (N: Positive :=1;        -- N bit register
      tLH: Time :=  0 ns;           -- rise inertial delay
      tHL: Time :=  0 ns;           -- fall inertial delay
      strn: STRENGTH := strn_X01;-- output strength
      tHOLD : Time := 0 ns;         -- Hold time
      tSetUp : Time := 0 ns;        -- Setup time
      tPwHighMin : Time := 0 ns;  -- min pulse width(high)
      tPwLowMin : Time := 0 ns);  -- min pulse width(low)
   port (Data: in STD_LOGIC_VECTOR (N-1 downto 0);
      Clock,                        -- clock input
      Reset: in STD_LOGIC;          -- reset input
      Output: out STD_LOGIC_VECTOR (N-1 downto 0);
      NOutput: out STD_LOGIC_VECTOR (N-1 downto 0));
end DFFREG;
entity ANDGATE is
   generic (N:    Positive := 2;-- number of inputs
      tLH: Time := 0 ns;            -- rise inertial delay
      tHL: Time := 0 ns;            -- fall inertial delay
      strn: STRENGTH := strn_X01);-- output strength
   port (Input: in STD_LOGIC_VECTOR (1 to N); -- inputs
      Output: out STD_LOGIC);    -- output
end ANDGATE;
entity INVGATE is
   generic (tLH: Time := 0 ns; -- rise inertial delay
      tHL: Time := 0 ns;            -- fall inertial delay
      strn: STRENGTH := strn_X01);  -- output strength
   port (Input: in STD_LOGIC;       -- input
      Output: out STD_LOGIC);       -- output
end INVGATE;
```

Figure 7.45   Model interface descriptions.

up counter circuit shown in Figure 7.36. Some basic questions that a designer might ask about this design are:

1. What is the maximum clock frequency at which the counter will work properly for a given set of delays?
2. What is the critical signal path?

This is actually a specific case of a more general problem shown in Figure 7.47. Here a signal IN is clocked into FF1 at the rise of CLK. The output of FF1 drives a combinational logic network, whose output OUT is also sampled by FF2 on the rise of CLK. For the circuit to work properly, the period (PER) of the clock must be greater than the sum of delay through FF1, after it is clocked (CLKDEL), and the delay through the combinational logic (CLDEL). Actually, the situation is worse than this, as we must also allow for the setup time of FF2, but for now we will ignore this problem.

In the case of the counter in Figure 7.36, the two high order counter flip-flops are FF1 and FF2. The combinational logic is the single AND gate. A race condition can develop in this circuit as follows: Assume the counter is in the state CNT(2) CNT(1) CNT(0) = 010. In this state,

```
configuration PARTS1 of TWO_CONSECUTIVE is
  for STRUCTURAL
    for all: EDGE_TRIGGERED_D use entity work.DFFREG(A)
      port map(Clock => CLK, Data(0) => D, Reset => NCLR,
                         Output(0) => Q, Noutput(0)=> QN);
    end for;
    for all: INVG use entity INVGATE(A)
      port map(Input => I, Output => O);
    end for;
    for all: AND3G use entity ANDGATE(A)
      generic map(N => 3)
      port map(Input(1) => I1, Input(2) => I2,
               Input(3) => I3, Output => O);
    end for;
    for all: OR2G use entity ORGATE(A)
      generic map(N => 2)
      port map(Input(1) => I1, Input(2) => I2, Output => O);
    end for;
  end for;
end PARTS1;

configuration PARTS2 of TWO_CONSECUTIVE is
  for STRUCTURAL
    for all: EDGE_TRIGGERED_D use entity work.DFFREG(A)
      generic map(TLH=> 5 ns, THL=> 6 ns)
      port map(Clock => CLK, Data(0) => D, Reset => NCLR,
                         Output(0) => Q, Noutput(0)=> QN);
    end for;
    for all: INVG use entity INVGATE(A)
      generic map(TLH=> 1 ns, THL=> 2 ns)
      port map(Input => I, Output => O);
    end for;
    for all: AND3G use entity ANDGATE(A)
      generic map(N => 3, TLH=> 3 ns, THL=> 4 ns)
      port map(Input(1) => I1, Input(2) => I2,
               Input(3) => I3, Output => O);
    end for;
    for all: OR2G use entity ORGATE(A)
      generic map(N => 2, TLH=> 2 ns, THL=> 3 ns)
      port map(Input(1) => I1, Input(2) => I2, Output => O);
    end for;
  end for;
end PARTS2;
```

Figure 7.46   Two configurations binding to library components.

the output of the AND gate (signal C01) is a '0'. If then clocked, the Counter will change, after CLKDEL, to the value 011. Following this, the output of the AND gate C01 will change to ',' after gate delay (DEL). In this case then, CLKDEL + CLDEL must be less then HI_TIME + LO_TIME, the oscillator period. Figure 7.48 shows a configuration FAST of CTR where this condition is just met. As shown in Figure 7.49, the signal change on C01 arrives 1 ns before the

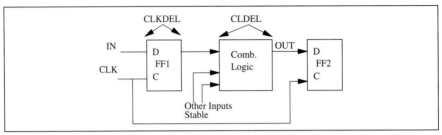

Figure 7.47   A basic timing problem.

```
configuration FAST of CTR is
  for STRUCTB
    for C0: OSC use entity work.COSC(WAIT_DEL)
      generic map(11 ns, 10 ns);
    end for;
    for C1: UP_CNT use entity work.UP_CNT(STRUCT)
      generic map(10 ns, 15 ns);
      for STRUCT
        for all: T use entity work.TFF(ALG)
          generic map(10 ns, 15 ns);
        end for;
        for C4: AND2G use entity work.AND2(BEH)
          generic map(5 ns);
        end for;
      end for;
    end for;
  end for;
end FAST;

configuration TOO_FAST of CTR is
  for STRUCTB
    for C0: OSC use entity work.COSC(WAIT_DEL)
      generic map(10 ns, 9 ns);
    end for;
    for C1: UP_CNT use entity work.UP_CNT(STRUCT)
      generic map(10 ns, 15 ns);
      for STRUCT
        for all: T use entity work.TFF(ALG)
          generic map(10 ns, 15 ns);
        end for;
        for C4: AND2G use entity work.AND2(BEH)
          generic map(5 ns);
        end for;
      end for;
    end for;
  end for;
end TOO_FAST;
```

Figure 7.48   Configurations used to model race conditions.

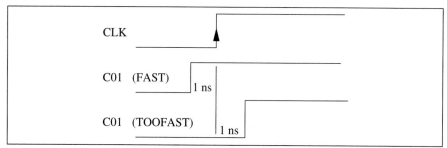

Figure 7.49   Counter race condition.

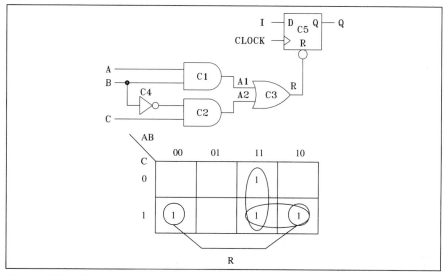

Figure 7.50   Circuit with hazards.

rise of the clock. However, for the configuration TOOFAST (Figure 7.48), the condition is violated and the signal C01 switches to '1' 1 ns too late. As a result, the counter counts 000,001,010,011,000,001 etc., instead of the normal 8-state counting sequence. The use of circuit configurations makes the studying of such timing problems easy.

The accurate timing of gate-level models in configuration declarations also allows realistic modeling of timing hazards. Figure 7.50 shows a circuit with potential hazards. A, B, and C are inputs into a circuit, which develops the reset signal R for the flip-flop. The Karnaugh map for the logic function corresponding to the signal R is also shown in the figure. Consider the case, where the input combination ABC is initially 111 but then switches to 101, i.e., B switches low. Examination of the K-map shows that both input combinations 111 and 101 produce an output 1, but when switching, particular gate and inverter delays can cause the output to go to 0 momentarily.

```
library SYNOPSYS;
use SYNOPSYS.TYPES.all;
entity RGLITCH is
   port(A,B,C,I,CLK: in MVL7; Q: out MVL7);
end RGLITCH;
architecture STRUCT of RGLITCH is
   signal A1,A2,NB,R : MVL7;
   component DFF
      port(R,CLK,D: in MVL7; Q: out MVL7);
   end component;
   component AND2
      port(I1,I2: in MVL7; O: out MVL7);
   end component;
   component OR2
      port(I1,I2: in MVL7; O: out MVL7);
   end component;
   component INV
      port(I: in MVL7; O: out MVL7);
   end component;
begin
   C1: AND2
      port map(I1 => A, I2  => B, O => A1);
   C2: AND2
      port map(I1 => NB, I2 => C, O => A2);
   C3: OR2
      port map(I1 => A1, I2 => A2, O => R);
   C4: INV
      port map(I => C, O => NB);
   C5: DFF
      port map(R => R, CLK => CLK, D => I, Q => Q);
end STRUCT;
```

Figure 7.51   Unbound structural architecture.

To study which combinations of delays produce a hazard, it is useful to create an unbound structural model of the circuit, then experiment with configuration bodies containing different delay sets to determine under which conditions the hazard causes a spike. Figure 7.51 gives a structural model for the circuit. Note: All components that are declared are unbound.

Figure 7.52 shows a configuration of the circuit, which will produce a "glitch." In this configuration, the data path for signal B through the inverter C4 and in the AND gate C2 is longer than the upper path for signal B through C1. Figure 7.53 shows the time response of the model when the input combination changes from ABC=111 to ABC=101, i.e., input B changes from 1 to 0. The output of the upper AND gate, signal A1, switches low 3 ns before the output of the lower AND gate, signal A2, switches high. A 3 ns 0 level glitch is produced on signal R, which will reset the flip-flop Q.

The hazard we have just analyzed is called a logic hazard. It can be eliminated by adding the term AC to the Boolean expression for R. This term will hold the OR gate output at logic 1, as B changes from 1 to 0, as shown by the dotted couple on the K-map in Figure 7.50. In general, it has been shown that logic hazards can always be eliminated by including all prime implicants of a function in its expression.

```
configuration GLTCH of RGLITCH is
  for STRUCT
    for all: DFF use entity WORK.DFFREG(A)
      generic map(TLH => 5 ns, THL => 6 ns)
      port map(CLOCK => CLK, DATA(0) => D, RESET => R,
               OUTPUT(0) => Q);
    end for;
    for all: INV use entity SYNOPSYS.INVGATE(A)
      generic map(TLH => 2 ns, THL => 3 ns)
      port map(INPUT => I, OUTPUT => O);
    end for;
    for C1 : AND2 use entity SYNOPSYS.ANDGATE(A)
      generic map(N => 2, TLH => 2 ns, THL => 3 ns)
      port map(INPUT(1) => I1, INPUT(2) => I2, OUTPUT => O);
    end for;
    for C2 : AND2 use entity SYNOPSYS.ANDGATE(A)
      generic map(N => 2, TLH => 4 ns, THL => 5 ns)
      port map(INPUT(1) => I1, INPUT(2) => I2, OUTPUT => O);
    end for;
    for all: OR2 use entity SYNOPSYS.ORGATE(A)
      generic map(N => 2, TLH => 2 ns, THL => 3 ns)
      port map(INPUT(1) => I1, INPUT(2) => I2, OUTPUT => O);
    end for;
  end for;
end GLTCH;
```

Figure 7.52 Glitch-producing configuration of the circuit.

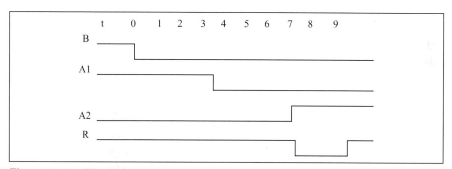

Figure 7.53 Effect of a logic hazard.

Another type of hazard is a function hazard. Here, a momentary incorrect output is also induced when switching from one input combination to another; however, addition of terms to the Boolean expression cannot correct this type of hazard. A function hazard for signal R is caused by switching from the input combination ABC = 001 to the combination ABC = 110. This circuit is sensitive to this fault when the paths for signals A and B through AND gate C1 are slow, relative to the path for signal C through AND gate C2. Figure 7.54 gives a configuration that causes this condition. Figure 7.55 gives the circuit response. In this case, the rise on A1 lags the fall on A2, causing a spike on R, which resets the flip-flop.

```
configuration GLTCH2 of RGLITCH is
  for STRUCT
    for all: DFF use entity WORK.DFFREG(A)
      generic map(TLH => 5 ns, THL => 6 ns)
      port map(CLOCK => CLK, DATA(0) => D, RESET => R,
               OUTPUT(0) => Q);
    end for;
    for all: INV use entity SYNOPSYS.INVGATE(A)
      generic map(TLH => 2 ns, THL => 3 ns)
      port map(INPUT => I, OUTPUT => O);
    end for;
    for C1 : AND2 use entity SYNOPSYS.ANDGATE(A)
      generic map(N => 2, TLH => 4 ns, THL => 5 ns)
      port map(INPUT(1) => I1,INPUT(2) => I2,OUTPUT => O);
    end for;
    for C2 : AND2 use entity SYNOPSYS.ANDGATE(A)
      generic map(N => 2, TLH => 1 ns, THL => 2 ns)
      port map(INPUT(1) => I1,INPUT(2) => I2, OUTPUT => O);
    end for;
    for all: OR2 use entity SYNOPSYS.ORGATE(A)
      generic map(N => 2, TLH => 1 ns, THL => 2 ns)
      port map(INPUT(1) => I1,INPUT(2) => I2, OUTPUT => O);
    end for;
  end for;
end GLTCH2;
```

Figure 7.54   Another glitch-producing configuration.

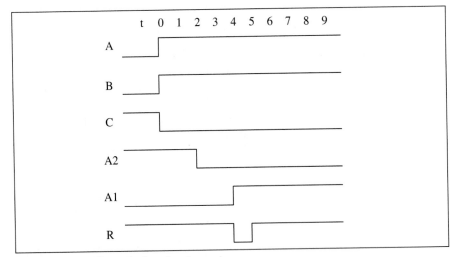

Figure 7.55   Effect of a function hazard.

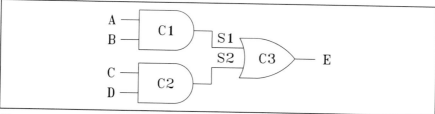

Figure 7.56 Three-gate circuit.

It is interesting that these two hazards are sensitized by two totally different delay situations. As can be seen, the use of different configuration bodies is a very effective means of sensitizing a circuit to different hazards.

7.6 APPROACHES TO DELAY CONTROL

In this section we illustrate some different approaches to delay control. Consider the circuit shown in Figure 7.56. Three likely delay modeling situations are:

1. All gates have the same delay.
2. All gates of a given type have the same delay.
3. Each gate might have a different delay.

We illustrate these three situations for the three-gate circuit, using the VHDL code shown in Figure 7.57. Gate models AND2G and OR2G are the basic gate models employed, each having a generic delay parameter DEL. Architecture AND_OR of entity THREE_GATES gives a structural model of the circuit in which the component declarations for AND2 and OR2 are unbound. The three delay mapping situations listed previously are accomplished by configurations ONE, TWO, and THREE of architecture AND_OR. In configuration ONE, generic GATE_DEL of entity THREE_GATES is mapped to the generic DEL for each gate; thus, all gates will have the same 2 ns delay. In configuration TWO, all AND gates are given the same delay (2 ns) and all OR gates the same delay (3 ns). Configuration THREE assigns a 2 ns delay to AND gate C1, 3 ns delay to AND gate C2, and 4 ns delay to OR gate C3.

It should be noted that it is not necessary to use configuration declarations to do these types of mappings. They could be done with configuration specifications in the structural architecture itself. However, the use of configuration declarations is an orderly way to do the different mappings, which does not require reanalysis of the architecture in order to change delay characteristics.

PROBLEMS

7.1 Develop a VHDL model for an SN7400 2-input NAND gate. Use generics to represent TPHL and TPLH. Simulate the gate using typical values.

7.2 Develop a VHDL model for a 2-input NAND gate, which incorporates input delay due to metal runs. Assume that the NAND gate is used in the gate circuit shown in Figure 7.58. Assume also that the length of a metal run is proportional to the length of lines on the drawing.

```
entity  OR2G is
  generic(DEL: TIME);
  port(I1,I2: in BIT; O: out BIT);
end OR2G;
architecture GNR_DEL of OR2G is
begin
  O <= I1 or I2 after DEL;
end GNR_DEL;

entity  AND2G is
  generic(DEL: TIME);
  port(I1,I2: in BIT; O: out BIT);
end AND2G;
architecture GNR_DEL of AND2G is
begin
  O <= I1 and I2 after DEL;
end GNR_DEL;

entity THREE_GATES is
  generic(GATE_DEL: TIME:= 2 ns);
  port(A,B,C,D: in BIT; E: out BIT);
end THREE_GATES;
architecture AND_OR of THREE_GATES is
  signal S1,S2: BIT;
  component  OR2
    port(I1,I2: in BIT; O: out BIT);
  end component;
  component  AND2
    port(I1,I2: in BIT; O: out BIT);
  end component;
begin
  C1: AND2
    port map(I1 => A, I2 => B, O => S1);
  C2: AND2
    port map(I1 => C, I2 => D, O => S2);
  C3: OR2
    port map(I1 => S1, I2 => S2, O => E);
end AND_OR;

configuration ONE of THREE_GATES is
  for AND_OR
    for all: AND2 use entity work.AND2G(GNR_DEL)
      generic map(DEL => GATE_DEL);
    end for;
    for all: OR2 use entity  work.OR2G(GNR_DEL)
      generic map(DEL => GATE_DEL);
    end for;
  end for;
end ONE;
```

Figure 7.57   Delay mapping.

```
configuration TWO of THREE_GATES is
  for AND_OR
    for all: AND2 use entity work.AND2G(GNR_DEL)
      generic map(DEL => 2 ns);
    end for;
    for all: OR2 use entity  work.OR2G(GNR_DEL)
      generic map(DEL => 3 ns);
    end for;
  end for;
end TWO;

configuration THREE of THREE_GATES is
  for AND_OR
    for C1: AND2 use entity work.AND2G(GNR_DEL)
      generic map(DEL => 2 ns);
    end for;
    for C2: AND2 use entity work.AND2G(GNR_DEL)
      generic map(DEL => 3 ns);
    end for;
    for C3: OR2 use entity  work.OR2G(GNR_DEL)
      generic map(DEL => 4 ns);
    end for;
  end for;
end THREE;
```

Figure 7.57   Delay mapping (continued).

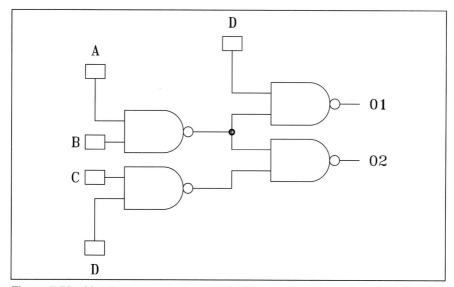

Figure 7.58   Metal runs.

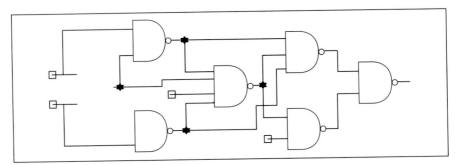

Figure 7.59   Circuit fanout.

Consult documentation on a gate array family and derive delay values vs. length of metal run. Use these values to simulate the network. The internal gate delay values can come from the gate array family also.

7.3   The purpose of this problem is to become familiar with the EIA Timing and Environmental Specification. Assume a linear model for ldelay. Next, use laboratory measurements and manufacturers data to develop an expression for T'_{pd} for an SN7400 2-input NAND gate. Include the effects of power supply voltage and temperature, if possible.

7.4   Develop a NAND gate model with the following characteristics:
a.   A general N input model with N being a generic in the entity description.
b.   A delay function that is fanout sensitive and where each gate calculates its own fanout when instantiated into a specific network. Use this model to simulate the logic circuit shown in Figure 7.59.

7.5   Using the Timing Control Package given in this chapter, model the gate network shown in Figure 7.59 using delta delay, minimum, typical, and maximum timing.

7.6   Figure 7.50 shows a circuit with a hazard, whose output feeds the reset input of a flip-flop. The flip-flop used in this example was a library flip-flop. Develop your own flip-flop model with spike detection on its set and reset inputs, and simulate the hazard situation.

7.7   Fundamentally, logic gates are two-valued, but realistic modeling of the circuit effects associated with them requires the use of multivalued logic. Make a list of the modeling situations that require special values and list the commonly used symbol for those values.

7.8   Develop a VHDL model for an SN74LS240 buffer with tristate output. Incorporate timing that accurately models data and enable delays, including the switching to and from the high-impedance state.

7.9   Model and simulate an NMOS inverter using the nine-valued logic system in Figure 7.60 (not the IEEE nine-valued system).

7.10   Implement a NAND gate model using the IEEE nine-valued system. For the RS flip-flop shown in Figure 7.60, assume that both gate outputs are initially U, and that the two inputs $\overline{R}$ and $\overline{S}$ are initially 0 and switch to 1 simultaneously. Simulate the circuit when:
a.   Both gates have identical delays.
b.   The two gates have different delays. Compare your results.

7.11   The use of more logic values is supposed to result in less pessimistic simulation results, i.e., the number of X's (unknown) values is less. Design and simulate an experimental circuit that tests this hypothesis.

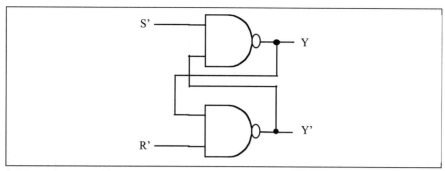

Figure 7.60   R S flip-flop.

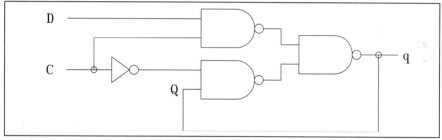

Figure 7.61   D flip-flop.

7.12 Multivalued logic can be used to uncover instability in sequential circuits. Consider the circuit of a simple D flip-flop shown in Figure 7.61. Assume all elements have delta delay and that initially $D = C = Q = 1$.

    a. Simulate the circuit when at t=0, C switches to X, and then at t= 10 ns, C switches to 0. Verify the result that q = X. (This simulation procedure is known as Eichelberger's technique. Consult books on switching theory and logic design to learn more about it.)

    b. Assign delays to the circuit, which will first eliminate and then produce the X output.

    c. Develop a Boolean expression and Karnaugh map for q in terms of D, C, and Q. Eliminate any static hazards found on this K-map by adding appropriate terms to the expression. Repeat part a to verify that the circuit is stable, independent of delays.

7.13 Using techniques from abstract algebra, prove that the WiredX resolution function is associative and commutative.

7.14 Following is the value system for system HILO.

```
              X
          U       D
       T     W     B
    P     R     F     N
  1     H     Z     L     0
```

Give the meaning of the following values: 1, H, Z, L, 0, F, U.

7.15 Suppose a 12-valued system is developed as follows:

    a. Three states: 0, 1, and X. These states are defined as in the text.

    b. Four strengths: F, R, Z, and U. F, R, and Z are defined in the text. U is an unknown strength.

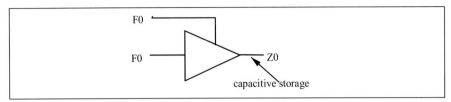

Figure 7.62   High impedance buffer.

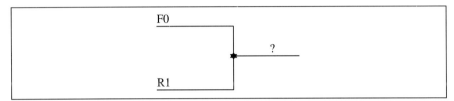

Figure 7.63   Bus resolution.

Thus, we have both an unknown state (X) and an unknown strength (U).

    a. Show the 12 values in the system.

    b. Classify the system as being a state strength or interval system.

    c. Consider the high-impedance buffer shown in Figure 7.62. Assume it has capacitive storage on the output, and that the initial states of the data inputs, outputs, and the enable are as shown. Assume the enable switches from F0 to FX. What will the output value switch to? Explain.

7.16 Difficulties arise when two interconnected VHDL models use different value systems. Develop a mapping between MVL4 and the IEEE nine-valued system. Discuss information loss in doing the mapping. Using principles of abstract algebra, discuss the algebraic properties of the mapping.

7.17 This question refers to a multivalued system using a state-strength model. For the wired situation shown in Figure 7.63 what should the resultant value be?

7.18 Create a switch-level structural model for the PLA modeled as a single behavioral model in Chapter 4 (see Figure 4.29).

7.19 Using the Smith and Acosta models for MOS switches, simulate a small MOS network and evaluate your results.

7.20 Develop a structural model for the circuit shown in Figure 7.64 that is unbound. Use configuration bodies to bind the circuit to a model library in which the models have input spike detection. Use K-map analysis to determine static or function hazards for the circuit and which input combinations will produce those hazards. Use delays in the configuration bodies to sensitize the circuit to different hazard conditions. Simulate the circuit to validate that the hazards are detected.

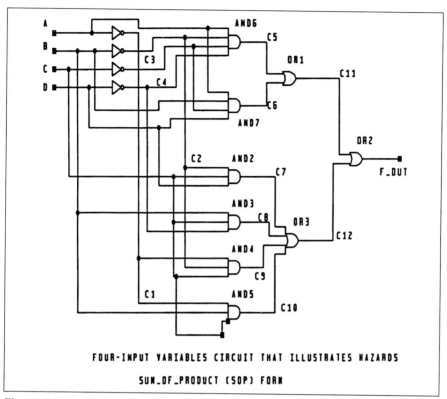

Figure 7.64   Circuit with hazard.

7.21   This problem is a repeat of Problem 4.19. In that problem a structural model for a counter was developed from an oscillator, JK flip-flops, and an AND gate. Delays were specified in terms of oscillator high time and low time and gate and flip-flop TPLH and TPHL. Develop an unbound structural model for the counter. Then develop configuration bodies, which can be used to assign values to these delays. For a given set of values of TPLH and TPHL for the flip-flops and gate, determine the maximum possible oscillator frequency. What is the critical path? Express the maximum oscillator frequency in terms of delays in this critical path. Simulate to validate your results.

CHAPTER 8

HDL-Based Design Techniques

$\mathbf{I}$n this chapter, we discuss structured design techniques for combinational and sequential logic. We integrate the use of a hardware description language with Karnaugh maps and state tables to develop design techniques for combinational and sequential logic that range from algorithmic descriptions to gate-level implementations. The chapter concludes with an integrated approach to the design of microprogrammed control units.

8.1 DESIGN OF COMBINATIONAL LOGIC CIRCUITS

In this section, the design of combinational logic circuits is discussed. Before the techniques of this section can be applied, the designer must have made an initial decision to use combinational logic in the system design. This decision must be based on an analysis of the design specifications and requires judgment on the part of the designer. For example, consider the device whose block diagram is shown in Figure 8.1.

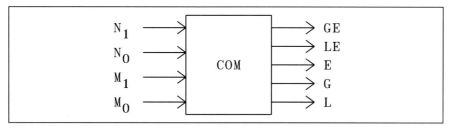

Figure 8.1 Block diagram for binary comparator.

Example 8.1: Specification for Binary Comparator (COM). A device must compare two binary numbers and determine which is larger. The inputs are 2-bit binary numbers denoted

```
--
-- Device to compare two binary inputs.
--
entity COM is
  generic (D:time);
  port (N1, N0, M1, M0: in BIT;
        GE, LE, E, G, L: out BIT);
end COM;
```

Figure 8.2   Entity specification for device COM.

Table 8.1   Binary output signals.

Signal	Description
GE	Binary output that is true when N is greater than or equal to M.
LE	Binary output that is true when N is less than or equal to M.
E	Binary output that is true when N is equal to M.
G	Binary output that is true when N is strictly greater than N.
L	Binary output that is true when N is strictly less than M.

as $N = N_1 N0$ and $M = M_1 M_0$. The outputs are binary signals GE, LE, E, G, and L, which are defined in Table 8.1.

From the problem specification and block diagram, the entity declaration for device COM is easily derived and is shown in Figure 8.2. The generic variable D is used to specify the time delay for the device from input to output. There are four input ports (N1, N0, M1, and M0) that represent, respectively, the binary inputs N_1, N_0, M_1, M_0. There are five output ports (GE, LE, E, G, and L) that represent the five binary outputs of device COM.

After reading this specification, it should be clear that successive outputs are entirely independent. Therefore, it should be possible to implement this device using combinational logic. After the decision is made to explore combinational logic implementations, the techniques to be discussed in this section can be used. The existence of a combinational logic implementation for a circuit does not prohibit the existence of a sequential circuit implementation for the same circuit. In fact, if the inputs are held in registers, this device has a sequential circuit implementation, which is developed later in this chapter.

8.1.1   Combinational Logic Design at the Algorithmic Level

If the problem size permits, the first step in the design of combinational logic is to create a truth table. The truth table lists the values of the outputs for each combination of values on the inputs. The truth table for this problem is shown in Figure 8.3.

In this section, we describe two ways to model truth tables at the algorithmic level in the behavioral domain. The ARRAY model can be mapped directly to a ROM implementation. The CASE model can be mapped directly to a multiplexer implementation. By using optimization procedures, several alternative multiplexer implementations can be designed.

N_1	N_0	M_1	M_0	GE	LE	E	G	L
0	0	0	0	1	1	1	0	0
0	0	0	1	0	1	0	0	1
0	0	1	0	0	1	0	0	1
0	0	1	1	0	1	0	0	1
0	1	0	0	1	0	0	1	0
0	1	0	1	1	1	1	0	0
0	1	1	0	0	1	0	0	1
0	1	1	1	0	1	0	0	1
1	0	0	0	1	0	0	1	0
1	0	0	1	1	0	0	1	0
1	0	1	0	1	1	1	0	0
1	0	1	1	0	1	0	0	1
1	1	0	0	1	0	0	1	0
1	1	0	1	1	0	0	1	0
1	1	1	0	1	0	0	1	0
1	1	1	1	1	1	1	0	0

Figure 8.3　Truth table for 2-bit comparator.

8.1.1.1 ARRAY Model

This method uses an array of arrays to represent the truth table as shown in Figure 8.4. Package TRUTH4x5 is used as a convenience to define the data types. Constant NUM_OUTPUTS is the number of outputs, which is five in device COM. The outputs of COM are GE, LE, E, G, and L. Constant NUM_INPUTS is the number of inputs which is four in device COM. The inputs are N_1, N_0, M_1, M_0. Constant NUM_ROWS is the number of rows in the truth table, which is computed as 2^{num_inputs}. For device COM, this is 16. For each row in the truth table, the five outputs can be represented as an array of bits called a WORD. For example, the WORD for row 0 in the truth table is (1,1,1,0,0). The elements of the array are referenced by subscripts numbered from 4 down to 0. The whole truth table (TRUTH) is then an array of WORDS (type MEM). Note: The index values for this array are decimal integers from 0 to 15. The whole array is initialized in the package declaration for TRUTH. The reader should verify these statements:

```
TRUTH(0)=(11100) and TRUTH(6)=01001.
TRUTH(0,4) = TRUTH(0,3) = TRUTH(0,2) = 1.
TRUTH(0,1) = TRUTH(0,0) = 0.
```

Because the input combination associated with a row is an array of bits (type ADDR) and the index value that defines a row of array TRUTH is an integer, we must convert the binary label of the row to an integer. For example, row 0101 is row 5. This conversion is defined by the function INTVAL.

The actual architecture (TABLE) consists of a single process that is activated when any of the device inputs change. The process is a simple table lookup. First, the array of input values is converted to a decimal index (INDEX), using function INTVAL. This decimal index is used to select a WORD from the constant TRUTH that holds the desired values of the five outputs (WOUT). The five outputs are then assigned the desired values after a time delay of D.

```
-- Standard model for truth table using ARRAY structure.
-- This model can be directly translated into a ROM
-- implementation.
package TRUTH4x5 is
  constant NUM_OUTPUTS: INTEGER:=5;
  constant NUM_INPUTS: INTEGER:=4;
  constant NUM_ROWS: INTEGER:= 2 ** NUM_INPUTS;
  type WORD is array(NUM_OUTPUTS-1 downto 0) of BIT;
  type ADDR is array(NUM_INPUTS-1 downto 0) of BIT;
  type MEM is array (0 to NUM_ROWS-1) of WORD;
  constant TRUTH: MEM :=
                ("11100", "01001", "01001", "01001",
                 "10010", "11100", "01001", "01001",
                 "10010", "10010", "11100", "01001",
                 "10010", "10010", "10010", "11100");
  function INTVAL(VAL:ADDR) return INTEGER;
end TRUTH4x5;
package body TRUTH4x5 is
  function INTVAL(VAL: ADDR) return INTEGER is
    variable SUM: INTEGER:=0;
  begin
    for N in VAL'LOW to VAL'HIGH loop
      if VAL(N) = '1' then
        SUM := SUM + (2 ** N);
      end if;
    end loop;
    return SUM;
  end INTVAL;
end TRUTH4x5;
-- Description of COM using table lookup.
use work.TRUTH4x5.all;
architecture TABLE of COM is
begin
  process (N1,N0,M1,M0)
    variable INDEX: INTEGER;
    variable WOUT: WORD;
  begin
    INDEX := INTVAL (N1&N0&M1&M0);
    WOUT := TRUTH (INDEX);
    GE <= WOUT(4) after D;
    LE <= WOUT(3) after D;
    E  <= WOUT(2) after D;
    G  <= WOUT(1) after D;
    L  <= WOUT(0) after D;
  end process;
end TABLE;
```

Figure 8.4   VHDL model for device COM using the ARRAY method.

The ARRAY model can be synthesized into a ROM implementation in which the contents of the ROM is specified by the array constant. The inputs of device COM are connected to the

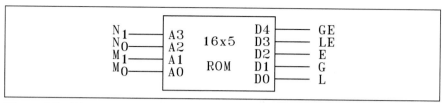

Figure 8.5   Direct translation of ARRAY model into a ROM.

address inputs of the ROM. The data outputs of the ROM are connected to the outputs of the device. For example, a ROM to implement device COM would require 16, 5-bit words (a 16x5 ROM). The pictorial representation of this implementation is shown in Figure 8.5. As shown in Chapter 10, automatic synthesis tools, generally, do not implement in full ROM form, but determine the inferred logic circuits and implement minimal versions of these.

8.1.1.2   CASE Model

In this model, each row of the truth table is associated with one choice in a *case* statement. Figure 8.6 shows a VHDL CASE model for the COM device.

The architecture is named MUX because its structure closely resembles the structure of a hardware multiplexer. The architecture consists of a single process that is executed anytime one of the device inputs changes. The process contains one large *case* statement that assigns logic values to each output for each of the 16 possible input combinations. These values are derived directly from the problem specification. For example, if $N_1N_0M_1M_0 = 0101$, i.e., N = 01 and M = 01, then the numbers are equal. Therefore, GE=LE=E=1 and L=G=0. Similarly, if $N_1N_0M_1M_0 = 0111$, i.e., N = 01 and M = 11, then N < M. Therefore, L=LE=1 and GE=E=G=0. For each combination of input variables, the value of each output is derived directly from the specifications.

This model can be directly translated into a hardware multiplexer. Figure 8.7 shows a multiplexer implementation of the output signal GE. The device inputs, N_1, N_0, M_1, M_0, are connected directly to the address inputs of the MUX, with N_1 connected to the high-order address input, and M_0 connected to the low-order address input. The logic constant connected to data input Di is the value of GE for CASE i. For example, input D5=1 because GE=1 for CASE $N_1N_0M_1M_0 = 0101$. Output signals LE, E, G, and L can also be implemented using four additional multiplexers.

8.1.1.3   Optimization Procedures at the Algorithmic Level

The translation in the previous section is the most straightforward way to produce a MUX implementation. However, there are other more efficient MUX implementations. These can be explored by applying transformations to the VHDL models and translating the resulting models into hardware. Such transformations are referred to as optimization steps. Such steps usually entail intelligent decisions and, therefore, require some form of artificial intelligence techniques to automate the procedures. In this section, we explore optimizations that apply to MUX style implementations.

A relatively straightforward optimization can be achieved by arbitrarily selecting one input variable to be eliminated from the multiplexer address set. To demonstrate the algorithm,

```
-- Standard model for combinational logic,using CASE statement.
-- This model can be directly translated
-- into a MUX implementation.
architecture MUX of COM is
begin
  process(N1,N0,M1,M0)
  begin
    case N1&N0&M1&M0 is
      when "0000" => GE <= '1' after D; LE <= '1' after D;
        E <= '1' after D; G <= '0' after D; L <= '0' after D;
      when "0001" => GE <= '0' after D; LE <= '1' after D;
        E <= '0' after D; G <= '0' after D; L <= '1' after D;
      when "0010" => GE <= '0' after D; LE <= '1' after D;
        E <= '0' after D; G <= '0' after D; L <= '1' after D;
      when "0011" => GE <= '0' after D; LE <= '1' after D;
        E <= '0' after D; G <= '0' after D; L <= '1' after D;
      when "0100" => GE <= '1' after D; LE <= '0' after D;
        E <= '0' after D; G <= '1' after D; L <= '0' after D;
      when "0101" => GE <= '1' after D; LE <= '1' after D;
        E <= '1' after D; G <= '0' after D; L <= '0' after D;
      when "0110" => GE <= '0' after D; LE <= '1' after D;
        E <= '0' after D; G <= '0' after D; L <= '1' after D;
      when "0111" => GE <= '0' after D; LE <= '1' after D;
        E <= '0' after D; G <= '0' after D; L <= '1' after D;
      when "1000" => GE <= '1' after D; LE <= '0' after D;
        E <= '0' after D; G <= '1' after D; L <= '0' after D;
      when "1001" => GE <= '1' after D; LE <= '0' after D;
        E <= '0' after D; G <= '1' after D; L <= '0' after D;
      when "1010" => GE <= '1' after D; LE <= '1' after D;
        E <= '1' after D; G <= '0' after D; L <= '0' after D;
      when "1011" => GE <= '0' after D; LE <= '1' after D;
        E <= '0' after D; G <= '0' after D; L <= '1' after D;
      when "1100" => GE <= '1' after D; LE <= '0' after D;
        E <= '0' after D; G <= '1' after D; L <= '0' after D;
      when "1101" => GE <= '1' after D; LE <= '0' after D;
        E <= '0' after D; G <= '1' after D; L <= '0' after D;
      when "1110" => GE <= '1' after D; LE <= '0' after D;
        E <= '0' after D; G <= '1' after D; L <= '0' after D;
      when "1111" => GE <= '1' after D; LE <= '1' after D;
        E <= '1' after D; G <= '0' after D; L <= '0' after D;
    end case;
  end process;
end MUX;
```

Figure 8.6   VHDL model for device COM using the CASE method.

we arbitrarily select variable M0 for elimination from the address set. Any other variable could have been selected. The new VHDL description after applying the optimization step is shown in Figure 8.8.

The transformation is performed as follows. In Figure 8.6, consider the two cases "0000" and "0001". We will replace these two cases with a single case "000" in the optimized descrip-

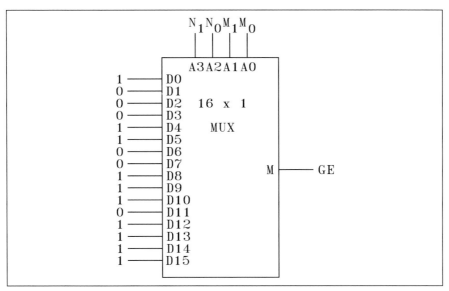

Figure 8.7 Direct translation of CASE model into MUX implementation.

```
-- Improved CASE description. This model directly
-- corresponds to an 8x1 MUX implementation.
architecture MUX3 of COM is
begin
 process (N1, N0, M1, M0)
 begin
  case N1&N0&M1 is
    when "000" => GE <= not M0 after D; LE <= '1' after D;
     E <= not M0 after D; G <= '0' after D; L <= M0 after D;
    when "001" => GE <= '0' after D; LE <= '1' after D;
     E <= '0' after D; G <= '0' after D; L <= '1' after D;
    when "010" => GE <= '1' after D; LE <= M0 after D;
     E <= M0 after D; G <= not M0 after D; L <= '0' after D;
    when "011" => GE <= '0' after D; LE <= '1' after D;
     E <= '0' after D; G <= '0' after D; L <= '1' after D;
    when "100" => GE <= '1' after D; LE <= '0' after D;
     E <= '0' after D; G <= '1' after D; L <= '0' after D;
    when "101" => GE <= not M0 after D; LE <= '1' after D;
     E <= not M0 after D; G <= '0' after D; L <= M0 after D;
    when "110" => GE <= '1' after D; LE <= '0' after D;
     E <= '0' after D; G <= '1' after D; L <= '0' after D;
    when "111" => GE <= '1' after D; LE <= M0 after D;
     E <= M0 after D; G <= not M0 after D; L <= '0' after D;
  end case;
 end process;
end MUX3;
```

Figure 8.8 An improved CASE-style description for device COM.

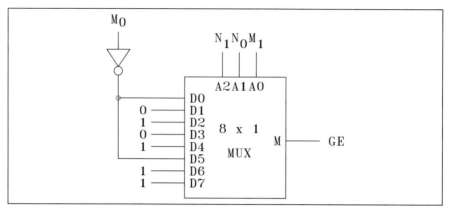

Figure 8.9 Direct translation of improved CASE model into MUX implementation.

tion, based on the variables N1, N0, and M1 as shown in Figure 8.8. The signal assignments for the variables are obtained as follows. For each output variable (GE, LE, E, G, and L), compare the required signal assignments in the original VHDL description to the values of input signal M0. There are four possibilities:

1. If the output signal is equal to variable M0 in both cases, assign the output signal to be M0.
2. If the output signal is the NOT of variable M0 in both cases, assign the output signal to be "not M0".
3. If the output signal is logic '0' in both cases, assign the output signal to be '0'.
4. If the output signal is logic '1' in both cases, assign the output signal to be '1'.

In the original VHDL description, variable GE is assigned '1' in case "0000" (where M0=0) and is assigned '0' in case "0001" (where M0=1). Therefore, assign GE to be equal to "not M0" for case "000" in the optimized description. Also, since LE='1' for both cases "0000" and "0001" in the original VHDL description, assign LE='1' in case "000" in the optimized description. Because G='0' in both cases, assign G='0' in case "000". Finally, as L='0' in case "0000" (where M0='0') and L='1' in case "0001" (where M0='1'), assign L=M0 in the case "000" in the optimized VHDL description. By examining all pairs of cases that differ only in variable M0, the reduced description shown in Figure 8.8 can easily be derived.

    The optimized VHDL description can then be translated directly into a multiplexer implementation. Figure 8.9 shows the improved multiplexer implementation for output GE. This translation is accomplished by mapping the CASE variables (N1, N0, and M1) to the multiplexer address inputs and by using the signal assignment statements to define the data inputs to the multiplexer. For example, "not M0" is connected to data input D0 because the signal assignment statement for case "000" is defined as "not M0". Similarly, D6 is connected to logic "1" because the signal assignment statement for case "110" is a constant 1. Similar implementations for LE, E, G, and L exist.

    Since the algorithm is defined in terms of the VHDL description, it is possible to automate the algorithm by writing a program that examines the CASE-style VHDL description for a device and produces an optimized VHDL description for the device.

8.1.2   Design of Data Flow Models of Combinational Logic in the Behavioral Domain

As indicated in the design methodology flow chart in Figure 1.6, the designer may translate the algorithmic behavioral model into a data flow behavioral model. This translation usually is accompanied by an optimization procedure. Designers frequently use Karnaugh maps, a pictorial representation, to optimize designs with a small number of variables (3-8) and programmed methods, such as the Quine-McCluskey procedure, a textual representation, to optimize problems with an intermediate number of variables (7-20). The translation also uses intelligent analysis to identify relationships among the variables that will simplify the results.

8.1.2.1   Decomposition

Design decomposition can frequently be employed to partition the original device into less complex components. To illustrate design decomposition, consider the comparator of Example 8.1. Observe that the five outputs are by no means independent. By carefully studying the five outputs, we conclude that outputs E, G, and L can be generated from the outputs GE and LE as follows:

```
E = GE and LE
G = GE and not LE
L = LE and not GE
```

Therefore, it is only necessary to find explicit realizations of GE and LE.

8.1.2.2   Synthesis of Data Flow Descriptions

It is possible to obtain data flow descriptions for combinational logic that can be easily translated into gate-level structural domain descriptions directly from the truth table description by using canonical sum of products or canonical product of sums standardized forms. However, these implementations would be very inefficient. Karnaugh maps (K-maps) are frequently used to optimize problems with six or fewer variables. To demonstrate the process, we will use K-maps to optimize the data flow implementation of device COM.

Figure 8.10 shows the optimized two-level sum of products realizations of $GE(Z_0)$ and $LE(Z_1)$, and Figure 8.11 shows the optimized two-level product of sums realizations. Clearly, the product of sums yields fewer gates and fewer gate inputs than the sum of products form. Therefore, we select the POS form for further development.

The VHDL optimized data flow model follows directly from the logic equations by using a signal assignment statement for each binary variable as shown in Figure 8.12. (See Chapter 10 for further discussion of automated synthesis techniques.)

Many other data flow models can be developed directly from the SOP or POS model. For example, the POS model can be transformed into a NOR model by using these rules:

1. Convert each OR term into a NOR term by inserting a "not" operator in front of each parenthesized OR term.
2. Replace each "and" operator with an "or" operator and insert a "not" operator in front of the entire expression.

Figure 8.13 shows the result of applying these transformation rules to the example in Figure 8.12. This model should be simulated to verify the functional integrity.

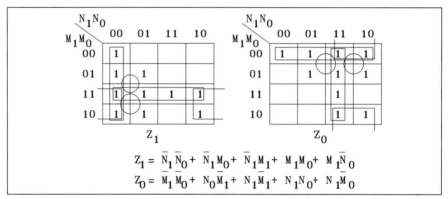

$$Z_1 = \overline{N}_1\overline{N}_0 + \overline{N}_1 M_0 + \overline{N}_1 M_1 + M_1 M_0 + M_1 \overline{N}_0$$

$$Z_0 = \overline{M}_1\overline{M}_0 + N_0\overline{M}_1 + N_1\overline{M}_1 + N_1 N_0 + N_1\overline{M}_0$$

Figure 8.10 Optimum two-level sum of products realizations of $GE(Z_0)$ and $LE(Z_1)$.

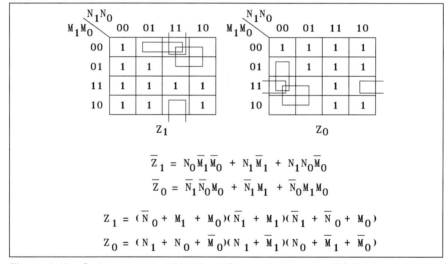

$$\overline{Z}_1 = N_0\overline{M}_1\overline{M}_0 + N_1\overline{M}_1 + N_1 N_0\overline{M}_0$$

$$\overline{Z}_0 = \overline{N}_1\overline{N}_0 M_0 + \overline{N}_1 M_1 + \overline{N}_0 M_1 M_0$$

$$Z_1 = (\overline{N}_0 + M_1 + M_0)(\overline{N}_1 + M_1)(\overline{N}_1 + \overline{N}_0 + M_0)$$

$$Z_0 = (N_1 + N_0 + \overline{M}_0)(N_1 + \overline{M}_1)(N_0 + \overline{M}_1 + \overline{M}_0)$$

Figure 8.11 Optimum two-level product of sums realizations of $GE(Z_0)$ and $LE(Z_1)$.

The optimization techniques discussed here are implemented in most industrial synthesis tools, which we discuss in Chapter 10.

8.1.3 Synthesis of Gate-Level Structural Domain Combinational Logic Circuits

In this section, the translation of combinational logic data flow descriptions to gate-level structural descriptions is discussed. This process is quite straightforward and easily automated.

In general, a given data flow model can be translated into several different gate-level structural domain models. For example, the data flow model, based on the sum of products form, can be translated into a two-level AND-OR circuit, two-level NAND-NAND circuit, two-level OR-

```
      -- Device to compare two binary inputs.
      --
      entity COM is
        generic (D:time);
        port (N1, N0, M1, M0: in BIT;
              GE, LE, E, G, L: out BIT);
      end COM;
      --
      -- Optimum two-level product of sums data flow model.
      --
      architecture POSDF of COM is
        signal Z1,Z0: BIT;
      begin
        Z1 <= (not N0 or M1 or M0) and (not N1 or M1) and
              (not N1 or not N0 or M0);
        Z0 <= (N1 or N0 or not M0) and (N1 or not M1) and
              (N0 or not M1 or not M0);
        LE <= Z1 after D;
        GE <= Z0 after D;
         E <= Z1 and Z0 after D;
         G <= Z0 and not Z1 after D;
         L <= Z1 and not Z0 after D;
      end POSDF;
```

Figure 8.12 VHDL description of data flow model for device COM.

NAND circuit, or any number of multilevel circuits in a straightforward manner. Similarly, the data flow model, based on the product of sums form, may be directly translated into a two-level OR-AND circuit, a two-level NOR-NOR circuit, a two-level AND-NOR (sometimes called AND-OR-INVERT), or any number of multilevel circuits.

To illustrate the general procedure, we will transform the product of sums model for the comparator shown in Figure 8.12 into several gate-level structural models. The sum of products models could be transformed into dual circuits.

Starting with the VHDL description in Figure 8.12, we begin by simply replacing each "and" operator by a 2-input AND gate and each "or" operator by a 2-input OR gate. The result is shown in pictorial form in Figure 8.14.

To verify the function of the pictorial circuit shown in Figure 8.14, we will create a VHDL structural model for simulation. The model shown in Figure 8.15a and Figure 8.15b was simulated to verify the gate-level design. The specification of the gate entities is shown in Figure 8.16a and in Figure 8.16b. This simulation can also be used to verify the timing assumptions made in the algorithmic model. Note: The timing information has been moved from the entity COM to the gate entities (AND, OR, etc.). The simulation at this level provides more accurate timing information. The aggregate delay from the structural simulation can be back annotated into the algorithmic model to provide accurate timing at that level of abstraction. The algorithmic model can then be used as a component in other systems, taking advantage of its relative simplicity in place of the more complex structural-level model. The ability to use higher-level models with accurate information derived from more detailed models allows the simulation of

```
-- Device to compare two binary inputs.
--
entity COM is
  generic (D:time);
  port (N1, N0, M1, M0: in BIT;
        GE, LE, E, G, L: out BIT);
end COM;
--
-- Optimum two-level product of sums data flow model.
-- Converted to two-level-nor data flow model.
--
architecture NORDF of COM is
begin
  process (N1, N0, M1, M0)
    variable Z1,Z0: BIT;
  begin
    Z1 := not(not(not N0 or M1 or M0) or not(not N1 or M1)
          or not(not N1 or not N0 or M0));
    Z0 := not(not(N1 or N0 or not M0) or not(N1 or not M1)
          or not(N0 or not M1 or not M0));
    LE <= Z1 after D;
    GE <= Z0 after D;
    E  <= not(not Z1 or not Z0) after D;
    G  <= not(not Z0 or Z1) after D;
    L  <= not(not Z1 or Z0) after D;
  end process;
end NORDF;
```

Figure 8.13 Another VHDL description of a data flow model for device COM.

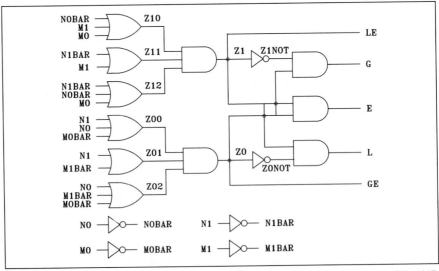

Figure 8.14 Direct translation of COM to gate level, resulting in a two-level OR-AND implementation.

```
      use work.all;
      -- Device to compare two binary inputs.
      entity COM is
        port (N1, N0, M1, M0: in BIT;
              GE, LE, E, G, L: out BIT);
      end COM;
      -- Two level OR-AND implementation derived from POS data
      -- flow model
      architecture TWO_LEVEL_OR_AND of COM is
        signal Z10,Z11,Z12,Z00,Z01,Z02: BIT;
        signal N0BAR,N1BAR,M0BAR,M1BAR: BIT;
        signal Z0,Z1,Z0NOT,Z1NOT: BIT;
        component NOT2G
          generic (D: TIME);
          port (I: in BIT; O: out BIT);
        end component;
        for all: NOT2G use entity NOT2(BEHAVIOR);
        component AND2G
          generic (D: TIME);
          port (I1, I2: in BIT; O: out BIT);
        end component;
        for all: AND2G use entity AND2(BEHAVIOR);
        component AND3G
          generic (D: TIME);
          port(I1,I2,I3: in BIT; O: out BIT);
        end component;
        for all: AND3G use entity AND3(BEHAVIOR);
        component OR2G
          generic (D: TIME);
          port(I1,I2: in BIT; O: out BIT);
        end component;
        for all: OR2G use entity OR2(BEHAVIOR);
        component OR3G
          generic (D: TIME);
          port (I1,I2,I3: in BIT; O: out BIT);
        end component;
        for all: OR3G use entity OR3(BEHAVIOR);
        component WIREG
          port (I: in BIT; O: out BIT);
          end component;
        for all: WIREG use entity WIRE(BEHAVIOR);
      begin
       C1: NOT2G
           generic map (2 ns)
           port map (N0, N0BAR);
       C2: NOT2G
           generic map (2 ns)
           port map (N1, N1BAR);
```

Figure 8.15a   VHDL structural model for COM using two-level OR-AND form.

```
       C3: NOT2G
           generic map (2 ns)
           port map (M0, M0BAR);
       C4: NOT2G
           generic map (2 ns)
           port map (M1, M1BAR);
       C5: OR3G
           generic map (2 ns)
           port map (N0BAR, M1, M0, Z10);
       C6: OR2G
           generic map (2 ns)
           port map (N1BAR, M1, Z11);
       C7: OR3G
           generic map (2 ns)
           port map (N1BAR, N0BAR, M0, Z12);
       C8: AND3G
           generic map (2 ns)
           port map (Z10, Z11, Z12, Z1);
       C9: OR3G
           generic map (2 ns)
           port map (N1, N0, M0BAR, Z00);
       C10:OR2G
           generic map (2 ns)
           port map (N1, M1BAR, Z01);
       C11:OR3G
           generic map (2 ns)
           port map (N0, M1BAR, M0BAR, Z02);
       C12:AND3G
           generic map (2 ns)
           port map (Z00, Z01, Z02, Z0);
       C13:NOT2G
           generic map (2 ns)
           port map (Z1, Z1NOT);
       C14:NOT2G
           generic map (2 ns)
           port map (Z0, Z0NOT);
       C15:AND2G
           generic map (2 ns)
           port map (Z0, Z1, E);
       C16:AND2G
           generic map (2 ns)
           port map (Z0, Z1NOT, G);
       C17:AND2G
           generic map (2 ns)
           port map (Z1, Z0NOT, L);
       C18:WIREG
           port map (Z0, GE);
       C19: WIREG
           port map (Z1, LE);
   end TWO_LEVEL_OR_AND;
```

Figure 8.15b VHDL structural model for COM using two-level OR-AND form (continued).

large complex systems that are not possible using the detailed models that require excessive simulation time.

Another translation is possible by recognizing that two-level OR-AND implementations can always be implemented by replacing each gate with a NOR gate of the appropriate size. This is left as an exercise for the reader in the PROBLEMS section at the end of the chapter.

8.1.4 Summary of Design Activity for Combinational Logic Circuits

The diagram in Figure 8.17 summarizes combinational logic design activities for product of sums models and shows the use of VHDL models in the design process. A similar set of activities could be developed for sum of product models. See Problem 8.6.

From a word description of the device, a truth table is constructed. This is a pictorial representation. The truth table is verified by using algorithmic-level VHDL models in the behavioral domain. Two such models are described. The ARRAY model can be directly translated into a ROM implementation, and the CASE model can be directly translated into a MUX implementation. The MUX implementation can be simplified by using an optimization step.

The algorithmic-level behavioral model is translated into a data flow level behavioral model using Karnaugh maps. Other methods, such as Quine-McCluskey, could also be used. The data flow model is verified through simulation before proceeding.

The data flow model is translated into a gate-level structural model in a straightforward manner. The gate-level structural model is then simulated to verify the functional operation of the circuit at that level. The timing information obtained from the gate-level structural model can be back annotated into the algorithmic and data flow models to allow accurate high level timing simulations.

8.2 DESIGN OF SEQUENTIAL LOGIC CIRCUITS

In this section, we discuss the integration of a hardware description language with the use of state tables to design sequential logic circuits. The design methodology ranges among several levels of representation.

We begin the design process with a word description for an example device.

Example 8.2. Design a serial to parallel converter. Figure 8.18 shows the block diagram. Input CLK is the clock that controls all operations in the system. Reset signal R is synchronous. If R=1 at the end of any clock period, the device must enter the reset state. Input A is asserted for exactly one clock period prior to the arrival of serial data on input D. For the next four clock periods, data arrives serially on line D. The device must collect the four bits of serial data and output them in parallel at output Z, which is a 4-bit vector. During the clock period when the parallel data is present at Z, signal DONE is asserted. The outputs Z and DONE must remain asserted for one full clock period. DONE alerts the destination device that data is present on Z. Figure 8.19 shows the timing. During the clock period when the parallel data is present on Z, the device may receive another pulse on line A indicating that new data will begin arriving on line D during the following clock period. If so, it must be prepared to receive that data. If not, the device goes to the reset state after sending out the parallel data and waits for new data to arrive.

```
entity NOT2 is
  generic (D: TIME);
  port(I: in BIT; O: out BIT);
end NOT2;
--
architecture  BEHAVIOR of NOT2 is
begin
  O <= not I after D;
end BEHAVIOR;
--
entity AND2 is
  generic (D: TIME);
  port (I1,I2: in BIT; O: out BIT);
end AND2;
--
architecture BEHAVIOR of AND2 is
begin
  O <= I1 and I2 after D;
end BEHAVIOR;
--
entity AND3 is
  generic (D: TIME);
  port(I1,I2,I3: in BIT; O: out BIT);
end AND3;
--
architecture BEHAVIOR of AND3 is
begin
  O <= I1 and I2 and I3 after D;
end BEHAVIOR;
--
entity OR2 is
  generic (D: TIME);
  port(I1,I2: in BIT; O: out BIT);
end OR2;
--
architecture BEHAVIOR of OR2 is
begin
  O <= I1 or I2 after D;
end BEHAVIOR;
--
entity OR3 is
  generic (D: TIME);
  port (I1,I2,I3: in BIT; O: out BIT);
end OR3;
--
architecture BEHAVIOR of OR3 is
begin
  O <= I1 or I2 or I3 after D;
end BEHAVIOR;
```

Figure 8.16a Specification of gate entities.

```
entity NOR2 is
  generic (D: TIME);
  port (I1,I2: in BIT; O: out BIT);
end NOR2;
--
architecture BEHAVIOR of NOR2 is
begin
  O <= I1 nor I2 after D;
end BEHAVIOR;
--
entity NOR3 is
  generic (D: TIME);
  port(I1,I2,I3: in BIT; O: out BIT);
end NOR3;
--
architecture BEHAVIOR of NOR3 is
begin
  O <= (I1 or I2) nor I3 after D;
end BEHAVIOR;
--
entity WIRE is
  port (I: in BIT; O: out BIT);
end WIRE;
--
architecture BEHAVIOR of WIRE is
begin
  O <= I;
end BEHAVIOR;
```

Figure 8.16b Specification of gate entities (continued).

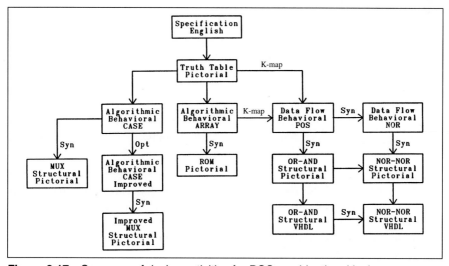

Figure 8.17 Summary of design activities for POS combinational logic.

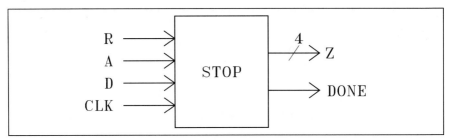

Figure 8.18 Block diagram for serial to parallel converter.

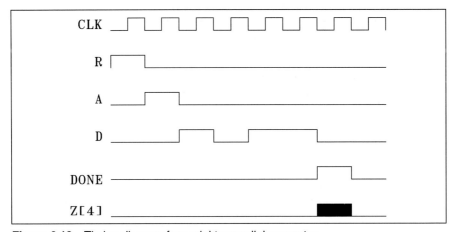

Figure 8.19 Timing diagram for serial to parallel converter.

8.2.1 Moore or Mealy Decision

The first step in the design of a finite state machine (FSM) is to decide whether a Moore or a Mealy circuit should be designed. The reader should already be familiar with the concept of a FSM and with the properties of Moore and Mealy devices. Therefore, we will only briefly review the issues as they relate to state table construction. By definition, the outputs from a Moore machine depend only on the state of the machine and are independent of the values of input signals. In a Mealy machine, the outputs depend on the state and on the input values. Functionally, it is always possible to build any specification as either a Moore or a Mealy machine. The primary difference is in the timing of the outputs. There are three practical effects to consider.

1. In a Moore machine, the outputs settle to their final values a few gate delays after the active edge of the clock. The outputs are constant for the remainder of the clock period, even if the inputs happen to change during the clock period. However, since the outputs are independent of the current inputs, any effect that the current input may have is deferred until the next clock period. One benefit of a Moore device is that it isolates outputs from inputs.

2. In a Mealy machine, since the outputs are a function of the inputs, an output may change in the middle of a clock period if an input changes. This allows the Mealy machine to respond one clock period earlier than the Moore machine to input changes, but also allows the output to follow spurious input changes. Noise on the input lines may be transferred to the outputs.

3. A Moore machine may require more states than the corresponding Mealy machine.

A specification may require a Moore, or a Mealy, or may allow either to be designed. Frequently, the choice involves judgment by the designer.

For the serial to parallel converter of Example 8.2, the output must be present during the clock period following the last input. Since the last input is no longer available, the outputs cannot depend on the inputs. Also, since the outputs are specified to be constant during the entire clock period, a Mealy machine cannot be used. Therefore, we must design a Moore machine.

8.2.2 Construction of a State Table

Having selected a Moore machine as the target device, we can begin state table construction. State table construction requires full utilization of the creative talents of the designer. There are many potential state tables that will meet the requirements of most specifications. Although there is no formal algorithm for state table construction, designers have a greater chance of success if they follow a structured methodology.

Designers often use two tools to simplify state table construction: state diagrams and transition lists. We illustrate both techniques. The state diagram provides a graphical and easily understandable description of the operation of a FSM, but is limited to relatively small devices. The transition list technique is used when the problem is too complex for a state diagram to be constructed.

8.2.3 Creating a State Diagram

Because the serial to parallel converter is a relatively small device, a state diagram is appropriate. To construct a state diagram, start with a state that is easily described in words. If there is a reset state, this is always a good place to start. We recommend writing a complete word description of each state as it is created. This permits easy reference as the design proceeds and provides documentation of the final design. The process is iterative in nature. A state description may evolve as the design progresses, and the final description may be different from that written early in the design process. However, it may be possible to revise a state description written earlier to accommodate new conditions, rather than creating a completely new state.

For the serial to parallel converter, we start with the reset state. We use the label S0 for this state and use this written description.

State S0, the Reset State. The device enters this state at the end of any clock period in which input R=1 regardless of the values of any other input. The device stays in this state until there is a logic 1 on line A. When in state S0, DONE=0, which means the data on line Z is ignored by the destination. This means we are free to place anything at all on output Z.

The next step is to decide what to do when the device is in state S0 for various conditions that may exist on the device inputs. If the device is in state S0, it stays in state S0 if R=1 regard-

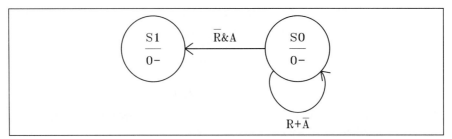

Figure 8.20   First partial state diagram for serial to parallel converter.

less of the values on any other input. When R=0, the device also stays in state S0 when A=0. The complete condition for the device to stay in state S0 is:

$$R + \overline{R}\,\overline{A} = R + \overline{A}$$

If R=0 and A=1 during some clock period, the device must get ready to receive data on line D during the next four clock periods. This means it must enter a new state at the end of the clock period. If S1 denotes the new state, a possible word description of the new state is:

State S1.   The device enters state S1 from state S0 when R=0 and A=1. When the device is in state S1, the first data value is present on line D and must be saved at the end of the clock period for later output. When in state S1, output DONE=0 and Z is unspecified.

As one develops the state descriptions, it is convenient to draw a state diagram as an aid in visualizing the device operation. The state diagram is a directed graph that has a node for each state in the device. The node label appears inside the node symbol. For a Moore machine, the outputs also appear inside the node symbol. There is a directed arc from node X to node Y if the device ever transitions from state X to state Y. Each arc has a label that defines the input condition that causes the state transition, which occurs in coincidence with the clock rise (or fall). If one constructs the state diagram while writing the state descriptions, it becomes a valuable aid in the design process. The state diagram for the part of the serial to parallel converter designed so far appears in Figure 8.20.

In the node for state S0, the notation 0- means that output DONE=0, and output Z is unspecified. The outputs when in state S1 are also 0-. The arc from node S0 to node S1 with the label $\overline{R}$ & A indicates that the device will transition from state S0 to state S1 at the end of any clock period when R=0 and A=1. The arc from node S0 to node S0 with label $R + \overline{A}$ indicates that the device will transition from state S0 to state S0 at the end of any clock period when R=1 or A=0.

Each new state must be analyzed to determine the required state transitions for all possible input conditions. These transitions are added to the state diagram by adding arcs between existing nodes or adding new nodes if necessary. This process continues until all nodes are analyzed. Hopefully, the process will terminate with a finite number of states in the state diagram. This is the reason for the name FSM (finite state machine). It is certainly possible to write a problem specification that requires an infinite number of states. Determining when this might happen is a very difficult problem that is not addressed in this book.

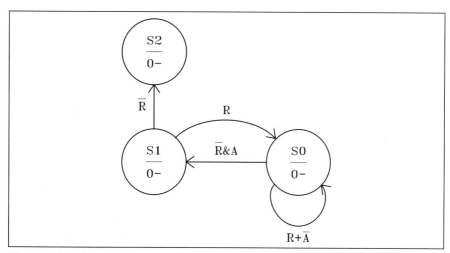

Figure 8.21 Second partial state diagram for the serial to parallel converter.

Analysis of state S1 for the serial to parallel converter proceeds as follows. If R=1 when the device is in state S1, the circuit should reset to state S0. This is indicated on the state diagram by adding an arc from S1 to S0 with label R. If R=0, the device should go to a new state to read in the second of the four input values on line D. Call the new state S2 with this description:

State S2. State S1 transitions to state S2 when R=0. When in state S2, the second data value is present on line D and must be saved at the end of the clock period. In state S2, the output DONE=0, and Z is unspecified.

Therefore, a new node is added to the state diagram for state S2. An arc from S1 to S2 with label $\overline{R}$ indicates the desired state transition. The state diagram now appears as shown in Figure 8.21.

Figure 8.22 shows the final state diagram for device STOP that is created by following the prescribed methodology to completion. The word descriptions for the additional states are described.

State S3. State S3 is entered from state S2 when R=0. When the device is in state S3, the third input is present on line D and must be saved at the end of the clock period. Output DONE=0, and Z is unspecified.

State S4. State S4 is entered from state S3 when R=0. When the device is in state S4, the fourth input is present on line D and must be saved at the end of the clock period. Output DONE=0, and Z is unspecified.

State S5. State S5 is entered from state S4 when R=0. During state S5, the device must output the four received bits in parallel and assert the DONE signal. Output DONE=1, and output Z is set equal to the four received bits.

Notice that a transition from state S5 to state S1 occurs when R=0 and A=1. Input A=1 means that the device must get ready to receive another four bits serially on line D starting in the very next clock period. At this time, we recognize that state S1 can be used again instead of cre-

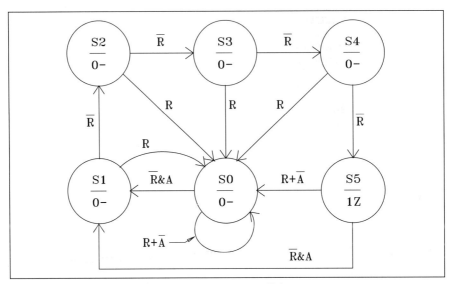

Figure 8.22   Final state diagram for serial to parallel converter.

ating a new state. This closes the state diagram and completes the design. Since we are using state S1 again, we must modify the original word description to reflect the new use. We can only use a state again when its new use is completely compatible with its previous use. The revised description for state S1 is:

Revised State S1.   State S1 is entered from state S0 or from state S5 when R=0 and A=1. When the device is in state S1, the first data value will be present on line D and must be saved at the end of the clock period for later output. When in state S1, output DONE=0, and Z is unspecified.

State S0 should also be revised to include the transition from state S5. This is left as an exercise for the reader.

8.2.4   Transition List Approach

The state diagram is a visual aid that helps one understand relationships between the states. Given the sequence of inputs, it is very easy to trace the sequence of state transitions. However, for large circuits with many states, the diagram becomes messy and difficult to draw in an easily usable form. A textual form of description known as the transition list is frequently used instead. It is created in the same way as the state diagram. Instead of drawing arrows in the state diagram to indicate state transitions, the transitions are listed in a table. Figure 8.23 shows the state transition list for the serial to parallel converter.

Principle of Mutual Exclusion.   The principle of mutual exclusion can be used as an aid in the design process and to check state diagrams for errors. The logic expressions on arcs leaving any node must be pairwise mutually exclusive. That is, no two expressions on different arcs leaving the same node can be true simultaneously. If two such expressions were both logic 1

Present State	Transition Expression	Nest State	Data Transfers	Output
S0	$R + \overline{A}$	S0	None	DONE=0, Z unspecified
S0	$\overline{R}$ & A	S1		
S1	$\overline{R}$	S2	Store bit 1	DONE=0, Z unspecified
S1	R	S0		
S2	$\overline{R}$	S3	Store bit 2	DONE=0, Z unspecified
S2	R	S0		
S3	$\overline{R}$	S4	Store bit 3	DONE=0, Z unspecified
S3	R	S0		
S4	$\overline{R}$	S5	Store bit 4	DONE=0, Z unspecified
S4	R	S0		
S5	$\overline{R}$ & A	S1	None	DONE=1, Z = parallel data out
S5	$R + \overline{A}$	S0		

Figure 8.23 Transition list for serial to parallel converter.

simultaneously, the machine would try to go to two different next states—clearly, that cannot happen.

The test for mutual exclusion is that the logical AND of any two expressions on different arcs leaving a node must be logical 0. For example, for node S0,

$$(\overline{R}A)(R + \overline{A}) = (\overline{R}AR) + (\overline{R}A\overline{A}) = 0 + 0 = 0$$

Therefore, the two logical expressions are mutually exclusive.

This principle is used to check state diagrams during their development. Clearly, transition lists must also satisfy this principle. The different logic expressions listed for a given state must all be mutually exclusive in pairs.

8.2.5 Creating a VHDL Model for State Machines

Either the state diagram or the transition list can be used to create a VHDL model. This model can be used to verify the functional operation of the circuit and thereby, discover any logical errors before proceeding to the next lower level of design.

The model assumes the system is partitioned into a data unit, control unit, and output unit as shown in Figure 8.24. The same clock synchronizes operations in both the control unit and the data unit. In a Moore machine, the outputs are a function of both the control state and logic values stored in the data unit. For a Mealy machine, the outputs may also depend on the inputs as shown by the dashed line.

From the state diagram or the transition list, the VHDL model shown in Figure 8.25 can be easily constructed. Entity declaration STOP defines the inputs and outputs as usual. The inputs R, A, D, and CLK are type BIT; output Z is a 4-bit vector with subscripts numbered from 3 down to 0, and output DONE is type BIT.

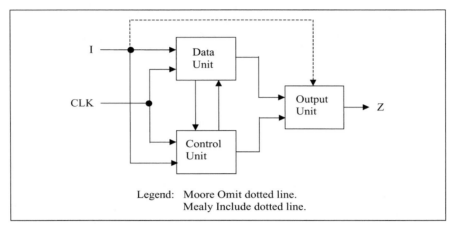

Figure 8.24 Block diagram for model of a state machine.

In the architecture (FSM_RTL), enumeration type STATE_TYPE defines the set of states. The elements in the type list are the names of the states taken directly from the state diagram or the transition list. Signal STATE holds the value of the current state of the device.

At this time we must decide how to save the bits that arrive serially on line D. There are many different ways to do this. Each leads to a different hardware configuration with a different cost. The choice of how to model the saving of the bits has a definite effect on the complexity and cost of the synthesized circuit. In this case, we decide to use a shift register to save the bits. The shift register is declared as a signal of type BIT_VECTOR with the same range as the final output Z.

The architecture includes two processes. Process STATE updates the state of the system at the end of each clock period when CLK rises from logic 0 to logic 1. Therefore, input CLK must be in the activity list for the process. The "if CLK='1' then" statement checks for a rising edge on CLK. If a rising edge is detected, the state is updated. The new state is computed in a *case* statement depending upon the current state. Each choice in the *case* statement corresponds to one possible value for the current state. For example, choice S0 in the *case* statement corresponds to state S0 in the state diagram. The state processing has two parts, the data section and the control section. The word statement for each state defines the data section, and the arrows leaving the state in the state diagram define the control section. For state S0, there is no data section operation. The state diagram implies that the next state should be S0 under the condition that R=1 or A=0. The VHDL model accomplishes this action via the statement "if R=1 or A=0 then STATE <= S0". The state diagram also implies the next state should be S1 if the current state is S0 and the condition "R=0 and A=1" is true. The VHDL model uses the statement "elsif R='0' and A='1' then STATE <= S1" to cause the desired state transition.

The remaining choices in the *case* statement define the other state transitions. The only additional effect is that in some states, such as S1, a data transfer must occur at the same time as the state transition. For example, the statement "SHIFT_REG <= D & SHIFT_REG(3 downto 1)" is added to the data section for state S1 to define the saving of the bit. Recall that the "&" symbol means concatenation. This statement implements a right shift of the register SHIFT_REG.

```
-- Serial to Parallel Converter
entity STOP is
  port (R, A, D, CLK: in BIT;
        Z: out BIT_VECTOR(3 downto 0);
        DONE: out BIT);
end STOP;

-- State Machine Description
-- for Serial to Parallel Converter (STOP)
architecture FSM_RTL of STOP is
  type STATE_TYPE is (S0, S1, S2, S3, S4, S5);
  signal STATE: STATE_TYPE;
  signal SHIFT_REG: BIT_VECTOR (3 downto 0);
begin
-- Process to update state at end of each clock period.
STATE: process (CLK)
  begin
    if CLK='1' then
      case STATE is
        when S0 =>
          -- Data Section
          -- Control Section
            if R='1' or A='0' then
              STATE <= S0;
            elsif R='0' and A='1' then
              STATE <= S1;
            end if;
        when S1 =>
          -- Data Section
          -- Shift in the first bit
            SHIFT_REG <= D & SHIFT_REG(3 downto 1);
          -- Control Section
            if R='0' then
              STATE <= S2;
            elsif R='1' then
              STATE <= S0;
            end if;
        when S2 =>
          -- Data Section
          -- Shift in the second bit
            SHIFT_REG <= D & SHIFT_REG(3 downto 1);
          -- Control Section
            if R='0' then
              STATE <= S3;
            elsif R='1' then
              STATE <= S0;
            end if;
```

Figure 8.25   VHDL model for serial to parallel converter.

The same information can be obtained from the transition list of Figure 8.23. The set of transitions from each state defines the control part of the *case* statement for that state. The word

```
-- Continuation of architecture FSM_RTL of STOP
--
      when S3 =>
        -- Data Section
        -- Shift in the third bit
          SHIFT_REG <= D & SHIFT_REG(3 downto 1);
        -- Control Section
         if R='0' then
           STATE <= S4;
         elsif R='1' then
           STATE <= S0;
         end if;
      when S4 =>
        -- Data Section
        -- Shift in the fourth bit
          SHIFT_REG <= D & SHIFT_REG(3 downto 1);
        -- Control Section
         if R='0' then
           STATE <= S5;
         elsif R='1' then
           STATE <= S0;
         end if;
      when S5 =>
        -- Data Section
        -- Control Section
           if R='0' and A='1' then
             STATE <= S1;
           elsif R='1' or A='0' then
             STATE <= S0;
           end if;
     end case;
   end if;
 end process STATE;
--
-- Output process
--
  OUTPUT: process (STATE)
  begin
    case STATE is
      when S0 to S4 =>
        DONE <= '0';
      when S5 =>
        DONE <= '1';
        Z <= SHIFT_REG;
    end case;
  end process OUTPUT;
end FSM_RTL;
```

Figure 8.25   VHDL model for serial to parallel converter (continued).

descriptions of the states define the data section part. The reader should examine the transition list and verify that the VHDL *case* statement could be constructed by using the information in the list.

Process OUTPUT defines the logic values on the device output signal lines. Since this is a Moore device, the only signal in the activity list is STATE. For a Mealy device, the activity list would include the state signal and the input signals. As a result, the output from the Moore machine only changes when the state changes; whereas, the output from a Mealy machine changes if either the state changes or an input changes. Since this is a Moore machine, the outputs can be defined using a "case STATE" structure. The outputs come from the state diagram (or from the transition list). For example, from the state diagram, the DONE output is logic 1 in state S5. The VHDL program uses the signal assignment statement "DONE <= '1'" to cause this output change. Also, the output Z is defined only during state S5. In the VHDL *case* statement, Z is assigned only during state S5. Due to the semantics of a VHDL signal, the value of Z remains unchanged in the other states, thereby, keeping its value until the machine enters S5 again. According to the problem specification, Z was unspecified at other times. The simulation will, therefore, correctly meet the specifications. However, in any automatic synthesis procedure the simulation of lower-level descriptions must match the simulated behavior at this higher level. Therefore, all lower-level simulations must cause Z to retain its value between assignments. Consequently, the synthesizer will probably create a latch to store Z. Chapter 10 contains detailed discussions of synthesis effects.

To take advantage of the fact that Z is unspecified during states other than S5, we would need to use multivalued logic. See Section 5.3.1.6 for a description of the IEEE nine-valued logic system that includes a mechanism for handling unspecified situations.

8.2.5.1   *Use of Aggregates in VHDL Models of State Tables*

We now illustrate a straightforward way to create a VHDL model of a state machine directly from a state table for the machine. This method was first proposed by Ken Bakalar while at Compass Design Systems. It makes use of records to represent state transitions and uses aggregates to represent the state table directly when the state table exists in the classical format.

The approach is to:

1. Represent the states with an enumeration type.
2. Encode the state table as an array aggregate that implements the next state and output function.
3. Access the array aggregate as required to model the operation of the finite state machine.

To illustrate this process we will use the state machine for the running example in Chapter 1, the system that detects two or more consecutive 1's or 0's on its input. Figure 1.4 gives the state table. Note that the machine implied is a Mealy machine.

An algorithmic model of this state table is shown in Figure 8.26. Note that the enumeration type STATE has the three values S0, S1, and S2, and the state of the machine is stored in signal FSM_STATE, which has type STATE. A record type, TRANSITION, is defined whose two fields are OUTPUT and NEXT_STATE. This is followed by a type declaration for a two dimensional array of TRANSITION records (type TRANSITION_MATRIX). Note that the two

```
entity TWO_CONSECUTIVE is
  port(CLK,R,X: in BIT; Z: out BIT);
end TWO_CONSECUTIVE;
--
architecture FSM of TWO_CONSECUTIVE is
  type STATE is (S0,S1,S2);
  signal FSM_STATE: STATE := S0;
  type TRANSITION is record
    OUTPUT: BIT;
    NEXT_STATE: STATE;
  end record;
  type TRANSITION_MATRIX is array(STATE,BIT) of TRANSITION;
  constant STATE_TRANS: TRANSITION_MATRIX :=
    (S0 => ('0' => ('0',S1), '1' => ('0',S2)),
     S1 => ('0' => ('1',S1), '1' => ('0',S2)),
     S2 => ('0' => ('0',S1), '1' => ('1',S2)));
begin
  process(R,X,CLK,FSM_STATE)
  begin
    if R = '0' then -- Reset
      FSM_STATE <= S0;
    elsif CLK'EVENT and CLK ='1' then -- Clock event
      FSM_STATE <= STATE_TRANS(FSM_STATE,X).NEXT_STATE;
    end if;
    if FSM_STATE'EVENT or X'EVENT then -- Output Function
      Z <= STATE_TRANS(FSM_STATE,X).OUTPUT;
    end if;
  end process;
end FSM;
```

Figure 8.26   Algorithmic model of a state table using records and aggregates.

indices for the array are of types STATE and BIT. The state table is implemented by defining a constant of type TRANSITION_MATIX, initialized with an aggregate that implements the state table. Note that the first index of type STATE selects a row in the array, while the second index of type BIT selects an element within that row. Each array element is a record aggregate of type TRANSITION.

The activities carried out by the FSM occur within a process. Signals R, X, CLK, and FSM_STATE trigger the process. There are three activities within the process: Reset, clock event, and output function. Reset overrides the clock event. The clock event and output function activities access the array and select the appropriate record field. The output function activity is triggered whenever either the input or the machine state changes.

The use of aggregates to represent tables is a powerful and practical approach. The data is organized in a form that allows automation of algorithmic model development from a graphical tabular input.

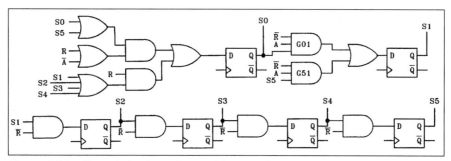

Figure 8.27 Control circuit synthesized from VHDL description for the serial to parallel converter.

To State	From State	Condition	To tate	From State	Condition
S0	S0	$R + \overline{A}$	S1	S0	$\overline{R}$ & A
S0	S1	R	S1	S5	$\overline{R}$ & A
S0	S2	R	S2	S1	$\overline{R}$
S0	S3	R	S3	S2	$\overline{R}$
S0	S4	R	S4	S3	$\overline{R}$
S0	S5	$R + \overline{A}$	S5	S4	$\overline{R}$

Figure 8.28 Control section synthesis list for serial to parallel converter.

8.2.6 Synthesis of VHDL State Machine Models

The model of a finite state machine can be synthesized in a fairly straightforward manner. The model assumes the general form shown in Figure 8.24. We begin by considering the synthesis of the control section.

By assigning each state to a control flip-flop, the synthesizer can specify a complete control unit design by examining the "if ... then" statements in each *when* clause of the *case* statement in the next state process. The control unit shown in Figure 8.27 could be produced by such a synthesizer.

The logic equation for the D input to each flip-flop is obtained by scanning the VHDL *case* statement and creating a list of all conditions that require a transition into the state. Figure 8.28 shows the list for the serial to parallel converter.

The control circuit in Figure 8.27 follows directly from the data in Figure 8.28. For example, consider the logic driving the D input to flip-flop S1. The 3-input AND gate G01 causes the transition to S1 from S0 under the condition $\overline{R}$A listed as the top entry in the right column of Figure 8.28. Similarly, the requirement of a transition to state S1 from state S5 under the condition $\overline{R}$A leads to the 3-input AND gate G51. An OR gate is selected to compute the logical OR of these two functions, and the OR gate output is connected to the D input of flip-flop S1. The outputs of all control flip-flops are available to the data unit and to the output unit in addition to being used in the logic expressions for the D inputs to the control flip-flops.

Data Transfer	State	Condition
SHIFT_REG <= D & SHIFT_REG(3downto 1)	S1	1
SHIFT_REG <= D & SHIFT_REG(3downto 1)	S2	1
SHIFT_REG <= D & SHIFT_REG(3downto 1)	S3	1
SHIFT_REG <= D & SHIFT_REG(3downto 1)	S4	1

Figure 8.29   Synthesis list for data unit for serial to parallel converter.

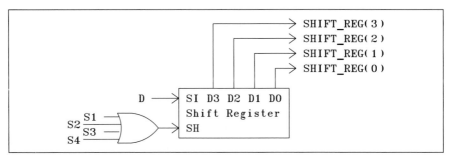

Figure 8.30   Data unit for serial to parallel converter.

This type of control unit design (one flip-flop per state) is called a "one hot" control unit because only one flip-flop is set at any one time. In Chapter 10 we see that this is a favored approach for synthesis with FPGAs.

The data unit can be synthesized from the *case* statement in the next state process by a similar procedure. First, the synthesizer scans the *case* statement and makes a list of all data transfers along with the required conditions. Figure 8.29 shows the list for the serial to parallel converter.

The control signals for each data transfer is obtained by logically ANDing each state with its corresponding condition (in this case, all conditions are 1) and then, logically ORing the results. From the table, the following control expression would be obtained for the shifting operation:

$$\text{SHIFT} = (S1)(1) + (S2)(1) + (S3)(1) + (S4)(1) = S1 + S2 + S3 + S4$$

Figure 8.30 shows the data unit design.

The VHDL code does not contain any information on how to design a shift register. The design of a shift register is another project. Usually, one has already been designed, and the synthesizer merely gets it from a library.

The output unit can be designed by scanning the output process and creating a logic equation from the *case* statement. For the serial to parallel converter, the logic equation for DONE is:

$$\text{DONE} = S5$$

The design for Z is more complex because the original specification only required that Z have a specific value in state S5. However, the VHDL signal semantics dictate that values be held on the Z line indefinitely until they are changed. As mentioned previously, the synthesis tool imple-

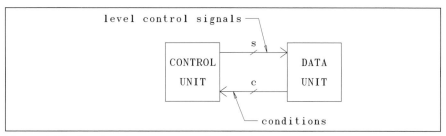

Figure 8.31   Interface between control unit and data unit.

ments Z as a latch (see Chapter 10) that is updated during S5 with the value of the shift register. Thus, the behavioral simulations and the synthesized circuit simulations will agree.

8.3   DESIGN OF MICROPROGRAMMED CONTROL UNITS

This section develops a systematic design procedure for microprogrammed control units. A hardware description language is combined with other design techniques to provide an integrated approach for multiple levels of design.

8.3.1   Interface Between Controller and Device

Figure 8.31 shows the interface between the control unit and the data unit for a typical device. The control unit generates s level control signals, labelled S0 through S(s-1), that are then used by the data unit to control data transfers. The data unit sends c condition signals, labeled C0 through C(c-1), to the control unit. The control unit uses these conditions to select an appropriate sequence of control steps to execute. The selected sequence defines the future data transfers. Both the controller and data unit must be timed by the same clock signal. The clock may be multiphase to allow several sequential steps to be executed during one, master clock period.

8.3.2   Comparison of Hardwired and Microprogrammed Control Units

A hardwired control unit is a custom circuit that includes flip-flops and logic gates. The controller defines a sequence of steps to be performed. Figure 8.32 is a section of a control unit that shows one way to accomplish a conditional transfer of control from step 5 (S5=1) to step 6 (S6=1) if condition A is true. Signal A represents some combination of condition signals from the control unit and signals generated by other control flip-flops. The S5 and S6 signals are used directly by the data unit as control signals. In a hardwired control unit, each control signal is individually derived using control unit signals and data unit condition signals.

In a microprogrammed control unit, the values of all control signals are read from an appropriate address location in ROM. The contents of each address in ROM is called a *control word*. During each clock period, the appropriate level control signals are read from ROM instead of being generated by logic circuitry. In principle this is a simple concept. Figure 8.33 shows a general block diagram for a microprogrammed control unit that includes an a-bit address register (MAR), a w-bit instruction register (MIR), a ROM with 2^a w-bit words, and address generation logic (AGL). At any given time, the address generation logic computes the address from

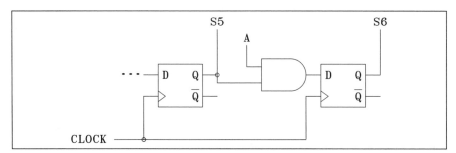

Figure 8.32   Typical control sequence in hardwired control unit.

which the next set of control signals is to be read. Information that is needed to compute the next address is also read from ROM. The address generation logic also uses the condition signals from the data unit to help define the next address. Generation of the ROM address at each time step can be difficult. The address generation logic varies considerably from design to design and is a major contributor to the constraints imposed by the controller.

Timing signals for the control unit and the data unit must be closely coordinated. As illustrated in Figure 8.33, to read a set of control signals from the control ROM, two sequential steps are required. The ROM address must be loaded into the MAR, and the control signal values must then be loaded into the MIR. These operations cannot be done in parallel because the data in the MIR is used to compute the next address. The control unit must perform two sequential data transfers to prepare for one set of data transfers in the data unit. Figure 8.34 shows how control unit operations may be overlapped with data unit operations. On the rising edge of the system clock new values of control signals are loaded into the MIR. On the falling edge of the system clock, the next address is loaded into the MAR. It is also possible to do data transfers in the data unit at this time if the delays in the data unit circuitry are sufficiently small to guarantee that the control signals have time to propagate through the logic of the data unit so that all setup times on data registers are satisfied. On the next rising edge of the system clock, a second set of data transfers may be done in the data unit coincident with the loading of new control signal values in the MIR. Using this arrangement, there could be two sets of sequential operations in the data unit during each major clock cycle.

One could argue that the MIR is not really needed if the ROM output driver circuits can provide sufficient electrical power to accommodate the load generated by the data unit and address generation logic. At first glance, one might be tempted to conclude that the clock speed could be nearly twice as fast. However, this is not the case because the clock period must now satisfy the following constraints:

$$\text{PER\_without\_MIR} > \text{MAR\_delay} + \text{ROM\_delay} + \text{AGL\_delay} + \text{MAR\_setup\_time}$$

In the original design, with the MIR included:

$$\frac{\text{PER}}{2} > \text{MAR\_delay} + \text{ROM\_delay} + \text{MIR\_setup\_time}$$

and

$$\frac{\text{PER}}{2} > \text{MIR\_delay} + \text{AGL\_delay} + \text{MAR\_setup\_time}$$

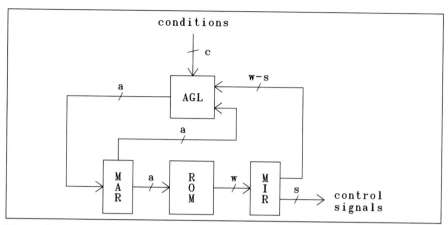

Figure 8.33 Block diagram for a microprogrammed control unit.

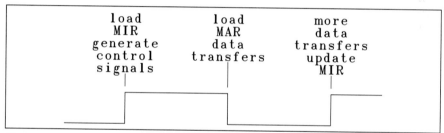

Figure 8.34 Timing in a microprogrammed control unit.

because the events were divided between the low clock and the high clock. Therefore, the whole clock period must satisfy:

$$(PER\_with\_MIR) > MAR\_delay + ROM\_delay + MIR\_setup\_time + MIR\_delay + AGL\_delay \\ + MAR\_setup\_time.$$

Clearly, the difference between the two clock periods is:

$$(PER\_with\_MIR) - (PER\_without\_MIR) = MIR\_setup\_time + MIR\_delay.$$

This is obviously not equal to half of the total clock period. In practice, the clock period might be shortened by 10 to 20 percent by eliminating the MIR.

However, since many ROM chips have low power drivers, which at most can handle one normal load, an output register buffer might be needed anyway. It is comforting to know that the clock period must only be stretched by 10 to 20 percent. Also, if the MIR is eliminated, the data unit delays have a lower tolerance because the ROM delay is inserted in the control signal path. With the MIR in place, the control signals flow from the MIR directly to the data unit without passing through ROM. Without the MIR,

$$CONTROL\_SIGNAL\_path\_delay = MAR\_delay + ROM\_delay + DATA\_UNIT\_delay.$$

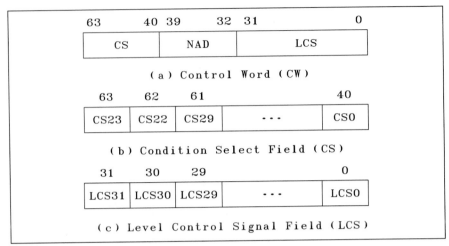

Figure 8.35   Organization of a control word in BMCU.

With the MIR, the control signal path delay is:

$$\text{CONTROL\_SIGNAL\_path\_delay} = \text{MIR\_delay} + \text{DATA\_UNIT\_delay}.$$

Assuming that MAR and MIR have similar delays, the primary difference is the ROM delay, which is typically quite large compared to gate or register delays.

Because the elimination of MIR does not gain that much and several problems are introduced as a result of eliminating MIR, we include the MIR in our presentation. The primary reason that microprogrammed control units are inherently slower than hardwired control units is the presence of ROM. Assuming that the control path delays in the data unit are comparable to the delays in the address generation logic, the total clock period must be stretched by an amount that approximates the ROM delay. A similar analysis would be performed by considering elimination of MAR. See Problem 8.35.

8.3.3   Basic Microprogrammed Control Unit

In this section, we describe an elementary microprogrammed control unit called the basic microprogrammed control unit (BMCU). We will use a simple address generation process that is known as two-way branching. Each control word (CW) contains 64 bits of data organized as shown in Figure 8.35a. Therefore, $w=64$.

The condition select (CS) field (CW(63:40)) contains information needed to select the input condition signal that is used to compute the address of the next instruction (see Figure 8.35b). In the BMCU, we use a "one-hot" code to select the condition signal. Bit CS_i will be logic '1' when condition C_i is selected, and all other bits in the CS field will be logic '0'. Due to the placement of the CS field in the control word, the CS_i select bit is in position CW(40+i) where i ranges from 23 downto 0. In this design, the maximum number of condition signals is 24 ($c=24$). If $C_i=1$ and C_i is selected ($CS_i=1$), the address in the Next ADdress (NAD) field (CW(39:32)) is the address of the next control word to be fetched. If C_i is selected and $C_i=0$, the current memory address (contents of MAR) is incremented to compute the next address.

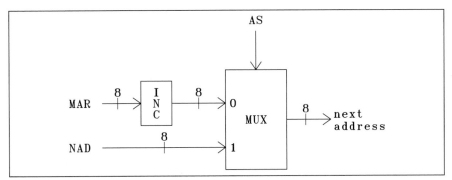

Figure 8.36 Address generation logic for BMCU.

The level control signal (LCS) field (CW(31:0)) contains the values of the control signals LCS31 to LCS0, respectively, as shown in Figure 8.35c. Therefore, in this design, the maximum number of level control signals is 32 ($s=32$). The value of control signal LCSi is equal to bit CW(i) for i ranging from 31 to 0. Because all addresses in ROM must be directly accessible through the NAD field, and the NAD field has 8 bits ($a=8$), the maximum ROM size is 256 words (2^8).

The address generation logic can be designed using a vector MUX as shown in Figure 8.36. The address selection signal (AS) must be logic 1 if the selected condition signal is TRUE. The logic equations for the AS control signal are:

$$AS = (CS23)(C23) \ + (CS22)(C22) \ + ... + (CS0)(C0)$$

or

$$AS = (CW(63))(C23) + (CW(62))(C22) + ... + (CW(40))(C0)$$

Therefore, if the selected condition is true, the contents of NAD will be used as the next address; if the selected condition is false, MAR+1 will be used as the next address.

8.3.4 Algorithmic-Level Model of BMCU

Used in high-level simulations, an algorithmic-level model for BMCU will now be developed. From the previous description, the process model graph shown in Figure 8.37 is easily derived. Note that a RESET signal has been added to initialize the BMCU.

Process MARP represents the MAR. When RESET is active, the MAR will be reset to address X"0000". On the falling edge of the system clock (CLK), a new address (NMAR) will be loaded into the MAR. Process ROMP represents the control ROM. When a new address is loaded into the MAR, a new set of control signals (CWORD) will be read from the control ROM. On the rising edge of the system clock, CWORD will be transferred into the MIR by process MIRP. When RESET is active, MIR will be initialized to the ZERO vector. Process DECODEP simply partitions the CWORD into its constituent parts (CS, NAD, and LCS). Process AGL computes the next address from which the next control word is to be read according to the rules described in the last section. If the selected condition bit is TRUE, the next address is NAD; otherwise it is MAR+1. Signal NMAR represents the computed address. On the falling

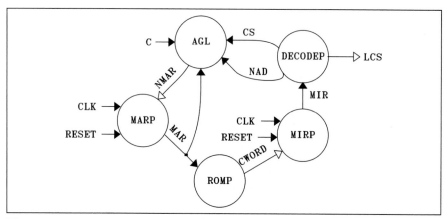

Figure 8.37 Process model graph for BMCU.

edge of the system clock, this new address is loaded into the MAR, and the whole sequence starts again.

From the process model graph, the algorithmic level VHDL model shown in Figure 8.38a and in Figure 8.38b is easily derived using the techniques described in Chapter 5. Note that generic delays for each component were added to allow high-level timing information to be represented.

Package BMCU_FUNCTIONS contains function definitions to convert BIT_VECTORS to INTEGER and to compute MAR+1. These are shown in Figure 8.39.

Notice that the functions are completely general and can be used on vectors of any size or index range. Attributes VEC'right and VEC'left are compared to determine whether the index range is ascending or descending. Each case is treated separately.

Computation of MAR+1 is done, recognizing that the low-order bit is always complemented when a binary vector is being incremented. Further, if there is a string of 1's on the low-order side of the vector, all these 1's are also complemented. The first 0 in the vector that is reached by scanning from right to left is also complemented. The remaining bits are not changed. Variable CARRY is used to determine when to stop complementing bits. As long as the AND of all bits so far is TRUE, then the scan has only found 1 bits so far. When the AND of all bits scanned becomes FALSE, at least one 0 bit has been found. CARRY is initialized to 1 to force complementation of the first bit. This algorithm is illustrated with two examples:

```
      0101 0111        1010 1010
  +            1    +            1
      0101 1000        1010 1011
```

8.3.5 Design of Microprogrammed Controllers for State Machines

Frequently it is possible to directly design microprogrammed controllers from algorithmic-level descriptions of devices. To illustrate the method, we will develop a systematic approach for the design of state machines that are represented using the VHDL modeling style described earlier in this chapter.

```
-- Basic MicroCoded Control Unit
entity BMCU is
  generic (AGL_DELAY, MAR_DELAY, ROM_DELAY, MIR_DELAY: TIME);
  port (C: in BIT_VECTOR (23 downto 0)--Cond. from data unit.
          :=B"0000_0000_0000_0000_0000_0000";
        CLK,RESET: in BIT:='0';  -- System clock and reset.
        LCS: out BIT_VECTOR (31 downto 0));--Level Control
end BMCU;                              --Signals to data unit.
--
-- Algorithmic level VHDL description of BMCU.
use work.BMCU_FUNCTIONS.all;
architecture ALGORITHMIC of BMCU is
  -- MAR is the Memory Address Register for the ROM.
  -- MIR is the Memory Instruction Register for the ROM.
  signal MAR,NMAR: BIT_VECTOR (7 downto 0);-- NMAR is nxt val
  signal CWORD: BIT_VECTOR (63 downto 0);-- Control word
  signal MIR: BIT_VECTOR (63 downto 0);
  signal NAD: BIT_VECTOR (7 downto 0);-- Branch address
  signal CS: BIT_VECTOR (23 downto 0);-- Condition Select
begin
  -- Memory Address Register process.
  MARP: process (CLK, RESET)
  begin
    if RESET='1' then
      MAR <= B"0000_0000" after MAR_DELAY;
    elsif CLK'event and CLK='0' then
      MAR <= NMAR after MAR_DELAY;
    end if;
  end process MARP;
  -- Memory Instruction Register Process
  MIRP: process (CLK, RESET)
  begin
    if RESET='1' then MIR <=
B"00000000_00000000_00000000_00000000_00000000_00000000_00000000_00000000"
      after MIR_DELAY;
    elsif CLK'event and CLK='1' then
      MIR <= CWORD after MIR_DELAY;
    end if;
  end process MIRP;
-- Model code continued in next figure.
```

Figure 8.38a Algorithmic-level VHDL description for BMCU.

A systematic process to design a microprogrammed controller for a state machine from an algorithmic level description in the register transfer style consists of these steps.

Procedure 1.   *Design of microprogrammed controller for a state machine.*

1. Prepare a table showing all transitions from each state and the conditions that define each transition by scanning the statements in the VHDL description.
2. Make a list of the conditions from step 1. After the assignment of ROM addresses to states, some of these conditions may not be needed.
3. Identify a reset state. Frequently, the specifications provide for a reset state. Select an arbitrary reset state if none is defined in the specification.

```
-- ROM process
  ROMP: process (MAR)
    type MEM_TYPE is array (0 to 255
      of BIT_VECTOR (63 downto 0);
    constant MEM: MEM_TYPE:=
  --|<-------- CS --------->|<-NAD->|<------------- LCS ----------->|
    (0=>B"0000_0000_0000_0000_0000_0000_0000_0000_0000_0000_0000_0000_0000_0000_0000_0000",
     1=>B"0000_0000_0000_0000_0000_0000_0000_0000_0000_0000_0000_0000_0000_0000_0000_0000",
     2=>B"0000_0000_0000_0000_0000_0000_0000_0000_0000_0000_0000_0000_0000_0000_0000_0000",
      others => (others => '0'));
  begin
    CWORD <= MEM(BIT_VECTOR_TO_INT(MAR)) after ROM_DELAY;
  end process ROMP;
  -- Decode the MIR to obtain NAD,CS, and LCS.
  DECODEP: process (MIR)
  begin
    NAD <= MIR(39 downto 32);
    LCS <= MIR(31 downto 0);
    CS  <= MIR(63 downto 40);
  end process DECODEP;
  -- Address generation process.
  AGL: process (MAR, NAD, C, CS)
    variable AS: BIT;  -- AS is TRUE if selected condition is TRUE.
  begin
AS:=(CS(23) and C(23)) or (CS(22) and C(22)) or (CS(21) and C(21))
 or (CS(20) and C(20)) or (CS(19) and C(19)) or (CS(18) and C(18))
 or (CS(17) and C(17)) or (CS(16) and C(16)) or (CS(15) and C(15))
 or (CS(14) and C(14)) or (CS(13) and C(13)) or (CS(12) and C(12))
 or (CS(11) and C(11)) or (CS(10) and C(10)) or (CS(9) and C(9))
 or (CS(8) and C(8))   or (CS(7) and C(7))   or (CS(6) and C(6))
 or (CS(5) and C(5))   or (CS(4) and C(4))   or (CS(3) and C(3))
 or (CS(2) and C(2))   or (CS(1) and C(1))   or (CS(0) and C(0));
    case AS is
      when '0' => NMAR <= INC(MAR) after AGL_DELAY;
      when '1' => NMAR <= NAD after AGL_DELAY;
    end case;
  end process AGL;
end ALGORITHMIC;
```

Figure 8.38b Algorithmic-level VHDL description for BMCU.

4. Assign a ROM address (or ROM addresses) to each state. The assignment of ROM addresses will significantly affect the cost and performance of the controller. There is no known efficient algorithm to optimize the address assignment. The optimum assignment is related to finding maximum length paths in the state diagram, which is a difficult graph theoretical problem that is beyond the scope of this book.

5. Remove the redundant signals from the condition list created in step 2, and assign the condition signals to the condition inputs in an arbitrary manner. This assignment will not affect the cost or performance of the controller. Of course, each condition needs to be passed only once although it may be used in many different branching situations.

6. Create a list of transfers during each state and a list of outputs required during each state. Include the conditions required for the transfers and outputs in this table.

7. From the list of transfers and outputs created in step 6, generate a list of control signals needed to time the transfers and outputs. Generally, a separate control signal is needed

```
-- Package of functions for BMCU.
--
package BMCU_FUNCTIONS is
  function BIT_VECTOR_TO_INT(VEC: BIT_VECTOR) return INTEGER;
  function INC(A: BIT_VECTOR) return BIT_VECTOR;
end BMCU_FUNCTIONS;
--
package body BMCU_FUNCTIONS is
  -- Convert a BIT_VECTOR to INTEGER.
  function BIT_VECTOR_TO_INT(VEC: BIT_VECTOR)return INTEGER is
    variable SUM, WT: INTEGER:=0;
  begin
    if VEC'right <= VEC'left then
      for N in VEC'right to VEC'left loop
        if VEC(N)='1' then
          SUM := SUM + (2 ** WT);
        end if;
        WT := WT+1;
      end loop;
    else
      for N in VEC'right downto VEC'left loop
        if VEC(N)='1' then
          SUM := SUM + (2 ** WT);
        end if;
        WT := WT+1;
      end loop;
    end if;
    return SUM;
  end BIT_VECTOR_TO_INT;
-- Increment a BIT_VECTOR.
  function INC(A: BIT_VECTOR) return BIT_VECTOR is
    variable CARRY: BIT:='1';
    variable RESULT: BIT_VECTOR (A'range);
  begin
    if A'right <= A'left then
      for N in A'right to A'left loop
        if CARRY = '1' then
          RESULT(N) := not A(N);
        else
          RESULT(N) := A(N);
        end if;
        CARRY := CARRY and A(N);
      end loop;
    else
      for N in A'right downto A'left loop
        if CARRY = '1' then
          RESULT(N) := not A(N);
        else
          RESULT(N) := A(N);
        end if;
        CARRY := CARRY and A(N);
      end loop;
    end if;
    return RESULT;
  end INC;
end BMCU_FUNCTIONS;
```

Figure 8.39 Package of functions for BMCU algorithmic-level model.

PS	NS	Condition	Type
S0	S0	$R + \overline{A}$	B
	S1	$\overline{R}A$	I
S1	S2	$\overline{R}$	I
	S0	R	B
S2	S3	$\overline{R}$	I
	S0	R	B
S3	S4	$\overline{R}$	I
	S0	R	B
S4	S5	$\overline{R}$	I
	S0	R	B
S5	S1	$\overline{R}A$	B
	S0	$R + \overline{A}$	B

Figure 8.40   Table of transitions created by executing step 1 of Algorithm 1.

for each transfer-condition pair and for each output-condition pair. However, some simplifications are possible if relationships among the control signals can be recognized by an optimizing procedure. Assign the control signals to control outputs in an arbitrary manner.

8. Draw a block diagram showing the condition and control signals produced. Include the reset and clock signals.

9. Determine the ROM program using the information in the lists produced in steps 1-8. This procedure is similar to that performed by an assembler program.

As an example, consider the serial to parallel converter (STOP) in Section 8.2. As defined in step 1, scan the VHDL code statements in Figure 8.25a and in Figure 8.25b to produce the table in Figure 8.40. Ignore the column "Type" for now. It will be added in step 4.

The list of conditions produced by step 2 is:

$$R + \overline{A}, \overline{R}A, \overline{R}, R$$

For step 3, the reset state S0 is defined in the problem specification.

Although the optimum assignment of addresses in step 4 is a difficult problem, the following simple approach will produce a reasonable assignment in a short time. Note: This procedure is very similar to the address assignment in the first pass of an assembler. In fact, programming ROM for a controller is often accomplished using a tool similar to an assembler.

Procedure 2.   *Assigning ROM addresses.*

1. Let variable P be the current state. Initialize P equal to the reset state. Let LC represent the location counter, which contains the next available memory address. Initialize LC=0.

2. Assign state P to address LC. If all states are assigned to memory addresses, then stop. Otherwise, go to step 3.

State	ROM Address
S0	0
S1	1
S2	2
S3	3
S4	4
S5	5

Figure 8.41 ROM addresses assigned to states in step 4 of Procedure 1.

3. To determine the number of ROM addresses needed to implement state P, let NSP be the number of next states for state P.

   a. If NSP=1, set LC = LC +1. Only one ROM address is needed. If the next state is not yet assigned, set P equal to the next state, mark the next state with an I (for increment), and go to step 2. If the next state is already assigned to a ROM location, then mark the next state with B (for branch), arbitrarily select any unassigned state for P, and go to step 2.

   b. If NSP>1, and if at least one of the next states of P are not yet assigned to a ROM address, then arbitrarily set Q equal to one of the unassigned next states of P. This is where the optimality breaks down. An optimal algorithm would try to make an optimum choice for Q. As we show later, NSP-1 memory locations are required for state P. Therefore, mark Q with I, mark all other next states of P with B, set LC = LC + NSP -1, set P=Q, and go to step 2.

   c. If NSP>1, and if all next states are already assigned to ROM addresses, then NSP memory locations are required to implement state P. Mark all next states with B, set LC=LC+NSP, arbitrarily select any unassigned state for P, and go to step 2.

It is convenient to place the marks for B and I in the table already constructed in step 1. Execution of Procedure 2 for the example problem produces the column "Type" in Figure 8.40. Figure 8.41 shows the assignment of states to ROM addresses produced by Procedure 2.

The next available address would be address 7 because state S5 requires two memory locations.

Given the address assignments in Figure 8.41 and the information in Figure 8.40, the ROM program can be easily constructed in a manner similar to pass 2 of an assembler. Before doing that, we need to define the condition and control signals that are required.

In step 5 of Procedure 1, the condition signals that must be passed from the data unit to the control unit are those that correspond to lines of Figure 8.40 that are marked with a "B". Figure 8.42 shows an arbitrary assignment of conditions to the condition inputs of the control unit. Note that condition $\overline{R}$ is not needed because all next states requiring condition $\overline{R}$ are marked with I indicating that the transitions will be accomplished by incrementing the MAR.

In step 6 of Procedure 1, scan the VHDL statements again to create a list of transfers and outputs. There is no reason why the lists of steps 1 and 6 cannot be done during the same scan of

Input	Condition
C23	R
C22	$R + \overline{A}$
C21	$\overline{R}A$

Figure 8.42 Assignment of conditions made in Step 5 of Procedure 1.

State	Transfer	Cond	DONE	Cond	Z	Cond
S0	-	-	0	1	-	-
S1	Shift SR	1	0	1	-	-
S2	Shift SR	1	0	1	-	-
S3	Shift SR	1	0	1	-	-
S4	Shift SR	1	0	1	-	-
S5	-	-	1	1	SR	1

Figure 8.43 Table of transfers and outputs generated at step 6 of Procedure 1.

Output	Signal	Transfer/Output
LCS31	SHIFT	$SR \leftarrow D \ \& \ SR[3:1]$
LCS30	DONE_CONTROL	DONE=1, Z=SR

Figure 8.44 Table of control signals generated at step 7 of Procedure 1.

the VHDL statements. We separated them into two steps for clarity only. Figure 8.43 shows the table that results from executing step 7 of Procedure 1. Note that a condition of 1 means an unconditional transfer or output. In this example, all transfers and outputs are unconditional.

Figure 8.44 shows the list of control signals that must be passed from the controller to the data unit. Because there are only two distinct output situations, the outputs can be controlled by one control signal called DONE_CONTROL. There is only one transfer required, indicated by control signal SHIFT. Figure 8.44 shows an arbitrary assignment of control signals to the output ports of the control unit.

Now that we have identified all the signals that pass between the controller and data unit, we can construct the block diagram for the data unit (step 8). Figure 8.45 shows the block diagram for the STOP device. Note that all unused inputs are connected to a constant logic '0' (represented by symbolic name F). Using the information generated by Procedure 1, it is fairly straightforward to fill in the contents of the control ROM (step 9). Figure 8.46 shows the final contents of ROM.

The contents of ROM address 0 follows from the information in the tables. From Figure 8.40, the condition for the branch operation (the next state labeled B) is C22 ($R + \overline{A}$); therefore,

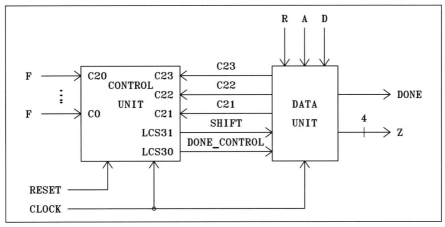

Figure 8.45 Block diagram for STOP showing intermodule signals.

State	Add	CS				NAD	LCS			CWORD bit
		23	22	21	20...0	7...0	31	30	29...0	
		63	62	61	60...40	39...32	31	30	29...0	
S0	0	0	1	0	0...0	00000000	0	0	----	
S1	1	1	0	0	0...0	00000000	1	0	----	
S2	2	1	0	0	0...0	00000000	1	0	----	
S3	3	1	0	0	0...0	00000000	1	0	----	
S4	4	1	0	0	0...0	00000000	1	0	----	
S5	5	0	0	1	0...0	00000001	0	1	----	
S5	6	0	1	0	0...0	00000000	0	1	----	

Figure 8.46 ROM contents for control ROM of STOP.

the CS field has a single 1 in column CS22. Since the branch is to state S0 when condition C22 is TRUE, the NAD field has the address of S0 (assigned to be address 0) in the NAD field. The 0 in SHIFT (LCS31) means that the shift operation is not performed during this state. The 0 in DONE_CONTROL (LCS30) means that DONE=0, and Z is undefined. When C22 is FALSE, the ROM address is incremented, and the device enters state S1 at address 1. The contents of addresses 1-4 are similarly determined.

The programming of the control signals for state S5 is more complex because the branching is not the two-way branch built into the controller. There are two direct branches instead of one direct branch and one increment branch. When there is more than one direct branch, the controller must perform them by examining the conditions one at a time. This requires more than one ROM address. In the present case, at ROM address 5, if condition C21 ($\overline{R}A$) is TRUE, the device branches to address 1 (state S1) as indicated by the NAD field. If C21 is FALSE, ROM goes to address 6 next, where condition C22 ($R + \overline{A}$) is checked. If C22 is TRUE, a branch to address 0 (state S0) is executed. In effect, the execution of state 5 is stretched over two microinstruction cycles. SHIFT (LCS31) is 0 because we do not want to shift during this cycle. DONE_CONTROL (LCS30) is 1 to indicate that the result is present on the Z outputs during this state. The stretching of state 5 over two microinstruction cycles will change the timing, relative to the problem specification. See Section 8.36 on limitations of the design for a discussion of this issue and also the list of problems at the end of this chapter.

Defining the ROM contents completes the design of the controller. Since the controller hardware is already designed, our only design task is to program a ROM chip and insert it into the controller board. The VHDL language can be used to verify the design prior to ROM programming with little additional work. The algorithmic model of the BMCU is modified by inserting a declaration for the ROM contents and by changing the name of the architecture from ALGORITHMIC to STOP. The entity declaration is not changed. Figure 8.47 shows the resulting architecture for the control unit.

The data unit still needs to be designed. This is a general design process that is independent of whether the control unit is microprogrammed or hardwired. Figure 8.48 shows a data flow model of the data unit.

Figure 8.49 shows a test bench that was used to verify the correct operation of the design. The control unit and the data units are represented as components in the test bench. The components are interconnected according to the block diagram of Figure 8.45. Because all inputs to a module must be connected to something, the unused condition inputs to the controller are connected to a constant logic 0 (signal F). Unused outputs are left open. This is consistent with good design practice in actual hardware. Notice that typical delays are included to simulate actual hardware delays.

8.3.6   Generalities and Limitations of Microprogrammed Control Units

One of the primary advantages of microprogrammed control units is that the same controller board, or controller chip, can be used for many different designs. In this section, we discuss the limits of applications by using the BMCU as an example.

The BMCU is limited to examining 24 conditions in the data unit because there are only 24 condition inputs built into the device. Since there are only 32 control outputs available, the controller is limited to controlling 32 independent operations in the data unit. Since the NAD field has 8 bits, the ROM program can have at most 256 control words. All these limitations arise because of a fixed size for various fields in the control word. Similarly, any standard microprogrammable controller will be limited by the size of the fields in its control word.

Perhaps the most severe limitation of a microprogrammed design is the limit imposed by the method used to generate the next address. The BMCU uses a very primitive two-way branching scheme by selecting one of the condition inputs, executing a conditional branch to an arbitrary location in ROM if the condition is TRUE and branching to the next sequential ROM location if the condition is FALSE. Although it is possible to execute a three-way branch, or an m-way branch, by executing multiple two-way branch instructions, this adversely affects the timing in the controller. For example, to execute a three-way branch involving two conditions and a default increment, two complete microinstruction cycles are required. In general, to execute an m-way branch would require m-1 microinstruction cycles. As a result, selecting the next address consumes a variable amount of time depending on the complexity of the branching oper-

```
-- Algorithmic level VHDL description of BMCU.
use work.BMCU_FUNCTIONS.all;
architecture STOP of BMCU is
  -- MAR is the Memory Address Register for the ROM.
  -- MIR is the Memory Instruction Register for the ROM.
  signal MAR, NMAR: BIT_VECTOR (7 downto 0);-- NMAR is next value.
  signal CWORD: BIT_VECTOR (63 downto 0);-- Control word from ROM.
  signal MIR: BIT_VECTOR (63 downto 0);--Word read from ROM.
  signal NAD: BIT_VECTOR (7 downto 0); -- Branch address.
  signal CS: BIT_VECTOR (23 downto 0); -- Condition Select Field
begin
  -- Memory Address Register process.
  MARP: process (CLK, RESET)
  begin
    if RESET='1' then
      MAR <= B"0000_0000" after MAR_DELAY;
    elsif CLK'event and CLK='0' then
      MAR <= NMAR after MAR_DELAY;
    end if;
  end process MARP;
  -- Memory Instruction Register Process
  MIRP: process (CLK, RESET)
  begin
    if RESET='1' then MIR <=
      B"00000000_00000000_00000000_00000000_00000000_00000000_00000000_00000000"
      after MIR_DELAY;
    elsif CLK'event and CLK='1' then
      MIR <= CWORD after MIR_DELAY;
    end if;
  end process MIRP;
  -- ROM process
  ROMP: process (MAR)
    type MEM_TYPE is array (0 to 255) of BIT_VECTOR (63 downto 0);
    constant MEM: MEM_TYPE:=
-----|<------- CS ------->|<-NAD->|<---------- LCS ---------->|
  (0=>B"01000000_00000000_00000000_00000000_00000000_00000000_00000000_00000000",
   1=>B"10000000_00000000_00000000_00000000_10000000_00000000_00000000_00000000",
   2=>B"10000000_00000000_00000000_00000000_10000000_00000000_00000000_00000000",
   3=>B"10000000_00000000_00000000_00000000_10000000_00000000_00000000_00000000",
   4=>B"10000000_00000000_00000000_00000000_10000000_00000000_00000000_00000000",
   5=>B"00100000_00000000_00000000_00000001_01000000_00000000_00000000_00000000",
   6=>B"01000000_00000000_00000000_00000000_01000000_00000000_0000_000_00000000",
      others => (others => '0'));
  begin
    CWORD <= MEM(BIT_VECTOR_TO_INT(MAR)) after ROM_DELAY;
  end process ROMP;
-- Code is continued on next page.
```

Figure 8.47 VHDL model for STOP control unit.

```
-- Decode the MIR to obtain NAD and S.
  DECODEP: process (MIR)
  begin
    NAD <= MIR(39 downto 32);
    LCS <= MIR(31 downto 0);
    CS  <= MIR(63 downto 40);
  end process DECODEP;
  -- Address generation process.
  AGL: process (MAR, NAD, C, CS)
    variable AS: BIT;  -- Selected condition is TRUE.
  begin
    AS := (CS(23) and C(23)) or (CS(22) and C(22)) or
          (CS(21) and C(21)) or (CS(20) and C(20)) or
          (CS(19) and C(19)) or (CS(18) and C(18)) or
          (CS(17) and C(17)) or (CS(16) and C(16)) or
          (CS(15) and C(15)) or (CS(14) and C(14)) or
          (CS(13) and C(13)) or (CS(12) and C(12)) or
          (CS(11) and C(11)) or (CS(10) and C(10)) or
          (CS( 9) and C( 9)) or (CS( 8) and C( 8)) or
          (CS( 7) and C( 7)) or (CS( 6) and C( 6)) or
          (CS( 5) and C( 5)) or (CS( 4) and C( 4)) or
          (CS( 3) and C( 3)) or (CS( 2) and C( 2)) or
          (CS( 1) and C( 1)) or (CS( 0) and C( 0));
    case AS is
      when '0' => NMAR <= INC(MAR) after AGL_DELAY;
      when '1' => NMAR <= NAD after AGL_DELAY;
    end case;
  end process AGL;
end STOP;
```

Figure 8.47   VHDL model for STOP control unit (continued).

ation. In effect, certain microinstruction cycles will be "stretched" to allow time to compute the next address. This stretching of the microinstruction cycle might cause timing problems at the device primary inputs and outputs, in addition to requiring more execution time.

For the STOP device, the fact that two microinstructions are required to execute the branching in state 5 means that if the A input pulse occurs during the second of these microinstruction cycles, it will be missed because the controller is not looking at the A signal at that time. In general, such problems can only be solved by using time-consuming handshaking protocols between devices. Note that there is usually no problem between the controller and data unit as we are designing both devices. However, the primary input and output signals are often impossible to implement according to original specifications if fixed timing relationships are required between the current device and other devices. For example, the STOP device specified that the A pulse would be present for only one clock period, and that data would immediately follow during the next clock period. It is probably impossible to use the BMCU as a microcontroller to accommodate complex fixed timing relationships among input signals. In fact, the addressing scheme of any microprogrammed controller will impose timing constraints on the implemented system.

```
-- Data Unit for Serial to Parallel Converter
--
entity DATASTOP is
  generic (SHIFT_DELAY, GATE_DELAY: TIME);
  port (R, A, D, CLK, SHIFT, DONE_CONTROL: in BIT;
        Z: out BIT_VECTOR(3 downto 0);
        DONE, C23, C22, C21: out BIT);
end DATASTOP;
--
architecture DATAFLOW of DATASTOP is
  signal SHIFT_REG: BIT_VECTOR (3 downto 0);
begin
  --
  -- Shift register
  --
  SHIFT_REG <= D & SHIFT_REG(3 downto 1) after SHIFT_DELAY
    when CLK'event and CLK='1' and SHIFT='1'
    else SHIFT_REG;
  --
  -- Condition signals needed in the control unit
  --
  C23 <= R;
  C22 <= R or not A after GATE_DELAY;
  C21 <= not R and A after GATE_DELAY;
  --
  -- Output signals
  --
  DONE <= DONE_CONTROL;
  Z <= SHIFT_REG;
end DATAFLOW;
```

Figure 8.48   VHDL model for STOP data unit.

8.3.7 Alternative Condition Select Methods

In the BMCU, one bit of the CS field was devoted to each condition. This method is called linear select, or one-hot. There are a large number of codes that might be used to select the condition. The major trade-off is the cost of decoding vs. the number of bits required in the control word. For linear select, there is no cost for decoding, but the number of conditions is limited to the number of bits in the condition select field of the control word. This is one end of a spectrum of choices.

At the other end of the spectrum, the CS field can be fully decoded. The 16 bits in the CS field of the BMCU can be decoded by a 16-input, 65,536-output decoder circuit. This represents the maximum number of conditions that can be selected using a 16-bit data field and also represents the maximum cost for decoding since such a decoder would be very expensive. From another perspective, the control word for BMCU could be shortened from 64 bits to 52 bits by using a 4-bit CS field and a 4x16 decoder to generate the condition select signal. The ROM space savings almost certainly offsets the cost of the decoder in this case since a 4x16 decoder requires only sixteen 4-input gates and four inverters.

```
use work.all;
entity TEST_BENCH is
end TEST_BENCH;
--
architecture BMCU_TEST of TEST_BENCH is
   signal R,A,D,CLK,INIT,RESET: BIT;
   signal C23, C22, C21: BIT;
   signal SHIFT, DONE_CONTROL: BIT;
   signal DONE: BIT;
   signal Z: BIT_VECTOR (3 downto 0);
   signal X: BIT_VECTOR (3 downto 1);
   signal F: BIT; -- Constant false signal.
   --
   component DATA_UNIT
      generic (SHIFT_DELAY, GATE_DELAY: TIME);
      port (R, A, D, CLK, SHIFT, DONE_CONTROL:  in BIT;
            Z: out BIT_VECTOR(3 downto 0);
            DONE, C23, C22, C21: out BIT);
   end component;
   --
   component MICRO_CONTROL_UNIT
    generic (AGL_DELAY, MAR_DELAY, ROM_DELAY, MIR_DELAY,
            MIR_SETUP, MAR_SETUP: TIME);
    port (C: in BIT_VECTOR (23 downto 0); -- Conditions.
          CLK, RESET: in BIT;  -- System clock and reset.
          LCS: out BIT_VECTOR (31 downto 0));-- Level CS.
   end component;
   --
   -- Component configuration statememts.
for L1: DATA_UNIT  use entity DATASTOP(DATAFLOW);
   for L2: MICRO_CONTROL_UNIT use entity BMCU(STOP);
begin
   L1: DATA_UNIT
     generic map (20 ns, 10 ns)
     port map (R, A, D, CLK, SHIFT, DONE_CONTROL, Z,
               DONE, C23, C22, C21);
   L2: MICRO_CONTROL_UNIT
     generic map (50 ns, 20 ns, 50 ns, 20 ns, 5 ns, 5 ns)
     port map (C(23) => C23, C(22) => C22, C(21) => C21,
          C( 0) => F, C( 1) => F, C( 2) => F, C( 3) => F,
          C( 4) => F, C( 5) => F, C( 6) => F, C( 7) => F,
          C( 8) => F, C( 9) => F, C(10) => F, C(11) => F,
          C(12) => F, C(13) => F, C(14) => F, C(15) => F,
          C(16) => F, C(17) => F, C(18) => F, C(19) => F,
          C(20) => F,
          RESET => RESET, CLK => CLK,
          LCS(31) => SHIFT, LCS(30) => DONE_CONTROL);
   -- Code continued on next page.
```

Figure 8.49   VHDL model for STOP system.

```
          -- Clock process
            process
            begin
              CLK <= '0';
              wait for 200 ns;
              CLK <= '1';
              wait for 200 ns;
            end process;
            --
            -- Process to provide inputs.
            F <= '0';
            RESET <= '1', '0' after 30 ns;
            process (X)
            begin
              R <= X(3);
              A <= X(2);
              D <= X(1);
            end process;
            --
            -- Assignment of values to input X
            X <= "100" after    50 ns, "010" after   650 ns,
                 "001" after  1050 ns, "000" after  1450 ns,
                 "001" after  1850 ns, "001" after  2250 ns,
                 "000" after  2650 ns, "000" after  3050 ns,
                 "010" after  3450 ns, "000" after  3850 ns,
                 "001" after  4250 ns, "001" after  4650 ns,
                 "000" after  5050 ns, "010" after  5450 ns,
                 "001" after  5850 ns, "001" after  6250 ns,
                 "011" after  6650 ns, "010" after  7050 ns,
                 "010" after  7450 ns, "000" after  7850 ns,
                 "101" after  8250 ns, "001" after  8650 ns,
                 "001" after  9050 ns, "000" after  9450 ns,
                 "001" after  9850 ns, "000" after 10250 ns,
                 "001" after 10650 ns, "000" after 11050 ns,
                 "000" after 11450 ns;
          end BMCU_TEST;
```

Figure 8.49   VHDL model for STOP system (continued).

There is a wide variety of alternatives between the two extremes. For example, a k-out-of-n code is particularly easy to decode and provides for more conditions than the linear select code. In a k-out-of-n code, each condition is specified by a code word that has exactly k 1's and $n-k$ 0's. Clearly, the linear select code is a 1-out-of-n code. We will use a 6-bit CS field to illustrate these concepts.   The linear select method requires no decoding logic, but accommodates only six independent conditions. Full decoding allows $2^6 = 64$ conditions and requires sixty-four 5-input gates and six inverters to decode. For a 2-out-of-6 code, there are 15 ways to specify exactly two 1's and four 0's. This corresponds to the number of ways to select two items out of a set of six items. Each select signal requires a two-input AND gate. Therefore, the total cost is fifteen 2-input AND gates and 15 conditions are available. A 3-out-of-6 code requires twenty 3-input AND gates and provides 20 condition signals. Figure 8.50 summarizes the spectrum of

Code	Maximum Number of Conditions	Cost
Linear	6	None
2-out-of-6	15	15 Two-input AND gates
3-out-of-6	20	20 Three-input AND gates
Full	64	64 Five-input AND gates

Figure 8.50 Partial spectrum of choices for decoding a 6-bit field.

6362 61	56 55	48 47	40 39	32 31	0
AT CS	NAD1	NAD2	NAD3	LCS	

Figure 8.51 Control word for flexible branching microcontroller.

choices discussed so far. Many other codes are available with diverse properties. Discussion of these are beyond the scope of this book.

8.3.8 Alternative Branching Methods

There are many ways to modify the address generation logic that will reduce the limitations imposed on the design. The major trade-offs are (1) additional branching capability at a cost of additional bits in the control word and/or (2) additional hardware cost for the address generation logic. Figure 8.51 shows a control word for a microprogrammable controller that has four different branching methods. The 2-bit AT field selects one of four address generation types. The 6-bit CS field will be fully decoded to select one of 64 condition inputs. The three 8-bit fields labeled NAD1, NAD2, and NAD3 will be used to compute the target address. The ROM address will be a 16-bit address, so ROM programs can be 64K words in size. The 32-bit LCS field contains 32 level control signals that can be used to control data transfers in the control unit.

When AT=00, the controller unconditionally branches to the 16-bit ROM address computed by concatenating fields NAD1 and NAD2. The CS and NAD3 fields are not used for this addressing mode. Figure 8.52(a) illustrates the format for an unconditional branch operation.

When AT=01, the controller executes a two-way branch exactly as in the BMCU. The CS field is fully decoded to select one of sixty-four condition inputs. If the selected input condition is TRUE, the controller branches to the 16-bit address that is computed by concatenating fields NAD1 and NAD2. If the selected condition is FALSE, then the controller branches to address MAR+1. Figure 8.52(b) illustrates the two-way branch format.

When AT=10, the controller executes a three-way branch. The user must select two consecutive condition inputs, with the first of the two conditions being an even number (for example, 10 and 11). The CS field is fully decoded to obtain the lower of the two consecutive numbers denoted by CDL. The controller then complements the low-order bit in the CS field and fully decodes the result to select the higher of the two consecutive numbers denoted by CDH. Of course, the user must be responsible for ensuring that the two conditions cannot both be

63 62	61 56	55 40	39 32	31 0
00	X	NAD	X	LCS

(a) Unconditional Branch

63 62	61 56	55 40	39 32	31 0
01	CS	NAD	X	LCS

(b) Two-way Branch

63 62	61 56	55 48	47 40	39 32	31 0
10	CS	NADH	NADL1	NADL2	LCS

(c) Three-way Branch

63 62	61 56	55 48	47 40	39 32	31 0
11	CS	NADL1	NADL2	NADL3	LCS

(d) Four-way Branch

Figure 8.52   Control word formats for extended branching methods.

TRUE at the same time. If CDL is TRUE, the controller branches to the address computed by concatenating NAD1 with NAD2. If CDH is TRUE, the controller branches to the address computed by concatenating NAD1 with NAD3. If both CDL and CDH are FALSE, the controller branches to MAR+1. Figure 8.52(c) illustrates the three-way branch format. Notice that both branch addresses must have the same high-order eight bits (denoted as NADH). The low-order half of the address is NADL1 if condition CDL is TRUE and is NADL2 if condition CDH is TRUE. This method was chosen, so we would not have to add more bits to the control word. If two, complete 16-bit addresses were contained in the control word, the size of the control word would have to be increased to 72 bits. We are reducing word size at the cost of adding constraints to the branch addresses.

If AT=11, the controller executes a four-way branch. The user must select three consecutive condition inputs with the first of the three being an even multiple of four, i.e., the right-most two bits of the lower condition must both be logic 0. For example, the user might select conditions 24, 25, and 26 with binary representations respectively of 011000, 011001, and 011010. The lower number is encoded by the programmer in the CS field. The controller then fully decodes the CS field to obtain CD1 (in the example, CD1=24), complements the low-order bit and fully decodes the result to obtain CD2 (in the example, CD2=25), and finally, complements the second bit from the right and fully decodes the result to obtain CD3 (in the example, CD3=26). The user must assure that the three condition signals are mutually exclusive. If CD1=1, the next address is computed by concatenating the high half of MAR with the NAD1. If CD2=1, the next address is computed by concatenating the high half of MAR with NAD2. If CD3=1, the next address is computed by concatenating the high half of MAR with NAD3. If CD1=CD2=CD3=0, the next address is MAR+1. Note that this method restricts all three branch addresses to be on the same 256-word page as the current address. Figure 8.52(d) illustrates the four-way branch address format. All three direct branch addresses must share the high-order half of its address with the current high-order half in MAR. Again, we accept these restrictions in order to keep the control word length at 64 bits. To have three complete branch addresses in the control word would require 24 additional bits (88 total).

Using this combined address generation scheme, it should be possible to program most applications without stretching the clock to allow multiple way branches. Of course, five-way or higher branches cannot be directly implemented. Also, certain restrictions must be followed for three-way and four-way branches, but they are not that difficult to meet in most applications.

The cost of the additional flexibility in branching is much more complicated address generation logic (see Problem 8.28).

PROBLEMS

8.1   Develop an improved CASE-style VHDL description for device COM from the basic CASE-style model of Figure 8.6 by eliminating each of the following variables from the multiplexer address set. Simulate each model.
   a. N1
   b. N0
   c. M1

8.2   Consider the improved CASE-style VHDL models developed in Problem 8.1. Translate each model into an improved MUX implementation.

8.3   Using the rules given in this chapter, translate each of the following VHDL statements developed for a POS model into a VHDL statement appropriate for a NOR model:
   a. X <= (not A or B or not C) and (not B or E or not C or not E) and (A or E or F)
   b. Y <= (not A or not C) and (B or C) and (A or not B or not C)
   c. Z <= A and (not B or C or D) and not E
   d. P <= not A and not B

8.4   Consider the VHDL data flow model of Figure 8.13.
   a. Develop rules for automatically translating the VHDL data flow model of Figure 8.13 into an implementation using only NOR gates. Plan to translate each assignment statement into a two-level (or less) NOR gate configuration. (NOT gates may be used to invert input variables, but this is not counted as a level.)
   b. Using the rules developed above, translate the VHDL data flow model of Figure 8.13 into a network of NOR and NOT gates.
   c. Create a structural VHDL model of the gate-level circuit created as a result of the above translation. Simulate the model to verify its accuracy.

8.5   Consider the VHDL statements generated in Problem 8.3. Using the rules developed in Problem 8.4.a, translate each VHDL statement into a two-level network of NOR gates. (NOT gates may be necessary to invert input variables, but do not count this as an additional level.)

8.6   The summary of design activities for combinational logic shown in Figure 8.17 emphasizes POS standard forms and NOR gates. Draw a similar pictorial representation for design activities for combinational logic that emphasizes sum of products (SOP) standard forms and NAND gates.

8.7   Write a VHDL model for the optimum SOP representation of device COM shown in Figure 8.10. Simulate the model to verify its accuracy.

8.8   Develop a set of rules to translate a sum of products VHDL model into a NAND VHDL model. Hint: Apply the principle of duality to the rules for translating the product of sums VHDL model into a NOR VHDL model given in the text.

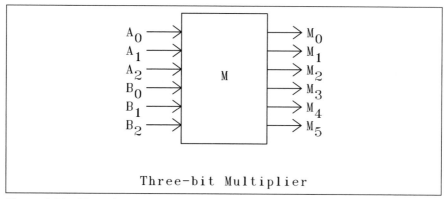

Figure 8.53   Three-bit binary multiplier.

8.9   Consider the rules developed in Problem 8.8. Create a NAND VHDL model for device COM by applying these rules to the SOP VHDL model created in Problem 8.7. Simulate the model to verify its accuracy.

8.10 Consider the NAND VHDL data flow model for device COM created in PROBLEM 8.9. Perform the following tasks:

   a. Develop rules for automatically translating the NAND VHDL data flow model into an implementation using only NAND gates. Plan to translate each assignment statement into a two-level (or less) NAND gate configuration. (NOT gates may be used to invert input variables, but this is not counted as a level.)

   b. Using the rules developed in part a, translate the NAND VHDL data flow model for device COM into a network of NAND and NOT gates.

   c. Create a structural VHDL model of the gate-level circuit created as a result of the above translation. Simulate the model to verify its accuracy.

8.11 Design a hardware multiplier circuit (M) that will compute the product of two, positive 3-bit binary numbers. Figure 8.53 shows a block diagram for device M. $A_2A_1A_0$ represents one, 3-bit number, $B_2B_1B_0$ represents the second 3-bit number, and $M_5M_4M_3M_2M_1M_0$ represents the 6-bit product. For example, if $A_2A_1A_0 = 110$ and $B_2B_1B_0 = 011$, then $M_5M_4M_3M_2M_1M_0 = 010010$. Use the path through the combinational design methodology chart of Figure 8.17, which starts with the English specification and ends with the ROM implementation.

   a. Document the translation process from each representation to the next representation along the path.

   b. Fully simulate all VHDL representations along the path.

   c. Show a block diagram of the final implementation with all signals named.

   d. If a hardware simulator is available, fully simulate the final hardware implementation.

8.12 Repeat the design of the 3-bit multiplier described in Problem 8.11. Use the path through the combinational design methodology chart of Figure 8.17, which starts with the English specification and ends with the standard MUX structural representation.

8.13 Repeat the design of the 3-bit multiplier described in Problem 8.11. Use the path through the combinational design methodology chart of Figure 8.17, which starts with the English specification and ends with the improved MUX structural representation.

8.14 Electrical sensors may experience serious transients as the sensor outputs change from one value to the next-higher or next-lower value. For example, if a 4-bit standard binary BCD code is used at the output of some electrical sensor, and the value changes from 7 (0111) to 8 (1000), then all four output signals must change. Since it is impossible for all four signals to change simultaneously, the output sequence could be:

0111
0110
0100
0000
1000

The physical quantity being measured (say, temperature) may stay on the boundary between value 7 and 8 for a significant amount of time, and in fact, may undergo many changes between 7 and 8 as the temperature hovers halfway in between values 7 and 8. In this case, a computer that is sampling the sensor output (say, every second) may read an erratic sequence of data values ranging from 0 to 15 when the actual value is between 7 and 8. Use of a Gray code will prevent such transients. The following chart shows a decimal Gray code and a binary BCD code for comparison:

Decimal Digit	Gray Code	BCD Code
0	0000	0000
1	0001	0001
2	0011	0010
3	0010	0011
4	0110	0100
5	1110	0101
6	1010	0110
7	1011	0111
8	1001	1000
9	1000	1001

Note that in the Gray code, 7 is represented by 1011, and 8 is represented by 1001. Because the only bit that is different is the third bit from the left, a computer that samples the sensor output will either read 1011 (7) or 1001 (8), as the value hovers between 7 and 8, instead of the erratic response of the BCD output. However, after the signal is sampled, it may be necessary to convert the code to the equivalent BCD code for internal processing within the computer. Design a hardware device that can translate the sampled signal from Gray code to BCD code. The device inputs will be the four bits of the Gray code, and the device outputs will be the four bits of the equivalent BCD code. The illegal input combinations (0100, 0101, 0111, 1100, 1101, 1111) may be treated as don't care situations. Use one of the following paths through the design methodology chart of Figure 8.17.

   a. Start with the English specification and end with ROM pictorial representation.
   b. Start with the English specification and end with the standard MUX structural pictorial representation.
   c. Start with the English specification and end with the improved MUX structural pictorial representation.
   d. Start with the English specification and end with the OR-AND structural pictorial representation.
   e. Start with the English specification and end with the OR-AND structural VHDL representation.
   f. Start with the English specification and end with the NOR-NOR structural pictorial representation.

g. Start with the English specification and end with the NOR-NOR structural VHDL representation.

h. Start with the English specification and end with the AND-OR structural pictorial representation using the dual concepts to those shown in the chart.

i. Start with the English specification and end with the AND-OR structural VHDL representation using the dual concepts to those shown in the chart.

j. Start with the English specification and end with the NAND-NAND structural pictorial representation using the dual concepts to those shown in the chart. Use the results of Problems 8.8. and 8.10(a).

k. Start with the English specification and end with the NAND-NAND structural VHDL representation using the dual concepts to those shown in the chart.

For each case, perform these steps:

a. Document the translation process from each representation to the next representation along the path.

b. Fully simulate all VHDL representations along the path.

c. Show a schematic diagram of the final implementation with all signals named.

d. If a hardware simulator is available, fully simulate the final hardware implementation.

8.15 Consider the Gray code described in Problem 8.14. Design an illegal code detector that has a single output (E) that will be logic 1 if and only if the input code is one of the illegal input codes (0100, 0101, 0111, 1100, 1101, 1111). Use one of the design paths described in Problem 8.14. Perform the same design steps.

8.16 Design an Equality Tester that interfaces with a computer. The tester should receive a sequence of 4-bit data items from the computer and decide when the last three data items are equal. The data items are transmitted over a 4-bit data bus. The tester and computer are both synchronous devices, but do not operate from a common clock. The tester should work correctly independent of whether the computer is faster or slower than the tester. Figure 8.54 shows the interface and timing for the Equality Tester. Signal READY is asserted whenever the tester is finished processing the current data and is ready for the next data item. Signal READY remains asserted until the computer responds by asserting the DAV line and simultaneously placing four bits of data on the DATA line. After the DAV line is asserted, the tester waits for at least one full clock period before transferring the four bits of data into an internal register. The clock period following the transfer of data, the tester drops the READY line indicating to the computer that the data has been successfully transferred. The computer holds the DATA line active until it sees the READY line drop. The computer then drops the DAV line and removes the data from the DATA line. At the same time that the READY line is dropped, the tester asserts the EQ output if the last three data items are all equal. When the computer sees the READY line drop, it knows that the EQ line has the desired result. It stores the value on the EQ line before dropping the DAV line. When the tester sees the DAV line drop, it can remove the EQ value, and assert the READY line. However, for safety, it will hold the EQ line active one full clock period before changing it and asserting the READY line again. Follow the design methodology described in the text.

a. Create a state table.

b. Develop an algorithmic-level VHDL model to verify the state table.

c. Simulate the algorithmic-level VHDL model. Include a sequence of inputs that will fully validate the functional specification.

d. Transform (by hand) the VHDL model into a hardware control unit and data unit. Fully document the transformation process and the resulting hardware circuit.

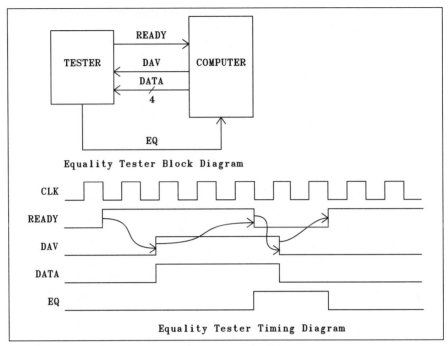

Figure 8.54 Interface and timing for the Equality Tester.

    e. Discuss the possibility of automation for the hardware translation. If an automated synthesizer is available, use it to synthesize the algorithmic-level model and compare with the hand simulation.

    f. If a hardware simulator is available, simulate the control and data unit and compare the simulation output to that produced by the algorithmic-level simulation.

8.17 Synthesize a microprogrammed control unit for the Equality Tester described in Problem 8.16. Perform the following design activities:

    a. Use the synthesis procedure described in this chapter to derive ROM program for a BMCU for the Equality Tester.

    b. Show a block diagram for the Equality Tester with blocks for the microprogrammed control unit and data unit.

    c. Develop a system-level VHDL model of the microprogrammed control unit and data unit similar to that shown in Figures 8.47a through 8.49b, inclusive.

    d. Simulate the system-level VHDL model. Include a sequence of inputs that will fully validate the functional specification.

    e. Translate the system-level VHDL model into hardware by hand. Document the translation process and discuss the possibilities for automation. If an automated synthesizer is available, try it.

    f. If a hardware simulator is available, simulate the hardware produced in the previous step. Use the same sequences used to validate the system-level VHDL model and compare the resulting outputs. Document and explain any differences. If there are any differences in outputs, do the differences constitute violations of specifications or are the differences just artifacts of the different representations?

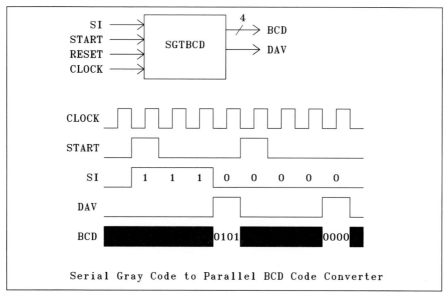

Figure 8.55   Gray code to BCD converter.

8.18   Modify the VHDL code for the device STOP in Figure 8.25a to formally take into account the unspecified behavior of output Z. Use package MVL4 from the CD-ROM. Refer to Chapter 5 for information on MVL4.

8.19   Using the state table approach, design a sequential circuit that converts a 4-bit Gray code into a 4-bit BCD code. A description of the Gray code may be found in Problem 8.14. The inputs and outputs are timed by the same system clock. Assume the device receives a START pulse coincident with the left-most bit of the Gray code. The remaining bits of the Gray code are received one bit at a time (from left to right). At the end of each 4-bit Gray code, the device outputs the corresponding 4-bit BCD code in parallel, along with a DAV signal. The next input could start during the clock period following the last input bit or at any time after that. The device should have a RESET input that initializes the device to the correct starting state. Figure 8.55 shows a block diagram and sample timing diagram.

8.20   Design a decimal Gray code counter that counts in the Gray code sequence shown in Problem 8.14. The counter should advance from 9 to 0 at the end of the sequence. Make the counter synchronous with a system clock.

8.21   Using the state table approach, design a Mealy sequence detector that detects a sequence of four consecutive 1 inputs. The detector has a single binary input, X, and a single binary output, Z. Signal Z should be logic 1 if and only if the last four inputs were all logic 1. Here is an example input-output sequence:

   X   0101111111011101011110
   Z   00000011110000000000010

Notice that the 1 output occurs simultaneously with the fourth consecutive 1 input.

8.22   Using the state table approach, design a Mealy sequence detector that has a single binary input, X, and a single binary output, Z. The output Z should be logic 1 if and only if the last four inputs were 1100 or 1001. Here is a sample input output sequence:

X   11001001101100111000
Z   00011001000000110010

Notice that the 1 output occurs simultaneously with the fourth bit in the detected sequence.

8.23 Using the state table approach, design a Moore sequence detector that has a single binary input, X, and a single binary output, Z. The output Z should be logic 1 if and only if the last four inputs were 1101 or 0110. Here is a sample input output sequence:

X   1101101101000110100-
Z   0000101101100001100

Notice that the 1 output is delayed one clock period with respect to the end of the detected sequence.

8.24 Four-bit data words are to be transmitted in parallel along a 4-bit data bus in blocks of two to fifteen words. Each data word, DAT(3 downto 0), consists of three bits of information and a parity bit. The parity bit, DAT(3), is selected to ensure an even number of 1's in each 4-bit word. The last word in each block is a parity word. The bits in the parity word are chosen to make the number of 1's in each column of the block an even number. This example illustrates the format:

	pddd
data word0	0011
data word1	1001
data word2	1111
data word3	0000
parity word	0101

In word0, the data is 011. The first bit (parity bit) is 0 because the number of 1's in the data part is an even number. In word1, the data is 001. The parity bit is 1 because the number of 1's in the data part is odd. Word4 is the parity word. Notice that the total number of 1's in each column is always even.

When a block of data is received, the parity information is used to detect and correct transmission errors. If no error occurs, the parity in each received word is even. The parity in each column is also even. If a single bit is received in error anywhere in the transmitted block, that error can be detected and corrected because it will cause exactly one row to have odd parity and exactly one column to have odd parity. The bit at the intersection of the odd parity row and the odd parity column can be corrected. For example, let the bit in column 1 of word2 in the data block shown previously be received in error as shown in the following table. At the right of each row is a binary variable that is logic 0 if the parity is even and logic 1 if the parity is odd. Similarly, at the bottom of each column is the value of a binary variable that is logic 1 if the total number of 1's in the column is odd and logic 0 if the total number of 1's in the column is even. The values of these computed variables are called the syndrome of the received data.

		Syndrome Column
received word0	0011	0
received word1	1001	0
received word2	1101	1
received word3	0000	0
received word4	0101	0
syndrome row	0010	

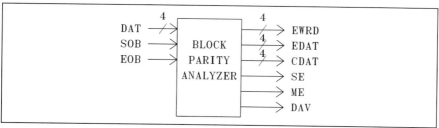

Figure 8.56a Block diagram for the BLOCK PARITY ANALYZER.

Since the only row with odd parity is that for word2, and since the only column with odd parity is column 1, the erroneous word can be corrected by performing a bit-wise XOR operation between word2 and the computed column parity:

faulty word2	1101
computed column parity (syndrome row)	0010
corrected word	1111

It should be clear that all single-bit errors have different syndromes and can, therefore, be distinguished from each other. Furthermore, if exactly two bits are received in error, either two rows will have odd parity, or two columns will have odd parity, or both. Try some examples to convince yourself of this fact. Thus, it is possible to distinguish all 2-bit errors from all 1-bit errors. However, not all 2-bit errors are distinguishable from one another. Can you find two, 2-bit errors that produce the same syndrome? If more than two bits are received in error, this fact can sometimes be detected and sometimes not, depending on exactly how many erroneous bits are present and where the erroneous bits are located. Try to find an example of a 3-bit error that produces the same syndrome as a single-bit error. Find a 4-bit error that cannot be detected at all because it produces the same syndrome as no errors.

Design a BLOCK PARITY ANALYZER device. Figure 8.56a shows a block diagram for the BLOCK PARITY ANALYZER. DAT is the 4-bit data input vector. SOB is a Start of Block level signal that lasts one clock period. EOB is an End of Block level signal that lasts one clock period. All inputs are synchronized with the system clock. If a single-bit error has occurred, then EWRD is a 4-bit output vector that indicates which word was in error. If word2 was in error, then EWRD=0010. EDAT is the 4-bit word that was received in error. CDAT is the corrected 4-bit vector. SE is a signal that indicates that exactly one error has occurred. If a multiple error can be detected, then ME=1. Multiple errors are detected if more than one word shows odd parity or if more than one column shows odd parity. DAV=1 when the other outputs contain valid data. All outputs are unspecified, except during the single clock period when DAV=1. The timing diagram in Figure 8.56b indicates the timing relationships for a three-word sequence. The three dots in the clock sequence represent an unknown number of clock periods needed to compute the required output information. After the clock period following the DAV pulse, the circuit returns to a state that waits for another SOB pulse. Internal registers should be initialized for another computation. The SOB pulse may be present during the very next clock period following DAV=1.

    a. Develop an algorithmic-level VHDL model for the BLOCK PARITY ANALYZER.
    b. Simulate the algorithmic-level VHDL model.
    c. Transform (by hand) the VHDL model into a hardwired control unit and data unit. Fully document the transformation process and the resulting hardware circuit.

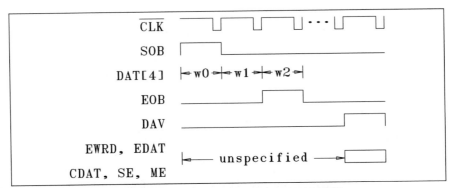

Figure 8.56b   Timing example for BLOCK PARITY ANALYZER.

    d. Discuss the possibility of automation for the hardware translation.

    e. If a hardware simulator is available, simulate the control and data unit and compare the simulation output to that produced by the algorithmic-level simulation.

8.25 Design a microprogrammed control unit for the BLOCK PARITY ANALYZER described in Problem 8.24. Use the following design steps:

    a. Use the synthesis procedure described in this chapter to derive the ROM program for a BMCU for the BLOCK PARITY ANALYZER.

    b. Show a block diagram for the BLOCK PARITY ANALYZER with blocks for the microprogrammed control unit and data unit.

    c. Develop a system-level VHDL model of the microprogrammed control unit and data unit similar to that shown in Figures 8.47a through 8.49b, inclusive.

    d. Simulate the system-level VHDL model. Include a selection of input sequences of various lengths that will fully validate the functional specification.

    e. Translate the system-level VHDL model into hardware. Document the translation process and discuss the possibilities for automation. If an automated synthesizer is available, try it.

    f. If a hardware simulator is available, simulate the hardware produced in the previous step. Use the same sequences used to validate the system-level VHDL model and compare the resulting outputs. Document and explain any differences. If there are any differences in outputs, do the differences constitute violations of specifications or are the differences just artifacts of the different representations?

8.26 Consider the implementation of state 5 of device STOP shown in Figure 8.46. Two clock periods were required because of the restriction on address generation logic to two-way branches. Discuss the effects of this implementation for state 5 on both input and output timing.

    a. For example, suppose that either R or A changes during the second half of the branching operation corresponding to the last entry in ROM. What problems could this cause?

    b. In what circumstance could the "stretched" micro-operation clock cycle cause problems in the output timing?

    c. Are there any other problems that might be caused by the "stretched" cycle?

    d. For each of the problems identified in part c, suggest possible solutions that could be specific to the STOP device.

    e. For each of the problems identified in (c) above, suggest possible solutions that could be applicable in general.

8.27 Suppose a microcontroller has an 8-bit condition select field. Make a table similar to that in Figure 8.50 that shows the number of conditions available and the cost of the decoding circuit for all possible k-out-of-n codes and for a fully decoded field. Is there any advantage to a 5-out-of-8 code that is not present in a less costly k-out-of-n code?

8.28 For the microcontroller with four branching methods (4BMCU) defined in Figures 8.51 and 8.52, do the following:

    a. Draw a block diagram for the controller similar to that for the BMCU shown in Figure 8.33.

    b. Develop an algorithmic-level VHDL description for 4BMCU that can be used as a template for designing devices with this type of control unit. Hint: Figures 8.38 and 8.39, inclusive, for the BMCU can be used to generate ideas.

    c. Design the address generation logic for 4BMCU.

8.29 Using 4BMCU instead of BMCU, determine the ROM contents for a microcontroller for device STOP described in the text.

8.30 Repeat Problem 8.17. Use 4BMCU instead of BMCU.

8.31 Repeat Problem 8.25. Use 4BMCU instead of BMCU.

8.32 The following is a state table of a Moore finite state machine, which performs serial parity checking on an input X. Below the machine is the partially completed code for a VHDL algorithmic model of the machine. Complete the code by (a) filling in the array aggregate, which represents the state table and (b) writing the executable statements in the model. Assume that when R = '1', the machine is reset asynchronously to S0.

Present State	Input		Output
	X=0	X=1	Z
S0	S0	S1	0
S1	S1	S0	1
	Next State		

```
entity MOORE is
  port(CLK,R,X: in BIT; Z: out BIT);
end MOORE;

architecture ALG of MOORE is
  type STATE is (S0,S1);
  signal FSM_STATE: STATE := S0;
  type TRANSITION is record
    OUTPUT: BIT;
    NEXT_STATE: STATE;
  end record;
  type TRANSITION_MATRIX is array(STATE,BIT)
      of TRANSITION;
  constant STATE_TRANS: TRANSITION_MATRIX :=

  --Insert state table here.

begin

  --Insert code here.

end ALG;
```

8.33　Figure 8.26 shows a concise algorithmic description style for a Mealy finite state machine. Although the same storage format could be used for a Moore machine, it is possible to modify the format, so the data structures are more efficient, by using separate arrays to store the next state and output values. Show how to implement the following Moore machine using the more efficient data structures.

Present State	X		Z
	0	1	
A	B	D	0
B	C	B	0
C	B	A	0
D	B	C	0
	Next State		

8.34　Implement an algorithmic model of a Mealy finite state machine. The state table of the machine is:

Present State	Input			
	0	1	2	3
S1	S2,1	S4,0	S2,1	S3,1
S2	S1,0	S3,1	S3,1	S3,0
S3	S1,1	S3,1	S2,0	S5,0
S4	S2,1	S1,0	S2,1	S2,1
S5	S2,1	S1,0	S2,1	S3,1
	Next State, Output			

Your model is to employ (1) an enumeration data type to represent the states, (2) a record type to represent the state transition and output information, and (3) array aggregates to encode the table. Shown below is a package to use with your description and the entity declaration for the finite state machine. It is up to you to develop the architectural body.

```
package INT_4 is
   type INT4 is range 0 to 3;
end INT_4;
use work.INT_4.INT4;
entity  FSM is
   port(CLK,R: in BIT; X: in INT4;
        Z: out BIT);
end FSM;
```

Simulate the application of a minimum touring sequence, i.e., a minimum length input sequence, which causes all states to be visited at least once.

8.35　Consider the possibility of eliminating the MAR from the micro programmed control unit shown in figure 8.33. Do a timing analysis and discuss advantages and disadvantages similar to that done for MIR in Section 8.3.2.

ASICs and the ASIC Design Process

$\mathbf{I}$n this chapter, we explain what an Application Specific Integrated Circuit (ASIC) is and describe the different technologies used to implement it. Next, we give a general description of the ASIC design process, emphasizing in detail the modeling and synthesis aspects. We describe how to use the Synopsys synthesis tools to design ASICs using a Standard Cell Library and how to use Xillinx tools to implement ASICs using Field Programmable Gate Arrays (FPGAs).

9.1　WHAT IS AN ASIC?

In the 1960s and 1970s design was done with standard parts, which we define as:

Standard Part.　A part produced in large quantities with functionality suitable for a wide range of applications.

The standard parts used in those eras are best represented by the 7400 TTL family of parts first developed by Texas Instruments and subsequently *second sourced* by other vendors. The family first consisted of parts at a Small Scale of Integration (SSI), e.g., gates and flip-flops. Soon parts at a Medium Scale of Integration were added to the TTL family, such as, decoders, multiplexers, and registers. SSI and MSI chips were mounted on a printed circuit board and interconnected by electro-plated metal runs. In the early 1970s, integrated circuit technology had developed to the point where Larger Scale Integrated (LSI) circuits could be developed. Companies, such as Intel and Motorola developed single chip microprocessors and single chip RAM and ROM memories. Other support chips were soon added: parallel ports, serial interfaces (UARTs), and interrupt controllers. Thus new families of LSI chips were developed that could be used for microcomputer design. Note that design with chip families is the strongest example of a bottom-up design methodology (discussed in Chapter 1).

However, major problems remained. For many applications, microprocessors were too slow, so many sections of *random logic* still had to be implemented with TTL parts. Also, the printed circuit interconnection approach was expensive and soldered connections created reliability problems. The major semiconductor vendors had the technology to integrate this random logic into a chip, but the cost to do so for small volumes was prohibitive. Two industrial developments changed this situation. First, *silicon foundries* were developed. For integrated circuit designs delivered in a standard format, they could produce chips in small volume at a reasonable price. Next, gate arrays were developed, which, as we shall discuss, contained rows of prefabricated transistors. The user would develop a file representing the metal interconnect and send that to the gate array house (another silicon foundry), which added the metal interconnect to complete the chip. However, designing at the transistor level is tedious; thus, companies were started that developed libraries of pre-laid out cells, which correspond to SSI and MSI components. This allowed the designers to develop a design file at a higher level of abstraction for delivery to the silicon foundry. Finally, to support these processes, many CAD tools were developed. Schematic capture allowed rapid entry of structural models. By the late 1980s hardware description languages, such as VDHL and Verilog were developed, which provided for the entry of designs behaviorally. High speed simulators and simulation engines allowed rapid validation of design models. Automatic checking and analyzers removed human error from the process. Finally, synthesis tools automated the process of translating behavioral HDL models to a structural model of the logic.

All these developments have supported the design of specialized, integrated circuits intended for a small market.

Application Specific Integrated Circuit (ASIC).   An integrated circuit produced for a specific application and (generally) produced in relatively small volumes.

This definition is technology independent. Thus, as we will see, a range of design technologies can be used to implement an ASIC: PLDs, Gate Arrays, FPGAs, Standard Cell, and Custom. However, like other generic definitions in the computer field, for example, RAM, the name *ASIC* has taken on a more specialized meaning. PLDs and FPGAs are referred to as *Programmable Logic*, while the term *ASIC* is reserved for Standard Cell and Gate Arrays where a chip is actually fabricated as opposed to being programmed at the customer's site. Thus, an engineer will say, "I'm not sure whether to use an FPGA or ASIC." The Custom approach is almost exclusively reserved for standard part design.

The range of application for a specific ASIC or for specific programmable logic is typically narrow, such as, Digital Signal Processing, Motor Control, and Network Routing, although, this is not always the case. Random logic in a general purpose computer can be placed on a chip using standard cell technology. And ASICs are not always sold in small volume. The aforementioned computer chip wouldn't be sold in small volume. Also, a DSP chip developed for a specific application may prove to be generally useful and be sold like a standard part. The essential point is that ASIC (and Programmable Logic) design technology puts the control of chip design under the user's control.

A recent development that pushed ASIC design to a new level was the emergence of the concept of *Intellectual Property (IP)*. IP refers to a working design that one company sells to another. The design is expressed as an HDL model, accompanied by test benches, and synthesis

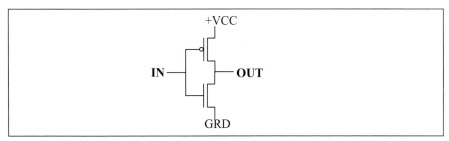

Figure 9.1   CMOS inverter.

constraints or script files to control synthesis. IP can be RAM memory or a microprocessor. It represents a new level of ASIC library development and should have a profound effect on design productivity.

The advent of ASICs and Programmable Logic has given the system designer a set of three design paradigms that can be used and mixed in system design:

1. The Von Neumann Machine Architecture: CPU, memory, and I/O. The designer translates system requirements into program code for the machine.
2. ASICs: For fixed logic where high speed is required.
3. FPGAs: For reconfigurable logic that allows a rapid change of function.

Using mixtures of these paradigms in a single design is an exciting prospect. An IP microprocessor could be placed on the same chip as the ASIC. This chip and the FPGA chip could then be bonded to the same substrate to form a Multi-Chip Module (MCM).

9.2 ASIC Circuit Technology

Almost all ASICs and Programmable Logic chips today use CMOS technology.

Figure 9.1 shows a CMOS inverter. When IN is low the NMOS transistor (the pull-down device that is connected to GRD) is OFF, and the PMOS transistor (the pull-up device that is connected to +VCC) is ON, making OUT=+5. When IN is high, the NMOS transistor is ON, and the PMOS transistor is OFF, forcing OUT=GRD. Therefore, statically, the DC power is very low because there is always high impedance from VCC to ground.

Other CMOS circuits, such as the NOR gate shown in Figure 9.2 exhibit similar low static power dissipation as either the pull-up circuit or the pull-down circuit is off. However, when the circuit switches, for a very short period of time, both the pull-up and the pull-down circuits are on, and a current spike occurs from VCC to ground. The dynamic (AC) power dissipated is given by the equation:

$$P_{AC} = CV_{CC}^2 f \tag{9.1}$$

where C is the load capacitance, V_{CC} is the power supply voltage, and f is the switching frequency. Although the CMOS AC power is much higher than the CMOS DC power, it is still very low, relative to other logic families.

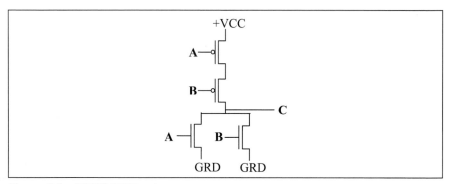

Figure 9.2   CMOS NOR gate.

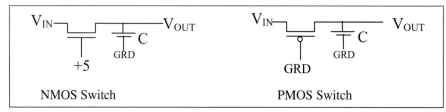

Figure 9.3   CMOS switches.

Advances in CMOS technology have reduced the feature size of CMOS circuits to as low as 0.15 microns. The combination of small feature size and low power has allowed the fabrication of high speed, low power, densely-packed complex circuits, which are the basis for VSLI technology.

9.2.1   CMOS Switches

The systems we are considering not only use inverters and gates, but also simple transistor switches (also called pass transistors). It is important that we consider the effectiveness of these switches in passing data values.

For the NMOS switch in Figure 9.3, if V_{IN} = 5v, the output will be a weak 1 (H in the IEEE logic system). If V_{IN} = GRD, the output will be a strong 0. For the PMOS device, a 5V input produces a strong 1 output, while a GRD input produces a weak 0 (L) output. Thus, the use of either switch alone produces signal degradation, especially if the switches are cascaded.

Figure 9.4 shows a "good" switch that corrects these deficiencies. Because of the parallel arrangement with both switches on, the NMOS switch passes the good 0, the PMOS switch the good 1. However, two transistors are now required for each switch, and for an interconnection structure that requires thousands of switches, such as is used in FPGAs, this might be a prohibitive price to pay.

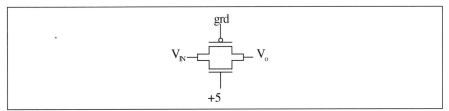

Figure 9.4   Good CMOS switch.

9.3   TYPES OF ASICS

In this section we describe the different approaches to ASIC and programmable logic design: PLDs, FPGAs, Gate Arrays, Standard Cell, and Custom.

9.3.1   PLDs

Programmable Logic Devices (PLDs) take the form shown in Figure 9.5. Input variables, for example, I_1 through I_8, and their complements are fed down the length of the chip on rails. These vertical rails can be connected to horizontal rails, which form AND terms of some subset of the input variables and their complements. Eight such AND terms are input to an OR gate. There are nine OR gates, each with eight AND inputs. The output of the OR gate is either connected to the input of a D flip or is directly connected to an output pin. Feedback occurs either directly from the OR gate or from the output of the flip-flop. All outputs are buffered. Complex PLD (CPLD) circuits contain a number of PLDs interconnected through switches.

PLDs are most effective for implementing logic where complex combinational logic functions terminate at a D flip-flop, such as, state machines. However, many systems have logic with a much higher ratio of flip-flops to gates. For these systems Field Programmable Gate Arrays (FPGAs) are more suitable.

9.3.2   Field Programmable Gate Arrays

Field Programmable Gate Arrays (FPGAs) are an extension of Mask Programmable Gate Arrays (MPGAs), commonly referred to as Gate Arrays. Gate arrays consist of cellular rows of NMOS and PMOS transistors with provisions for interconnecting the transistors within the cells to form gates and for routing signals between the cells. The interconnect is performed in a silicon foundry by using a mask to deposit the metal layer interconnect. An FPGA also consists of an array of cells, but the cells are more complex, for example, look-up tables or multiplexers. These elements, as well as the interconnections, are programmable at the user's site thus, providing a rapid, economical way to implement logic.

Figure 9.6 shows the essential elements of an FPGA organized as a 2-D array of cells (1) a rectangular array of Configurable Logic Blocks (CLBs) capable of implementing a variety of logic functions, (2) wiring tracks to route signals between the cells, (3) Xbar switches to connect horizontal and vertical wires, and (4) input/output (I/O) pads for signal conditioning at the chip input and output pins. All these resources are programmable (and possibly) reprogrammable at the designer's site.

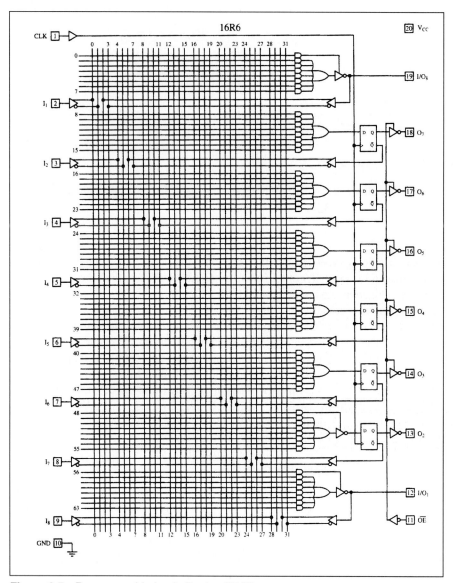

Figure 9.5 Programmable Logic Device (PLD).

Figure 9.7 illustrates a sea of modules architecture. Three layers of metal above the silicon surface are used to perform routing. This architecture is used in Actel Corporation's FPGAs. The cells are closely packed and use a high degree of local routing.

9.3.2.1 Configurable Logic Blocks (CLBs)

A major consideration in the selection of logic blocks is generality. Suppose, one wishes to implement the logic function shown in Table 9.1.

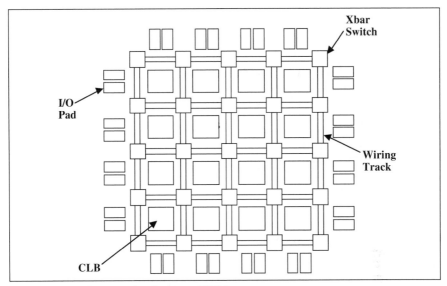

Figure 9.6   2-D array of cells for an FPGA.

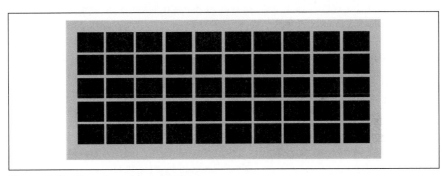

Figure 9.7   Sea of modules architecture.

Table 9.1   3-Input logic function.

A	B	C	F
0	0	0	0
0	0	1	1
0	1	0	1
0	1	1	0
1	0	0	1
1	0	1	0
1	1	0	0
1	1	1	1

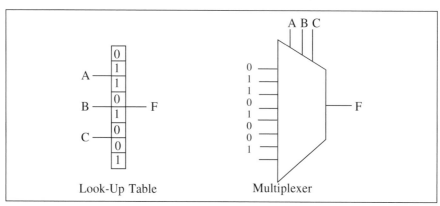

Figure 9.8   General logic cells.

Figure 9.8 shows two possible ways to implement this function using a general cell approach.

On the left is a look-up table (LUT) implementation. A, B, and C are address inputs. A specific combination of address inputs selects the table entry that contains the output corresponding to that particular input combination. Look-up tables are implemented using RAM. The LUT in Figure 9.8 would be implemented as an 8x1 RAM. On the right is a multliplexer implementation in which the values of F are applied to 8 multiplexer data inputs. A, B, and C are applied to the multiplexer select inputs. In general, an n-input LUT can implement any function of n variables. With constant inputs, the same holds for the MUX. However, if the $n+1$ variable and its complement are available, a MUX with n select inputs can implement any function of $n+1$ variables. The (n+1)st variable and its complement in combination with 1's and 0's are applied to the MUX inputs. See Section 8.1.1.2 for a detailed example of MUX design.

A basic question is: What is the ideal value of n? The temptation is to make it large, hoping to cover complex functions with a single cell. However, experience has shown that for look-up tables, a value of n=4 is optimal and for MUXes, a value of n=3 is optimal. For larger values, many LUT inputs are unused and much logic is wasted. An early FPGA implementation by Algotronix used a look-up table with n=2, that is, it could generate all 16 functions of two variables. However, for cells this small, even though the utilization of cell logic is high, many cells must be interconnected to implement even simple functions. The extra space needed for interconnections and the extra delay incurred proved unsatisfactory.

Although one can use combinational logic with feedback to implement sequential elements, this is a cumbersome approach. Thus, CLBs are typically equiped with flip-flop storage. They also contain multiplexers for data routing within the cell. Figure 9.9 shows such an arrangement. In this case, the CLB is used to implement a JK flip-flop. A and B are the J and K inputs respectively. The state of the flip-flop is fed back through the left-most MUX to LUT input C' by setting its control input to 1. A similar setting for the right-most lower MUX connects the flip-flop reset (R) to the cell input C. The cell output X is connected to the flip-flop output via the top-right multiplexer.

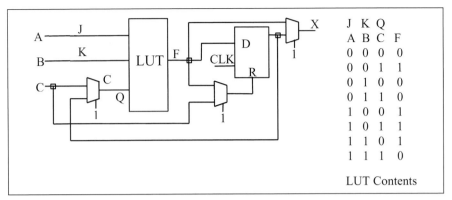

Figure 9.9 CLB with flip-flop storage.

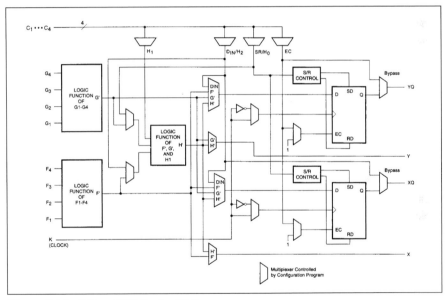

Figure 9.10 Xilinx 4000 Series CLB.

Figure 9.10 shows the CLB for the Xilinx 4000 series FPGAs. It contains two 4-input LUTs and one 3-input LUT which may be programmed independently or in combination. It has numerous MUXes for data selection and internal cell routing and two flip-flip storage elements. The 4000 Series CLB can implement:

- Any two 4-variable functions and any 3-variable functions—all involving different input variables.
- Any 5-variable function.
- Any function of four variables, together with some functions of six variables.
- Some functions of up to nine variables.

The two flip-flops or storage elements:

- Can be configured as a latch or edge-triggered flip-flop.
- Have a common clock, CLOCK, and clock enable (CE). The common clock is individually invertible for each storage element.
- Are controlled by a global reset, and are programmed to set or reset in response to it. (The global reset is not shown in the the figure.)

Note that there are numerous MUXes for controlling the inputs to LUTs and flip-flops, for selecting cell outputs, and for routing data through the cell.

Figure 9.11 shows the logic cells for the Actel SX Family of FPGAs. The R Cell is the sequential storage cell. The D flip-flop can be clocked by a hardwired clock or a routed clock and clock polarity is selectable. The flip-flop has an asynchronous set and reset. It can be preset from nearest neighbor and routed data. The C cell is the combinational logic cell. It can implement all 3-variable functions (note the presence of the inverter), and some 4- and 5-variable functions.

9.3.2.2   Interconnect

Switch technology.   The interconnection problem essentially consists of connecting two conductors together. Two main approaches are used, reprogrammable and one-time programmable.

Figure 9.12 shows a Programmable Interconnection Point (PIP). The square box represents a RAM location that controls the state of the switch. Because RAM contents can be quickly changed, the switch is quickly reprogrammable. EPROM and EEPROM technology can also be used for reprogrammable switches, but their reprogramming times are longer. The chip using EPROM must be removed from the board, erased, and then reprogrammed. EEPROM switches can be reprogrammed in place, but the reprogramming time is long, relative to the PIP. FPGAs using SRAM technology can be reprogrammed in milliseconds. However, from our previous discussion, we know that the CMOS switch degrades a signal. This degradation can be modeled in the form of an RC network, which loads and delays the signal as shown in Figure 9.13. Signal degradation can be eliminated by buffering the signal at various points, but this in turn causes additional delay.

An important type of connection switch is the *antifuse* as shown in Figure 9.14. Two conductors are separated by a dielectrical material, which normally exhibits high impedance. However, if a large enough voltage is applied across the dielectric, it breaks down, current flows, and a permanent, low resistance connection is made between the conductors. The process is irreversible. There are two main commercial implementations of antifuses. The Plice antifuse consists of a dielectric between polysilicon and n+ diffusion. The ViaLink antifuse has amorphous silicon between two layers of metal.

When selecting a switch technology, it is important to compare the following aspects of each technology:

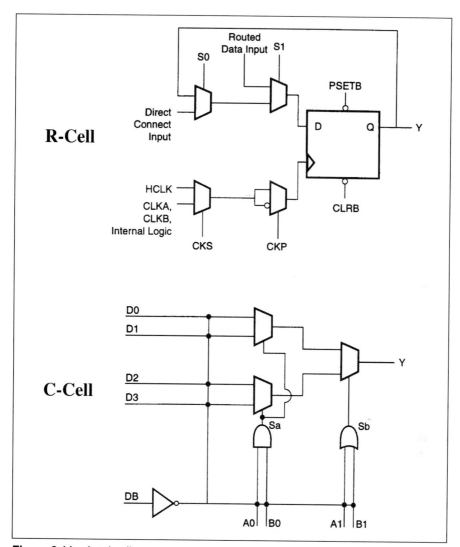

Figure 9.11 Actel cells.

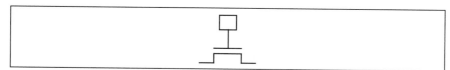

Figure 9.12 Programmable interconnection point (PIP).

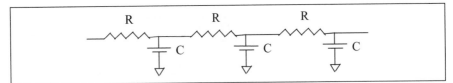

Figure 9.13   Switch-delay model.

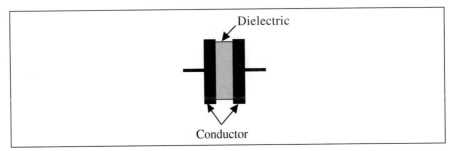

Figure 9.14   Antifuse.

1. Values of R and C for the Switch: Three switches in series results in the delay model shown in Figure 9.13.
2. Reprogrammability: Important if one wants to easily incorporate design changes or to change the function of the logic.
3. Volatility: What happens when the power is lost? How easily is the functionality restored?
4. Area of the switch: More switches make routing easier; fewer switches makes routing difficult and, in some cases, impossible.

Table 9.2   Characteristics of programming technologies.

Programming Technology	Volatility	Re-programability	Chip Area	R(ohm)	C(ff)
SRAM	Yes	In circuit	Large	1-2K	10-20
Plice Antifuse	No	No	Small antifuse, large prog. trans.	300-500	3-5
ViaLink Antifuse	No	No	Small antifuse, large prog. trans.	50-80	1.3ff
EPROM	No	Out of circuit	Small	2-4K	10-20ff
EEPROM	No	In circuit	2 x EPROM	2-4K	10-20ff

Table 9.2 compares different technologies. The two most popular choices are SRAM and antifuse. SRAM is volatile, but can be reprogrammed quickly. It's RC time constant is relatively high, causing a relatively slow interconnect. Antifuses are not volatile and can't be reprogrammed. But, they have the lowest RC time constants of all the options. It would appear that they could support the highest clock rate. Actel uses Plice antifuses. However, Actel logic cells

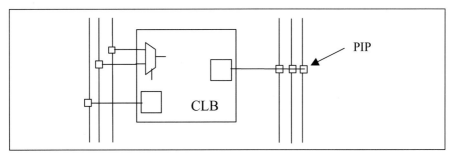

Figure 9.15   CLB wiring track connections.

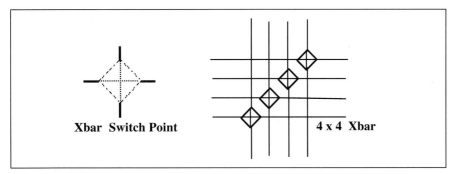

Figure 9.16   Xbar.

are smaller than Xilinx logic cells, and thus, more routing is required than with Xilinx. So, the clock rate advantage may not be pronounced. Actel FPGAs are superior in applications, where volatility is dangerous, for example, in man-rated systems. Xilinx FPGAs are the clear choice for reconfigurable systems.

Interconnect structures.   Figure 9.6 illustrated the interconnect structure of an FPGA organized as a 2-D array of cells. The interconnection mechanism for this array involves using wiring tracks and Xbar switches to interconnect CLBs. We will illustrate this for the Xilinx 4000 series FPGAs. Figure 9.15 shows how connections are made to wiring tracks.

PIPs (see Figure 9.12) are used to connect CLB inputs and outputs to wires. Multiplexers internal to the CLB, in some cases, select from a number of wired inputs. Figure 9.15 only shows connections to vertical wires. In fact, the Xilinx CLB has outputs on all four sides and can make connections to horizontal wires as well.

Figure 9.16 shows how the Xbar is constructed. On the left is the Xbar switch point. Each dotted line in the switch represents a PIP. On the right is a 4x4 crossbar. Each horizontal or vertical input can be connected straight through or can make a right or left turn. An input can also be connected to more than one output line. The switch is bidirectional. The switch is expensive in terms of hardware. An N x N crossbar requires 6N transistors.

We, thus, have the capability to route from one CLB to any other CLB. However, switch delay must be considered. In Figure 9.6, if one wanted to route from the lower-right CLB to the upper-right CLB, seven PIP delays would be encountered, five delays through crossbar switch PIPs and two PIP delays for CLB output and input connections. The simple FPGA in Figure 9.6 is a 4x4 array of CLBs. The Xilinx XC 4010XL has a 20x20 array of CLBs, so the routing delay could be much worse. To solve this problem, Xilinx has developed a hierarchy of wire types within the routing tracks:

1. Single length lines pass through every switch matrix in their direction of travel (Figure 9.6 implies this type of line). These lines are not suitable for long connections but are perfect for short, local connections.
2. Double length lines run past two CLBs before entering a switch matrix.
3. Quad length lines run past four CLBs before entering a switch matrix. Thus, double and quad length lines will have shorter delays than single length lines that connect CLBs that are far apart.
4. Direct interconnect—North, South, East, and West—connection to nearest neighbor CLBs. This is an unswitched connection, and thus, the fastest local interconnect.
5. Long lines run the entire length or width of the array. For high fanout, time-critical nets, or nets distributed over large distances.

One of the functions of the FPGA routing program is to select the best type of routing for a signal run.

Figure 9.7 showed the Actel architecture as a sea of modules architecture. No interconnect structure was shown. In the Actel SX family of FPGAs, the interconnect is handled by three layers of metal, separate from the cell layer. Figure 9.17 shows the layered structure. Antifuses are between metal layers 2 and 3. The structure has no specific routing channels but uses the whole surface area of the chip.

Figure 9.18 shows the details of the routing. R and C cells are formed into clusters (shaded gray area). The interconnect type hierarchy is:

1. Direct connect: Unswitched connection. Used for intracluster routing.
2. Fast connect: Enables horizontal and vertical routing between adjacent clusters. This routing incurs one antifuse delay.
3. Segmented routing: This provides for a variety of track lengths between clusters. The delay is typically two antifuses and a maximum of five antifuses.
4. High drive routing: This structure provides three clock networks. The first clock, HCLK, is hardwired from the central HCLK buffer to the clock select MUX in each R-cell. The two remaining clocks (CLKA, CLKB) are global clocks that can be sourced from external pins or from internal logic within the array. They can also be selected at each R-cell.

9.3.2.3   *Input/Output Blocks (IOBs)*

At the periphery of the FPGA are Input/Output Blocks (IOBs). Figure 9.19 shows the IOB for the Xilinx 4000 series FPGA. It has these characteristics:

1. Inputs (buffered) can be direct or clocked into a flip-flop. Delay can be inserted at the f/f input to increase the setup time and insure that the hold time is not negative.
2. Inputs can go into a "fast capture" latch, which then feeds the clocked flipflop.
3. Outputs (buffered) can be direct or first clocked into a flip-flop.

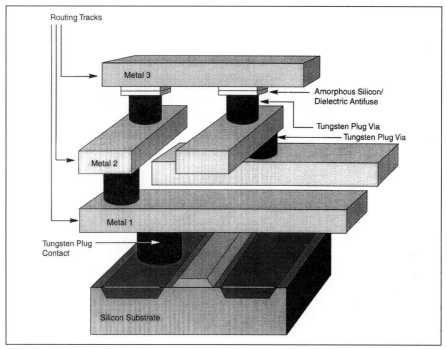

Figure 9.17 Actel multilayer interconnect.

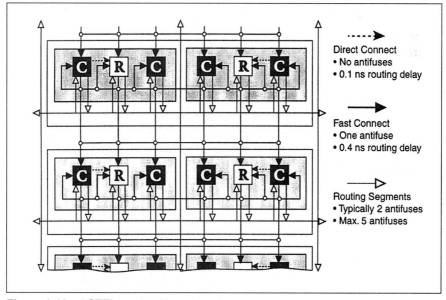

Figure 9.18 ACTEL routing hierarchy.

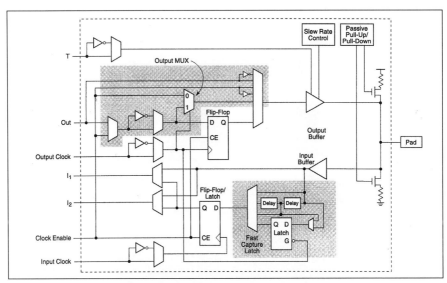

Figure 9.19   Xilinx 4000 series I/O block.

4. Outputs are slew rate controlled and are either TTL or CMOS tristate buffers. Slew rate control is used to minimize noise coupling when many outputs are switching simultaneously.

Actel FPGAs also contain IOBs, which can be configured as inputs, outputs, tristate outputs, or bidirectional pins. They contain no flip-flop storage but use normal R-cells for this purpose.

9.3.2.4   Special logic

Certain types of logic cannot be implemented by FPGA CLBs or are implemented very inefficiently, for example, (1) bussed tristate systems, (2) decoders, (3) oscillators, (4) fast adder logic, and (5) RAM cells. Xilinx 4000 series FPGAs have special circuits to implement wired AND and mutliplexed buses. Located on the edge of the chip are four programmable decoders. An onboard oscillator provides 8 MHz, 500 kHz, 16 kHz, 490 Hz, and 15 Hz clock signals. RAM can be implemented efficiently by utilizing the two, 4-input look-up tables in the CLB as two 16x1 RAM memories. Each CLB has special fast adder logic.

9.3.3   Gate Arrays

A gate array chip contains prefabricated adjacent rows of PMOS and NMOS transistors. The user creates a file which specifies the metal interconnect paths. Gate array design is accomplished using a CAD tool that maps schematics or HDL models to transistor configurations. Horizontal routing between transistors can be done (1) in horizontal channels between the rows of transistors and (2) using paths over unused transistors (sea of gates). This second method is now preferred because it allows a higher gate density than can be achieved with the channel routing method. Routing layer approaches include (1) single-layer metal, (2) single-layer metal and contacts, (3) double-layer metal, contacts, and vias, or (4) triple-layer metal, vias, and contacts. The file that the designer creates is sent to the gate array manufacturer (a silicon foundry),

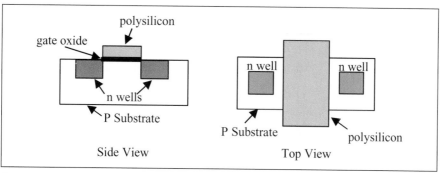

Figure 9.20 NMOS transistor.

where a set of masks is created, which are used to deposit the metal runs. Because of the mask creation process, gate arrays are sometimes called Mask Programmable Gate Arrays (MPGAs) to differentiate them from FPGAs. Thus, MPGAs are not programmable by the user, and are not programmable logic under the definition we presented at the outset, but are actually ASICs. MPGAs predated FPGAs, in fact, FPGA technology was derived from MPGA technology. MPGAs provide for greater gate densities and clock speeds than FPGAs and are non-volatile. Changes to an MPGA design are expensive and time-consuming because new masks must be created prior to a new manufacturing run.

To understand how gate arrays are fabricated, one has to understand the basic MOS gate fabrication process. Figure 9.20 shows the layout of an NMOS transistor. Two highly doped n wells are defused into the P Substrate to form the transistor source and drain. A layer of oxide is deposited on the surface of the silicon between the source and the drain. A layer of polysilicon is deposited on top of the gate oxide to form the gate connection. Thus, the basic principle of transistor construction is that wherever a polysilicon run crosses a structure like that shown in Figure 9.20, an NMOS transistor is formed. PMOS transistors are formed in a similar fashion except the wells are P-type diffusion in a large n well, which is defused into the P Substrate.

Figure 9.21 shows the gate array layout. Each row of transistors consists of a P diffusion strip and an N diffusion strip. Overlaying these strips are polysilicon gates. A transistor is formed by making metal connections (source and drain) on each side of the poly gate and applying power to the devices. The layout of the inverter in Figure 9.1 is illustrated in Figure 9.21.

The wiring of the inverter illustrated in Figure 9.21 shows the intracell wiring. Early gate arrays used a horizontal channel between rows of transistors to do horizontal intercell routing. Vertical routing was done on another wiring plane. More recent gate arrays use unused rows of transistors for horizontal routing, and there are no separate horizontal routing channels. Transistors are packed more densely than in the channel routed gate array. With this approach the array is referred to as a sea of gates. One disadvantage of gate arrays is that the transistors are all one size, and therefore, faster versions of gates are not immediately available. We will see that designing with standard cells addresses this problem.

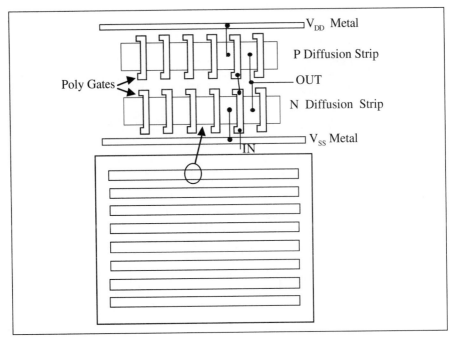

V_{DD} Metal

P Diffusion Strip

Poly Gates

OUT

N Diffusion Strip

V_{SS} Metal

IN

Figure 9.21   Gate array layout.

9.3.4 Standard Cells

Currently, most ASICs are designed using standard cells from a library. The chips produced are also referred to as Cell-Based Integrated Circuits (CBICs). At the designer's level, the library consists of logic elements of varying complexity: SSI logic, MSI logic, data path blocks, memories, and system-level blocks. The library contains a complete layout for each logic element. The user only deals with the logic block descriptions and is isolated from the details of the circuit layout.

Figure 9.22 shows an example of a standard cell layout. All diffusions, contacts, vias, and runs of poly and metal are completely defined. When this cell is used in a design, the cell layout is "pasted" electronically into a row of cells on the chip layout.

Figure 9.23 shows standard cell interconnect layout. The cells (gray boxes) are configured in rows. They are all equal height but vary in width, based on their complexity. Multilayers of metal (in this case two) are used to perform horizontal and vertical routing. These layers are routed over the top of the cells; thus, one has complete X-Y plane freedom. However, as the density of circuits increases (and feature size decreases), third, fourth, and even fifth layers of metal interconnect may be needed. To fabricate the chips, a full set of masks must be made for the integrated circuit. All layers are fabricated at the silicon foundry, in contrast to gate arrays, where only masks for the metal layers are needed, and only metal is run at the silicon foundry.

However, the logic designer is divorced from these details, and is concerned only with logic representation. Using schematic capture or an HDL input to a synthesis tool, the designer

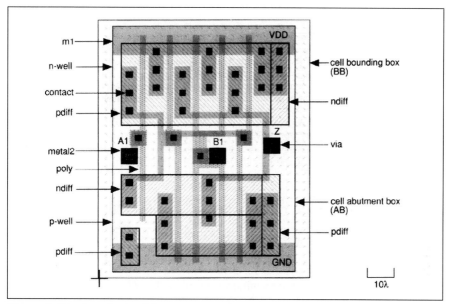

Figure 9.22   Standard cell layout.

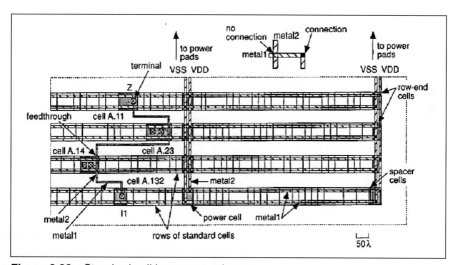

Figure 9.23   Standard cell interconnect layout.

specifies which library elements are used and their interconnect. The types of library elements available in order of increasing complexity are:

1. SSI logic: nand, nor, xor, aoi, oai, inverters, buffers and registers
2. MSI logic: decoders, encoders, parity trees, adders and comparators
3. Data path: alus, adders, register files, shifters, bus extractors, and inserters

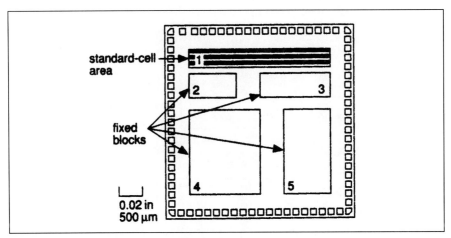

Figure 9.24   Mixing standard cells with fixed blocks.

4. Memory: RAM, ROM, and CAM
5. System level blocks: multipliers, micro-controllers, UARTs, and RISC cores

When one reaches the complexity level of memories and system-level blocks, the layout of the blocks for the cell layout approach shown in Figure 9.23 will no longer work. Instead the arrangement shown in Figure 9.24 is used.

Here standard cells (top rows) are mixed with fixed blocks from the memory and system-level block categories. They are also referred to as cores, and if purchased from another company, are often referred to as intellectual property (IP).

As an example of an ASIC library, we consider LSI Logic's 500 K library. Table 9.3 shows the combinational logic cells. In general, there are four power levels (a) low (l) and (b) normal and high (a,b,c), with a being the lowest of these and c being the highest. For circuits like AND-OR-INV, the number sequence 2-2-2 means that the input stage has three two-input AND gates. Table 9.4 shows the flip-flop cells and their options. These are basic cells. LSI offers CAD software, which can build macros from these basic cells. Examples of macros include: adders (many variations), counters, large MUXes, incrementers and decrementers, decoders, shifters, FIFOs, comparators, multipliers, barrel shifters, and ALUs. They also offer cores, such as microprocessors and blocks for mixed signal circuits (A/D and D/A).

In addition to the functionality of a cell, we need to know its delay. To compute a cell's delay, we need to know the wiring load for the cell. However, if the chip has not been laid out, one cannot measure the length of metal runs. Figure 9.25 contains an example wire load table. For a given wire area estimated in mm^2 and gate fanout, the table gives the capacitive load of the interconnect. The capacitive load is given in terms of the standard load, which is the input capacitance of an inverter, e.g., 0.032 pf. Figure 9.26 shows a D flip-flop data sheet. To calculate the propagation delay for a device, one uses this equation:

$$C_T = C_{wireload} + C_{inpload} \qquad (9.2)$$

Table 9.3 LSI Logic's 500 K library basic, combinational logic cells.

Function	No. of Inputs	Power Versions
NAND	2,3	a,b,c,l
NAND	4-8	a,c
AND	2,3	a,b,c,l
NOR	2,3	a,b,c,l
NOR	4-8	a,c
OR	2,3	a,b,c.l
OR	4	a,c
Exclusive NOR	2	a,b,c,l
Exclusive NOR	3	a,b,c
Exclusive OR	2	a,b,c,l
Exclusive OR	3	a,b,c,l
Inverting MUX	2	a,b,c,l
MUX	2	a,b,c,l
MUX	3,4,6,8	a,c
AND-OR-INV	2-1, 2-2, 2-1-1, 2-2-2	a,c
OR-AND-INV	2-1, 2-2, 2-1-1 2-2-2-2	a.c
AND-OR-OR-INV	2-2	a,b,c
OR-AND-AND-INV	2-2	a,b,c
Half Adder	2	a,b,c
Full Adder	3	a,b,c

Table 9.4 LSI 500 K basic, sequential cells

Flip-Flop Type	Options	Power
D	asyn clear	(a,c)
	asyn set and clear	
	asyn set	
	neg edge clock	
	neg edge clock and asyn clear	
	many scan options	
JK	async clear	(a.c)
	async set and clear	
	many scan options	

Size in mm2	Fanouts[1] 1	2	3	4	5	6	7	8	16	32	64	Slope[2]
0.5x0.5	0.593	1.186	1.779	2.372	2.965	3.558	4.151	4.744	9.488	18.976	37.952	0.593
1.0x1.0	0.649	1.298	1.947	2.596	3.245	3.894	4.543	5.192	10.384	20.768	41.536	0.649
2.0x2.0	0.763	1.526	2.289	3.052	3.815	4.578	5.341	6.104	12.208	24.416	48.832	0.763
3.0x3.0	0.877	1.754	2.631	3.508	4.385	5.262	6.139	7.016	14.032	28.064	56.128	0.877
4.0x4.0	0.992	1.984	2.976	3.968	4.960	5.952	6.944	7.936	15.872	31.744	63.488	0.992
5.0x5.0	1.106	2.212	3.318	4.424	5.530	6.636	7.742	8.848	17.696	35.392	70.784	1.106
6.0x6.0	1.220	2.440	3.660	4.880	6.100	7.320	8.540	9.760	19.520	39.040	78.080	1.220
7.0x7.0	1.334	2.668	4.002	5.336	6.670	8.004	9.338	10.672	21.344	42.688	85.376	1.334
8.0x8.0	1.448	2.896	4.344	5.792	7.240	8.688	10.136	11.584	23.168	46.336	92.672	1.448
9.0x9.0	1.562	3.124	4.686	6.248	7.810	9.372	10.934	12.496	24.992	49.984	99.968	1.562
10.0x10.0	1.676	3.352	5.028	6.704	8.380	10.056	11.732	13.408	26.816	53.632	107.264	1.676
12.0x12.0	1.904	3.808	5.712	7.616	9.520	11.424	13.328	15.232	30.464	60.928	121.856	1.904
14.0x14.0	2.133	4.266	6.399	8.532	10.665	12.798	14.931	17.064	34.128	68.256	136.512	2.133
16.0x16.0	2.361	4.722	7.083	9.444	11.805	14.166	16.527	18.888	37.776	75.552	151.104	2.361
18.0x18.0	2.590	5.180	7.770	10.360	12.950	15.540	18.130	20.720	41.440	82.880	165.760	2.590
20.0x20.0	2.817	5.634	8.451	11.268	14.085	16.902	19.719	22.536	45.072	90.144	180.288	2.817

1. Wire length in mm = (wlests * number_of_fanouts) + wlestc. For LCB500K: wlestc = 0; wlests = 0.026X + 0.122
2. Intercept = 0.

Figure 9.25 Equivalent standard load matrix for 2, and 3-layer metal.

where $C_{wireload}$ is derived from the wireload table in Figure 9.25 and $C_{inpload}$ is computed from the loads presented by the driven gates. For instance, the D input on the D flip-flop of Figure 9.26 presents a standard load of 0.6. Once C_T is computed, it is then used as an index into the delay table. For example, if $C_T = 10$, then, from the table in Figure 9.26, tpLH for version fd1qa is 0.94 ns (interpolating between 0.87 and 1.02).

After the chip is laid out, the actual length of metal runs can be measured, and the actual interconnect delays back annotated into the simulation model using the Standard Delay Format (SDF), an industry standard.

9.3.5 Full Custom Chips

In full custom chips, the designer has complete control over the design of the circuit within the design rules set up for the given technology, e.g., line spacing. Each transistor can be individually sized for optimum performance. However, the productivity of this approach is low, for example, figures, such as 6.17 transistors/day/person are quoted. This can be improved by electronically cutting and pasting sections from previous designs. But, with decreasing feature size, this straightforward reuse is not always possible. Thus, very large numbers of people are required to do a custom design. It is typically only used for standard parts design, such as microprocessors. One exception is the area of mixed signal design. ASICs used in communication circuits will have analog portions that are custom designed.

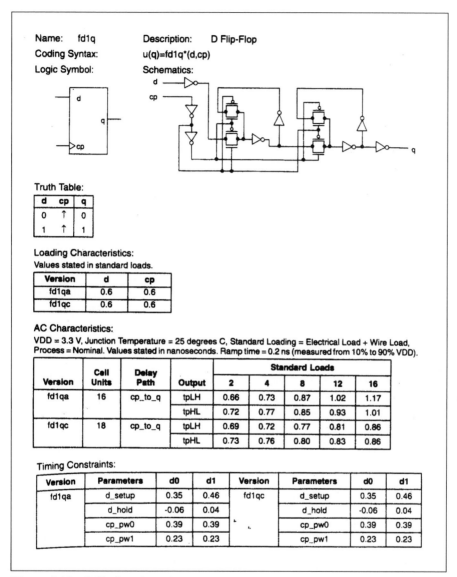

Figure 9.26 D flip-flop data sheet.

9.3.6 Relative Cost of ASICs and FPGAs

Designers must not only be concerned with the performance of the circuits but also their costs. Table 9.5 gives a summary of the design steps for MPGAs and standard cells vs. FPGAs. FPGAs require no chip fabrication or testing because the FPGA manufacturer does this. Thus, the Non-Recurring Engineering (NRE) costs for FPGAs are much smaller than for either MPGAs or for standard cells. Between MPGAs and standard cells, standard cells have the higher NRE. The

Table 9.5   Design steps for MPGAs and standard cell vs. FPGAs

MPGA and Standard Cell	FPGA
System Design	System Design
Logic Design	Logic Design
Place and Route	Place and Route
Timing Simulation	Timing Simulation
Test Pattern Generation	
Mask Making	
Wafer Fabrication	Download/Programming
Packaging	
Testing	
System Integration	System Integration

standard cell CAD process is more complicated. More masks must be made and more layers must be fabricated on the silicon surface. Some example data from *Application Specific Integrated Circuits* by Michael John Sebastian Smith (Addison Wesley, 1997) illustrates the situation. Assume one wishes to implement a 10,000 gate circuit. Typical NRE costs for the design would be:

1. FPGA - $21,800
2. MPBA- $86,000
3. CBIC - $146,000

On the other hand, the cost of an individual part shows the opposite trend:

1. FPGA - $39
2. MPGA - $10
3. CBIC - $8

These part costs result from the varying utilization of cells. FPGAs have the lowest utilization; MPGAs the next higher utilization, and CBICs the highest utilization. Thus, an FPGA to implement a 10,000 gate circuit would require a relatively large die size; whereas, an MPGA for a 10,000 gate circuit would require a much smaller die size than the FPGA, but still, much greater than a CBIC implementation. Larger die size means fewer dies per wafer and thus, a higher manufacturing cost.

Total Parts Cost is given by:

$$\text{Total Parts Cost} = \text{NRE} + \text{Cost/Part} \times \text{Sales Volume} \tag{9.3}$$

Figure 9.27 illustrates this relationship. It shows that for small volumes of less than 11,000 parts, FPGAs cost less. Between 11,000 parts and 40,000 parts, MPGAs cost less, while at volumes over 40,000 parts, standard cell CBICs cost less. Thus, the production volume is a critical variable in determining which approach to use.

Another important consideration is meeting the time window for product marketability. Figure 9.28 shows a situation, where a product introduction is delayed by one quarter. Conse-

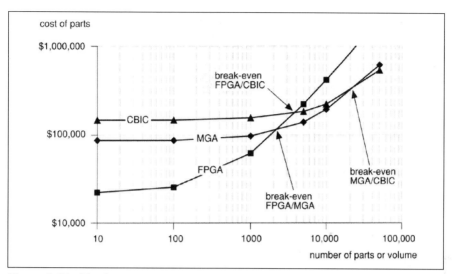

Figure 9.27   Total parts cost.

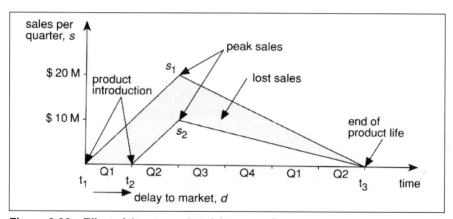

Figure 9.28   Effect of time-to-market delay on profits.

quently, a loss of $35 million in sales results. In terms of time to market, the ranking of the three methods, from shortest to longest is:

1. FPGAs
2. MPGAs
3. CBICs

Thus, this situation can have a profound effect on the choice of approach. Originally, FPGAs were thought of as strictly prototyping tools. However, their performance in the field has proven excellent, and they are a common choice for small and moderate volume products to reduce the time to market.

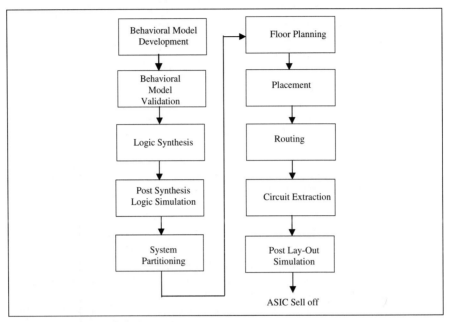

Figure 9.29   The ASIC design process.

9.4   THE ASIC DESIGN PROCESS

In this section we describe the ASIC design process with particular emphasis on HDL modeling and synthesis aspects. This process is applicable to all types of ASICs and programmable logic. We illustrate the process by describing how the Synopsys tools are used to synthesize standard cell ASICs and how the Xilinx tools are used to synthesize FPGAs.

Figure 9.29 shows the flow of the ASIC design process. The major steps are:

1. **Behavioral model development.**   In this step the requirements called out in the specification are translated into an HDL behavioral model. As part of this process, the system may be partitioned into a set of ASICs. Chapter 11 discusses the partitioning process in detail. The size of blocks in the functional decomposition is limited by the number of gates that can effectively be synthesized by existing tools. To use an automated synthesis tool, the behavioral models must be developed in a style acceptable to the tool. We discuss synthesis styles in Chapter 10. After the system is partitioned into a set of blocks, the ASIC design steps are repeated for each block.

2. **Behavioral model validation.**   Typically, a test bench is developed to validate the behavioral model. This is a complicated process as the test bench must reflect the requirements in the specification. One major issue is how exhaustively to test the model. For example, it is usually impossible to apply all possible input sequences to a model of a sequential circuit. We discuss these issues in detail in Chapter 10. The test bench is used to simulate the behavioral model of the ASIC. Simulation results are then examined to determine if the algorithms implied by the specification are correctly

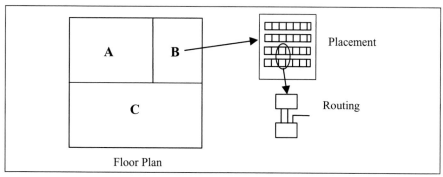

Figure 9.30   Floor planning, placement, and routing.

implemented. Gross, timing relationships can also be checked at this stage in the design process.

3. **Synthesis.**   A logic synthesis tool is used to transform the behavioral description into a structural gate-level circuit. The gate-level circuit is comprised of primitives from the ASIC library. The synthesis process may require many iterations as constraints on delay, area, and power consumption are applied in the synthesis process.

4. **Post synthesis logic simulation**.   In this step we simulate the synthesized circuit (1) to validate that the synthesis transformation is correct. For mature tools this is usually not a problem, but inevitably tool bugs will be uncovered, particularly if the user uses coding styles not tried before, and (2) to check the timing of the synthesized circuit. The timing for the synthesized circuit comes from an ASIC library and the structural interconnect, not from the behavioral model. Thus, the circuit must be checked for correct timing, relative to output delays, output hazards, register setup times, etc.

5. **System partitioning.**   In the development of the behavioral model, the system was divided into ASIC-sized chunks. However, the resultant gate-level circuit may be very large, requiring further subdivision within the ASIC. For example, in preparation for floor planning, the synthesized circuit might need to be partitioned into functional blocks. Such partitioning might require a repeat of the post synthesis logic simulation.

6. **Floor planning.**   In this step, we arrange the blocks of the net list on the chip. Figure 9.30 shows how three blocks (A, B, and C) might be arranged on a chip.

7. **Placement.**   In this step, standard cells are placed within each block to implement the function of the block. Figure 9.30 shows the cell placement within block B.

8. **Routing.**   In this step, wired connections are made between cells and blocks. Figure 9.30 illustrates the connections between two cells in block B.

9. **Circuit extraction.**   Determine the resistance and capacitance of the interconnect. This information is back annotated into the gate-level model, replacing the information derived from the wire load model. A Standard Delay Format (SDF) is used for this purpose.

10. **Post layout simulation.**   In this step, the gate-level circuit with accurate timing back annotated from the layout is resimulated to check circuit timing and to provide a final check on circuit functionality. Successful completion of these simulations is termed *ASIC sell off*. The design is now ready to be shipped to the silicon foundry.

9.4.1   Standard Cell ASIC Synthesis

In this section we describe the basic steps for synthesizing a standard cell ASIC. The discussion here is specific to Synopsys but other vendor tools require similar steps. The LSI 10K library is used in all our examples because of availability. Other libraries would show similar characteristics.

With the Synopsys system, one typically goes through the following steps to synthesize a circuit:

1. Analyze
2. Elaborate
3. Compile
4. Report
5. Save

Analyze.   The source VHDL file is analyzed for conformance to the VHDL synthesis subset. Models that will analyze successfully with a conventional VHDL simulator might fail this step if constructs are used, which the synthesis tool does not support. If the analysis is successful, the source VHDL file is translated into an intermediate format.

Elaborate.   Using the intermediate format generated in the analyze step, a design is built using generic components. This design is technology independent, that is, it is not tied to any specific ASIC library.

Compile.   This step translates the technology independent design to a library-based design. Compilation is frequently constrained in terms of the desired area, delay, or power of the synthesized design. The result of this step is a gate-level circuit expressed in terms of ASIC library macros.

Report.   This command documents the features of the synthesized circuit, such as cell area and the delay for the critical path. The command has many options, and its intelligent use is critical if one wants to understand the nature of the synthesized circuit.

Save.   This commands saves the results of synthesis. Two important file formats are used. The *.db format uses the tool's internal data structure. This file cannot be read by an editor tool. The file can be reloaded at a later time, and new compilation steps can be applied. Saving in the *.vhd format generates a VHDL structural model of the circuit, which can be used for simulation of the synthesized circuit to validate its function. This file can be read by an editor tool.

9.4.1.1   Synthesis Tool Use Strategies

Two important points can be made about synthesis tools. First, the documentation does not completely define the operation of the tool. This is in contrast to VHDL simulators, whose expected results are defined in the *VHDL Language Reference Manual*. Tool vendors realize a good deal of their revenue from teaching companies how to use their tools. Thus, all the useful tricks are not specified in the documentation. One must experiment with the tool to understand its operation. This is also true with simulators, but synthesis tools are an order of magnitude more difficult to use than a simulator, which leads to our second point. The operation of a synthesis tool is complex. There are many ways to constrain the design. After compilation, there are many report options that can be used to characterize it. We stated the necessity of experimenting with the

```
      library IEEE;
      use IEEE.std_logic_1164.all;
      use work.finc.all;
      entity SM_COUNT is
        port(CLK,CON,RESET: in std_logic;
             COUNT: inout std_logic_vector(3 downto 0));
      end SM_COUNT;

      architecture ALG of SM_COUNT is
        begin
        process(CLK,CON,RESET)
          begin
            if RESET = '1' then
              COUNT <= "0000";
            elsif CLK'EVENT and CLK='1' then
              if CON = '1' then
                COUNT <= INC(COUNT);
            end if;
            end if;
        end process;
      end ALG;
```

Figure 9.31   VHDL description of a counter circuit.

```
      analyze -format vhdl sm_count.vhd
      elaborate SM_COUNT -architecture ALG
      create_clock  -period 10 -waveform {0 5} CLK
      compile
      write -format vhdl -output SM_COUNT_1.VHD
      Write -output SM_COUNT_1.db
      report_timing > SM_COUNT_1.rep
      report_area  >> SM_COUNT_1.rep
```

Figure 9.32   Script for synthesis of the circuit.

tool. To learn from experiments, they must be repeatable. When one controls synthesis tools from menus, it is very difficult to remember command sequences used for a given experiment, thus, making it difficult to repeat. To handle this problem one must control the tool with scripts.

Figure 9.31 shows the VHDL description for a counter circuit. Figure 9.32 is a simple script which analyzes, elaborates, and compiles the circuit with a clock constraint. Next the design is saved in both *.db and *.vhd formats. Finally, a report is generated.

Figures 9.33a and 9.33b show sample reports. Figure 9.33a shows the timing report. In the script file (Figure 9.32), a clock with a period of 10 ns was created. This places a timing constraint on the design because the longest delay (called the critical path delay) in the next state logic must complete during a clock period. The timing report shows that the critical path between two registers begins at the CLK pin (CP) of COUNT_reg<0> and terminates at the TI input of COUNT_reg<3>. The logical elements along the path are also enumerated. The total path delay is 4.57 ns. The library cell used for COUNT_reg<3> (FJK2S) had a set up time of

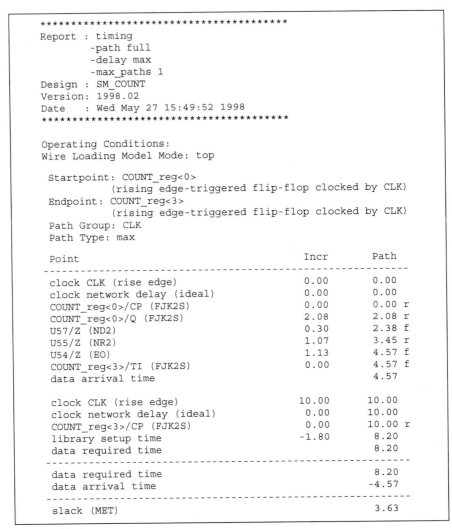

```
*******************************************
Report : timing
         -path full
         -delay max
         -max_paths 1
Design : SM_COUNT
Version: 1998.02
Date   : Wed May 27 15:49:52 1998
*******************************************

Operating Conditions:
Wire Loading Model Mode: top

 Startpoint: COUNT_reg<0>
            (rising edge-triggered flip-flop clocked by CLK)
 Endpoint: COUNT_reg<3>
            (rising edge-triggered flip-flop clocked by CLK)
 Path Group: CLK
 Path Type: max

 Point                                      Incr      Path
 ---------------------------------------------------------
 clock CLK (rise edge)                      0.00      0.00
 clock network delay (ideal)                0.00      0.00
 COUNT_reg<0>/CP (FJK2S)                     0.00      0.00 r
 COUNT_reg<0>/Q (FJK2S)                      2.08      2.08 r
 U57/Z (ND2)                                 0.30      2.38 f
 U55/Z (NR2)                                 1.07      3.45 r
 U54/Z (EO)                                  1.13      4.57 f
 COUNT_reg<3>/TI (FJK2S)                     0.00      4.57 f
 data arrival time                                     4.57

 clock CLK (rise edge)                      10.00     10.00
 clock network delay (ideal)                 0.00     10.00
 COUNT_reg<3>/CP (FJK2S)                      0.00     10.00 r
 library setup time                         -1.80      8.20
 data required time                                    8.20
 --------------------------------------------------------
 data required time                                    8.20
 data arrival time                                    -4.57
 --------------------------------------------------------
 slack (MET)                                           3.63
```

Figure 9.33a Timing report.

1.80 ns. Thus, the data must arrive in 8.2 ns. The path has a positive (good) slack of 3.63 ns. This path is the critical path because all other paths related to the clock are guaranteed to have a slack no worse than this. Thus, the circuit will work with this clock period. We discuss timing issues in more detail later in this chapter.

Figure 9.33b shows the area report, which provides the number of cells of each type used and shows the area of the synthesized circuit in equivalent 2-input NAND gates. Both combinational (14) and noncombinational area (52) are given. If a wire load model were specified, an estimate of the interconnect area would have been included. In this case no wire load model was specified. The combined timing and area report gives one the information to allow trade-offs between these two important design parameters.

```
*****************************************
Report : area
Design : SM_COUNT

Version: 1998.02
Date   : Wed May 27 15:49:53 1998
*****************************************

Library(s) Used:

  lsi_10k (File:
/software/synopsys/1998.02/libraries/syn/lsi_10k.db)

Number of ports:              7
Number of nets:              19
Number of cells:             12
Number of references:         5
Combinational area:       14.000000
Noncombinational area:    52.000000
Net Interconnect area:    undefined
 (No wire load specified)
Total cell area:          66.000000
Total area:               undefined
```

Figure 9.33b   Area report.

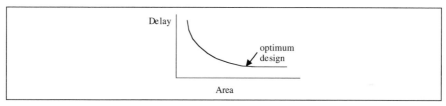

Figure 9.34   Ideal delay vs. area curve.

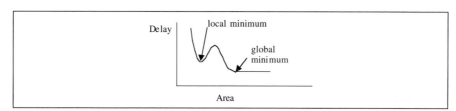

Figure 9.35   Realistic delay vs. area curve.

9.4.1.2   *Minimum Delay Circuits*

In this section we discuss how to synthesize circuits that have minimum delay. There is a trade-off between delay and area. Figure 9.34 shows an ideal delay vs. area curve. It is ideal because the delay is a monotonically decreasing function of area. The synthesis process can move along this curve until the flat region begins. The beginning of this region is the optimum design as an increase in area beyond this point does not yield any further reduction in delay.

Figure 9.35 shows a more realistic delay curve. This curve exhibits both a local and global minimum. Depending on the initial design point and the trajectory taken by the synthesis algorithm, it can get trapped in the area of the local minimum and never reach the global minimum. We illustrate this situation by means of an example. Consider this command script:

```
/*script 1*/
analyze -format vhdl sm_count.vhd
elaborate SM_COUNT -architecture ALG
create_clock  -period 10 -waveform {0 5} CLK
compile
```

The command:

create_clock -period 10 -waveform {0 5} CLK

specifies a 10 ns period, 50 percent duty cycle clock, and the corresponding constraint on data paths terminating at a clocked flip-flop. The circuit is then compiled. A subsequent timing report or highlight critical path command shows that the critical path is:

from COUNT_reg<0>/CP to COUNT_reg<3>/TI with a delay of 4.57 ns.

One can then constrain this path with the command:

set_max_delay 3 -from "COUNT_reg<0>/CP" - to "COUNT_reg<3>/TI"

which is then used to optimize the path. Recompilation produces another circuit with another critical path of 3.04 ns. Again a 3 ns constraint is applied and compilation performed. This process continues until the delay no longer changes. Table 9.6 summarizes the scenario.

Table 9.6   Scenario 1.

Action	Delay	Area
Execute script 1	4.45	66
Constrain critical path to 3 ns.	3.04	67
Constrain critical path to 3 ns.	3.04	65

Script 2 below applies another set of constraints, which develops a different initial design point:

```
/*script 2*/
analyze -format vhdl sm_count.vhd
elaborate SM_COUNT -architecture ALG
create_clock  -period 10 -waveform {0 5} CLK
set_max_delay 0 -to all_outputs() + all_registers(-data_pins)
```

Note the command:

set_max_delay 0 -to all_outputs() + all_registers(-data_pins)

specifies a maximum delay of zero for all paths terminating at registers or outputs. This command instructs the synthesis tool to exert maximum effort to minimize the delay along each path in the circuit. Script 2 produces a circuit with a maximum delay of 3.6 ns and an area of 69. Starting with this circuit, an additional constraint of 3 ns on the critical path produces a circuit with a delay of 3.11 ns and an area of 68. As shown in Table 9.7, additional synthesis runs produce little change.

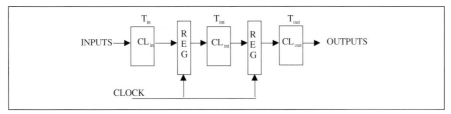

Figure 9.36 Synchronous system.

Table 9.7 Scenario 2.

Action	Delay	Area
Execute script 2	3.60	69
Constrain critical path to 3 ns.	3.11	68
Constrain critical path to 3 ns.	3.11	69
Constrain critical path to 3 ns.	3.11	68

Because scenario 2 produced a design with more area than scenario 1, one could conclude that scenario 2 caused the optimization algorithm to get trapped in a local minimum while scenario 1 produced an initial design point from which a global optimum could be developed. There are many tracks in the design space, and the initial point and the trajectory produced by the command sequence dictate what sort of minimum is achieved. Again, it is important to use scripts, so these experiments can be repeated and comparisons made.

9.4.1.3 Optimally Timed Synchronous Circuits

Figure 9.36 shows a block diagram of a synchronous system. Suppose, we wish to run the system at a clock frequency of f_{clk} and clock period $T_{clk} = 1/f_{clk}$, then it is clear that:

$$T_{int} < T_{clk} \tag{9.4}$$

for proper operation. Note that Tint includes register clock delays and register setup time. We would like to use this definition to define an *optimally timed synchronous circuit* (OTSC). However, in making such a definition, we should also take into account that the circuit shown in Figure 9.36 must be able to interface with other circuits of a similar nature. Thus, we add the additional requirements that:

$$T_{in} < \frac{T_{clk}}{2} \text{ and } T_{out} < \frac{T_{clk}}{2} \tag{9.5}$$

If conditions 9.4 and 9.5 are met, we are assured that the total delay between clocked registers will be less than T_{clk}, even if the registers are in different circuits.

Although one does not always need to satisfy condition 9.5, it is good design practice to include it, so we can be insured that the new component will work in all applications. If we will use a synthesis tool to achieve this, we have to be careful what we ask it to do. Figure 9.37 gives

```
library IEEE;
use IEEE.std_logic_1164.all;
use IEEE.std_logic_arith.all;
entity ADD is
  port(CLK,RESET: in std_logic;
       A: in signed(1 downto 0);
       C: buffer signed(1 downto 0));
end ADD;

architecture ALG of ADD is
begin
  process(CLK)
  begin
    if RESET = '1' then
      C <= "00";
    elsif CLK'EVENT and CLK = '1' then
      C  <= A + C;
    end if;
  end process;
end ALG;
```

Figure 9.37   VHDL description for registered adder.

the VHDL description for a two-bit registered adder. One can synthesize it initially with this script:

```
analyze -format vhdl add.vhd
elaborate ADD -architecture ALG
create_clock  -period 10 -waveform {0 5} CLK
compile
```

and then apply constraints on the critical internal path (CINTP). Table 9.8 shows the scenario. Figures 9.38, 9.39, and 9.40 show the resultant circuits. Note: Trial 2 improved the value of T_{int} but at the expense of T_{in}; whereas, Trial 3 reduced T_{in} and generated an OTSC with a clock period of 4 ns.

Table 9.8   Synthesis scenario for registered adder

Trial	Constraint	Critical Input Path (CINTP)	T_{int} (max)	Critical Input Path (CINPP)	T_{in} (max)	Area	Circuit
1	$T_c = 10$ ns	Creg<1>/CP to Creg<1>/D	4.27 ns	A to Creg<1>/D	2.25 ns.	31	add_int_scr.db Adder a
2	CINTP < 3 ns	Creg<0>/CP to Creg<0>/D	2.75 ns	A to Creg<1>/TE	3.57 ns	35	addb.db Adder b
3	CINTP < 2.5 ns	Creg<0>/CP to C<0>	1.63 ns	A to Creg<0>/TE	1.99 ns	36	addc.db Adder c

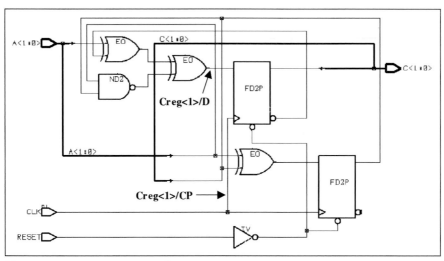

Figure 9.38   Adder A.

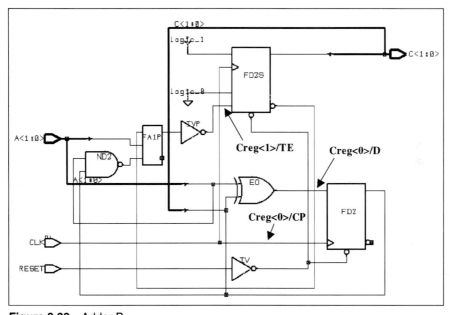

Figure 9.39   Adder B.

Thus, it is important to constrain T_{in} and T_{out} as well as T_{int}. We can do this by applying input delays, del_{in}, and output delays, del_{out}, to the circuit. The delays are placed in paths terminating (input) or originating (output) at a clocked register. In the context of the synthesis tool, they are tagged to the clock. Optimization will then optimize the input and output paths, including these delays relative to the clock period T_{clk}. Figure 9.41 illustrates this technique.

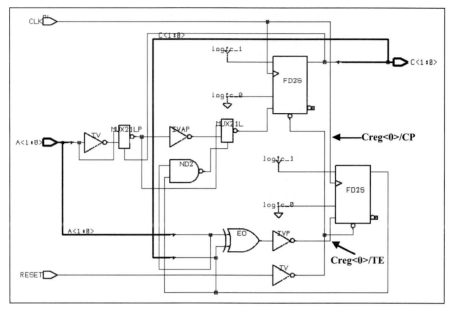

Figure 9.40   Adder C.

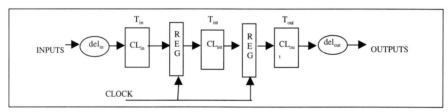

Figure 9.41   Optimizing with input and output delays.

```
analyze -format vhdl add.vhd
elaborate ADD -architecture ALG
create_clock  -period  5 -waveform {0 2.5} CLK
set_input_delay -clock CLK -max -rise 2.5 "A"
set_input_delay -clock CLK -max -fall 2.5 "A"
set_output_delay -clock CLK  -max -rise 2.5 "C"
set_output_delay -clock CLK  -max -fall 2.5 "C"
compile
report_timing -path end -delay max -max_paths 40 -nworst 5 >
add.rpt
```

Figure 9.42   OTSC script.

In general, the tool will try to achieve $T_{in} = T_{clk} - del_{in}$ and $T_{out} = T_{clk} - del_{out}$. For our case, to achieve $T_{in} < T_{clk}/2$ and $T_{out} < T_{clk}/2$, one sets $del_{in} = del_{out} = T_{clk}/2$. Figure 9.42 shows a script to achieve this. Figure 9.43 shows the synthesized circuit.

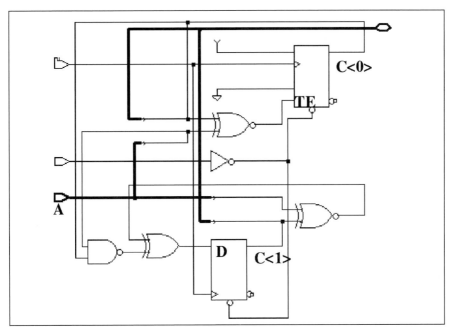

Figure 9.43   OTSC circuit.

Figure 9.44 shows the report file (add.rpt). The report gives the end points of paths in critical order, that is, the worst path is given first (in terms of its slack value -0.53), which means that the path is 0.53 ns too long; other paths come in order, based on their slack values. The first five paths are input paths, which terminate at the D input of an FD2 sequential element. (This is not evident from this report, but other reports, which give the end point of a path, show this.) From working with the library, one would know that the setup time for this cell input is 0.85 ns. For T_{CLK}= 5 ns, the required data arrival time (path required) is 4.15 ns. The first actual path is 4.68 ns, thus, the -0.53 ns slack. The next four paths are also input paths, which terminate at the TE input of an FD2S sequential element, which has a setup time of 1.25 ns and, thus, a data arrival time of 3.75 ns. The tenth path is an output path, and its requirement is that the data arrives 2.5 ns after the clock rise. The eleventh path is an internal path between registers, which terminates at the TE input and, thus, has a required data arrival time of 3.75 ns. The last three paths shown are output paths like path 10. To see both path end points, one can change the -path option of the report_timing command to: short. To see the full path, which shows the cells traversed, one can change the -path option to: full. The -max_paths option constrains the number of paths shown. The -nworst option controls how many paths terminating at a given endpoint are shown. In our case, max_paths = 40. Only fourteen paths are shown. However note that there are five paths each shown terminating at C_reg<1>/D and C_reg<0>/ TE; thus, these pins are "maxed out." Thus, one can conclude that at most 26 other paths could be displayed for max_paths = 40 if nworst was increased sufficiently. Increasing nworst allows one to see a variety of path types at the expense of repeating paths with the same endpoints.

```
*****************************************
Report : timing
        -path end
        -delay max
        -nworst 5
        -max_paths 40
Design : ADD
Version: 1998.02
Date   : Wed May 27 08:03:21 1998
*****************************************
Operating Conditions:
Wire Loading Model Mode: top


Endpoint                Path Delay      Path Required      Slack
-----------------------------------------------------------------
C_reg<1>/D (FD2)           4.68 f          4.15            -0.53
C_reg<1>/D (FD2)           4.68 f          4.15            -0.53
C_reg<1>/D (FD2)           4.50 r          4.15            -0.35
C_reg<1>/D (FD2)           4.50 r          4.15            -0.35
C_reg<1>/D (FD2)           4.50 f          4.15            -0.35
C_reg<0>/TE (FD2S)         3.69 f          3.75             0.06
C_reg<0>/TE (FD2S)         3.69 f          3.75             0.06
C_reg<0>/TE (FD2S)         3.58 r          3.75             0.17
C_reg<0>/TE (FD2S)         3.58 r          3.75             0.17
C<0> (inout)               1.63 r          2.50             0.87
C_reg<0>/TE (FD2S)         2.82 f          3.75             0.93
C<0> (inout)               1.53 f          2.50             0.97
C<1> (inout)               1.48 r          2.50             1.02
C<1> (inout)               1.47 f          2.50             1.03
```

Figure 9.44   Path report.

The results show that for a T_{ClK} of 5 ns we do not have an OTSC because the input delay is too long. However, if we change T_{CLK} to 5.53 ns, we do meet the requirements. Alternatively, one could constrain the delay from the input A to C_reg<1>/D and try another synthesis.

The above scenario suggests the possibility of using an iterative script for developing an OTSC. A high-level description of the script is:

1. Synthesize the circuit initially with a script of the type shown in Figure 9.42.
2. Determine T_{in}, T_{inp}, T_{out} of the critical path that has the worst slack.
3. Constrain the critical path with a value r times (T_{in}, T_{int}or T_{out}), where $0 < r < 1$.
4. Resynthesize the circuit.
5. If the circuit is an OTSC, stop. Otherwise, go back to step 2.

9.4.1.4   Validation of Synthesized Models

Synthesized models can be validated through (1) hand checking and (2) simulation.

Hand Checking.   Hand checking, although impractical for all but the smallest models, is a useful way to understand how synthesis transformations are made and how the basic ASIC cells operate. Thus, from an educational point of view, it is important to be able to examine sim-

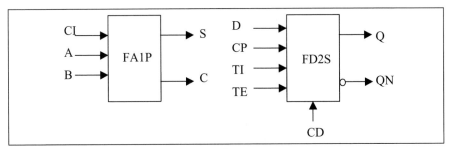

Figure 9.45   Pin-outs for FA1P and FD2S.

ple synthesized circuits and develop the circuit function. Figure 9.37 gives the VHDL description of a clocked two-bit adder. Figure 9.39 gives a circuit synthesized from the VHDL description (Adder B).

From the library file lsi_10k.lib, one can identify the functionality of the elements as (1) ND2—two input NAND, (2) EO—two input exclusive OR, (3) IV, IVP—inverters, (4) FD2S, FD2—clocked flip-flops, and (5) FA1P—full adder. The ND2, EO, IV, and IVP function as expected. The FD2 functions as a conventional clocked D flip-flop. One needs to know the pin-out and logic equations for the FD2S and the FA1P. Cell pin-out can be determined by turning on this layer using the View Style command in the Design Analyzer. Logic equations come from the .lib file.

Figure 9.45 shows the pin-outs for the FA1P and FD2S. We can now determine the functionality of the circuit. From Figure 9.39:

$$C<0>(D) = A<0> \text{ XOR } C<0>$$

which is correct for the low-order sum bit. From the *.lib file, the sum output (S) for the FA1P (Full Adder) component is:

$$S = CI \ A' \ B' + CI' \ A \ B' + CI' \ A' \ B + CI \ A \ B$$

Then, if we let CAR = A<0> C<0> and note that the inputs to the full adder are A = A<1>, B = CAR', and CI = C<1>', then:

$$S = \ C<1>'A<1>'CAR + C<1>A<1>CAR + C<1>A<1>'CAR' + C<1>'A<1>CAR'$$

which is the exclusive-OR of C<1>, A<1>, and CAR, the carry from the 0 position.

In the FD2S cell, the expression for the next state of Q is:

$$D \ (TE)' + (TI) \ (TE)$$

which for TI = 0 and D =1 equals (TE)'. Then, given that the output of the full adder passes though an inverter, the next state of C<1> will take on the value for S given above. Thus, C<1> is the correct value for the high-order sum bit.

9.4.2 Post Synthesis Simulation

After synthesis is complete, one simulates the generated structural model to verify that it is correct (ASIC sell-off). This is done for two reasons. First, one cannot completely trust the transformations performed by the synthesis tool. Their correctness is partially established by customer feedback. Second, one must verify the timing of the synthesized circuit. The timing in the synthesized circuit comes from the target ASIC library, not the original model that was the input to the synthesis tool. Synthesis tools ignore timing in the input model. Not only are path delays different, but the synthesized model, because it is a gate-level model, will exhibit such timing phenomena as hazards, which are not present in the original model. These differences must be taken into account when comparing the synthesized model to the behavioral model.

Assume we plan to synthesize the counter model shown in Figure 9.31. Before synthesis we test the model in a test bench (Figure 9.46), where we excite the counter with a 100 ns period clock with the counter enabled (CON =1). In this situation, because the model is a delta delay behavioral model, the counter output switches when the clock rises with delta delay.

Assume this model is synthesized and that the synthesized model is tested in the same test bench, that is, the counter runs with CON=1. Now, even though the clock switches at 100 ns intervals, the outputs of the counter (COUNT) switch later due to gate delays in the synthesized circuit. Not only that, the circuit exhibits hazards (normal). Figure 9.47 shows a situation where the output of the synthesized circuit is switching from 3 to 4. Ideally, the circuit should switch at 400 ns (400,000 ps), but switching doesn't begin until 1.45 ns (1450 ps) later, and exhibits a 20 ps hazard before settling out 1470 ps later. This situation has implications for validation of the model. One would hope to validate the synthesized model by comparing the simulation outputs of the synthesized model to the simulation outputs of the behavioral model. Thus, this delay would have to be accounted for when comparing the two models, that is, one would have to be sure that the output of the structural model had settled before comparing it to the output of the behavioral model. The best time to make the comparison is when the clock rises. At that time the behavioral model has not yet changed (it will change delta delay later), and the synthesized model has settled out from the previous clock transition.

Perhaps when simulating the synthesized model, one would like to reuse the same test bench (Figure 9.46). The behavioral source code has a file name sm_count.vhd. The entity name for the model is SM_COUNT. When storing the synthesized VHDL gate-level model, one should use a different UNIX file name, for example, SM_COUNT.vhd, so you don't overwrite the source. The synthesis tool uses the name SM_COUNT for the entity name in the gate-level model. Note in the test bench file, a default binding is used to bind SM_COUNT. The binding will use the most recently analyzed entity SM_COUNT. Thus, the test bench can be used to test either the behavioral or the synthesized model.

To analyze for behavioral simulation, one uses these two commands:

```
vhdlan -t ps -xsim sm_count.vhd
vhdlan -t ps  -xsim tb_count.vhd
```

and the behavioral model is bound into the test bench.

For simulation of the gate-level synthesized circuit, one uses those two commands:

```
vhdlan -t ps -xsim SM_COUNT.vhd
vhdlan -t ps  -xsim tb_count.vhd
```

and the structural model is bound in the test bench. The picosecond time scale is used, so hazards can be observed.

Another comment on how the test bench models are bound. In the configuration declaration statement, binding is carried out using the default binding. This statement uses the default

```
Library IEEE;
use IEEE.std_logic_1164.all;
entity TB_COUNT is
end;
architecture TESTBENCH of TB_COUNT is
  signal CLK,CON,RESET : std_logic := '0';
  signal COUNT: std_logic_vector(3 downto 0);
  component SM_COUNT
    port(CLK,CON,RESET: in std_logic;
         COUNT: inout std_logic_vector(3 downto 0));
  end component;
begin
  UUT :SM_COUNT
    Port Map (CLK,  CON,   RESET,   COUNT);
SignalSource : process
  begin
    CON <= '0', '1' after 60 ns;
    RESET <= '1', '0' after 40 ns;
    CLK <= '1', '0' after 50 ns, '1' after 100 ns,
           '0' after 150 ns,'1' after 200 ns,
           '0' after 250 ns, '1' after 300 ns,
           '0' after 350 ns, '1' after 400 ns,
           '0' after 450 ns, '1' after 500 ns,
           '0' after 550 ns,'1' after 600 ns,
           '0' after 650 ns, '1' after 700 ns,
           '0' after 750 ns, '1' after 800 ns,
           '0' after 850 ns, '1' after 900 ns,
           '0' after 950 ns, '1' after 1000 ns,
           '0' after 1050 ns, '1' after 1100 ns,
           '0' after 1150 ns,'1' after 1200 ns,
           '0' after 1250 ns, '1' after 1300 ns,
           '0' after 1350 ns, '1' after 1400 ns,
           '0' after 1450 ns, '1' after 1500 ns,
           '0' after 1550 ns, '1' after 1600 ns,
           '0' after 1650 ns, '1' after 1700 ns,
           '0' after 1750 ns, '1' after 1800 ns,
           '0' after 1850 ns, '1' after 1900 ns,
           '0' after 1950 ns, '1' after 2000 ns,
           '0' after 2050 ns, '1' after 2100 ns,
           '0' after 2150 ns;
    wait;
  end process;

end TESTBENCH;
configuration CFG_TB_COUNT of TB_COUNT is
  for TESTBENCH
    for UUT : SM_COUNT
    end for;
  end for;
end;
```

Figure 9.46 Counter test bench.

binding for either the original behavioral or the structural architecture, which is generated by the synthesis tool. This synthesized architecture contains no configurations specifications. Rather, the component declarations use the same names as the library elements, for example, ND2 for a

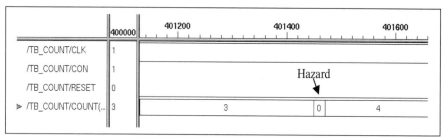

Figure 9.47 Synthesized circuit response.

two-input NAND gate. The default binding associates these component declaration names with model names in the cell simulation library. It binds at the top level, as well as the lower level, by looking for a model with the same entity name as the component declaration.

9.5 FPGA SYNTHESIS

In this section we describe FPGA synthesis. We illustrate this process using Xilinx software and a XC4010XL SRAM programmable FPGA chip. FPGA design differs somewhat from the ASIC design process shown in Figure 9.29. Figure 9.48 shows the FPGA design process.

Behavioral model development takes place as in the ASIC design process of Figure 9.29. The next step is logic synthesis, where the behavioral description is translated into a gate-level model. In this case the gate-level description has no delays, as the ultimate timing will come from the FPGA circuitry. This synthesis is carried out by the FPGA Express program which was developed by Synopsys and is included in the Xilinx Tool set. FPGA Express develops a *.xnf file that is forwarded to the Design Manager program, which carries out the remainder of the design activities.

Figure 9.49 shows the design flow carried out by Design Manager:

1. **Translate:** This process invokes a program NGDBUILD, which converts the .xnf file from FPGA Express to a Native Generic Database (NGD) file. There is some technology mapping in this step as primitives from the .xnf file are mapped into Xilinx library primitives.

2. **Map:** Mapping is the process of allocating CLBs, IOBs, or other Xilinx resouces to logical elements in the design.

3. **Place and Route:** The PAR program automatically performs optimal Place and Route on the mapped CLBs and IOBs in the design. This process can be subject to timing constraints specified by the user. Placement is performed first, so as to minimize the length of routing. Iteration might be required if routing fails. Routing is more difficult in FPGAs than ASICs because the routing resources are fixed. The result of Place and Route is an .ncd file, that is, a Native Circuit Description file.

4. **Timing:** Timing analysis on the placed and routed design is performed by the Trace program. Trace is a static timing analysis program that determines path delays in the design. It also checks to see if timing constraints imposed on the design have been violated. Reports are generated that can be used to modify the design if necessary.

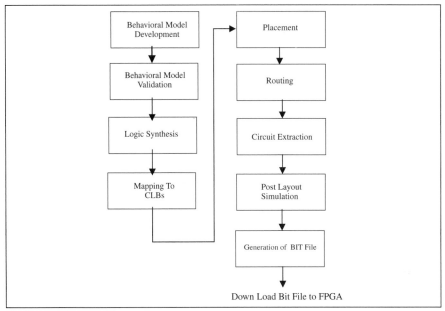

Figure 9.48 FPGA design process.

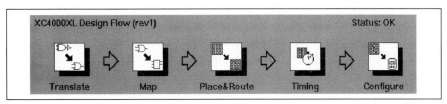

Figure 9.49 Design Manager design flow.

5. **Configure:** This function is performed by the BitGen program. It takes a fully routed .ncd file and produces a configuration bit stream (BIT file). The BIT file can then be downloaded in the FPGA's control memory cells or can be used to create a PROM, which will be used for initialization in an operational system.

9.5.1 FPGA Example

We now present an FPGA design example. The Xilinx chip employed is the XC4010XL mounted on an XESS XS40 Version 1.2 board with an XTEND board added. The combination of the two boards provides a variety of input and output devices to drive the chip, as well as a microcontroller, RAM, and Stereo Codec circuitry.

The example in Figure 9.50 has input logic to clocked flip-flops, logic between clocked flip-flops, logic between a clocked flip-flop and the output and a tristate output circuit. Figure 9.51 shows the general purpose I/O resources provided by the XESS/XTEND board (1) inputs: three pushbuttons, eight dip switches, and an oscillator and (2) outputs: three 7-segment displays

```
library IEEE;
use IEEE.STD_LOGIC_1164.all;
entity TIME_TEST is
Port( A,B,C,EN,CLK,RESET: in STD_LOGIC; F: out STD_LOGIC);
end TIME_TEST;

architecture ALG of TIME_TEST is
  signal  FF1, FF2: STD_LOGIC;
begin
 P1:process(RESET,CLK)
  begin
    if RESET = '1' then
      FF1 <=  '0';
    elsif ( CLK'EVENT and CLK = '1')
      FF1 <= A and B;
    end if;
  end process;
 P2: process(RESET,CLK)
  begin
    if RESET = '1' then
      FF2<=  '0';
    elsif ( CLK'EVENT and CLK = '1')
      FF2 <= FF1 and C;
    end if;
  end process;
  F <= FF2 and FF1 when EN = '1' else
       'Z' when others;
end ALG;
```

Figure 9.50 Example model (TIMECKT.VHD).

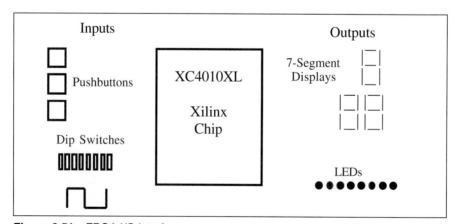

Figure 9.51 FPGA I/O interface.

and eight LEDs. A model of the circuitry to service this interface must be provided as part of the top-level structural model. Figure 9.52 shows a diagram of the top-level structural model.

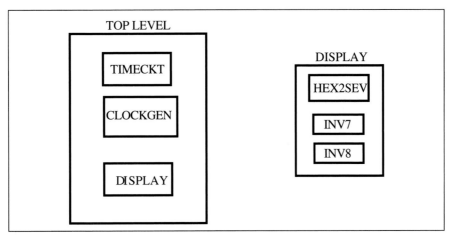

Figure 9.52   Top-level structural model.

The TOPLEVEL model instantiates three components (1) TIMECKT—the model being designed, (2) CLOCKGEN—this model counts down a 12 MHz clock to the desired clock frequency, (3) DISPLAY—conditions TIMECKT outputs for driving seven-segment and LED displays. DISPLAY is itself a structural model consisting of a hex to seven segment converter (HEX2SEV) and 7- and 8-bit inverters (INV7 and INV8). We leave the implementation of the other blocks as problems at the end of the chapter.

To place and route a design, the designer must prepare a User Constraints File (UCF). Figure 9.53 shows such a file as an example. The majority of the file defines the mapping of signal names to pin names. However, at the bottom of the file is a timing constraint section:

1. The first TIMESPEC command defines a constraint that says the delay between clocked flip-flops should not exceed 10 ns. FFS denotes the general signal group, flip-flops.
2. The second TIMESPEC command defines a constraint that says the delay between input PADS and clocked flip-flops should not exceed 4 ns. PADS denotes the general signal group, PADS.
3. The TNM command defines a signal group, CKTOUT, and places the signal F in that group.
4. The third TIMESPEC command defines a constraint that says the delay between FFS and the signal group (CKTOUT) should not exceed 17 ns.

Figure 9.54 shows a partial FPGA layout for the example circuit. The upper left CLB implements all the functionality of processes P1 and P2 (see Figure 9.50). Two 4-input LUTs and two flip-flops are utilized. The middle CLB uses one 4-input LUT to implement the AND of FF1 and FF2. The lower-right CLB inverts the input signal EN using a 4-input LUT. Where the signals converge on the lower right, a tristate buffer is used to implement the conditional assignment statement in the model. A fourth CLB (not shown) is used to invert the sense of the reset pushbutton and connect it to the start circuit.

```
# TIMECKT Configuration File
NET XS40_CLK    LOC=P13;
# TOP SEVEN SEGMENT CONNECTIONS(ACTIVE-LOW)
NET TOP_7S<6>   LOC=P25;
NET TOP_7S<5>   LOC=P26;
NET TOP_7S<4>   LOC=P24;
NET TOP_7S<3>   LOC=P20;
NET TOP_7S<2>   LOC=P23;
NET TOP_7S<1>   LOC=P18;
NET TOP_7S<0>   LOC=P19;
# Dip Switch Connections
NET DS<3>      LOC=P6;
NET DS<2>      LOC=P9;
NET DS<1>      LOC=P8;
NET DS<0>      LOC=P7;
#PUSHBUTTON SWITCHES (ACTIVE-LOW)
NET RESET_BUT  LOC=P37;
# LEFT SEVEN SEGMENT CONNECTIONS (ACTIVE-LOW)
NET LEFT_7S<6> LOC=P3;
NET LEFT_7S<5> LOC=P4;
NET LEFT_7S<4> LOC=P5;
NET LEFT_7S<3> LOC=P78;
NET LEFT_7S<2> LOC=P79;
NET LEFT_7S<1> LOC=P82;
NET LEFT_7S<0> LOC=P83;
# RIGHT SEVEN SEGMENT CONNECTIONS (ACTIVE-LOW)
NET RIGHT_7S<6>LOC=P59;
NET RIGHT_7S<5>LOC=P57;
NET RIGHT_7S<4>LOC=P51;
NET RIGHT_7S<3>LOC=P56;
NET RIGHT_7S<2>LOC=P50;
NET RIGHT_7S<1>LOC=P58;
NET RIGHT_7S<0>LOC=P60;
# LED CONNECTIONS (ACTIVE-LOW)
NET LED<7>      LOC=P10;
NET LED<6>      LOC=P80;
NET LED<5>      LOC=P81;
NET LED<4>      LOC=P35;
NET LED<3>      LOC=P38;
NET LED<2>      LOC=P39;
NET LED<1>      LOC=P40;
NET LED<0>      LOC=P41;
#Timing constraints
TIMESPEC TS01=FROM FFS TO FFS 10;
TIMESPEC TS02=FROM PADS TO FFS 4;
NET F TNM=CKTOUT;
TIMESPEC TS03=FROM FFS TO CKTOUT 17
```

Figure 9.53 UCF file.

Table 9.9 shows the back-annotated values from the placed and routed FPGA array. The constraints with * were not met.

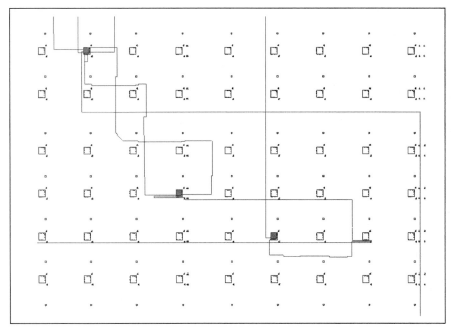

Figure 9.54 Partial FPGA layout as an example.

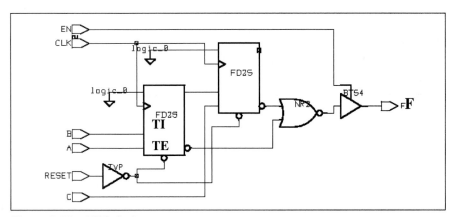

Figure 9.55 ASIC design.

Table 9.9 Timing results.

Constraint	Requested	Actual	Logic Levels
FFS to FFS	10 ns	4.01 ns	2
PADS to FFS *	4 ns	4.725 ns	2
FFS to F *	17 ns	22.2 ns	4

9.5.2   Comparison with an ASIC Design

Figure 9.55 shows an ASIC design of TIMECKT.VHD. The design takes advantage of the TI and TE inputs of the FD2S library elements to perform the ANDing with this element. Thus, there is no input logic, nor logic between sequential elements. The output delay to F is 3.18 ns. The FD2S element has a clock delay of 1.42 ns and a setup time of 1.25 ns, which gives a minimum clock period of 2.67 ns and a max clock frequency of 374.5 MHz. The TRACE program for the Xilinx FPGA design predicts a minimum clock period of 4.01 ns and a maximum clock frequency of 249 MHz. However, the LSI 10 K library is mature technology, while XC4010A is relatively new. A comparable sequential element from the LSI 500 K library would have a clock delay of 0.58 ns and a setup time of 0.61 ns, which yields a minimum clock period of 1.19 ns, and a clock frequency of 840 MHz, at least for the isolated design in the example. The ASIC library design has an area of 25 equivalent two input NAND gates. The result of the Xilinx mapping process gives an equivalent gate count of 39. Note that Xilinx uses two 4-input LUTs very inefficiently. One is used to implement a 2-input function, and one is used to implement an inverter.

PROBLEMS

9.1   Explain how standard parts design and ASIC design relate to top-down and bottom-up design as described in Chapter 1.

9.2   Search vendor literature and trade magazines. Find one example each of commercial products that are (1) standard parts, (2) ASIC design, and (3) a mixture of standard parts and ASIC design.

9.3   Give five examples of large cores (Intellectual Property) that can be obtained commercially.

9.4   List four CAD tool vendors that provide VHDL simulation and synthesis capabilities.

9.5   Give an example of a product that mixes Von Neumann Machine architecture, ASICs, and FPGAs.

9.6   Draw the transistor-level diagram of a CMOS three-input NAND gate.

9.7   What is the typical static power dissipation of a CMOS inverter? With a clock period of 10 ns, a VCC of 4.75 v, and a load capacitance of 100 pf, what is the dynamic power dissipation? If one wants to conserve power dissipation in a CMOS microprocessor during periods of no computation, what should one do?

9.8   Why does CMOS logic have low-static power dissipation?

9.9   Why is low-power dissipation important in the gates used to build complex chips?

9.10 What is the smallest feature size used in ASIC CMOS implementations today? In FPGA CMOS implementations?

9.11 Implement the state machine whose state diagram is given in Figure 8.22 of the text on a PLD. Simulate the design. Estimate the percentage of PLD resources required.

9.12 Implement the majority function of three inputs (see 1's Counter example in Chapter 3) using (1) an 8 to 1 multiplexer, (2) a 4 to 1 multiplexer and necessary inverters, and (3) table lookup. Compare the hardware complexity of the three circuits.

9.13 Figure 9.56a shows the block diagram of a "toggle when equal circuit." The circuit is a clocked sequential circuit. Output Z toggles when X1 and X2 are equal. When X1 and X2 are not equal, output Z does not change. The detector has a highly active, asynchronous reset, R, which sets the output Z to '0'. Figure 9.56b shows a CLB cell for an FPGA. The table in the cell can implement any function of three variables. Show how the CLB can implement the last

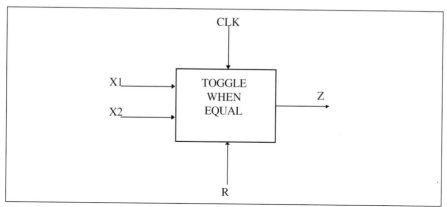

Figure 9.56a "Toggle When Equal" detector.

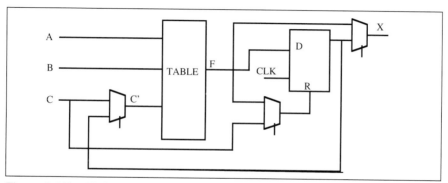

Figure 9.56b FPGA CLB.

agreement detector. Assume (1) A is the most significant address bit for the table, C' the least, (2) when a multiplexer control line is 0 the upper input is selected, and (3) the clock (CLK) is hardwired to the cell. The output Z should equal the present state of the circuit.

9.14 For this problem, use the figures in Problem 9.13. However, in this case Figure 9.56a is the block diagram of a "last agreement detector." The circuit is a clocked sequential circuit. Output Z is equal to the last value inputs X1 and X2 agreed upon. So, if the last time X1 equaled X2, X1 = X2 = '1', then Z = '1'. If the last time X1 equaled X2, X1 = X2 = '0', then Z = '0'. When X1 and X2 differ, Z does not change. The detector has a highly active, asynchronous reset, R, which sets the output Z to '0'. Figure 9.56b shows a CLB cell for an FPGA. The table can implement any function of three variables. Show how the CLB can implement the last agreement detector. Assume (1) A is the most significant address bit for the table, C' the least, (2) when the multiplexer control line is 0, the upper input is selected, and (3) the clock (CLK) is hardwired to the cell.

9.15 Determine the delay through the Switch Delay Model in Figure 9.14. Assume the input is a 5 v, 10 ns duration pulse. Using SPICE, determine the delay for two cases (a) SRAM interconnect and (b) ViaLink antifuse, and compare the results.

9.16 Give two reasons why one would choose FPGAs using fusible links over Xilinx SRAM FPGAs.

9.17 This question refers to the XC4010XL FPGA described in the text. Rank the following types of interconnect in terms of speed (1 for highest, 4 for lowest).

quad length lines _____

single length lines _____

direct interconnect _____

double length lines _____

9.18 Assume a CMOS chip designer has the following options in implementing her chip: FPGAs, Gate Arrays (MPGAs) and standard cell (CBICs). Assuming performance is not an issue, which option would be the least expensive for the following types of design? (Use data from the text.)

a. a) A chip with 85 gates, 300 chips made? _____

b. b) A chip with 15,000 gates, 9,000 chips made? _____

c. c) A chip with 50,000 gates, 200,000 chips made? _____

9.19 Compare present-day gate array and standard cell technologies in two areas, gate density and clock rate.

9.20 In Standard Cell libraries, one usually has low power versions of gates and high power versions of gates. Why would one use a high power version in place of a low power version?

9.21 Use integration to show that the sales lost in Figure 9.28 is indeed $35 million dollars.

9.22 One of the later steps in the ASIC design process is circuit extraction. What are the results of this step used for?

9.23 Using Synopsys Design Analyzer and the LSI 10 K library, synthesize the counter desscription in Figure 9.31 using the script in Figure 9.32. However, change the clock period to 5 ns. Compare area and time results with that achieved in the book. What conclusion can you draw?

9.24 Determine the OTSC with the highest clock frequency for the registered adder described in Figure 9.37.

9.25 Implement the counter (Figure 9.31) and registered adder (Figure 9.37) on a Xilinx PFGA. Compare achievable clock frequency with that using Synopsys and the LSI 10K library.

9.26 In this problem you will design and synthesize a triangle wave generator chip.

Specification: Design a chip which implements a Triangular waveform generator. The output of the chip is to have eight bits of precision. When enabled (EN =1), the output will ramp from 0 to 255_{10} and back to 0, and then repeat the cycle. The period of the ramp will be $510_{10}*T_{CLK}$, where T_{CLK} is the period of the clock. When the reset is low (R= 0), the counter is reset and counting is inhibited.

a. Develop a VHDL behavioral model for the chip. Use IEEE_STD logic packages, including those packages that overload + and -. Use overloaded + and - to increment and decrement the ramp count.

b. Validate the model with a VHDL system.

c. Synthesize the model with Xilinx. Generate two versions.

1. Slow clock version—drive with 1.4 Hz clock and observe response on LEDs.

2. Fast clock version—drive with 12 mHz and observe the response of the high order stage on the oscilloscope. For this version, use timing analysis from the Xilinx tools to determine what the maximum possible clock frequency is.

d. Synthesize the model on Synopsys. Perform timing analysis and determine the maximum clock frequency that the circuit can run at. In your report include a copy of the circuit schematic. Highlight the critical path on the schematic. Compare the timing analysis with c.2 above. Which circuit can be run faster? Is this what you expected? Explain. The

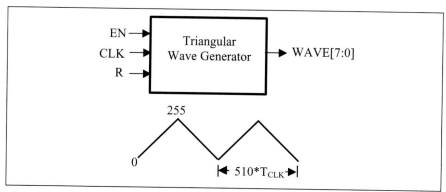

Figure 9.57 Triangular waveform generator.

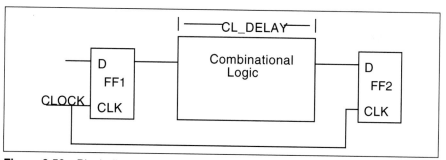

Figure 9.58 Block diagram for Problem 9.28.

ASIC library we are using with Synopsis is the LSI 10 K library, which is old CMOS technology, while the XC4010XL is a fairly new FPGA chip. So, the comparison is not actually "fair." The LSI 500 K library has gates whose delays are one-fifth of the corresponding gates in the LSI 10 K library. Repeat the comparison assuming you are using the LSI 500 K library.

9.27 Using the Synopsys system, develop an OTSC for the model developed in Problem 9.26

9.28 Given the following flip-flop data:

setup time	5 ns
hold time	2 ns
tplh(when clocked)	12 ns
tphl(when clocked)	14 ns

CL_DELAY is 35 ns, independent of the direction of signal change, as shown in Figure 9.58.

What is the maximum possible frequency for the signal CLOCK for which the circuit will operate properly?

Modeling for Synthesis

$\mathbf{I}$n this chapter, we present modeling styles appropriate
for synthesis. We also discuss synthesis tool commands that support a particular modeling style.
For each model that we present, we show the resultant circuit that is synthesized by Synopsys or
Xilinx software.

10.1 BEHAVIORAL MODEL DEVELOPMENT

In Figure 9.29, we showed the steps in the ASIC design process. The initial steps were the
development and validation of the behavioral model, followed directly by synthesis. The impli-
cation was that this initial behavioral model was in the proper form for synthesis. In this chapter,
we examine this process in more detail.

Figure 10.1 illustrates a common design paradigm. In many cases two models are actually
involved:

Model 1.   Model 1 is the behavioral model which reflects the interpretation of the specification. This model is sometimes referred to as the *executable specification* because it is an
immediate interpretation of the specification. We discuss this in detail later in the chapter. The
reader should also see Chapter 11 for an in-depth discussion of model development. Another
point to be made is that one does not typically synthesize a model representing an entire system.
Thus, Model 1 is usually a component in a system-level schematic. Chapter 11 also discusses
how to create such a schematic. The results of the model validation simulation are fed back to
the specification writer to answer the question: Is this what you meant? If the answer is Yes, one
wants to proceed directly to synthesis. But before synthesis is begun, one must check to see if
this behavioral model is in a style compatible with the synthesis tool. As we discuss in this chapter, not all constructs in the VHDL language can be synthesized—only those that have hardware
interpretations can be.

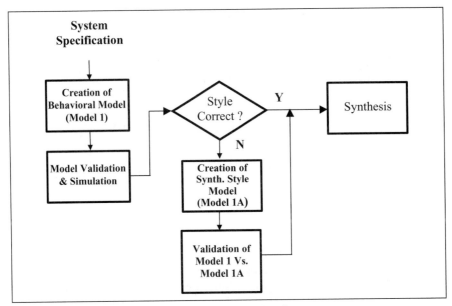

Figure 10.1   Modeling and simulation in the design flow.

Model 1A.   In many cases the initial behavioral model is not in the correct style and one has to create another model, which we refer to as Model 1A. It, hopefully, implements the same behavior as Model 1 but is coded in a style that is compatible with the synthesis tool. Having both a Model 1 and a Model 1A is problematic because additional simulation is required to show that that they exhibit equivalent behavior. The question is: Why can't they be the same model? The answer is that the goal of the developer of Model 1 is to make sure that the interpretation of the specification is correct. In many cases, for example, in DSP, the specification is embodied in a mathematical model that has been proven correct. Thus, the model developer tends to directly implement this mathematical model in VHDL using high-level constructs that might not have direct interpretation in hardware. He may also use high-level data types, such as Real, which are closely related to the mathematical model. Also, his main concern is the simulation semantics of the language. When developing the model for synthesis, the designer is thinking about the hardware implementation and the synthesis semantics of the language. Thus, for many systems it is difficult to use the same model for both purposes.

This problem has been alleviated somewhat by the development of *behavioral compilers*, which can translate some high-level descriptions into register transfer-level descriptions, although they cannot successfully transform all models.

10.1.1   Creation of the Initial Behavioral Model

The first step in the modeling process is to create the behavioral model that represents the first interpretation of the specification. The specification is expressed in terms of natural language (English), block diagrams, state diagrams, and timing diagrams. There have been efforts to auto-

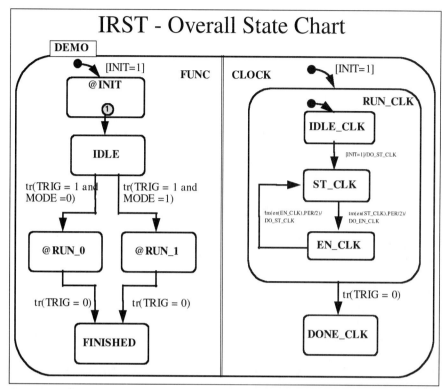

Figure 10.2 IRST overall state chart.

mate this process. Researchers at Virginia Tech have investigated a process whereby English from computer system specifications is first translated into conceptual graphs and then into VHDL model fragments. Typically, only model fragments are generated because the syntax and semantics of English are not exact. Correcting these deficiencies may require an English-oriented specification language. In industrial practice, the generation of behavioral models from specifications is performed manually, but we feel that, increasingly, graphics-based tools will be employed in preparing models. There are two types of graphics-based tools: *application-domain* and *language-domain* tools.

10.1.2 Application-Domain Tools

In application-domain tools, the graphical representation is dictated by the application area. For example, *state charts* can be used in control intensive models. Figure 10.2 shows the top-level state chart for a model used to generate two-dimensional frames of pixels for an infrared search and track (IRST) system. State charts allow for concurrent states and the modeling of concurrent control activity. In Figure 10.2 note that states FUNC and CLOCK are concurrent. Nesting of states is also allowed, thus preventing "state explosion." For example, states IDLE_CLK, ST_CLK, and EN_CLK are nested within the state RUN_CLK. States with their names preceded by an @ are super states, which imply an underlying state chart. For example,

```
ACTION DICTIONARY
Project: IRST

DO_ANG_CALC
Defined in chart: INIT
Definition: TAR_XCOORD:=XCOORD;
            TAR_YCOORD:=YCOORD;
            if(MODE_MOTION=1)then
                RAD_ANGLE:=0;
            end if;
            if(MODE_MOTION=2)then
                RAD_ANGLE:=(90*3.14)/180;
            end if;
            if(MODE_MOTION=3)then
                RAD_ANGLE:=(ANGLE*3.14)/180;
            end if;
            R_VECTOR:=VELOCITY;
```

Figure 10.3   Express VHDL code template example.

INIT, RUN_0, and RUN_1 are super states which themselves have underlying state charts. State transitions are triggered by events or elapsed time. In addition to the state charts, one typically has to fill in code templates, which define functions invoked when a transition is made.

Figure 10.3 shows a template for a function DO_ANG_CALC, which is invoked when the INIT state chart is invoked.

State chart models can be simulated at this graphics level. Code generation interfaces can be used to generate C or synthesizable VHDL or Verilog.

Data flow models can also be developed graphically. Figure 10.4 shows a model of a synthetic aperture radar developed on Cadence's Signal Processing Workbench (SPW). The model can be simulated at this graphics level and an FFT of the output can be used to validate model correctness. Again, a code-generation interface can then be used to produce an equivalent VHDL behavioral model. The complex signal sinks in Figure 10.4 generate data files that can be used as inputs to test lower-level models.

Figure 10.5 shows the received down-converted signal from the radar, and Figure 10.6 shows the FFT of the output signal. Note that it indicates a major response at a single frequency. The FFT is normalized to the sampling frequency, but an associated readout gives the peak frequency to which the range is proportional.

SPW is DSP specific. More general data flow systems can be modeled by activity charts as implemented in Ilogix Express VHDL. Graphical representations, such as flow charts and block diagrams are also used in such systems as Visual HDL from Summit.

The advantage of these graphics-based systems is that the model can be validated at a high level, and in the case of digital signal processing (DSP) models using proven system mathematics. Thus, the validation process does not require detailed knowledge of a hardware description language because the modeler is working strictly in the application domain. After the model is found to be correct, C, VHDL, or Verilog can be automatically generated.

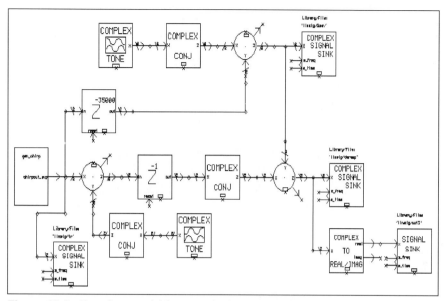

Figure 10.4 Data flow model for a synthetic aperture radar.

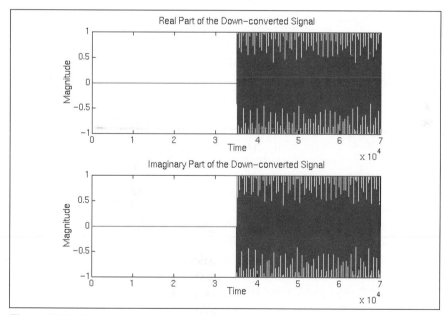

Figure 10.5 Received down-converted signal.

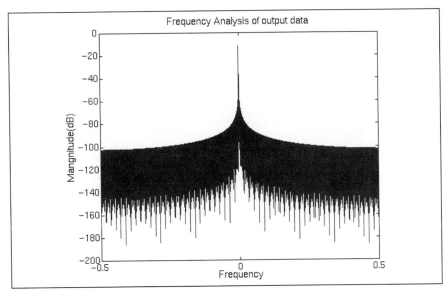

Figure 10.6   FFT of de-ramped signal.

10.1.3   Language-Domain Modeling

One can also create models in the language domain. This is the domain used by CAD engineers and designers, who want close control over model development. These individuals must be knowledgeable in the target language. They can use standard editors to prepare a source file that contains code written in the target language. Modelers may also employ *context sensitive editors* that check the syntax of the VHDL as it is entered. This straightforward editing approach is used in most cases, but has the disadvantage that in complex models consisting of many lines of code, it is easy to lose perspective. Thus, one would prefer a tool that works in the language domain but that provides a graphical representation for a model to help manage model complexity.

In VHDL, behavioral models are represented by a network of communicating processes represented by a Process Model Graph (PMG). The nodes of the graph represent processes and the arcs represent VHDL signals. A modeling tool known as the Modeler's Assistant uses this representation to develop VHDL behavioral models. The user first enters the PMG, then selects constituent processes from a library of process primitives. Models are constructed rapidly and are correct by construction. Figure 10.7 shows the Modeler's Assistant screen for a six-process model of the synthetic aperture radar (SAR) sensor using the process primitives GENCHIRP, DELAY, DOWNCONVERT, DERAMP, and DECIMATE from the SAR process primitive library.

The system provides for code reuse at the process level, a capability that is not available in conventional VHDL software. Modeling primitives can have a finer granularity than is provided by entity models. Rapid construction of models is only possible when there exists an available library of processes. To develop a process library that covers all models is an impractical goal, but for a given design group, it is quite practical to develop a process library, which covers most model functions. For example, for radar modeling one has processes, such as chirp generator

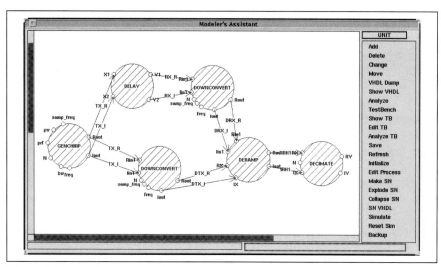

Figure 10.7   SAR sensor model.

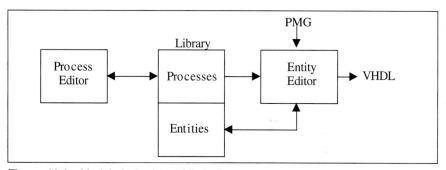

Figure 10.8   Modeler's Assistant block diagram.

(generates the radar transmitted signal), delay, down-convert, de-ramp, and decimate. To aid in this library construction, the Modeler's Assistant allows one to construct process *primitives,* which are generalized processes. They are generalized in the sense that process input and output port[1] names, bit widths, and the sense of process inputs are general and are specified when these primitive processes are used.

Figure 10.8 shows a block diagram of the Modeler's Assistant. Processes and process primitives are created and refined by the Process Editor and stored in the Processes section of the MODAS library. The Entity Editor uses processes from the process library and the Process Model Graph (PMG), which is input graphically. It constructs the entity model and stores it in

---

1.   Because processes are the modules from which one builds the Process Model Graph of the model, we define input and output ports for processes. These ports disappear when the model VHDL is generated.

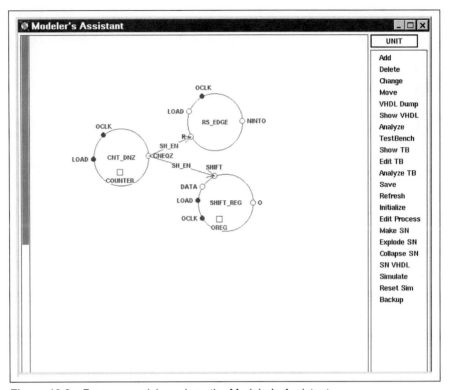

Figure 10.9   Process model graph on the Modeler's Assistant.

the Entity section of the library for later use and refinement. The Entity Editor also generates the VHDL code.

Figure 10.9 shows the menu from the Entity[2] Editor for a three process model of a parallel to serial conversion circuit. The parallel to serial conversion circuit is the output stage of a UART. Figure 10.10 shows the PMG for the full UART. The added nodes represent the rest of the UART: a serial to parallel converter (input stage) and buffer for gating data onto the data bus.

As can be seen, the Process Model Graph for the UART is complex. To reduce this complexity one can collapse groups of nodes into a *Super Node*. Figure 10.11 shows one way to collapse nodes for the UART model. Because PMGs for complicated systems can contain many processes, we developed this mechanism for collapsing a PMG into a Super Node, which can be used in a higher-level graph consisting of Super Nodes only or a mixture of processes and Super Nodes. The Super Node concept provides an efficient method for modeling behavioral hierarchy in the VHDL language. The program can also translate a higher-level graph consisting of only Super Nodes into a structural model. Thus, one can study partitioning of functionality by trying different Super Node configurations. The hierarchy implied by Super Nodes is graphical only—

---

2.   In the Modeler's Assistant an entity is referred to as a unit.

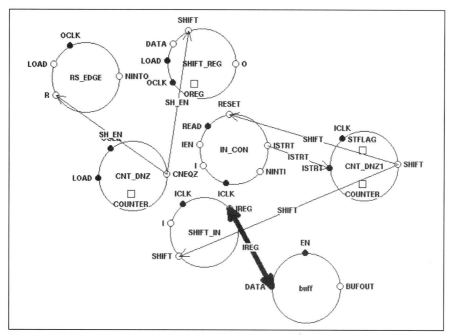

Figure 10.10 PMG of complete UART.

in VHDL one cannot nest a process within a process. The VHDL code resulting from the graph with Super Nodes is flat; that is, it is a conventional multi-process, single architecture behavioral model. In one study that we ran, this flat code was more efficient in a simulation and synthesis sense than the code generated by systems that convert a hierarchical graphic representation into a VHDL model, which maintains the hierarchy.

The coding style with Modeler's Assistant is process-oriented, but this fits well into the synthesis paradigm, which is also process-oriented.

10.1.4 Modeling and Model Efficiency

When one is creating a model with a tool instead of hand coding it, it is important to assess the efficiency of the model creation process. For the IRST model shown in Figure 10.2 (only the top state chart is shown) the complexity data are (1) state charts: 43 nodes and 56 arcs, (2) templates: 520 lines of text, (3) resultant VHDL (generated automatically): 948 lines, and (4) programming time: about a month because the modelers were learning the Ilogix software and the IRST specification. Assuming knowledge of both, a model of this complexity could be done in under a week. Therefore, in this case there was little coding gain, but model checkout is much faster at the high level.

One should also be concerned about the efficiency of code generated by high-level modeling tools. Code efficiency is measured in two domains (1) *simulation efficiency*, that is, how fast does the code simulate and (2) *synthesis efficiency*, that is, what is the cell count when synthesized? Results of a study at Virginia Tech showed that hand-coded models were as much as three

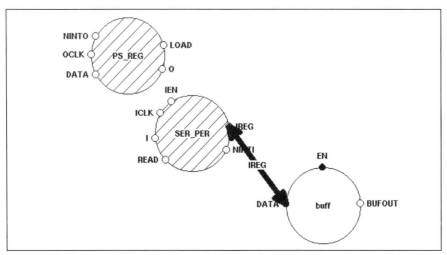

Figure 10.11   Super Node UART model.

times more simulation efficient and required 20 percent less cells when synthesized, as compared to models developed from high-level tools. Results vary depending on the tools used and the individual doing the hand coding. The important point is to monitor the efficiency of code developed from high-level tools.

10.1.5   Application-Domain vs. Language-Domain Modeling

Application-domain modeling is done by designers who do not want to be burdened by language details. These designers are familiar with the pictorial representation used in schematic capture. The graphic representation in the application domain (for example, state charts) is a semantic domain that designers are familiar with. The underlying mathematics of this domain is well known and well defined. They can validate their models quickly in this domain. However models developed in this domain tend to be less efficient than those that are hand coded. Also, because the application-domain models are developed from system mathematics, for example, DSP, they may contain language constructs that can not be automatically synthesized. However, tool vendors try to alleviate this problem by providing code interfaces that generate VHDL code in various styles compatible with synthesis tools.

Another issue is tool coverage of different application domains. The DSP domain is well covered by high-level modeling tools, which is fortunate because a great deal of ASIC design deals with signal processing. However, the tools available for general modeling are not as powerful in terms of their ability to support reasoning at a high level of abstraction. This problem, however, is being addressed. Some systems have combined control flow and data flow modeling capability. For example, Ilogix can also model data flow systems with activity charts. Visual VHDL from Summit Design Inc. also allows modeling of both control and data flow behavior with block diagrams, flow charts, state diagrams, and state tables.

With language-domain modeling, one must be familiar with the language constructs, such as VHDL processes. CAD engineers and designers trained in hardware description languages do language-domain modeling. Model development requires more time in this domain, but because language-domain modeling is a direct use of the language, models constructed in this domain tend to be more efficient in terms of both simulation and synthesis than those constructed in the application domain. The modeling style is controlled; thus, one can personally assure compatibility with the synthesis tool one wants to use.

10.2　THE SEMANTICS OF SIMULATION AND SYNTHESIS

In this section we discuss VHDL modeling for synthesis. The key point that we emphasize is that there is one set of VHDL semantics for simulation and another for synthesis. The VHDL semantics for simulation are defined in the *Language Reference Manual*. The VHDL semantics for synthesis are defined by a particular vendor's synthesis tool, although there is much consensus in the industry, and IEEE has a working group attempting to define a standard.

In section 10.1, we discussed development of a model that represents the system specification. This model has two main purposes. First, it represents the first interpretation of the specification, and its validation indicates whether the specification is correctly interpreted. Second, after validated, it can be used as input to a synthesis tool. The key question here is: Is the model in the correct style for synthesis? One can use constructs in a high-level model that will simulate quite properly but are difficult or impossible to synthesize. Then, there is the issue of data types. In the DSP application, starting at the system level, one uses Real number models to check out a filter algorithm. After this algorithm is validated, fixed-point integers are used to consider the effects of truncation and rounding on accuracy. Finally, these Integer data types must be replaced by bit vector data types.

The level of abstraction of the primitives in the model must also be considered. In the SAR SPW model shown in Figure 10.4, a complex conjugator is one of the primitive functions. This mathematical operation forms the complex conjugate of a complex number. In logic circuits, data operations are Boolean or can, at least, be quickly translated into Boolean operations, such as increment or addition. Figure 10.12 shows the model of a decimation block used to resample complex real-numbered data, thereby reducing the number of samples.

The decimation block mode uses high-level data types (Real and Integer) and the mathematical operation "rem" to accomplish the desired result. In contrast, consider the model of a counter shown in Figure 9.31 The data type used was the IEEE standard logic vector that is used to represent circuit states. Data operations include assigning bit patterns to objects and incrementing a counter (operations that have direct Boolean implementations).

Let us consider another example of the effects of modeling style. Figure 10.13 gives a model of a clocked parallel to serial converter.

When START is received, COUNT is initialized to 7, and DONE is set to FALSE. Then, for each subsequent rise of SHCLK, if DONE is FALSE, one bit of the parallel input vector is selected and transferred to the output SO. COUNT is then decremented. If COUNT is negative, DONE is set to TRUE. Note that COUNT is declared as an integer with restricted range. If type INTEGER is used instead, the synthesis tool will implement a counter with an unnecessarily large number of stages.

```
entity DECIMATE is
generic (DEC: INTEGER);
  port(RX,IX: in REAL:=0.0;
       N:   in REAL:=0.0;
       RY,IY: out REAL:=0.0;);
end DECIMATE;
architecture BEHAVIOR of DECIMATE is
begin
  process
    variable cnt: integer :=0;
  begin
    wait on N;
    cnt := integer(N) rem DEC;
    if cnt = 0 then
       RY <= RX;
       IY <= IX;
    end if;
  end process;
end BEHAVIOR;
```

Figure 10.12   High-level decimation block.

If one examines the model, one can see that the hardware implied by process P1 is that shown in Figure 10.14. A Counter (with output COUNT), driven by an external clock (SHCLK), counts 8 steps. By connecting the counter outputs to the address inputs of a MUX, a unique bit of register PAR_IN is routed to the output S0. This is straightforward and efficient hardware mapping.

Figure 10.15 shows another algorithmic implementation for the converter, which relies on internal clocking of the shifting process. In this case all the serial outputs are generated in one execution of the process; that is, the loop runs eight times and produces eight corresponding changes on the output SO. Figure 10.16 shows the simulation response of the two converters. START_CLK starts the clocked converter, and SO_CLK is its response; whereas, START_SCHED starts the scheduled converter, and SO_SCHED is its response. Other than the difference in timing of the start signals, the responses of the two models are the same. One could argue that the scheduled model is superior because it is self-contained and requires no external clock. However, let us now consider the hardware implied by this model. As shown in Figure 10.17, the implied hardware consists of a counter, which controls a scheduler that inserts eight events in a time queue. These events are subsequently processed by an event processor to produce SO. From a simulation point of view, this presents no problem, as the simulator handles the scheduling, time queue, and event processing functions. However, from the synthesis point of view, implementing these functions in hardware would be very complicated, unnecessarily so, in light of the straightforward mapping implied by the clocked model. As a practical matter, current synthesis tools could not process the scheduled converter correctly since the circuit timing is controlled by the delay in the *after* clause in the signal assignment statement, which assigns a scheduled value to the output SO.

Another difference between simulation and synthesis semantics involves how the sensitivity list of the process is used. For simulation, the sensitivity list of the process is used to deter-

```
library IEEE;
use IEEE.std_logic_1164.all;
entity PAR_TO_SER is
  port(START,SHCLK: in STD_LOGIC;
       PAR_IN: in STD_LOGIC_VECTOR(7 downto 0);
       SO: out STD_LOGIC);
end PAR_TO_SER;

architecture ALG1 of PAR_TO_SER is
begin
P1:process(START,SHCLK)
    variable COUNT: INTEGER range 7 downto -1 := 0;
    variable DONE: BOOLEAN;
  begin
    if  START = '1' then
      COUNT := 7;
      DONE := FALSE;
    elsif SHCLK'EVENT and SHCLK = '1' then
      if DONE = FALSE then
        SO <= PAR_IN(COUNT);
        COUNT := COUNT - 1;
      end if;
      if COUNT < 0 then
        DONE := TRUE;
      else
        DONE  := FALSE;
      end if;
    end if;
  end process;
end ALG1;
```

Figure 10.13   Clocked parallel to serial converter.

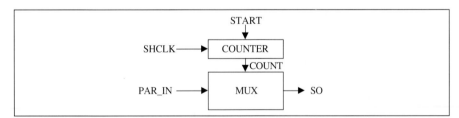

Figure 10.14   Clock driven parallel to serial converter.

mine which signals trigger execution of the process. For synthesis, the sensitivity list is used to list process inputs in a documentation sense, but has no effect on synthesis. We illustrate this situation with several models of a T flip-flop. Figure 10.18 shows the first model. Because the model needs to read its state to toggle it, an internal signal Q is used for the internal flip-flop state, while the signal QOUT is the circuit output. The value of Q is transferred to QOUT using a signal assignment statement outside the process. Figure 10.19 shows a model, which is identi-

```
use IEEE.std_logic_1164.all;
entity PAR_TO_SER_SCHED is
  generic(PERIOD: TIME);
  port(START: in STD_LOGIC;
       PAR_IN: in STD_LOGIC_VECTOR(7 downto 0);
       SO: out STD_LOGIC);
end PAR_TO_SER_SCHED;
architecture ALG2 of PAR_TO_SER_SCHED is
begin
P1:process(START)
    variable COUNT: INTEGER;
  begin
    if  START = '1' then
     COUNT := 7;
     while COUNT >= 0 loop
      SO <= transport PAR_IN(COUNT) after (7-COUNT)*PERIOD;
      COUNT := COUNT - 1;
     end loop;
    end if;
  end process;
end ALG2;
```

Figure 10.15   Scheduled parallel to serial converter.

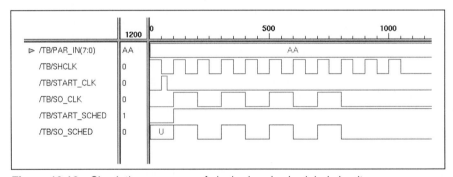

Figure 10.16   Simulation response of clocked and scheduled circuits.

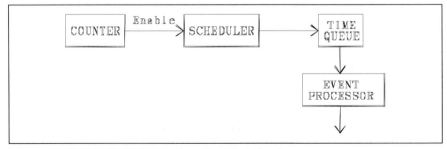

Figure 10.17   Hardware implied by scheduled circuit.

```
entity T_FF is
  port(RESET,T,CLK: in STD_LOGIC; QOUT: out STD_LOGIC);
end T_FF;
architecture ALG of T_FF is
  signal Q: STD_LOGIC;
begin
  process(RESET,T,CLK)
  begin
    if (RESET = '1') then
      Q  <= '0';
    elsif (CLK'EVENT and CLK = '1') then
      if T = '1' then
        Q  <=  not Q ;
      end if;
    end if;
  end process;
  QOUT <= Q;
end ALG;
```

Figure 10.18 T flip-flop model that simulates correctly.

```
entity T_FF2 is
  port(RESET,T,CLK: in STD_LOGIC; QOUT: out STD_LOGIC);
end T_FF2;
architecture ALG of T_FF2 is
  signal Q: STD_LOGIC;
begin
  process(RESET,T,CLK)
  begin
    if (RESET = '1') then
      Q  <= '0';
    elsif (CLK'EVENT and CLK = '1') then
      if T = '1' then
        Q <=  not Q ;
      end if;
    end if;
    QOUT <= Q;
  end process;
end ALG;
```

Figure 10.19 T flip-flop model that simulates incorrectly.

cal to the one in Figure 10.18, except that the transfer of Q to QOUT takes place inside the process. Figure 10.20 shows the simulation response of the two models. Q1 is the output of the model in Figure 10.18. Q2 is the output of the model in Figure 10.19. Note that the output Q2 differs from Q1 because, when the process in the second model is called due to a positive going clock change with T ='1', Q is scheduled to invert but does not do so until one delta cycle later. Thus, QOUT does not assume the same value as that in the first model until the next call of the process at the clock fall. The second model can be "fixed" by adding Q to the sensitivity list of the process. Q3 is the output for this modified model.

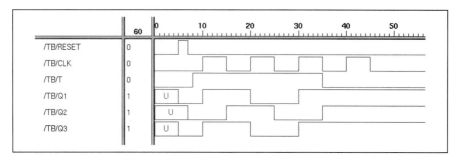

Figure 10.20   Simulation responses of T flip-flop models.

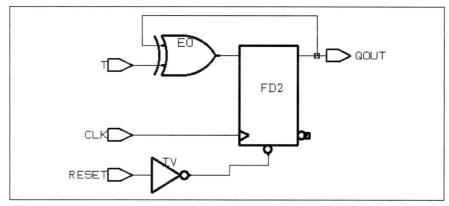

Figure 10.21   Synthesized circuit for T flip-flop.

Synthesis ignores the sensitivity list. Thus, when the descriptions in Figure 10.18 and Figure 10.19 are synthesized, both yield the circuit shown in Figure 10.21. The lesson to learn is that two models with different simulation responses can both synthesize to the same circuit. Thus, it is important to pay close attention to what signals are included in the process sensitivity list.

10.2.1   Delay in Models

In the previous section we discussed differences between simulation and synthesis semantics. An important aspect of modeling is model delay. As writers of behavioral code, we are used to writing:

$$C <= A \text{ and } B \text{ after } 10 \text{ ns};$$

which implies the AND function with a delay of 10 ns. This code will simulate properly but will cause great difficulty for a synthesis tool if one expects it to implement a 10 ns delay. Synthesis tools can implement functionality, including clocked behavior, but it is difficult for a synthesis tool to implement an exact or even approximate delay. All one can expect is that the tool will implement the functionality (AND). The delay will come from the delay of the ASIC library elements actually used to synthesize the device. In fact most synthesis tools require that you use only delta delay, and thus, you should write:

$$C <= A \text{ and } B;$$

Similarly, a statement "wait for 10 ns" makes no sense for synthesis. You must wait for events, for example, "wait until clock = '1';"

10.2.2 Data Types

Model data types should be restricted to STD_LOGIC, STD_LOGIC_VECTOR, and constrained integers. Most synthesis tools are designed to work with STD_LOGIC. Although many synthesis tools will accept types BIT and BIT_VECTOR in the source file, the synthesized circuit still use types STD_LOGIC and STD_LOGIC_VECTOR. Therefore, in order to validate the synthesized circuit, it is necessary to compare outputs with different data types. In addition, all the IEEE packages used to support modeling for synthesis use STD_LOGIC, STD_LOGIC_VECTOR, and constrained integers.

10.3 MODELING SEQUENTIAL BEHAVIOR

Modeling sequential behavior for synthesis is an interesting and critical problem. A number of approaches can be used. In some HDLs, for example AHPL, flip-flops and registers are declared as memory, and a special symbol (<=) is reserved for assignment to memory. All such assignments are assumed to be synchronized with the rise or fall of a system clock. Assignment to the output of a combinational logic block is called "connection" and is denoted by the symbol (:=). Thus, constructs built into the language identify what is sequential and what is combinational logic. This simplifies the synthesis task but at the expense of generality. For exmple, if all assignments to memory elements are clocked, how does one model asynchronous memory devices and latch behavior? Another possibility is to use naming conventions to identify synchronous elements. For example X_CLK signal would identify X as a clock signal. A process could be named A_FF. However, such an approach restricts the modeler and requires the imposition of a naming standard on top of the language itself. In VHDL synthesis, neither of these approaches is used. Rather, basic language constructs are combined in certain ways to imply sequential behavior.

There are three main concerns when synthesizing sequential behavior (1) identifying the clocking control mechanism, (2) determining what values have to be registered (stored), and (3) initializing the circuit. To identify the clocking control mechanism, VHDL models a clock edge as CLK'event and CLK = '1' (positive transition) or CLK'event and CLK = '0' (negative transition). In the STD_LOGIC_1164 package, RISING_EDGE and FALLING_EDGE functions are defined, which can be used to check for these conditions. The code for function RISING _EDGE is:

```
function RISING_EDGE (signal S: STD_ULOGIC)
  return BOOLEAN is
begin
  return (S'event and (To_X01(S) = '1')
              and (To_X01(S'last_value) = '0'));
end;
```

The function To_X01(S) maps the value of S to X, 0, or 1. The mapping is:

```
('X', 'Z', 'U', '-') map to 'X'.
('1', 'H') map to '1'.
('0', 'L') map to '0'.
```

Thus, the function RISING_EDGE returns TRUE for a transition from '0' or 'L' to '1' or 'H'.

The edge expression (or function call) is used in an *if* or *wait* statement. The *if* statement typically has the following form:

```
if (RESET = '1') then -- alternatively,(RESET = '0')
    -- perform asynchronous reset and initialization
elsif (CLK'event and CLK = '1') -- alternatively CLK = '0'
    -- load flip-flops and registers
end if;
```

The models in Figure 10.13, Figure 10.18, and Figure 10.19 use this approach. In general the asynchronous reset and clock activities can be as simple as loading a value into a register or may involve more complex data operations that conclude with register loading. Note that this template covers two of the concerns: identifying the clocking control mechanism and initializing the circuit. One important point is that most synthesis tools do not allow other conditions to be added to the clock edge condition, for example, CLK'EVENT and CLK = '1' and EN = '1' is invalid. Synthesis tools require that the clock edge condition be isolated. If there is an enable, EN, it should be tested within the *elsif* region. This prevents gated clocks, which is a poor design procedure. If EN is located within the *elsif* clause, then EN will be applied to the clock enable input of the flip-flop. Note that the T flip-flop model in Figure 10.18 follows this proper modeling procedure.

Let us consider the possibility of using a *wait* statement to detect the clock edge:

```
wait until (CLK'EVENT and CLK = '1');
-- load flip-flops and registers
```

This approach has no built-in reset mechanism and is, therefore, modeling strictly synchronous activity. Synchronous resets would have to rely on preceding code or on "upstream" logic to produce the reset condition on the data input. In fact what is frequently done is to have a process representing pure combinational logic drive a purely sequential process with a registering *wait* statement. Finally, recall that the *wait* statement approach can only be used in processes that have no sensitivity list.

To further illustrate the problems in using the *wait* statement, we consider a clocked sequential circuit, which performs a Toggle On Equal Function. Whenever the single bit input I has the same value for three consecutive clock periods, the output TEQ toggles. Figure 10.22 shows an initial model of the detector which uses the *wait* statement approach to clocked action. Previous values of the input are stored in the variables IBK1 and IBK2. This model simulates properly as long as the output TEQDET is initialized to '0' or '1' (which it is). If this is not done, the output begins with the value 'U' and continues to hold that value throughout simulation. When the model is synthesized, however, the synthesis tool ignores the initialization value in the port statement, and there is no way to solve this dilemma using a *wait* statement. The model needs an explicit reset statement. Figure 10.23 shows a model that uses an *if-then* structure to

```
entity EQDET is
  port(I,CLK: in STD_LOGIC; TEQDET: inout STD_LOGIC :='0');
end EQDET;

architecture ALG of EQDET is
begin
  process
    variable  EQ,IBK1,IBK2: STD_LOGIC;
  begin
    wait until (CLK'EVENT and CLK = '1');
    if(IBK1 =IBK2) and (IBK2 = I) then
      EQ := '1';
    else
      EQ := '0';
    end if;
    TEQDET <= (EQ xor TEQDET);
    IBK2 := IBK1;
    IBK1 := I;
  end process;
end ALG;
```

Figure 10.22   Initial equality detector model.

```
entity EQDET is
  port(RESET,I,CLK: in STD_LOGIC; TEQDET: inout STD_LOGIC);
end EQDET;

architecture ALG of EQDET is
begin
  process(RESET,CLK)
    variable  EQ,IBK1,IBK2: STD_LOGIC;
  begin
    if (RESET = '1') then
      IBK1 :=  '0';
      IBK2 :=  '0';
      TEQDET <= '0';
    elsif (CLK'EVENT and CLK = '1') then
      if (IBK1 = I) and (IBK1 = IBK2) then
        EQ := '1';
      else
        EQ := '0';
      end if;
      TEQDET <= (EQ xor TEQDET);
      IBK2 := IBK1;
      IBK1 := I;
    end if;
  end process;
end ALG;
```

Figure 10.23   Equality detector with reset in *if-then* structure.

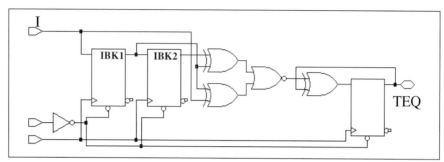

Figure 10.24   Synthesized detector.

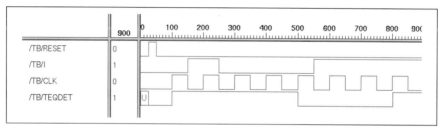

Figure 10.25   Detector simulation response.

initialize the circuit and the CLK'event construct to identify the clocking mechanism. This explicit reset solves the initialization problem. Figure 10.24 shows the synthesized circuit. Figure 10.25 shows its simulation response. Note: In Figure 10.24 the variables IBK1 and IBK2, which are delayed values of I, have been registered. In the *elsif* clause controlled by the clock, the synthesis tool registers all signal outputs and also all variables, which are read before updating. Thus, IBK1 and IBK2 are registered, but EQ is not. Note, also, that the order of updating IBK1 and IBK2 in the code is critical. If one were to change the order to:

```
IBK1 := I;
IBK2 := IBK1;
```

then, during one call of the process, IBK1 and IBK2 will take on the same value, and the synthesis tool will generate a circuit which clocks I directly into both flip-flops.

Figure 10.26 shows a signal-based model, where IBK1 and IBK2 are represented as signals. Here, the order of updating IBK1 and IBK2 is unimportant as individual signal assignment statements under control of the clock are automatically registered. The signal-based model (Figure 10.26) synthesizes to a circuit, which is identical to the variable-based model (Figure 10.23). Because the variable-based model simulates more efficiently at the behavioral level, it is preferred.

A further comment about resetting. Asynchronous resets are preferred because a desired initial state is easily attained. Synchronous reset is only effective if the input to the synthesized register element can be controlled. In the case of the detector circuit, this was not possible without modifying the model.

```
entity EQDET is
  port(RESET,I,CLK: in STD_LOGIC; TEQDET: inout STD_LOGIC);
end EQDET;

architecture ALG of EQDET is
  signal IBK1,IBK2: STD_LOGIC;
begin
  process(RESET,CLK)
    variable  EQ: STD_LOGIC;
  begin
    if (RESET = '1') then
      IBK1 <=  '0';
      IBK2  <=  '0';
      TEQDET <= '0';
    elsif (CLK'EVENT and CLK = '1') then
      if (IBK1 = I) and (IBK1 = IBK2) then
        EQ := '1';
      else
        EQ := '0';
      end if;
      TEQDET <= (EQ xor TEQDET);
      IBK1 <= I;
      IBK2  <= IBK1;
    end if;
  end process;
end ALG;
```

Figure 10.26 Signal-based model.

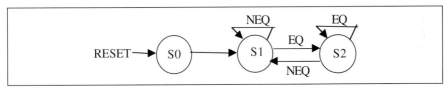

Figure 10.27 Detector state diagram.

Another approach to the design of the detector is to use a finite state machine (FSM). Figure 10.27 shows the state diagram. The state definitions are as follows: 1) S0: the intial state, 2) S1: entered after an input has been received, and 3) S2: entered when an input was received which was equal to the previous input. Transitions labeled EQ mean that the present input is equal to the previous input, those labeled NEQ the opposite condition. When S2 is entered the signal TEQ will toggle at the end of the clock period if the input equals the previous input.

Figure 10.28 shows a VHDL model for the FSM. States are represented using an enumerative type for the states. A *case* statement controls the clocked state movement. Actions in each state are defined by a *when* clause of the *case* statement. The *when* clause has two parts (1) next state movement and (2) data and output operations. Figure 10.29 shows the synthesized circuit. The left-most flip-flop stores the variable IBK1; the right-most stores the circuit output TEQ-

```
entity EQDET is
  port(RESET,I,CLK: in STD_LOGIC; TEQDET: inout STD_LOGIC);
end EQDET;
architecture FSM of EQDET is
begin
 P1:process(RESET,CLK)
    type STATE_TYPE is (S0,S1,S2);
    variable STATE: STATE_TYPE;
    variable IBK1: STD_LOGIC;
  begin
    if RESET = '1' then
       STATE := S0;
       IBK1 := '0';
       TEQDET <= '0';
    elsif (CLK'EVENT and CLK = '1') then
       case (STATE) is
         when S0 =>
           STATE := S1;
           IBK1 := I;
         when S1 =>
           if (IBK1 = I) then
             STATE := S2;
           else
             STATE := S1;
           end if;
           IBK1 := I;
         when S2 =>
           if (IBK1 = I) then
             STATE := S2;
             TEQDET <= not TEQDET;
           else
             STATE := S1;
           end if;
           IBK1 := I;
       end case;
    end if;
  end process;
end FSM;
```

Figure 10.28   Detector FSM model.

DET, and the two middle flip-flops implement the state machine. In this case the FSM implementation is more costly than the models in Figures 10.23 and 10.26, principally because one more flip-flop is required. In general though, the finite state machine approach is a very powerful technique for modeling control intensive circuits. This approach was discussed in detail in Chapter 8.

We now list other important synthesis modeling concerns for sequential circuits:

1. Do not assign signals and variables initial values in the declaration statements because initial values in declaration statements are ignored by most synthesis tools. Such initial values will be used by the simulator; therefore, a model may simulate correctly. How-

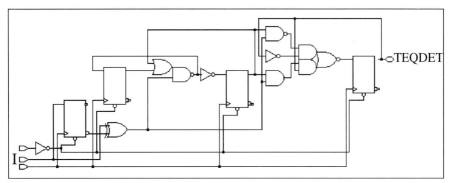

Figure 10.29   Detector FSM synthesized circuit.

ever, since the synthesis program ignores them, the resulting hardware will not be properly initialized. For example, do not use initialization statements, such as:

variable SUM:INTEGER:=0;

Instead, in the code body, make an explicit assignment to SUM:

SUM := 0;

2. Do not use unconstrained integers. For example, a modeler might want to use an integer signal to represent a counter, which they could declare:

signal COUNT: INTEGER;

Since unconstrained type INTEGER has an extremely large range of possible values that depends on the word length of the host machine, the synthesis tool will insert a large register to hold the value, typically 32 bits in modern machines. The cost is further increased since 32-bit data paths will be needed to transfer data into and out of the register. The solution is to use a constrained integer for COUNT:

signal COUNT: range 0 to 15;

Declaring a constrained integer allows the synthesis tool to reduce the size of the register and data paths to that needed to hold the expected range of values.

3. Use attributes to specify state encoding in finite state machine models. In the finite state machine model shown in Figure 10.28, the enumerative type state values are S0, S1, and S2. The question is, How should they be encoded by the synthesis tool? A binary encoding could be used: S0 - 00, S1- 01, and S2 -10, or a "one hot" code might be used: S0 - 001, S1 - 010, and S2 - 100. The implementation of the model in Figure 10.28 shown in Figure 10.29 uses binary encoding. Implementation of the FSM model in Figure 8.25 shown in 8.27 uses one-hot encoding. Comparing the two approaches, binary encoding uses fewer flip-flops, but the next state logic may be relatively complicated. One hot encoding uses more flip-flops, but the next state logic can be simpler. Because of their abundance of flip-flops, Xilinx 4000 FPGAs favor the one-hot approach. The default method for the Synopsys Design Analyzer seems to be binary

encoding. One can control the type of encoding used by the synthesizer through the use of user defined attributes:

```
type STATE_TYPE is (S0,S1,S2);
attribute ENUM_ENCODING: STRING;
attribute ENUM_ENCODING of STATE_TYPE is "00 01 10";
```

The statements above would force binary encoding. Changing the third statement to:

```
attribute ENUM_ENCODING of  STATE_TYPE is "001 010 100";
```

would force one-hot encoding.

4. Synthesizing RAM models. When coding a RAM, one typically makes the following declarations:

```
type MEMORY is array(0 to 1023) of
      STD_LOGIC_VECTOR(7 downto 0);
type MEM: MEMORY;
```

The question is, How will this 1024 x 8 RAM be synthesized? It could be synthesized as a 2D array of flip-flops. For an ASIC design this might be acceptable, but for an FPGA, if the two flip-flops in the CLB were used, this would require the use of 512 CLBs. The solution is to use a predefined RAM module provided by the synthesis vendor and instantiate it as a component in a structural model. For example, Xilinx provides various forms of 16 x 1 RAM block. A tool called LogiBLOX can be used to define larger RAM components. These predefined blocks take advantage of the ability to convert 4-input look up tables to 16 x 1 RAM components and results in efficient use of CLBs.

5. Synthesizing latches. This is the code for a latch circuit:

```
LATCH: process(GATE,DATA)
begin
  if(GATE = '1') then
    Q <= DATA;
  end if;
end process;
```

Because the value of Q is not specified when GATE is not equal to '1', a latch is inferred. Since ASIC libraries have efficient latch primitives, this presents no problem. In XC4000XC devices, the two flip-flops in the CLB can be configured as latches, so there is, again, no problem.

Figure 10.30 shows a gate-level latch circuit. If this had to be implemented with LUTs, it would be very expensive. Thus, for FPGAs without the built-in latch capability, their use should be avoided as much as possible. In many cases a clocked flip-flop can be used instead.

10.4 MODELING COMBINATIONAL CIRCUITS FOR SYNTHESIS

When modeling combinational circuit behavior for synthesis, one uses a process or another concurrent VHDL construct because such constructs are all equivalent to a process. All circuit inputs are modeled by signals in the sensitivity list of the process or equivalent process. One can

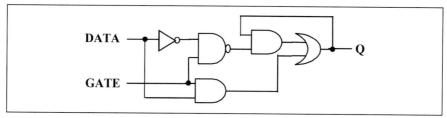

Figure 10.30 A gate-level latch circuit.

Table 10.1 Truth table for 1's counter.

A_2	A_1	A_0	C_1	C_0
0	0	0	0	0
0	0	1	0	1
0	1	0	0	1
0	1	1	1	0
1	0	0	0	1
1	0	1	1	0
1	1	0	1	0
1	1	1	1	1

choose the level of abstraction used in coding the processes. The level chosen controls the ease with which the synthesis tool translates the model into gate-level logic.

Table 10.1 shows the truth table for the 1's counter first introduced in Chapter 3. The two-bit output C gives the number of 1's in the three-bit input A. Figure 10.31 shows the entity description for the ones counter and five different architectures listed in increasing level of abstraction: data flow, MUX, ROM, algorithmic and subprogram.

Architectures DATA_FLOW, MUX, and ROM are synthesized by Synopsys in one optimization step, producing the efficient circuit shown in Figure 10.32. Architecture ALGORITHMIC represents an interesting case. A straightforward implementation is the sequential circuit shown in Figure 10.33. Surprisingly, a synthesis tool can deduce a combinational logic circuit from this description. Figure 10.34 shows the technology-independent circuit developed by the Synopsys tool. It consists of MUXes, adders, and associated gating. Figure 10.35 shows the inefficient circuit, realized using the LSI 10 K Library during the first synthesis optimization pass. The circuit has a cost of 17 cells and maximum path delay of 4.89 ns. Compare this to the efficient circuit shown in Figure 10.32, which has a cost of 9 cells and a maximum path delay of 2.49 ns. However, the efficient circuit can be produced from the inefficient one by applying two more optimization steps using the maximum effort setting on the Synopsis tool. The lesson to learn from this is that very abstract models can be synthesized but require more effort than those coded with a style, which has a more direct physical implementation.

Architecture PROC utilizes a procedure to implement another abstract implementation of the 1's counter. Incrementing is done by using the overloaded + operation for unsigned standard

```
library IEEE;
use IEEE.std_logic_1164.all;
use IEEE.std_logic_unsigned.all;
use IEEE.std_logic_arith.all;

entity ONES_CNT  is
  port (A: in STD_LOGIC_VECTOR(2 downto 0);
        C: out STD_LOGIC_VECTOR(1 downto 0));
end ONES_CNT;

architecture DATA_FLOW of ONES_CNT is
begin
  C(1) <= (A(1) and A(0)) or (A(2) and A(0))
          or (A(2) and A(1));
  C(0) <= (A(2) and not A(1) and not A(0))
          or (not A(2) and not A(1) and A(0))
          or (A(2) and A(1) and A(0))
          or (not A(2) and A(1) and not A(0));
end DATA_FLOW;

architecture MUX of ONES_CNT is
begin
  process(A)
  begin
    case A is
      when "000" => C<= "00";
      when "001"|"010"|"100" => C<= "01";
      when "011"|"101"|"110" => C<= "10";
      when "111" => C<= "11";
      when others => null;
    end case;
  end process;
end MUX;

architecture ROM of ONES_CNT is
begin
  process (A)
    type ROM_TABLE is array( 0 to 7)
      of STD_LOGIC_VECTOR(1 downto 0);
    constant ROM: ROM_TABLE :=
      (('0','0'),('0','1'),('0','1'),('1','0'),
       ('0','1'),('1','0'),('1','0'),('1','1'));
  begin
    C <= ROM(CONV_INTEGER(A));
  end process;
end ROM;
```

Figure 10.31   Architectures for 1's counter.

logic vectors from package STD_LOGIC_ARITH. The result, RES, is initialized to '0' using the type conversion function CONV_STD_LOGIC_VECTOR from package STD_LOGIC_

```
architecture ALGORITHMIC of ONES_CNT is
begin
  P1:process(A)
    variable NUM: INTEGER range 0 to 3;
  begin
    NUM := 0;
    for I in 0 to 2 loop
      if A(I) = '1'  then
        NUM := NUM + 1;
      end if;
    end loop;

case NUM is
    when  0 => C <= "00";
    when  1 => C <= "01";
    when  2 => C <= "10";
    when  3 => C <= "11";
    end case;
  end process P1;
end ALGORITHMIC;

architecture PROC of ONES_CNT is
  procedure ONES_CNT_PROC(X : in STD_LOGIC_VECTOR;
      Z_SIZE:INTEGER; signal Z: out STD_LOGIC_VECTOR) is
    variable  RES:  STD_LOGIC_VECTOR(Z_SIZE-1 downto 0);
  begin
    RES := CONV_STD_LOGIC_VECTOR(0,Z_SIZE);
    for I in  X'RANGE loop
      if (X(I) = '1') then
        RES := unsigned(RES) + 1;
      end if;
    end loop;
Z <= RES;
  end ONES_CNT_PROC;
begin
  ONES_CNT_PROC(A,2,C);
end PROC;
```

Figure 10.31 Architectures for 1's counter (continued).

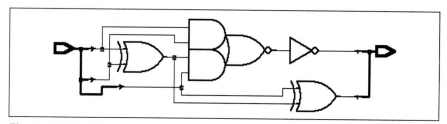

Figure 10.32 Efficient 1's counter circuit.

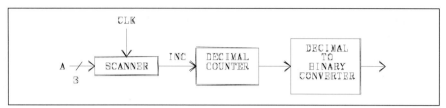

Figure 10.33   Implied sequential circuit.

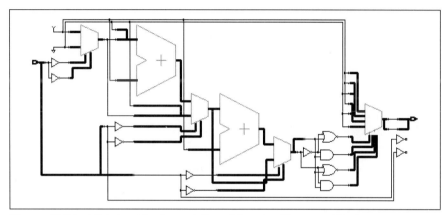

Figure 10.34   High-level implementation of the algorithmic architecture.

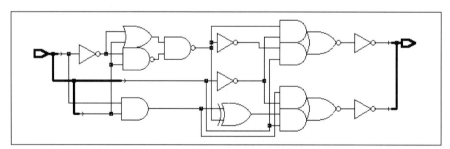

Figure 10.35   Inefficient implementation of 1's counter.

ARITH, which converts the integer 0 to type STD_LOGIC_VECTOR. This approach is required because the procedure accepts inputs of variable length. Because of the abstract nature of the approach, the Synopsis tool achieves the same result as that for the architecture ALGORITH-MIC; that is, three optimization steps are required to achieve the minimal circuit shown in Figure 10.32. However, the use of procedures and functions to model combinational logic circuits is a very powerful technique that we will further exploit.

```
        Library IEEE;
        use IEEE.STD_LOGIC_1164.all;
        use IEEE.STD_LOGIC_ARITH.all;
        entity SIMP_ADD is
          port(A,B: in STD_LOGIC_VECTOR(3 downto 0);
               CIN: in STD_LOGIC;
               C: out STD_LOGIC_VECTOR(3 downto 0);
               CAR_OUT: out STD_LOGIC);
        end SIMP_ADD;
        architecture ALG of SIMP_ADD is
        begin
          P1:process(A,B,CIN)
            variable  PADDED_CIN: STD_LOGIC_VECTOR(3 downto 0);
            variable A_UNSIGNED: UNSIGNED(3 downto 0);
            variable C_UNSIGNED: UNSIGNED(4 downto 0);
          begin
            A_UNSIGNED := UNSIGNED(A);
            PADDED_CIN :="000"&CIN;
            C_UNSIGNED := CONV_UNSIGNED(A_UNSIGNED,5) +
                          UNSIGNED(B) + UNSIGNED(PADDED_CIN);
            C  <= STD_LOGIC_VECTOR(C_UNSIGNED(3 downto 0));
            CAR_OUT  <= C_UNSIGNED(4);
          end process;
        end ALG;
```

Figure 10.36 Adder model.

10.4.1 Synthesis of Arithmetic Circuits

In this section we discuss synthesis of arithmetic circuits. Modeling of these circuits is made easier by the use of the IEEE library that not only contains the package STD_LOGIC_1164 but also a number of supporting packages that support arithmetic. Here, we use the Synopsys version of these packages. Specifically, we use the package STD_LOGIC_ARITH, which defines two new types :

```
type SIGNED is array(NATURAL range <>) of STD_LOGIC;
type UNSIGNED is array(NATURAL  range <>) of STD_LOGIC;
```

Type SIGNED is used for 2's complement arithmetic. Type UNSIGNED is used for arithmetic involving unsigned binary numbers. The package contains overloaded arithmetic and relational operators for the two new types, as well as numerous type conversion functions to convert SIGNED and UNSIGNED to Integer and STD_LOGIC_VECTOR and vice versa.

Figure 10.36 gives an adder model that receives two 4-bit operands and an input carry and generates a 4-bit sum and an output carry. Input and output data types are either STD_LOGIC or STD_LOGIC_VECTOR. Inputs A and B are converted to type SIGNED; the input carry is padded with three leading zeros and converted to type SIGNED. Adding the three operands (A, B, and CIN) will produce a five-bit result. In the addition operation, A is extended to five bits (a leading zero is inserted on the left) by the CONV_UNSIGNED function. The overloaded + automatically adjusts the length of the other operands. The low four bits of the result (C_UNSIGNED) become the sum output C. The left-most bit of C_UNSIGNED becomes

```
library IEEE,DW02;
use IEEE.STD_LOGIC_1164.all;
use IEEE.STD_LOGIC_arith.all;
use DW02.DW02_components.all;

entity MULT10 is
  port(DATA_IN: in STD_LOGIC_VECTOR(3 downto 0);
       PRODUCT: out STD_LOGIC_VECTOR(7 downto 0));
end MULT10;

architecture ALG of MULT10 is
begin
  process(DATA_IN)
    variable PROD_US: UNSIGNED(7 downto 0);
  begin
    PROD_US := UNSIGNED(DATA_IN)*CONV_UNSIGNED(10,4);
    PRODUCT <= STD_LOGIC_VECTOR(PROD_US);
  end process;
end ALG;
```

Figure 10.37   Multiply by 10 model.

Figure 10.38   Multiply by 10 and add module.

CAR_OUT. Note that the three intermediate variables are created because a type conversion function cannot determine the type of an operand without such a declaration.

Multiplication can also be overloaded. From a synthesis point of view, the multiply operator (*) implies the use of a combinational logic multiplier. In the Synopsis system, multipliers are components in the Design Ware Library. One has the option of specifying a Carry Save Array multiplier or a Wallace tree, the Carry Save Array being the default. One can also infer a signed or unsigned multiplication, depending on whether the operand data type is SIGNED or UNSIGNED. Figure 10.37 shows a multiply by 10 module. Note that the length of the result is twice the length of the operands—a normal and required condition for specifying a multiply operation. When synthesizing a multiplication by a constant, it is wasteful to implement a full multiplier. For example, to multiply by 10, one can multiply the multiplicand by 2 and by 8, which are left-shifted versions of the multiplicand, and then add the two terms. A synthesis tool most likely takes this approach. Shifting a fixed number of positions is implemented by wiring.

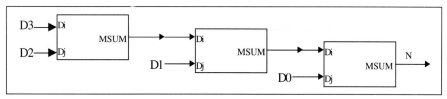

Figure 10.39 Three-digit decimal to binary converter.

10.4.2 Hierarchical Arithmetic Circuit: BCD to Binary Converter

Decimal to binary conversion is an important logic function in digital circuits. Horner's rule for polynomial evaluation provides an efficient way to implement decimal to binary conversion. The usual way to represent a base 10 number is:

$$N = D_3 x 10^3 + D_2 x 10^2 + D_1 x 10^1 + D_0$$

where the D_i are represented in the standard 4-bit BCD code. Horner would rewrite it in this form:

$$N = ((D_3 x 10 + D_2) x 10 + D_1) x 10 + D_0$$

Thus, BCD to binary conversion can be accomplished by repeated application of a "multiply by 10 and add" operator. Figure 10.38 shows a block symbol for a logic circuit which performs the "multiply and add" function (MSUM = $D_i x 10 + D_j$). Figure 10.39 shows the interconnection of three of these modules to calculate the expression for N in the above equation.

Figure 10.40 shows the VHDL model of the Multiply by 10 and add circuit. Again, in order to infer a multiply, it is necessary to have a variable PROD to hold the product and whose length is double that of the multiplication operands. Appropriate low-order bits of the variable PROD are then used in the subsequent add. The model is generic in the input size. This is necessary because the value of D_i grows as one moves right in the cascade. Table 10.2 shows the required D_i size. Note that it is sufficient that the output size of the circuit be larger by four, than the input size.

Table 10.2 Required D_i input size.

Stage	D_i Max	D_i Size
1	9	4
2	99	7
3	999	10
4	9999	14
5	99999	17
6	99999	20

Figure 10.41 shows a structural model for the BCD to binary converter, and Figure 10.42 shows its simulation response.

```
library IEEE,DW02;
use IEEE.STD_LOGIC_1164.all;
use IEEE.STD_LOGIC_arith.all;
use DW02.DW02_components.all;

entity MADD is
  generic(IN_WIDTH: NATURAL := 4);
  port(DI: in STD_LOGIC_VECTOR(IN_WIDTH-1 downto 0);
       DJ: in STD_LOGIC_VECTOR(3 downto 0);
       MSUM: out STD_LOGIC_VECTOR(IN_WIDTH+3 downto 0));
end MADD;

architecture ALG of MADD is
begin
  P1: process(DI,DJ)
    variable MSUM_US: UNSIGNED(IN_WIDTH+3 downto 0);
    variable PROD:UNSIGNED(2*IN_WIDTH-1 downto 0);
  begin
    PROD :=  UNSIGNED(DI)*CONV_UNSIGNED(10,IN_WIDTH);
    MSUM_US := PROD(IN_WIDTH+3 downto 0)+ UNSIGNED(DJ);
    MSUM <= STD_LOGIC_VECTOR(MSUM_US);
  end process;
end ALG;
```

Figure 10.40   VHDL model of the multiply by 10 and add circuit.

10.4.3 Synthesis of Hierarchical Circuits

To synthesize the structural model in Figure 10.41, MADD must be analyzed first, followed by entity BCDCONV. During elaboration of BCDCONV, the synthesis tool creates three instances of MADD. The generic values 4, 8, and 12 are, respectively, assigned to IN_WIDTH in the three instances. After elaboration, compilation of the structural model produces the synthesized circuit.

Figure 10.43 gives the hierarchy report for the synthesized BCD to binary converter. Shown at each level are the names of the cells used. At the highest level, level 3 are the MADD circuits. At the next level down (level 2) are multipliers and adders from the Synopsis Design Ware Library. Because multipliers contain adders, level 1 contains adders and basic primitives. Level 0, the lowest level, contains only basic primitives, such as AN2 (two input AND), and EO (exclusive OR).

When compiling hierarchical circuits, one can perform the *ungrouping* and *uniquify* operations:

Ungrouping operation. Removes the hierarchy in the circuit, so the circuit has a single level. This is sometimes referred to as *hierarchical flattening*. This type of flattening is not to be confused with the logic-level flattening, which we discuss below. Figure 10.44 shows the hierarchy report for the ungrouped (flattened) circuit. Note that the single level contains only primitive elements.

```
     library IEEE;
     use IEEE.STD_LOGIC_1164.all;
     entity BCDCONV is
     port(D0,D1,D2,D3: in STD_LOGIC_VECTOR(3 downto 0);
           BIN_OUT: out STD_LOGIC_VECTOR(15 downto 0));
     end BCDCONV;

     architecture STRUCTURAL of BCDCONV is
       component MADD
         generic(IN_WIDTH: NATURAL := 4);
         port(DI: in STD_LOGIC_VECTOR(IN_WIDTH-1 downto 0);
              DJ: in STD_LOGIC_VECTOR(3 downto 0);
              MSUM: out STD_LOGIC_VECTOR(IN_WIDTH+3 downto 0));
       end component;
       signal MSUM2: STD_LOGIC_VECTOR(7 downto 0);
       signal MSUM1:  STD_LOGIC_VECTOR(11 downto 0);
     begin
       C1: MADD
         generic map(4)
         port map(D3,D2,MSUM2);
       C2: MADD
         generic map(8)
         port map(MSUM2,D1,MSUM1);
       C3: MADD
         generic map(12)
         port map(MSUM1,D0,BIN_OUT);
     end STRUCTURAL;
```

Figure 10.41   Structural model of the BCD to binary converter.

		300	0	50	100	150	200	250
/TB/D3(3:0)		9	0		6	8		9
/TB/D2(3:0)		9	0		4	2		9
/TB/D1(3:0)		9	0		9	1		9
/TB/D0(3:0)		9	0		3	5		9
/TB/BIN_OUT(15:0)		270F	0000		195D	2017		270F

Figure 10.42   Simulation response of BCD to binary converter.

Flattening can result in significant reduction in area and can result in lower delay times. For the BCD to binary circuit, the initial compiled circuit had an area of 342 and a max delay of 30.34 ns. Ungrouping (flattening) the circuit results in an area of 309 and a maximum delay of 30.11 ns.

Uniquify operation.   If the same cell is referenced a number of times in the hierarchy, uniquify makes each one of those references a unique design that is compiled independently of the others.

```
*****************************************
Report : hierarchy
Design : BCDCONV
Version: 1998.02
Date   : Thu Jun 18 09:17:19 1998
*****************************************

BCDCONV
    MADD_IN_WIDTH4
        MADD_IN_WIDTH4_DW01_add_8_0
            AN2                                        lsi_10k
            EO                                         lsi_10k
            FA1A                                       lsi_10k
        MADD_IN_WIDTH4_DW02_mult_4_4_0
            AN2                                        lsi_10k
            EO                                         lsi_10k
            FA1A                                       lsi_10k
            MADD_IN_WIDTH4_DW01_add_6_0
                AN2                                    lsi_10k
                EO                                     lsi_10k
    MADD_IN_WIDTH8
        MADD_IN_WIDTH8_DW01_add_12_0
            AN2                                        lsi_10k
            EO                                         lsi_10k
            FA1A                                       lsi_10k
        MADD_IN_WIDTH8_DW02_mult_8_4_0
            AN2                                        lsi_10k
            EO                                         lsi_10k
            FA1A                                       lsi_10k
            MADD_IN_WIDTH8_DW01_add_10_0
                AN2                                    lsi_10k
                EO                                     lsi_10k
    MADD_IN_WIDTH12
        MADD_IN_WIDTH12_DW01_add_16_0
            AN2                                        lsi_10k
            EO                                         lsi_10k
            FA1A                                       lsi_10k
        MADD_IN_WIDTH12_DW02_mult_12_4_0
            AN2                                        lsi_10k
            EO                                         lsi_10k
            FA1A                                       lsi_10k
            MADD_IN_WIDTH12_DW01_add_14_0
                AN2                                    lsi_10k
                EO                                     lsi_10k
```

Figure 10.43   Hierarchy report.

Figure 10.45 shows the block diagram of a 12-bit adder constructed using three 4-bit adders. Figure 10.46 gives the VHDL model of the circuit. Note that the component SIMP_ADD is instantiated three times. If the hierarchy is not ungrouped then the uniquify command is used to generate three copies of SIMP_ADD, all with unique names. If entity ADD_12_REG is compiled, then Figure 10.47 shows the resultant hierarchy report. Note the cell names SIMP_ADD_0, SIMP_ADD_1, and SIMP_ADD_2 for the three instances of SIMP_ADD that have been created.

```
* * * * * * * * * * * * * * * * * * * * * * * * * * * * * * * * * * * * * * *
Report : hierarchy
Design : BCDCONV
Version: 1998.02
Date   : Thu Jun 18 10:27:04 1998
* * * * * * * * * * * * * * * * * * * * * * * * * * * * * * * * * * * * * * *


BCDCONV
     AN2                  lsi_10k
     AN3                  lsi_10k
     AO2                  lsi_10k
     AO4                  lsi_10k
     AO6                  lsi_10k
     EN                   lsi_10k
     EO                   lsi_10k
     EO1                  lsi_10k
     FA1A                 lsi_10k
     IV                   lsi_10k
     ND2                  lsi_10k
     ND3                  lsi_10k
     NR2                  lsi_10k
```

Figure 10.44   Hierarchy of ungrouped (flattened) circuit.

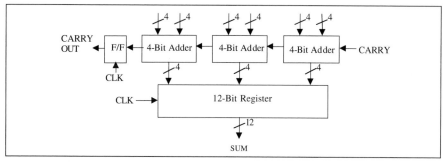

Figure 10.45   Block diagram of a 12-bit adder.

After uniquification has been performed, compilation of the model can take place in two ways, top down and bottom up. For top-down compilation, the synthesis tool compiles each sub-design, automatically taking into account the environment surrounding the subdesign. Subde-signs are recompiled as required to meet overall design constraints. The compilation, which resulted in the hierarchy in Figure 10.47, is an example of top-down compilation. Figure 10.48 shows the script for top-down compilation for the 12-bit adder.

For bottom-up optimization, subdesigns are first compiled separately or the design is ungrouped. Next, the overall design is compiled. For the subdesign compilation strategy, subde-signs are first characterized; then, the entire design is compiled. Figure 10.49 shows the script for the bottom-up compilation.

```
      entity ADD_12_REG is
        port(X,Y: in STD_LOGIC_VECTOR(11 downto 0);
              CI, CLK: in STD_LOGIC;
              Z:out STD_LOGIC_VECTOR(11 downto 0);
              CO: out STD_LOGIC);
      end ADD_12_REG;
      architecture STRUCTURAL of ADD_12_REG is
        component SIMP_ADD
          port(A,B: in STD_LOGIC_VECTOR(3 downto 0);
               CIN: in STD_LOGIC;
               C:out STD_LOGIC_VECTOR(3 downto 0);
               CAR_OUT: out STD_LOGIC);
        end component;
        component REG_12
          port(DATA_IN: in STD_LOGIC_VECTOR(11 downto 0);
               CLK: in STD_LOGIC;
               DATA_OUT: out STD_LOGIC_VECTOR(11 downto 0));
        end component;
        component REG
          port(DATA_IN: in STD_LOGIC;
               CLK: in STD_LOGIC;
               DATA_OUT: out STD_LOGIC);
        end component;
        signal LO_CAR,MID_CAR,HI_CAR:STD_LOGIC;
        signal Z_COMB: STD_LOGIC_VECTOR(11 downto 0);
      begin
        C1: SIMP_ADD
          port map(X(3 downto 0),Y(3 downto 0),CI,
                   Z_COMB(3 downto 0), LO_CAR);
        C2: SIMP_ADD
          port map(X(7 downto 4),Y(7 downto 4),LO_CAR,
                   Z_COMB(7 downto 4), MID_CAR);
        C3: SIMP_ADD
          port map(X(11 downto 8),Y(11 downto 8),LO_CAR,
                   Z_COMB(11 downto 8), HI_CAR);
        C4: REG_12
          port map(Z_COMB,CLK,Z);
        C5: REG
          port map(HI_CAR,CLK,CO);
      end STRUCTURAL;
```

Figure 10.46   12-bit adder model.

An alternative approach to bottom-up compilation is the "golden instance" approach. Uniquify is not used here. Rather, one instance of a multiply instantiated subdesign is characterized and compiled. This instance is considered golden because its environment is used for all instances. Next, the don't touch attribute is set on all instances, and the entire design is compiled. Figure 10.50 shows the golden instance script for the 12-bit adder. C2 is the golden instance.

Table 10.3 gives the comparative results of the compilation strategies. The top-down approach gives the lowest-cost circuit. The bottom up approach gives the fastest circuit. The

```
*****************************************
Report : hierarchy
Design : ADD_12_REG
Version: 1998.02
Date   : Wed Jul  1 14:01:42 1998
*****************************************

ADD_12_REG
   REG
         FD1                              lsi_10k
   REG_12
         FD1                              lsi_10k
   SIMP_ADD_0
      SIMP_ADD_0_DW01_add_5_0
            FA1A                          lsi_10k
   SIMP_ADD_1
      SIMP_ADD_1_DW01_addsub_5_1
            EO3P                          lsi_10k
            FA1A                          lsi_10k
   SIMP_ADD_2
      SIMP_ADD_2_DW01_add_5_1
            AN2P                          lsi_10k
            AO7                           lsi_10k
            EN                            lsi_10k
            EO                            lsi_10k
            EO3P                          lsi_10k
            IVP                           lsi_10k
            ND2                           lsi_10k
```

Figure 10.47   12-bit adder hierarchy report.

```
read -format vhdl add_12_reg.vhd
read -format vhdl reg.vhd
read -format vhdl reg_12.vhd
read -format vhdl sim_add_wc2.vhd
current_design  ADD_12_REG
create_clock -period 10 -waveform {0 5.0} CLK
uniquify
compile
```

Figure 10.48   Top-down script.

golden instance approach yields far from golden results. Choice of a different golden instance may yield better results.

10.5   INFERRED LATCHES AND DON'T CARES

When assignment to a signal or a variable is under the control of an *if* or *case* statement, the statement must specify what value the signal or variable gets under all conditions tested by the *if*

```
current_design  ADD_12_REG
characterize {C1 C2 C3 C4 C5}
current_design  SIMP_ADD_0
write_script >  SIMP_ADD_0.scr
compile
current_design  SIMP_ADD_1
write_script >  SIMP_ADD_1.scr
compile
current_design SIMP_ADD_2
write_script >  SIMP_ADD_2.scr
compile
current_design  REG_12
write_script >  REG_12.scr
compile
current_design REG
write_script >  REG.scr
compile
current_design  ADD_12_REG
set_dont_touch {C1 C2 C3 C4 C5}
compile
```

Figure 10.49   Bottom-up script.

```
read -format vhdl add_12_reg.vhd
read -format vhdl reg.vhd
read -format vhdl reg_12.vhd
read -format vhdl sim_add_wc2.vhd
current_design SIMP_ADD
compile
current_design  ADD_12_REG
create_clock -period 10 -waveform {0 5.0} CLK
characterize {C2}
current_design  SIMP_ADD
compile
current_design  ADD_12_REG
set_dont_touch {C1 C2 C3}
compile
```

Figure 10.50   Golden instance script.

Table 10.3   Hierarchical compilation results.

	Top-Down	Bottom-Up	Golden Instance
Area	225	277	254
Delay	8.84 ns	8.38 ns	11.19 ns

or *case* statement; otherwise, a latch is inferred. Consider the code in Figure 10.51. In process P_A_LATCHED, the signal A_LATCHED is assigned the value of IN_DAT when IN_EN = '1'.

The code does not say what happens to A_LATCHED when IN_EN='0'. From a simulation point of view, this is not a problem because signals have memory and will retain the last value assigned to them. From a synthesis point of view, however, a latch is inferred to insure that the value will be retained. Figure 10.52 shows the synthesized model. Note that the signal A_LATCHED is indeed latched. In process P_A_COMB, on the other hand, the *if* statement is symmetrical, that is, the *else* clause assigns '0' to the signal A_COMB when IN_EN = '0'. Refer again to Figure 10.52 and note that the signal A_COMB is the output of an AND gate.

A similar situation can exist for *case* statements. Consider process P_B_LATCHED in Figure 10.51. The *when* clauses must account for all conditions of the control variable. The *case* statement tests the signal SEL, which has four logic combinations of interest: "00","01","10", and "11". However, SEL is a STD_LOGIC_VECTOR of length 2. To account for the other vector values possible with the nine-valued logic system, the *when others* clause executes a null operation. Thus, from a simulation semantics point of view the statement is valid. However, note that for the case "11" a *null* operation is also coded. This implies that we want nothing to happen for this condition; that is, we want signal B_LATCHED to retain its pevious value when SEL="11". This requires a latch, as shown in Figure 10.52.

If a latch is not wanted, there are two possible alternatives. First, in process P_B_COMB_0, a value ('1') can be assigned under the "11" clause. Second, in process P_B_COMB_1, a default value can be specified for B_COMB_1 preceding the *case* statement. Note that the simulation semantics for each of these alternatives is correct. In the second case, the statement B_COMB_1 < = '1' will assign a default value to the signal driver which can be overridden during the same simulation cycle by a *case* statement clause. Note from Figure 10.52 that the signals B_COMB_0 and B_COMB_1 were both synthesized as the output of the same combinational circuit and were not latched.

In process P_B_COMB_2, under the *when* clause "11", the signal B_COMB_2 is assigned the value '-', which is the don't care value in the IEEE nine-valued logic system. From a simulation point of view, this value is handled just like any other enumeration type value; signal B_COMB_2 will receive the value '-' when SEL = "11". How synthesis treats the value is vendor dependent. For Synopsys tools, the don't cares are used to minimize the logic provided that the *flatten* directive is included. Note in Figure 10.52 that the signal B_COMB2 is the output of this minimized logic.

The effect of using the '-' value for combinational logic functions is better illustrated in Figure 10.53. Here, we have two ROM models of combinational logic functions. The functions are identical except that the second model has don't cares in two positions where the first model has 0's. Figure 10.54 shows the resultant circuits and illustrates that the second function has a lower cost implementation.

The main problem with the use of don't cares is the difference between simulation and synthesis semantics. Consider outputs first. If a '-' is assigned to an output, when you simulate the source model, this is the value you get. However, the synthesized model will produce a '0' or a '1'. Thus, the two models outputs are different, complicating validation. Testing for '-' on input signals produces even more difficult problems. For example, the statement:

if (A = '-') then

```
entity INFERRED is
  port(IN_DAT,IN_EN: in STD_LOGIC;
       SEL: in STD_LOGIC_VECTOR(1 downto 0);
       A_LATCHED, A_COMB, B_LATCHED, B_COMB_0, B_COMB_1,
         B_COMB_2: out STD_LOGIC);
--pragma dc_script_begin
--set_flatten true
--pragma dc_script_end
end INFERRED;

architecture ALG of INFERRED is
begin
  P_A_LATCHED: process(IN_DAT,IN_EN)
  begin
    if IN_EN = '1' then
      A_LATCHED <= IN_DAT;
    end if;
  end process;
  P_A_COMB: process(IN_DAT,IN_EN)
  begin
    if IN_EN = '1' then
      A_COMB <= IN_DAT;
    else
      A_COMB <= '0';
    end if;
  end process;
  P_B_LATCHED: process(IN_DAT,SEL)
  begin
    case (SEL) is
      when "00" => B_LATCHED <= IN_DAT;
      when "01" => B_LATCHED <= not IN_DAT;
      when "10" => B_LATCHED <= '0';
      when "11" =>  null;
      when others => null;
    end case;
  end process;
  P_B_COMB_0: process(IN_DAT,SEL)
  begin
    case (SEL) is
      when "00" => B_COMB_0 <= IN_DAT;
      when "01" => B_COMB_0 <= not IN_DAT;
      when "10" => B_COMB_0 <= '0';
      when "11" => B_COMB_0 <= '1';
      when others => null;
    end case;
  end process;
```

Figure 10.51 Model to illustrate inferred latches and don't cares.

will be interpreted as FALSE by the synthesis process in Synopsys tools, hardly the intended behavior. Thus, one should avoid testing for '-' in an *if* clause.

```
        P_B_COMB_1: process(IN_DAT,SEL)
        begin
          B_COMB_1 <= '1';
          case (SEL) is
            when "00" => B_COMB_1 <= IN_DAT;
            when "01" => B_COMB_1 <= not IN_DAT;
            when "10" => B_COMB_1 <= '0';
            when "11" => null;
            when others => null;
          end case;
        end process;
        P_B_COMB_2: process(IN_DAT,SEL)
        begin
          case (SEL) is
            when "00" => B_COMB_2 <= IN_DAT;
            when "01" => B_COMB_2 <= not IN_DAT;
            when "10" => B_COMB_2 <= '0';
            when "11" => B_COMB_2 <= '-';
            when others => null;
          end case;
        end process;
      end ALG;
```

Figure 10.51 Model to illustrate inferred latches and don't cares (continued).

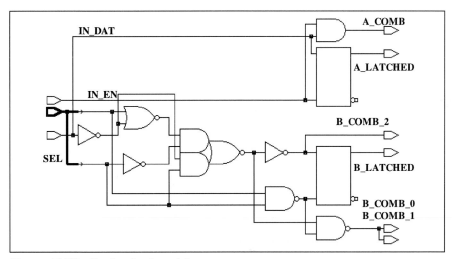

Figure 10.52 Synthesized model.

10.6 TRISTATE CIRCUITS

To model tristate circuits, two conditions must be met (1) the output of the tristate buffers must drive the same VHDL signal, implying a resolution function and (2) when the enable signal to a

```
library IEEE;
use IEEE.std_logic_1164.all;
use IEEE.std_logic_unsigned.all;
entity FUNCS is
  port(X: in STD_LOGIC_VECTOR(2 downto 0);
       Z1,Z2: out STD_LOGIC);
--pragma dc_script_begin
--set_flatten true
--pragma dc_script_end
end FUNCS;

architecture ROM of FUNCS is
  type ROM_1D is array(0 to 7) of STD_LOGIC;
begin
  FULLY_SPECIFIED: process(X)
    constant ROM1: ROM_1D:= "01101000";
  begin
    Z1 <=ROM1(CONV_INTEGER(X));
  end process;
  PARTIALLY_SPECIFIED: process(X)
    constant ROM2: ROM_1D:= "01101--0";
  begin
    Z2 <=ROM2(CONV_INTEGER(X));
  end process;
end ROM;
```

Figure 10.53   Two ROM models.

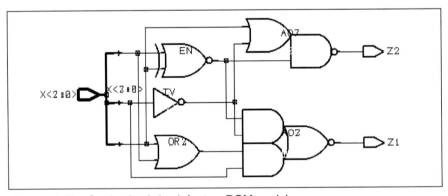

Figure 10.54   Synthesized circuit for two-ROM model.

tristate buffer goes to '0', the output of the buffer must receive the value 'Z'. If these two conditions are met, most synthesis tools will infer a tristate buffer. Figure 10.55 shows a two-process
model that meets these requirements. Data type STD_LOGIC has a resolution function attached
in the IEEE 1164 library. Figure 10.56 shows the synthesized circuit. Most synthesis tools allow
the use of directives to force a Wired AND or Wired OR bus, but lacking this, one gets a bus that
employs the IEEE 1164 resolution function.

```
library IEEE;
use IEEE.STD_LOGIC_1164.all;
entity TRISTATE is
  port(A,B,ENA,ENB: in STD_LOGIC; BUS_SIG: out STD_LOGIC);
end TRISTATE;

architecture   ALG of  TRISTATE is
begin
  PROCA: process(A,ENA)
  begin
    if (ENA = '1') then
      BUS_SIG <= A;
    else
      BUS_SIG <= 'Z';
    end if;
  end process;
  PROCB: process(B,ENB)
  begin
    if (ENB = '1') then
      BUS_SIG <= B;
    else
      BUS_SIG <= 'Z';
    end if;
  end process;
end ALG;
```

Figure 10.55 Tristate circuit model.

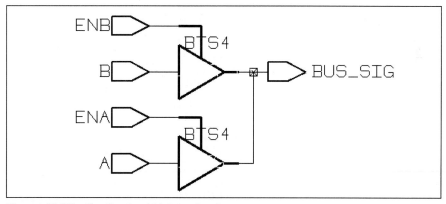

Figure 10.56 Synthesized tristate circuit.

10.7 SHARED RESOURCES

It is important, particularly when implementing arithmetic functions, to force the synthesis tool to share resources between the implementations of different statements. This can be done with synthesis directives or by coding style. We illustrate the second case here. Suppose, one is given

```
library IEEE;
use IEEE.STD_LOGIC_1164.all;
use IEEE.STD_LOGIC_ARITH.all;
entity SHARE_1 is
  port(A,B,C,D: in STD_LOGIC_VECTOR(3 downto 0);
       SEL_CD: in STD_LOGIC;
       F: out STD_LOGIC_VECTOR(3 downto 0));
end SHARE_1;
architecture DF of SHARE_1 is
  signal MUX1,MUX2:SIGNED(3 downto 0);
  signal SUM:SIGNED(3 downto 0);
begin
  MUX1 <= SIGNED(A)when SEL_CD = '0' else SIGNED(C);
  MUX2 <= SIGNED(B) when SEL_CD = '0' else SIGNED(D);
  SUM <= MUX1 + MUX2;
  F <= STD_LOGIC_VECTOR(SUM);
end DF;
```

Figure 10.57 Multiplex then add model.

the following specification for an arithmetic circuit with data inputs A, B, C, and D and control signal SEL_CD. When control signal SEL_CD is '0', the output of the circuit is A+B; otherwise, the output is C+D. There are two basic approaches to the design of this circuit (1) *multiplex then add* or (2) *add then multiplex*. The approach taken has a significant effect on the size and delay of the circuit. Coding style can be used to control which approach is used in synthesis. Figure 10.57 gives a *multiplex then add* model. Figure 10.58 gives an *add then multiplex* model. Figure 10.59 gives the high-level circuit for the *multiplex then add* model. Figure 10.60 gives the high-level circuit of the *add then multiplex* model. (The high-level circuit is the library independent circuit generated by elaboration in the Synopsys tool. It is useful for seeing which major logic blocks were used by the tool.) Table 10.4 gives the cost/delay data for the two circuits. As one might expect, the multiplex then add circuit is less complex than the add then multiplex circuit because adders cost more than multiplexers. However, the add then multiplex circuit has a lower delay than the multiplex then add circuit because the inverter delay occurs in parallel with the adder delay. Thus, a trade-off is involved.

Table 10.4 Cost/delay comparison.

Circuit	Area	Delay
Multiplex then add	51	8.47
Add then multiplex	73	7.09

10.8 FLATTENING AND STRUCTURING

When performing logic design, designers are aware that two-level sum of products or product of sums circuits are the fastest, while factored (mongrel) forms can result in lower-cost designs albeit slower. Logic flattening (as contrasted to hierarchical flattening) is used to achieve two-level circuits. Structuring yields factored circuits. Through the use of synthesis directives, one

```
library IEEE;
use IEEE.STD_LOGIC_1164.all;
use IEEE.STD_LOGIC_ARITH.all;
entity SHARE_2 is
  port(A,B,C,D: in STD_LOGIC_VECTOR(3 downto 0);
       SEL_CD: in STD_LOGIC;
       F: out STD_LOGIC_VECTOR(3 downto 0));
end SHARE_2;
architecture DF of SHARE_2 is
  signal ADD1,ADD2:SIGNED(3 downto 0);
begin
  ADD1 <= SIGNED(A) + SIGNED(B);
  ADD2 <= SIGNED(C) + SIGNED(D);
  F <= STD_LOGIC_VECTOR(ADD1) when SEL_CD = '0' else
       STD_LOGIC_VECTOR(ADD2);
end DF;
```

Figure 10.58   Add then multiplex model.

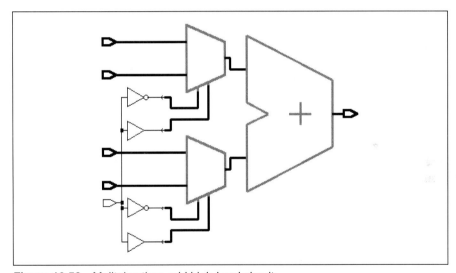

Figure 10.59   Mulitplex then add high-level circuit.

can control which type of circuit is synthesized. Figure 10.61 shows a model that implies a three-level circuit. Figure 10.62 shows the synthesized, flattened circuit. Figure 10.63 shows a model which implies a product of sums implementation. Figure 10.64 gives a circuit which is a factored equivalent.

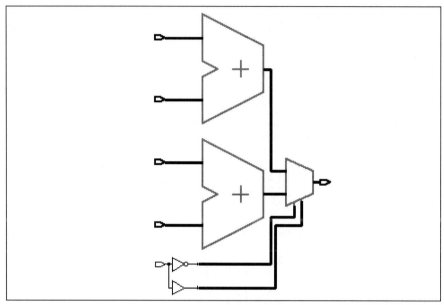

Figure 10.60   Add then multiplex high-level circuit.

```
        entity LOG_FUNC is
        port(A, B, C, D: in STD_LOGIC; F: out STD_LOGIC);
        --pragma dc_script_begin
        --set_flatten true
        --pragma dc_script_end
        end LOG_FUNC;

        architecture DF of LOG_FUNC is
          signal S1,S2: STD_LOGIC;
        begin
          S1 <= A or B;
          S2 <= S1 and C;
          F <= S2 and D;
        end DF;
```

Figure 10.61   Three-level circuit.

10.9 EFFECT OF MODELING STYLE ON CIRCUIT COMPLEXITY

We now consider some additional effects of modeling style on circuit complexity.

10.9.1   Effect of Selection of Individual Construct

In many cases, the choice of an individual language construct can have a profound effect on the synthesized circuit complexity. As an example, consider the two implementations of a 4-to-1

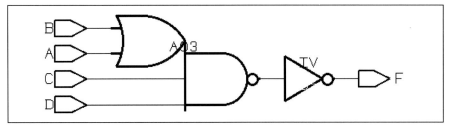

Figure 10.62 Flattened circuit.

```
entity LOG_FUNC is
   port(A,B,C,D: in STD_LOGIC; F1,F2: out STD_LOGIC);
--pragma dc_script_begin
--set_structure -timing true
--pragma dc_script_end
end LOG_FUNC;

architecture DF of LOG_FUNC is
begin
   F1 <= (A and B) or (A and D);
   F2 <= (B and C) or (C and D);
end DF;
```

Figure 10.63 Two-level model.

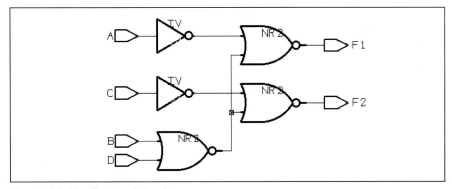

Figure 10.64 Factored circuit.

multiplexer shown in Figure 10.65. Figures 10.66 and 10.67, respectively, show the synthesized versions of the IF_STATEMENT multiplexer and the CASE_STATEMENT multliplexer using LSI 10 K library components. Table 10.5 summarizes the cost/delay data. for the two synthesized circuits. The CASE MUX is superior in both categories. Synthesis with Xilinx FPGAs gives similar results. Clearly, using the *case* statement generates lower cost and faster implementations than using the *if-then* construct. The lesson to learn from this example is that the selection of language construct can affect both the cost and delay in the resulting synthesized circuit.

```
library IEEE;
use IEEE.STD_LOGIC_1164.all;
entity MUX is
  port (SEL: in STD_LOGIC_VECTOR(1 downto 0);
        A,B,C,D: in STD_LOGIC;
        MUX_OUT: out STD_LOGIC);
end MUX;
architecture IF_STATEMENT of MUX is
begin
  process(SEL,A,B,C,D)
  begin
    if (SEL = "00") then MUX_OUT <= A;
    elsif (SEL = "01") then MUX_OUT <= B;
    elsif (SEL = "10") then MUX_OUT <= C;
    elsif (SEL = "11") then MUX_OUT <= D;
    else  MUX_OUT <= '0';
    end if;
  end process;
end IF_STATEMENT;

architecture CASE_STATEMENT of MUX is
begin
  process(SEL,A,B,C,D)
  begin
    case SEL is
      when "00" => MUX_OUT <= A;
      when "01" => MUX_OUT <= B;
      when "10" => MUX_OUT <= C;
      when "11" => MUX_OUT <= D;
      when others => MUX_OUT <= '0';
    end case;
  end process;
end CASE_STATEMENT;
```

Figure 10.65   *If* and *case* implementation of a muliplexer.

Table 10.5   Cost delay comparison.

Circuit	Area	Delay
if MUX	11	2.75 ns
case MUX	6	1.35 ns

10.9.2   Effect of General Modeling Approach

In many cases, the general approach to modeling can have a profound effect on the circuit complexity. For example, consider the design of a circuit that performs division and multiplication by two. Possible ways to model the circuit include (1) use of shift-left (shl) and shift-right (shr) operators from the IEEE library, (2) use of an overloaded multiplier operator from the IEEE library, which instantiates a combinational multiplier block, (3) use of multiplication and divi-

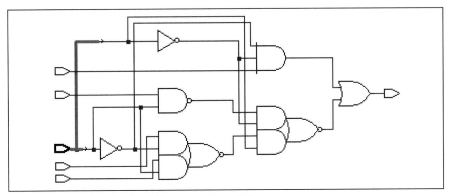

Figure 10.66 Synthesized *if* statement multiplexer.

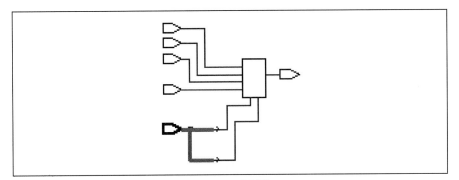

Figure 10.67 Synthesized *case* statement MUX.

sion algorithms, or (4) use of a clocked shift register and the concatenation operator. The first three general approaches imply large blocks of combinational logic, whereas the fourth approach implies a shift register and some control logic. In general, it is most efficient to use a model that has an evident, simple hardware implementation. Determining which of these three approaches is left as an exercise in the Problems section.

PROBLEMS

10.1 Implement a low pass digital filter in tools, such as SPW from Cadence or System View from Elanix. Simulate the filter in the tool. Save inputs and outputs from simulation. Have the tool generate either a VHDL model or a Xilinx model and simulate the model with the inputs saved from the tool and compare the results of the high-level simulation.

10.2 Select a network protocol that can be modeled by two state machines, one at each end of the channel. Model the system using state charts and simulate the protocol operation. Generate a VHDL model. Compare the efficiency of this model with one you could develop by hand coding, using the techniques given in Chapter 8.

10.3   Figure 3.9 shows the test bench for the entity ONES_CNT developed in Chapter 3. In this test
       bench, a process is used to assign inputs to ONES_CNT at 1 ns intervals. Explain how this
       process works. Could it be synthesized? Explain your answer.

10.4
```
     library IEEE;
     use IEEE.std_logic_1164.all;
     entity CKT is
        port(CLK,I: in std_logic; DOUT: out std_logic);
     end CKT;

     architecture ALG of CKT is
        signal X: std_logic;
     begin
        process(CLK)
        begin
           if (CLK'EVENT and CLk = '0') then
               X  <=  I;
           end if;
               DOUT <= X;
        end process;
     end ALG;

     library IEEE;
     use IEEE.std_logic_1164.all;
     entity tb is
     end tb;

     architecture TEST of tb is
        component CKT
           port(CLK,I: in std_logic; DOUT: out std_logic);
        end component;
        for C1:CKT use entity work.CKT(ALG);
        signal CLK,I,DOUT: std_logic;
     begin
        C1: CKT
           port map(CLK,I,DOUT);
        CLK <= '0' after 50 ns,  '1' after 100 ns,
               '0' after 150 ns, '1' after 200 ns,
               '0' after 250 ns, '1' after 300 ns,
               '0' after 350 ns, '1' after 400 ns,
               '0' after 450 ns;
        I <=  '0', '1' after 125 ns, '0' after 225 ns;
     end TEST;
```

       Given the above model and test bench, assume that the code is compiled and that the test
       bench is simulated. Plot the simulation results; that is, carefully plot: I, X (internal signal in
       component CKT), CLK, and DOUT. Check your results on a VHDL simulator.

10.5
```
     library IEEE;
     use IEEE.STD_LOGIC_1164.all;
     entity CKT1 is
        port(I,CLK,RESET: in STD_LOGIC; DOUT: out STD_LOGIC);
     end CKT1;
```

```
architecture  BEHAV1 of CKT1 is
  signal D: STD_LOGIC;
begin
  process(CLK,RESET)
  begin
    if RESET = '1' then
      D <= '0';
    elsif (CLK = '1' and CLK'EVENT) then
      D <= D xor I;
      Dout <= D;
    end if;
  end process;
end BEHAV1;

library IEEE;
use IEEE.STD_LOGIC_1164.all;
entity CKT2 is
  port(I,CLK,RESET: in STD_LOGIC; DOUT: out STD_LOGIC);
end CKT2;
architecture  BEHAV2 of CKT2 is
  signal D: STD_LOGIC;
begin
  process(CLK,RESET)
  begin
    if RESET = '1' then
      D <= '0';
    elsif (CLK = '1' and CLK'EVENT)then
      D <= D xor I;
    end if;
  end process;
  Dout <= D;
end BEHAV2;

library IEEE;
use IEEE.STD_LOGIC_1164.all;
entity CKT3 is
 port(I,CLK,RESET: in STD_LOGIC; DOUT: out STD_LOGIC);
end CKT3;

architecture  BEHAV3 of CKT3 is
begin
  process(CLK,RESET)
    variable D: STD_LOGIC;
  begin
    if RESET = '1' then
      D := '0';
    elsif (CLK = '1' and CLK'EVENT) then
      D := D xor I;
      Dout <= D;
    end if;
  end process;
end BEHAV3;
```

Shown above are three behavioral models with seemingly similar behavior.

    a. Simulate all three models and compare their responses. Differentiate between initialization behavior and run behavior.

    b. Synthesize all three models, and then simulate them. Use a test bench similar to that shown in Figure 9.46 to test the models. Is the behavior of the synthesized models the same as that of the behavioral models? Explain.

    c. Print the schematics of all three models.

    d. What general conclusions can you draw from this experiment?

10.6   The style for synthesis prohibits the use of time expressions in signal assignment statements and *wait* statements. In light of this, how does one incorporate delay into a model?

10.7   A modeler wishes to model a counter with an integer variable. Should he use type Integer or a constrained integer? Why?

10.8   For each of the following devices (1) create a behavioral model, (2) simulate the behavioral model, (3) synthesize the behavioral model, and (4) simulate the synthesized circuit. All devices should use a positive edge triggered (CLOCK). Comment on any problems you encounter.

    a. JK flip-flop with low active asynchronous set and reset.

    b. D flip-flop with high active asynchronous set and reset.

    c. Parallel load, shift-right, shift-left register (8 bits wide).

    d. 8-bit wide, 6-deep FIFO.

10.9   Assuming you are using the Synopsys synthesis style, predict the logic implied by the following code; then, synthesize the code. Compare the actual synthesized circuit with what you predicted.

```
entity SYNQ is
  port(A: in BIT; B: INTEGER range 0 to 3;
       C: BIT_VECTOR(0 to 3);
       D: out BIT);
end SYNQ;
architecture ALG of SYNQ is
  signal X: BIT;
begin
  X <= C(B);
  process(X,A)
  begin
    if  A ='1' then
      D <= X;
    end if;
  end process;
end ALG;
```

10.10 Repeat Problem 10.9 for this code:

```
library IEEE;
use IEEE.std_logic_1164.all;
entity  CKT is
  port(A,B,C: in STD_LOGIC; Q1,Q2: out STD_LOGIC);
end CKT;
architecture ALG of CKT is
begin
  process(A)
    variable  V :STD_LOGIC;
  begin
    if V = '1' then
      Q2 <= C;
    end if;
    if A'EVENT and A = '1' then
      Q1 <= B;
      V := not B;
    end if;
  end process;
end ALG;
```

10.11 Repeat Problem 10.9 for this code:

```
use work.PRIMS.all;
entity LOGIC is
  port(CON: in BIT_VECTOR(1 downto 0);
       A,B: in BIT_VECTOR(7 downto 0);
       CLK: in BIT;
       FOUT: out BIT_VECTOR(7 downto 0));
end LOGIC;

architecture ALG of LOGIC is
  signal F: BIT_VECTOR(7 downto 0);
begin
  FUNCTIONS:process(CON,A,B)
  begin
    case CON is
      when "00" => F <= A;
      when "01" => F <= not A;
      when "10" => F <= INC(A);
      when "11" => F <= A and B;
    end case;
  end process FUNCTIONS;
  STORE:process(CLK)
  begin
    if CLK = '1' then
      FOUT <= F;
    end if;
  end process STORE;
end ALG;
```

10.12 Repeat Problem 10.9 for this code:

```
use work.PRIMS.all;
entity TEST_QUESTION is
  port(CLK,W,X,Y,Z: in BIT; O: out BIT);
end TEST_QUESTION;

architecture BEHAV of TEST_QUESTION is
  signal  C: BIT_VECTOR(0 to 1);
begin
  BLK: block(not CLK'STABLE and CLK='1')
  begin
    C <= guarded INC(C);  --INC is an increment function
  end block BLK;
  O <= W when C="00" else
       X when C="01" else
       Y when C="10" else
       Z;
end BEHAV;
```

10.13 Repeat Problem 10.9 for this code:

```
use work.funcs.all;
entity SYS is
  port(C: in BIT; COM: BIT_VECTOR(0 to 1);
       INP: in BIT_VECTOR(0 to 7));
end SYS;
architecture CODE of SYS is
  signal X,Y: BIT_VECTOR(0 to 7);
begin
  process(C)
  begin
    if C='1' then
      case COM is
        when "00" => X <= INP;
        when "01" => Y <= INP;
        when "10" => X <= ADD8(X,Y);
        when  "11" => X <= ADD8(X,INC8(not(Y)));
      end case;
    end if;
  end process;
end CODE;
```

10.14 Figure 10.68 shows a block diagram and Table 10.6 shows a state table for a module. The module is a clocked synchronous circuit with an asynchronous reset. The state table gives the next state of module for each present state and also specifies which data operation is to be carried out in each state.

a. Write a single architecture VHDL behavioral model for the module. Write it in a style that is compatible with the Synopsys Synthesis Tool. Assume the functions inc and dec are in a package called FUNCPAC.

b. Hand draw a logic diagram, which represents a circuit you think would be synthesized from the model developed in (a). Divide your diagram into two areas, control unit and data unit. In the control unit area, use gates and flip-flops. In the data unit area, use boxes to represent logic functions.

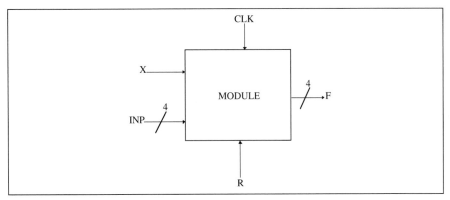

Figure 10.68 Block diagram for Problem 10.14.

Table 10.6 State table for a device module for Problem 10.14.

	Input X		
Present State	**0**	**1**	**Data Operation In Present State**
S0	S0	S1	F <= not INP
S1	S2	S1	F <= inc(INP)
S2	S2	S3	F <= dec(INP)
S3	S0	S3	F <= INP
	Next State		

    c. Synthesize the model using both Synopsys and Xilinx. Experiment with types of state machines for the control unit. Compare results with your hand drawing in part (b).

10.15 In the Xilinx FPGAs, some types of logic are not implemented efficiently using look-up tables. To improve this situation, Xilinx has added special logic. Which type of special logic could be used to implement a stack?

10.16 When synthesizing a finite state machine for a Xilinx FPGA, such as the XC 4010XL, which type of state encoding should you use?

10.17 Attempt to synthesize the Controlled Counter model from Chapter 5. (Figures 5.15 and 5.16). This model has asynchronous handshaking between processes. Is this acceptable coding style? If not, how would you change it to make it acceptable? Discuss any other problems.

10.18 Consider the design of a circuit that performs division and multiplication by two. Possible approaches to modeling include (1) use of shl and shr operators from the IEEE library, (2) use of an overloaded multiplier operator from the IEEE library, which instantiates a combinational multiplier block, (3) use of multiplier and division algorithms, and (4) Use of a clocked shift register and concatenation operator. The first three approaches imply large blocks of combinational logic. The fourth process implies a shift register and some control logic. In general, it is most efficient to use a model that has an evident, simple hardware implementation. Synthesize each of these approaches using Synopsys (ASIC) and Xilinx (FPGA). Compare complexity and speed among the approaches.

10.19 The structural model that is used for the XESS XS40 Board (shown in Figure 9.52) has three components (1) model being designed, (2) DISPLAY model for handling the FPGA inputs and outputs, and (3) CLOCKGEN model that counts down the 12 MHz board clock to a value suitable for this model. Study the XESS board and implement these models.

10.20 Using the Xilinx XC4010XL chip on the XESS XS40 board, develop and synthesize a model that reads the eight dip switches and transfers the output to the eight LEDs.

10.21 (Euclid's Algorithm) An algorithm for the computation of the greatest common divisor of two unsigned binary numbers is:

```
Registers    X(6:0), Y(6:0)
Input:
  X <= Xi
  Y <= Yi
Calculation:
  while (X /= Y) loop
    if (X < Y) then
      Y <= Y-X
    else
      X <= X-Y
  end loop
Output:
  OUT <= X
```

Note: The algorithm is given in pseudo code form, which must be translated into legal VHDL.
   a. Translate the algorithm into a single-process algorithmic model adhering to VHDL style restrictions for synthesis. Simulate this algorithmic model to validate it.
   b. Synthesize the algorithmic model developed in (a) using Synopsys (ASIC). Simulate the resultant structural model to validate it. Draw an equivalent block diagram of the circuit.
   c. Synthesize the model using the Xilinx XC4010XL chip on the XESS XS40 card. Check out the Xilinx model on the board.
   d. Compare the complexity and speed of the ASIC and FPGA implementations.

10.22 Design a simple calculator.

Overview. Design a simple calculator and implement it on a Xilinx 4010 FPGA. First, develop a behavioral model of the calculator and simulate it. Be sure to apply synthesis constraints on style, so it can be synthesized. Then, synthesize the circuit using Xilinx Foundation Express and download the synthesized circuit onto the XTEND board.

Specification. The calculator receives a sequence of inputs from the dip switches. The most significant bit (DS(1)) indicates the format of each data byte. If DS(1) = 1, the byte is an operator. If DS(1) = 0, the byte is an operand. If the current byte is an operator, then the least significant two bits (DS(7)-DS(8)) specify the operation to be performed according to Table 10.7 below. If the current byte is an operand, bits DS(2)-DS(8) specify the value of the operand (DS(2) is the sign bit) as a seven-bit number in 2's complement format.

Table 10.7

DS(1)	DS(7)-DS(8)	Operator	Note
0	---	---	DS(2)-DS(8) is the operand
1	00	+	Addition
1	01	-	Subtraction
1	10	*	Multiplication (Multiplier is a power of 2)
1	11	/	Division (Divisor is a power of 2)

Examples:

Table 10.8

DS(1)-DS(8)	Interpretation
1000 0010	Multiplication operator
0000 1010	(+10)
0111 1101	(-3)

The user enters an operand or operator by setting the bit values on the dip switches and pressing and releasing the SPARE button. The calculator reads the value into a buffer and decodes it. If it is an operand, the data value is sign-extended, pushed onto the stack, and displayed in the 7-segment displays. If it is an operator, the top two items on the stack are removed, the operation is performed, and the result is pushed back onto the stack and displayed on the 7-segment displays. Table 10.9 shows the results for the operations.

Table 10.9

Second from Top	Top	Operator	Operation
A	B	+	A+B
A	B	-	A-B
A	B	*	A*B
A	B	/	A/B

The arithmetic expression for the calculator is in reverse Polish notation (RPN). For example, the following expression (given in in-fix notation)

$$3*4 + 6/2$$

is expressed in RPN as:

$$3\ 4 * 6\ 2 / +$$

The use of RPN allows the expression to be evaluated by scanning the expression from left to right and saving values on a push-down stack. If a value is scanned, that value is pushed onto the top of the stack. If an operator is scanned, the top two items are removed from the stack, the specified operation is performed, and the result is pushed onto the stack. Table 10.10 shows how the RPN expression:

$$8\ 7\ 2 / 2 * - 5 +$$

is evaluated.

Table 10.10

Item Input	DS(1)	DS(2)-DS(8) DS(7)-DS(8)	Display after Operation	Stack after Operation
RESET	-	-	0	Empty
8	0	8	8	8
7	0	7	7	8,7
2	0	2	2	8,7 ,2
/	1	3	3	8,3
2	0	2	2	8,3 ,2
*	1	2	6	8,6
-	1	1	2	2
5	0	5	5	2,5
+	1	0	7	7

The calculator should produce the correct result as long as the maximum stack depth of 5 is not exceeded or an arithmetic overflow does not occur.

Detailed Specifications:

> Display: Three 7-segment displays.
>
> Maximum stack depth: 5.
>
> Range of input operand values: 7-bit 2's complement numbers. Calculate the range of values and include it in your report.
>
> Result range: 12-bit 2's complement numbers. Calculate the range of values and include it in your report.
>
> No overflow will occur.
>
> There will be no errors in the input expression.

Current dip switch settings should always be displayed on the LEDs as a check on the inputs *before* the pushbutton is depressed.

Procedure. Prepare an algorithmic VHDL model of the calculator, and embed it into a strucural model that includes the calculator entity and various display entities. The system inputs are: SPARE_PUSH_BUTTON, RESET_PUSH_BUTTON, DS(7:0), and XS40_CLK. The outputs of the system are: LED(7:0), TOP_7S(6:0), LEFT_7S(6:0), and RIGHT_7S(6:0). The file debounce.vhd from the book CD contains sample code for debouncing switches. Use a test bench to simulate the structural model. You may want to simulate individual components separately before assembling them into the final structural model. Simulate the following operation sequences (given in in-fix notation). The RPN form for the first expression is given. You will need to translate the second two expressions into RPN notation. Note: {-5} means that you enter a negative number already in correct 2's complement format.

1. (21+5*2)/2 – 16   [21 5 2 *+2 / 16 -]
2. 10*(5-(2+(3*10)))
3. (60*32-41/16)*{-64}

Implement the calculator in a Xilinx 4010XL chip.

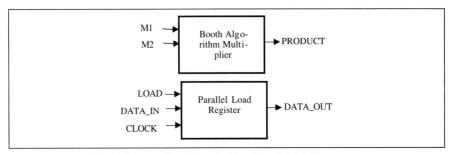

Figure 10.69 Block diagram for a Booth Algorithm Multliplier.

10.23 Design and synthesize a Booth Algorithm Multiplier.

Specification. Design a chip that implements a Booth Algorithm Multiplier. Inputs M1 and M2 are 8-bit signed numbers using the 2's complement number system. The output, PRODUCT, is a 16-bit signed number that represents the product of M1 and M2. This is a combinational device.

Procedure. Use these two files from the book CD, "booth.ucf" and "debounce.vhd." "Booth.ucf" is a user constraints file suitable for use in this project. Develop a VHDL behavioral model for the chip. Use IEEE standard logic packages, including those packages that overload + and -. Use overloaded + and - to implement the Booth multiplication algorithm. Validate the model with a VHDL system.

Develop a VHDL model for an 8-bit parallel load register, whose block diagram is shown in Figure 10.69. DATA_IN is an 8-bit parallel input. DATA_OUT is an 8-bit parallel output. LOAD is a control signal. CLOCK is a system clock. When LOAD is active high, the value on DATA_IN is loaded into the register on the low to high transition of CLOCK. DATA_OUT always displays the current value stored in the register. Validate the model with a VHDL system.

Create a structural VHDL model that interfaces your multiplier to the XSTEND board. Design the interface, so a user can set one of the operand values into the dip switches and push one of the pushbuttons to enter the operand into an internal register on the Xilinx chip. The user then sets the second operand value into the dip switches and pushes the second pushbutton to enter the second value into a second internal register. Bounce on the pushbutton switches may prove to be a problem. The project file "bounce.vhd" contains VHDL code for implementing a software debouncing operation. Display the high-order eight bits of the product in the LEDs. Display the low-order eight bits in the LEFT AND RIGHT 7-segment displays as two hexadecimal characters. Synthesize the model using Xilinx tools.

Replace your Booth multiplier component with another multiplier component generated, using the overloaded multiply operator from the IEEE package. Synthesize the new multiplier on Xilinx. Compare the two implementations as to efficiency of design (amount of chip resources used) and maximum frequency.

Integration of VHDL into a Top-Down Design Methodology

T he VHDL language was designed to support a top-down design methodology. This chapter describes how VHDL can be integrated into all abstraction levels of such a methodology.

11.1 TOP-DOWN DESIGN METHODOLOGY

The objective of this chapter is to describe a top-down design methodology in which a design is successively refined. Typically, a design evolves from word specification to system behavioral description, to system decomposition, to RTL models, to gate-level models and finally, to the circuit that can be physically fabricated. This is a complete design cycle in which a design evolves from a high abstraction level to a low abstraction level. As the design evolves and approaches the physical implementation, more and more detail is introduced.

VHDL models are used at all levels to specify and document hardware designs. They are also used for testing and simulation purposes. The initial function and requirements of a system/subsystem are captured in VHDL as an executable specification, which can be verified by a VHDL simulator. The initial executable specification may need to be refined to eliminate ambiguities in the written specification. The executable specification can be inserted into a test bench to verify the system function and to provide requirements for lower-level designs. Synthesis tools can generate gate-level models automatically from higher-level models.

During the life cycle of a design, refinements and upgrades occur periodically. Modifications are made to achieve functional improvements, correct design errors, increase speed of operation, reduce power consumption, cost, or weight, or to replace an obsolete part with a newer technology device. The next version of a design should be based on the current design. It is not cost effective to start completely from scratch. If the design of the previous version can be reused as the starting point for the new design, the redesign cycle can be significantly shortened.

It is, therefore, important for the design methodology to emphasize techniques that promote reuse of the VHDL models from earlier versions of the design. As the design progresses down the hierarchy, it is also important to be able to reuse test benches from higher levels to test models developed for lower levels. Issues of timing and data types become important.

After a system is decomposed into a set of modules, different design groups can be assigned to design individual modules. These design groups might be located in different geographical locations and even associated with different companies. It is, therefore, important for the design methodology to provide seamless integration of the various modules into a final system model for testing and evaluation.

In this chapter, we create an executable model from a written natural language specification, iteratively decompose the top-level model into components until the leaf nodes are RTL-level models, and finally use a synthesis tool to translate the RTL models into gate-level models. See Chapter 1 for a detailed discussion of abstraction hierarchy. The design time, cost, and efficiency are directly affected by the design methodology.

In top-down design, a design moves from the top to the bottom in the abstraction hierarchy. There is no generally accepted standard for either abstraction hierarchies or for design methodology. We present the following methodology as typical. Our methodology has been developed to highlight the use and reuse of VHDL models at multiple abstraction levels. Figure 11.1 shows a flow chart of a typical design process.

The first step is to create an executable VHDL model of the system that accurately reflects the system specifications. There are many ways to accomplish this step. Three parallel paths in Figure 11.1 describe three methods designers can use to create the top-level model. As indicated in the left-most path, experienced VHDL designers can directly generate this model from the written specifications. For those with less experience, the center path and the right-most path describe using tools to aid in the model construction activity.

The center path describes the use of a schematic capture tool to create a requirements repository for the system. Important system parameters appear as generics in this model. It is important that the schematic capture tool be able to automatically generate a VHDL code shell. The user must then insert a behavioral model for the system into the VHDL shell and simulate the complete model. Using the tool provides documentation for the system specifications and a convenient method to track changes in system specifications that occur during the design process. The designer must still be fluent in VHDL to write the behavioral model that is inserted into the VHDL shell. An example of this type of tool, Synopsys SGE, is described later in the chapter.

To further insulate designers from the details of the VHDL language, application-specific tools can be used to develop the VHDL executable specification. The right-most path in Figure 11.1 describes this alternative. For example, the Signal Processing Workbench (SPW) tool can be used to generate executable models of signal processing systems. This tool provides primitive functions, typical in signal processing applications, such as fast Fourier transforms (FFT) and finite impulse response (FIR) filters. The algorithm is modeled using a schematic capture technique to interconnect the primitive signal processing function modules. Simulation in SPW verifies the correctness of the algorithm. The SPW tool will then automatically generate a VHDL output file that contains a VHDL model of the SPW algorithm. This model is then simulated by a VHDL simulator to verify that the two models produce the same outputs. This method pro-

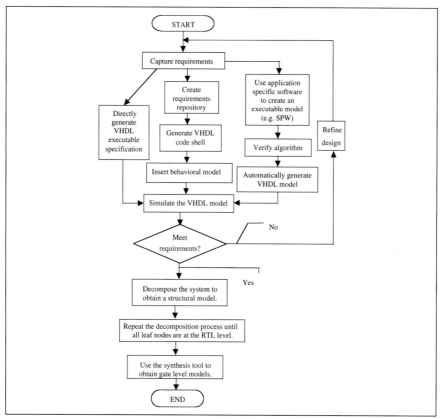

Figure 11.1   Top-down design methodology.

vides an audit trail for the design and also for any redesign. To make a design change, one simply makes the change in SPW, verifies that the new design meets system specifications by simulation in SPW, and requests SPW to generate a new VHDL model. This scenario does not require that the designer be an experienced VHDL programmer.

Sometimes, the creation of an executable specification uncovers errors in the written specification. The original specification might be incomplete or inconsistent. The written specification and the executable specification are then refined in an iterative fashion, as shown in Figure 11.1, until the executable specification matches the design requirements.

When the executable specification is acceptable, the system is decomposed into several smaller components at the same abstraction level or at the next lower abstraction level. This results in a structural model of the system. The components can themselves be further decomposed into smaller and smaller components. The decomposition process is repeated until the components are at the RTL level or are synthesizable. Then, a synthesis tool is applied to obtain a gate-level model.

We will now illustrate the steps in the design methodology by starting with a written specification and using the design methodology to create a series of VHDL models that terminate

Figure 11.2   Image processing system.

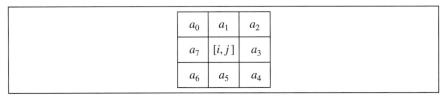

Figure 11.3   Typical arrangement of a window of pixels for an image.

with a gate-level design. The next section provides background information on a common, signal processing application that will be used as the running example.

11.2   SOBEL EDGE DETECTION ALGORITHM

Image processing is a technique frequently used to track moving objects in a succession of images. Figure 11.2 illustrates an image processing system that receives a continuous stream of images from a sensor and selects objects to track, based on the processed input image data. The raw image is enhanced, and then segmented to locate objects or regions of interest. After an object or region has been located, it is examined for identifying features that can lead to final classification of the object. Any identified object can then be tracked through subsequent images to determine its speed and direction of motion.

Before an image can be segmented, the objects in that image must be detected and roughly classified as to shape and boundary characteristics (edges). Edges are significant local changes in the image and are important features for analyzing images. Edge detection is frequently the first step in recovering information from images. Many edge detectors, such as Roberts Operator, Sobel operator, Prewitt Operator, Laplacian Operator, etc., have been developed in the last two decades. Because the Sobel operator is one of the most commonly used edge detectors in image processing systems, we use this operator to illustrate our design methodology.

The Sobel algorithm convolves the incoming image data with four 3x3 masks, that is, Sobel operators, weighted to measure the differences in intensity along the horizontal, vertical, and left and right diagonal directions. This process is commonly called filtering.

Consider an arrangement of pixels about an arbitrary pixel [i,j] as shown in Figure 11.3. Each window consists of nine elements, the current pixel [i,j] and its eight nearest neighbors which are represented by $a_0, a_1, ..., a_7$.

Intensity variations along the horizontal, vertical, and left and right diagonal directions can be computed as follows:

$$E_H = (a_0 + ca_1 + a_2) - (a_4 + ca_5 + a_6) \tag{11.1}$$

$$E_V = (a_2 + ca_3 + a_4) - (a_0 + ca_7 + a_6) \tag{11.2}$$

$$E_{DL} = (a_1 + ca_2 + a_3) - (a_7 + ca_6 + a_5) \tag{11.3}$$

$$E_{DR} = (a_1 + ca_0 + a_7) - (a_3 + ca_4 + a_5) \tag{11.4}$$

We will use c=2 as a typical value. E_H, E_V, E_{DL}, and E_{DR} can be implemented using these convolution masks:

$$H = \begin{array}{|c|c|c|} \hline 1 & 2 & 1 \\ \hline 0 & 0 & 0 \\ \hline -1 & -2 & -1 \\ \hline \end{array} \qquad V = \begin{array}{|c|c|c|} \hline -1 & 0 & 1 \\ \hline -2 & 0 & 2 \\ \hline -1 & 0 & 1 \\ \hline \end{array}$$

$$DL = \begin{array}{|c|c|c|} \hline 0 & 1 & 2 \\ \hline -1 & 0 & 1 \\ \hline -2 & -1 & 0 \\ \hline \end{array} \qquad DR = \begin{array}{|c|c|c|} \hline 2 & 1 & 0 \\ \hline 1 & 0 & -1 \\ \hline 0 & -1 & -2 \\ \hline \end{array}$$

These four parameters, E_H, E_V, E_{DL}, and E_{DR}, are then combined to compute the gradient magnitude and direction. A commonly used estimate for the gradient magnitude is:

$$Magnitude = Max\left[|E_H|, |E_V|, |E_{DR}|, |E_{DL}| \right] + \frac{1}{8}\lfloor |E_\perp| \rfloor \tag{11.5}$$

where $|E_H|, |E_V|, |E_{DR}|, |E_{DL}|$ are the absolute values of E_H, E_V, E_{DR}, and E_{DL}, respectively, and $E_\perp$ is the intensity variation in the direction, perpendicular to the direction of the maximum intensity variation. The notation $\lfloor \ \rfloor$ means "the greatest integer less than or equal to". If the gradient magnitude exceeds a specified threshold, the pixel in the center of the window is declared to be part of an edge.

The direction of the edge, also called the phase of the edge, is represented by a 3-bit vector as shown in Figure 11.4 and Table 11.1. The edge direction is defined to be the direction associated with the maximum intensity variation. For example, if the horizontal filter output has the maximum intensity and has a positive value, the direction is 000. If the horizontal filter output has the maximum intensity and the value is negative, the direction is 100.

Table 11.1 Codes for edge directions.

Direction (Phase)	Code
Positive Horizontal	000
Negative Horizontal	100
Positive Vertical	010
Negative Vertical	110
Positive right diagonal	001
Negative right diagonal	101
Positive left diagonal	011
Negative left diagonal	111

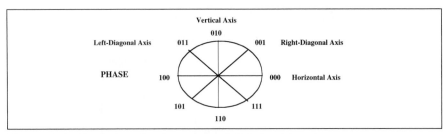

Figure 11.4   Representation for quantized direction (phase).

To illustrate the magnitude and phase computations, consider this 3x3 window:

$$12 \quad 01 \quad 08$$
$$05 \quad 09 \quad 03$$
$$40 \quad 03 \quad 10$$

The four Sobel masks are applied to this image to compute the intensity variations along the four directions as follows :

$$E_H = (-34) \quad E_V = (-38) \quad E_{DR} = (+4) \quad E_{DL} = (-68)$$

Then, $Max\left[|-34|, |-38|, |4|, |-68|\right] = 68$. The maximum value corresponds to the left-diagonal direction. Since the value is negative, the direction of maximum intensity variation is 111. Since a right diagonal is perpendicular to a left diagonal, $E_{\perp} = E_{DR} = 4$.
Therefore, the gradient magnitude is:

$$Magnitude = Max\left[|-34|, |-38|, |4|, |-68|\right] + \frac{1}{8}\lfloor |+4| \rfloor = 68$$

If the threshold is < 68, the pixel in the center of the example window is declared to be part of a left-diagonal edge with direction 111.

Frequently, the gray level of a pixel is assigned to be a value in the range from 0 to 255, with 0 corresponding to black, 255 corresponding to white, and shades of gray distributed over the middle values. Thus, eight bits are required to represent the gray level, which is always a positive integer. Since the filter outputs are calculated by Equations (11.1)–(11.5), we need two more bits to be able to represent the partial summation magnitude. The subtraction operation can create a negative number, so we need one more bit to represent the sign of the final result. Thus, we need 11 bits to represent the filter output. In order to use a standard bus, 12-bit 2's complement vectors are used to represent the four filtering outputs.

After edge detection, each edge pixel is assigned to the foreground value (255), while non-edge pixels are assigned to the background value (0). In the previous example, the center pixel in the window was declared to be part of an edge and its value set to be the foreground value (255). In the processed image, edges will appear as white lines on a black background.

11.3   SYSTEM REQUIREMENTS LEVEL

The first step in the design process is to formally capture the system level requirements from a specification written in a natural language, such as English. This step requires some original thinking by the designer. Specifications written in a natural language are often ambiguous and incomplete. One of the primary purposes of formally capturing the system requirements is to eliminate such ambiguity.

11.3.1   Written Specifications

A written specification uses natural language and possibly a block diagram and/or a timing diagram to describe a system. The written specification describes the input and output signals and the function of the system, as well as basic knowledge about the system, such as whether the system is synchronous or asynchronous, and whether the signal ports are serial or parallel. It is the designer's first introduction to the details of the system. However, as described above, the written specification might be ambiguous at times. As the design progresses, the designer will frequently refine the written specification to eliminate ambiguities. The following is a written specification of the Sobel edge detection system created by the authors:

> The Sobel edge detection system detects edges in images using the Sobel operator. The image pixels will be supplied in raster scan order (i.e., rows are scanned left to right and top to bottom.) The edge detector will serially output data which contains edge information. The system should be synchronized by a system clock. The system should be designed in such a way that images of various size can be processed.

The next section shows how to formally capture the written specifications.

11.3.2   Requirements Repository

This section describes how to capture the system requirements using a schematics capture tool, such as the Synopsis Graphical Environment (SGE) tool. This activity formally documents the system requirements in a format that allows automatic generation of a shell for a VHDL model.

The SGE model is called a *requirements repository* because it stores all the information about the system requirements in an organized manner. The system interface will be formally defined. The designer will decide the necessary ports for data and control signals, which will communicate with the outside world to fulfill the system function. Also, the parameters of the system, such as timing delays and data types, will be declared. This design stage is usually related to the eventual executable specification level in the abstraction hierarchy. After the system interface has been designed, the designer can develop a behavioral architecture for the system to verify the functionality. There are six steps that must be followed to create the requirements repository using the SGE tool.

Step 1.   From the written specification, define the interface signals.

This step requires some originality on the part of the designer. One way to satisfy the written specifications for the Sobel edge detector is to declare the following set of port signals. This is not necessarily a unique choice.

EDGE_START: An input signal of type STD_LOGIC that causes the data process-

Figure 11.5   Sobel block diagram.

ing to start when it assumes the value '1'. This signal also initializes all registers and control flip-flops to the desired starting state. Type STD_LOGIC is the IEEE 1164 standard nine-valued logic type.

CLOCK: An input signal of type STD_LOGIC that serves as the system clock. This signal is needed because the system is specified to be synchronous.

INPUT: An input signal of type PIXEL. The image data is input through this port as a continuous stream of pixel values in raster scan order. The pixels in each row are input from left to right, and the rows are input from top to bottom. Type *pixel* is a constrained integer data type with values between 0 and 255.

THRESHOLD: An input signal of type FILTER_OUT that defines the threshold value for determining edges in the Sobel algorithm. This parameter is provided as an input port instead of a generic because it is anticipated that the threshold value might vary with external system conditions. In this case, another module would compute the current value for the threshold based on some algorithm. The data type *filter_out* is a constrained Integer data type with values in the range from -2048 to +2047.

OUTPUT: A bidirectional signal of type PIXEL that specifies whether the current pixel is an edge pixel. Recall that the Sobel algorithm assigns a value of 255 (white) to an edge pixel and a value of 0 (black) to a pixel that is not an edge pixel. The result is an image of white edges on a black background. The output image pixel values will be present on this pin as a continuous sequence of pixel values, delayed by a fixed amount relative to the sequence of input pixel values. The actual delay value will be determined later in the design process when more detailed design information is available.

DIR: A bidirectional signal of type DIRECTION that specifies the direction of the line that contains an edge pixel. It is computed by the Sobel algorithm. Data type DIRECTION is a STD_LOGIC_VECTOR data type with values in the range from 000 to 111 representing the digitized orientation of the line, as described in an earlier section.

Step 2.   Using the SGE schematic capture tool, create a block diagram that represents the design entity with input and output ports attached.

Figure 11.5 shows such a block diagram for the Sobel edge detection system.

Step 3.   Using the SGE Pin Attributes tool, specify the port names, port data types, port mode, and the default values to be assigned to each port during simulation.

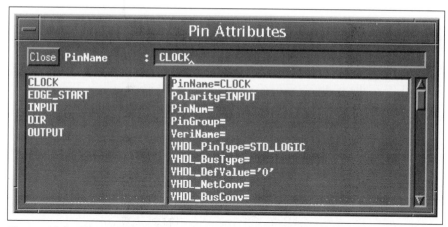

Figure 11.6 Pin attributes for the Sobel edge detection block diagram.

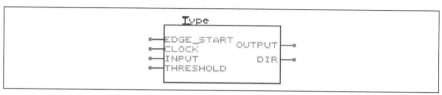

Figure 11.7 Final SGE symbol for Sobel edge detector.

Figure 11.6 shows a Pin Attributes window used to specify the attributes of the signal CLOCK for the Sobel edge detection system. The signal name (PinName) is CLOCK. The signal mode (Polarity) is INPUT. The data type (VHDL_PinType) for signal CLOCK is STD_LOGIC and the initial value (VHDL_DefValue) assigned to signal CLOCK during simulation is '0'. Figure 11.7 shows the appearance of the SGE symbol after defining the pin attribute information.

Step 4. Using the Symbol Attributes tool in SGE, specify the generics for the device.

Figure 11.8 shows the Symbol Attributes window used to specify the generics for the Sobel edge detector system. Note that we elected to use the IEEE_DEFAULTS library. The written specification stated that the size of the image should be variable. Therefore, we elected to specify image size by defining two generic parameters (NUM_ROWS and NUM_COLS) that specify the number of rows and columns, respectively, in the image. The designer can then use the module to process an image of any size by specifying values for the generics at compile time. The data type is Natural.

Step 5. Using the SGE tool, specify the name for the component.

The name for the Sobel edge detection block diagram is specified to be EDGE_DETECTOR. This name will be the entity name in the VHDL code automatically generated by SGE.

Step 6. Using the SGE tool, generate a shell for a VHDL model.

Figure 11.9 shows the code generated by SGE for the Sobel edge detection system.

Figure 11.8 Symbol attributes for the Sobel edge detection block diagram.

```
--VHDL Model Created from SGE Symbol for the
--Sobel edge_detector.sym May 31 14:32:43 1997
library IEEE;
  use IEEE.std_logic_1164.all;
  use IEEE.std_logic_misc.all;
  use IEEE.std_logic_arith.all;
  use IEEE.std_logic_components.all;
entity EDGE_DETECTOR  is
  generic( NUM_ROWS: NATURAL;
  NUM_COLS: NATURAL);
  Port (CLOCK:       In  STD_LOGIC := '0';
        EDGE_START: In  STD_LOGIC := '0';
        INPUT:       In  PIXEL:=0;
        THRESHOLD:  In  FILTER_OUT:=0;
        DIR:         InOut  DIRECTION :=0;
        OUTPUT:      InOut  PIXEL:=0 );
end EDGE_DETECTOR;

architecture BEHAVIORAL of EDGE_DETECTOR is
begin
end BEHAVIORAL;

configuration CFG_EDGE_DETECTOR_BEHAVIORAL of EDGE_DETECTOR is
  for BEHAVIORAL
  end for;
end CFG_EDGE_DETECTOR_BEHAVIORAL;
```

Figure 11.9 VHDL code generated by SGE for Sobel edge detector.

First, the IEEE library and the four packages (std_logic_1164, std_logic_misc, std_logic_arith and std_logic_components) are made visible. Then, the interface for the design entity is declared. The entity name is given as EDGE_DETECTOR as described before. The entity parameters (NUM_ROWS and NUM_COLS) are declared as generics. The port declarations are derived from the pin attributes in the SGE diagram. The architecture and configuration are empty. The designer must insert either a behavioral or structural description of the model in the architecture. The empty configuration could serve as a default binding, or the designer could insert a specific binding. Since this block diagram has no schematic related to it, it is treated as a leaf component and a behavioral model is assumed. Thus, BEHAVIORAL and CFG_EDGE_ DETECTOR_ BEHAVIORAL are assigned as the architecture and configuration names, respectively, by the SGE tool. Of course, the designer is free to edit this code and change the architecture and configuration names if desired. However, such changes eliminate the automatic links from the SGE model to the VHDL code and, therefore, make the code less reusable and more difficult to maintain. If default bindings are used, then no changes need to be made to this part of the code, thus, preserving the automatic links.

11.4 SYSTEM DEFINITION LEVEL

At the system definition level, a designer creates an executable specification for the system. This is the most abstract model in the design hierarchy that can be verified by a VHDL simulator. Any necessary refinements are applied to this model until a complete executable specification is obtained. Several iterations may be necessary to eliminate ambiguities in the written specification. Typically, this is a behavioral description of the system that directly implements the system requirements. Experienced VHDL programmers frequently create this model directly from the written specification. However, the VHDL code generated by the SGE tool can be edited to develop the model. Alternatively, application-specific software packages, such as SPW, can be used (refer to the three paths in Figure 11.1). In this section, we utilize the SGE shell developed in the previous section.

11.4.1 Executable Specification

The first step in the development of the executable specification is to edit the VHDL shell generated by SGE (Figure 11.9). For convenience, we divided the code into three sections. Figure 11.10 shows Section 1. After designing the top-level behavioral model, we discovered that we only needed one of the IEEE libraries that were automatically inserted by the SGE tool. We, therefore, deleted the three unused packages. In addition, we later discovered that a separate library and package to hold our design dependent constructs was convenient. Figure 11.10 also shows the declaration for this dedicated library and the *use* statement for the dedicated package.

Figure 11.11 shows Section 2 of the modified SGE code, which was unchanged. Figure 11.11 also shows Section 3 of the modified SGE code. We will insert the behavioral description in Section 3 later. Although we could have simulated the configuration that was generated by the SGE tool, we decided that this would introduce an unnecessary complication. Instead, we elected to simulate the architecture directly and therefore deleted the unused configuration declaration.

11.4.1.1 Library Structure

Libraries are used to organize and manage the models and analysis results for the models. At the system definition level, two libraries are used. One is the built-in IEEE library that is automati-

```
-- Executable Specification of the Sobel edge detector
------------------------------------------------------------
--   SECTION 1:  LIBRARY AND PACKAGE DECLARATION      --
------------------------------------------------------------
library IEEE;
use IEEE.STD_LOGIC _1164.all;

---   deleted unused packages
  -- use IEEE.std_logic_misc.all;
  -- use IEEE.std_logic_arith.all;
  -- use IEEE.std_logic_components.all;

-- added a user defined library and package
library BEH_INT;
use BEH_INT.IMAGE_PROCESSING.all;
```

Figure 11.10   Section 1 of the edited SGE code.

cally inserted by the SGE tool. In this library, only the STD_LOGIC_1164 package is made visible because we do not need the other three packages at this time. The second library is a user-defined library (BEH_INT). In this design library, we store all models and packages developed for use at the system definition level, as well as the simulation results at this level.

11.4.1.2 Packages

It is tedious and unnecessary to repeat commonly used declarations and function definitions. A package provides a way to collect such items in a single source that can be referenced by *use* statements. See Chapter 3 for details on how to use this statement. Figure 11.12 shows the part of the image processing package for the Sobel edge detection system, which contains the type and constant declarations. The two constants, BACKGROUND and FOREGROUND, define the allowable range of pixel values. In this example, we use values that can be represented by eight bits. Subtype PIXEL is a constrained INTEGER data type that represents pixel values.

Type PIX3 is a linear array of three pixels that is used to represent an arbitrary row in the 9-pixel window shown in Figure 11.13. Type PIXEL3_3 is a two-dimensional array of nine pixels used to represent an arbitrary window. Figure 11.13 shows the relationship between the PIXEL3_3 data type definition and the pixel window described in the section on the Sobel algorithm. Subtype FILTER_OUT is used to represent the outputs of the four Sobel filters. If this model were to be synthesized, this type should be constrained to reduce the number of bits assigned by the synthesis tool. Subtype DIRECTION is a 3-bit STD_LOGIC_VECTOR that is used to represent the eight individual orientations of Figure 11.4.

Figure 11.14 shows the function and procedure declarations in package IMAGE_PROCESSING. Function WEIGHT computes the weighted sums of three parameters as needed in Equations 11.1 through 11.4. The code for WEIGHT is:

```
function WEIGHT ( X1,X2,X3: PIXEL)
  return FILTER_OUT is
begin
  return X1+ 2*X2 + X3;
end WEIGHT;
```

```
-------------------------------------------------------
--    SECTION 2:   INTERFACE DECLARATION           --
-------------------------------------------------------
entity EDGE_DETECTOR is
   -- the generics were declared as symbol attributes
   -- in the SGE model
   generic( NUM_ROWS: NATURAL;   -- the number of rows
            NUM_COLS: NATURAL);  -- the number of columns
   -- the ports were declared as port attributes
   -- in the SGE model
   port (CLOCK: in  STD_LOGIC := '0'; -- the system CLOCK
         EDGE_START: in  STD_LOGIC := '0'; -- the START
                     -- signal for the image  processing
         INPUT: in  PIXEL:=0;        -- input image pixel
         THRESHOLD: in  FILTER_OUT:=0;  -- threshold to
                            -- determine edge pixels
         DIR: inout  DIRECTION :=0;   -- edge direction
         OUTPUT: inout  PIXEL:=0);  -- output image pixel
end EDGE_DETECTOR;
-------------------------------------------------------
--   SECTION 3:   ARCHITECTURE BODY DECLARATION      --
-------------------------------------------------------
architecture BEHAVIORAL of EDGE_DETECTOR is
     -- behavioral model declarations will be added here
begin
   SOBEL: process    -- Sobel process will be added here
end process SOBEL;
end BEHAVIORAL;
---- delete unused configuration part
-- configuration CFG_EDGE_DETECTOR_BEHAVIORAL of
-- EDGE_DETECTOR is
--    for BEHAVIORAL
--    end for;
-- end CFG_EDGE_DETECTOR_BEHAVIORAL;
```

Figure 11.11 Sections 2 and 3 of the modified SGE code.

The four filter functions, HORIZONTAL_FILTER, VERTICAL_FILTER, DIAGONAL_-L_FILTER, and DIAGONAL_R_FILTER implement Equations 11.1 through 11.4, respectively. As an example, the code for the HORIZONAL_FILTER function is:

```
function HORIZONTAL_FILTER ( A: PIXEL3_3)
   return FILTER_OUT is
begin
 return  WEIGHT( A(1,1), A(1,2), A(1,3) )
      - WEIGHT( A(3,1), A(3,2), A(3,3) );
end HORIZONTAL_FILTER;
```

The code for the other three filter functions is similar.

The procedure COMPARE determines which of the four filter outputs (H,V,LD,RD) has the maximum absolute value. Output X is a positive integer equal to the absolute value of the magnitude of the maximum filter output. Output DIR represents the direction of the maximum

```
package IMAGE_PROCESSING is
------ declare two constants --------------------------
constant FOREGROUND:INTEGER:=255;
    -- define the value for the foreground pixel
constant BACKGROUND:INTEGER:=0;
    -- define the value for the background pixel

------ declare types and subtypes ---------------------
subtype PIXEL is INTEGER range BACKGROUND to FOREGROUND;
type PIX3 is array (1 to 3) of PIXEL;
type PIXEL3_3 is array(1 to 3, 1 to 3) of PIXEL;
subtype FILTER_OUT is INTEGER;
subtype DIRECTION is STD_LOGIC_VECTOR(2 downto 0);

---- declare 8 direction constants --------------------
constant EAST:DIRECTION:="000";
constant NORTHEAST:DIRECTION:="001";
constant NORTH:DIRECTION:="010";
constant NORTHWEST:DIRECTION:="011";
constant WEST:DIRECTION:="100";
constant SOUTHWEST:DIRECTION:="101";
constant SOUTH:DIRECTION:="110";
constant SOUTHEAST:DIRECTION:="111";
```

Figure 11.12   Types and constants in package IMAGE_PROCESSING.

a_0	a_1	a_2
a_7	$[i,j]$	a_3
a_6	a_6	a_4

(1,1)	(1,2)	(1,3)
(2,1)	(2,2)	(2,3)
(3,1)	(3,2)	(3,3)

Sobel window   PIXEL3_3 data type

Figure 11.13   Relationship between Sobel window and PIXEL3 data type.

intensity variation for the current Sobel window. It is computed by examining the outputs of the four filters, selecting the one with the maximum magnitude, and looking up the appropriate three-bit code as defined in Figure 11.4. Output Y is a positive integer that represents the magnitude of the filter whose orientation is perpendicular to the direction of the maximum magnitude filter. The code for procedure COMPARE can be found on the CD ROM that accompanies this text.

The function MAGNITUDE computes the Sobel magnitude gradient of Equation 11.5. Input A is the maximum filter output and input B is the filter output perpendicular to the direction of the maximum filter. Both inputs are positive integers. The code for MAGNITUDE is:

```
function MAGNITUDE (A,B: FILTER_OUT)
  return FILTER_OUT is
begin
  return (A + (B/8));
end MAGNITUDE;
```

```
------- declaration of functions and procedures -------
function WEIGHT (X1,X2,X3: PIXEL)
    return FILTER_OUT;
function SHIFT_LEFT (A: PIX3; B: PIXEL)
    return PIX3;
function HORIZONTAL_FILTER (A: PIXEL3_3)
    return FILTER_OUT;
function VERTICAL_FILTER (A: PIXEL3_3)
    return FILTER_OUT;
function DIAGONAL_L_FILTER (A: PIXEL3_3)
    return FILTER_OUT;
function DIAGONAL_R_FILTER (A: PIXEL3_3)
    return FILTER_OUT;
procedure COMPARE (H,V,LD,RD :in FILTER_OUT;
                   X,Y: out FILTER_OUT;
                   DIR:out DIRECTION);
function MAGNITUDE (A,B:  FILTER_OUT)
    return FILTER_OUT;
function INT_TO_STDLOGIC8 (A: INTEGER)
    return STD_LOGIC_VECTOR;
function STDLOGIC_TO_INT (S:STD_LOGIC_VECTOR)
    return INTEGER;
function INT_TO_STDLOGIC12 (A: INTEGER)
    return STD_LOGIC_VECTOR;
function STDLOGIC_TO_BIT (A: STD_LOGIC_VECTOR)
     return BIT_VECTOR;
end IMAGE_PROCESSING;
```

Figure 11.14   IMAGE_PROCESSING package declarations.

Function SHIFT_LEFT shifts a 3-pixel linear array (A) one position to the left. The right-most position is filled with the value on input B. The remainder of the functions are type conversion functions of various kinds. The complete code for all of the functions can be found on the CD ROM that accompanies this text.

11.4.1.3   *Behavioral Model for the Sobel Edge Detection System*

This step requires considerable originality and ingenuity. There are many ways to implement the Sobel edge detection algorithm. We illustrate one implementation, but make no claim that this is the optimal implementation.

Recall that the pixels are arriving in raster scan order. Because we need to examine arbitrary 3x3 arrays of pixel values, we will elect to copy three full rows of pixels into an internal memory. The first few lines in the architecture for the model are:

```
architecture BEHAVIOR of EDGE_DETECTOR is
   type MEMORY_ARRAY is array(1 to 3, 1 to NUM_COLS)
       of PIXEL;
begin
```

Type MEMORY_ARRAY represents the internal pixel memory. It is an array with three rows. The number of columns is the number of pixels in each row of the image (NUM_COLS). Our Sobel edge detector will store three full rows of image pixels in this internal memory. The memory is used in a round robin fashion. That is, the top three rows of pixels are initially stored in the array. The fourth row of pixels to arrive will be stored in the first row of the array, the fifth row in the second row of the array, and so on. We will keep track of the location of the pixel image rows in the internal memory by a system of index pointers. For example, after three full rows of image pixels have arrived at the inputs to the Sobel edge detection system, the internal memory contents will be:

Row 1 of internal memory array. First row of image pixels.
Row 2 of internal memory array. Second row of image pixels.
Row 3 of internal memory array. Third row of image pixels.

After the fourth row of image pixels arrive at the inputs, the internal memory array will contain this set of pixels:

Row 1 of internal memory array. Fourth row of image pixels.
Row 2 of internal memory array. Second row of image pixels.
Row 3 of internal memory array. Third row of image pixels.

Note that the fourth row of image pixels replaces the first row of image pixels. Similarly, the fifth row of image pixels replaces the second row of image pixels in the internal memory array.

We will arbitrarily decide to use one process (SOBEL) to represent the Sobel edge detection system. The following variables will be used in the process. The purpose of each variable is indicated by a comment in the declaration section.

```
SOBEL: process
  variable BUSY1,BUSY2: std_logic:='0';
    -- BUSY1='1' while input pixels are arriving.
    -- BUSY2='1' while edge pixels are being sent out
  variable A: PIXEL3_3:=((0,0,0),(0,0,0),(0,0,0));
    -- The current 3x3 window of pixels being processed.
  variable TEMP: PIXEL;
    -- Temporary storage for the output edge pixel
  variable E_H:  FILTER_OUT;
    -- Temporary storage for output of horizontal filter
  variable E_V:  FILTER_OUT;
    -- Temporary storage for output of  vertical filter
  variable E_DL:  FILTER_OUT;
    -- Temporary storage for output of left_diagonal filter
  variable E_DR:  FILTER_OUT;
    -- Temp storage for output of right_diagonal filter
  variable M:  FILTER_OUT;
    -- Temp storage for max absolute value of the filters
  variable P:  FILTER_OUT;
    -- Temp storage for filter output perpendicular to
    -- the direction of maximum filter value.
  variable MAG:  FILTER_OUT;  -- output edge pixel value
  variable PHASE: DIRECTION;  -- direction of the edge
  variable X1:NATURAL:=1;
    -- Row index for internal memory WRITE operation
  variable Y1:NATURAL:=1;
    -- Column index for internal memory WRITE operation
  variable X2:NATURAL:=1;
    -- Row index for internal memory READ operation
```

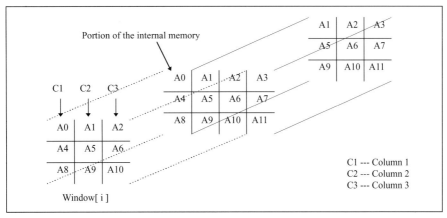

Figure 11.15 Updating for the current window.

```
variable Y2:NATURAL:=1;
  -- Column index for internal memory READ operation
variable COUNT_R1,COUNT_R2: INTEGER:=0;
  --COUNT_R1 is the row number for input pixels.
  --COUNT_R2 is column number for input pixels
variable MEMORY: MEMORY_ARRAY;
  -- Internal memory variable.

----- begin the SOBEL process -----
begin
```

Variable A is the current pixel window being processed. Figure 11.15 illustrates how this window is updated during each clock period. Note that each row of the window is shifted left and a new pixel is appended at the right end of the row.

During each clock cycle, one new input pixel is read from port INPUT and stored in the internal memory array. The current window is updated and a new pixel is analyzed to determine whether it is part of an edge. If it is part of an edge, the output pixel magnitude at port OUTPUT is given the value 255 and the DIR port is set to indicate the direction of the edge. If the current pixel is not part of an edge, port OUTPUT is set to value 0, and DIR is undefined. In steady state, the edge output pixels occur in a steady stream, one each clock cycle. Since we must have a window of nine pixels in order to look for an edge, we will not start the window processing until we start reading in the third row of the input image.

Two internal variables, BUSY1 and BUSY2, are initialized to '0' by declaration. A pulse on port EDGE_START provides a start signal for the edge detector. This event causes internal variable BUSY1 to be set to value '1'. BUSY1 remains set until the entire input image is processed. Internal signal BUSY2 is set to '1' when it is time to start shifting columns of data into the active window (i.e., the first pixel of row three is arriving at port INPUT). Counter COUNT_R1 keeps track of the rows, with the first row being row 0. The code for controlling BUSY1 and BUSY2 is:

```
wait until rising_edge(CLOCK);
  --  Set the internal busy variable ----------
if EDGE_START ='1' then
  BUSY1:='1';
end if;

-- Store the input image pixel in the internal memory array
if BUSY1 ='1' then
  MEMORY(X1,Y1):=INPUT;

    -- set internal busy variable BUSY2 and start to
    -- generate the address for the READ operation
    if COUNT_R1=2 and Y1=1 then
      BUSY2:='1';
    end if;
```

Note that variables X1 and Y1 keep track of the internal memory address into which the current pixel value is to be stored. Variable X1 keeps track of the row, and variable Y1 keeps track of the column. Both X1 and Y1 are initialized to 1 and, therefore, have a range from 1 to NUM_COLS, so we want to start loading data into the active window when the first pixel (Y1=1) of the third row (COUNT_R1=2) arrives.

We now calculate the memory address into which the next pixel should be written:

```
    -- calculate the internal memory location for the next
    -- WRITE operation
    Y1:=Y1+1; -- update the column in the same row

    -- start a new row if we are at the end of the row
    if Y1=NUM_COLS+1 then
      Y1:=1;
      X1:=X1+1;
      COUNT_R1:=COUNT_R1+1;
    end if;

    -- reuse first row of internal memory -----
    if X1=4 then
      X1:=1;
    end if;

    -- Reset BUSY1 if all rows have been read in --
    if COUNT_R1=NUM_ROWS then
      BUSY1:='0';
    end if;
end if; -- End for if BUSY=1 clause
```

To update the window variable, we shift each row left and insert the new value on the right, as shown in Figure 11.15. Since we are reusing the rows of the internal memory in a round robin fashion, we must take this into account in the window updating algorithm. The *mod3* operator performs this task. We also look for the end of the last row of the input image, so we can reset BUSY2 to stop window processing.

```
   -- Update the 3 x 3 buffer window A --
   if BUSY2='1' then

     ---- move columns 2 and 3 left one position -----
     for J in 1 to 2 loop
       for I in 1 to 3 loop
         A(I, J):= A(I, J+1);
       end loop;
     end loop;

     -- insert new values for column 3 -----
     for I in 1 to 3 loop
       X2:= (COUNT_R2 +I-1)mod 3+1;
       A(I, 3):=memory(X2,Y2);
     end loop;

     -- calculate the internal memory column from which the
     -- next memory READ operation will occur ----
     Y2:=Y2+1;
     if Y2=NUM_COLS+1 then
       COUNT_R2:=COUNT_R2+1;
       Y2:=1;
     end if;

     -- Reset the internal variable BUSY2 if we reach the
     -- last row of the input image.
     if COUNT_R2 = NUM_ROWS-2 then
       BUSY2 := '0';
     end if;
```

We now apply the four filtering functions and determine whether the current pixel is an edge pixel. The outputs are updated accordingly.

```
     -- apply the filtering function to the current window -
     E_H  := HORIZONTAL_FILTER(A);
     E_V  := VERTICAL_FILTER(A);
     E_DL := DIAGONAL_L_FILTER(A);
     E_DR := DIAGONAL_R_FILTER(A);

     -- determine the output edge pixels and directions --
     COMPARE(E_H,E_V,E_DL,E_DR,M,P,PHASE);
     MAG:= MAGNITUDE(M,P);
     if MAG >= THRESHOLD then
       TEMP := FOREGROUND;
     else
       TEMP := BACKGROUND;
     end if;

     -- update output signals ----
     OUTPUT <= TEMP;
     DIR <= PHASE;
   end if; -- This is the end of the BUSY1 section.

 end process SOBEL;   -- end the process
end BEHAVIORAL; -- end the architecture
```

The full code can be found on the CD-ROM.

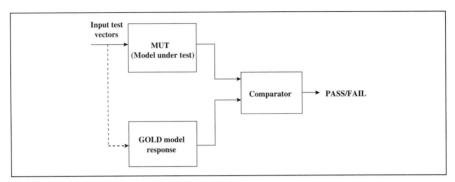

Figure 11.16   Basic test bench block diagram.

11.4.2   Test Bench Development for Executable Specifications

VHDL models are tested using an enclosing model called a test bench. A VHDL test bench can be defined as an executable VHDL model that instantiates a model under test (MUT) and provides the capability to drive the MUT with a set of test vectors, comparing the response with the expected response. By combining an efficient test bench methodology with a top-down design methodology, designers can boost productivity and gain increased confidence in the accuracy of their designs. The test bench is at the top level in the model hierarchy. It is the entity that is simulated when testing a model. Test benches provide the user with the capability to test the MUT thoroughly through simulation.

A VHDL test bench typically consists of an architecture body containing an instance of the component to be tested and processes that generate sequences of values on signals connected to the component instance. The architecture body may also contain processes that verify that the component instance produces the expected values on the output signals. Alternatively, we can use the monitoring facilities of the simulator to observe the outputs visually.

Figure 11.16 shows the block diagram of a basic test bench. A set of input test vectors are applied to the model under test (MUT), and its output response is compared with the output response of the GOLD model in a comparator, which generates a PASS/FAIL indication. The GOLD model can be an independently validated model, or it can be a file that contains the expected output sequence. If the GOLD model is a file, that file may be prepared by the designer or obtained from an external source.

At a minimum, the VHDL test bench must serve the following purposes:

1. Simulate the MUT under a variety of test conditions.
2. Automatically verify that the MUT meets the specifications, and log all the errors if it does not meet these specifications. This can be achieved by comparing the output response of the MUT with that of a GOLD model and reporting success or failure for each test. Online verification checks the performance of the MUT accurately and also consumes less time than manual verification, which is very tedious, cumbersome, and inefficient. Automation also enables a reduction in time for future maintenance effort, because it enables fast and reliable reverification of the model when changes are introduced.

3. The test bench should be capable of performing regression tests. These tests are used to ensure that a design change in the system does not adversely affect the performance of the system.

At higher levels of abstraction, the primary purpose of simulation is to verify the functionality of the system specification. At lower levels of abstraction, simulation is used to verify that a set of tests will detect common hardware faults and to compare the outputs of lower-level models with the outputs of higher-level models. Test sets developed at higher levels of abstraction should be reused to test the models at lower levels. However, additional test sets are usually needed to detect specific hardware faults in the lower-level models. Testing is accomplished by exercising the design using a number of stressing test cases. Because of the complexity of the tests, it is good practice to identify the goals of the verification before the actual simulation activity and devise a test plan to ensure that the test bench meets the desired goals.

A good set of test vectors must be developed. Although exhaustive testing can be adopted to test models of lesser complexity, this might be an impractical method for testing large systems. A common strategy for testing the model at the behavioral level of abstraction is to select tests that exercise all control flow paths in the model. Paths involving nested control constructs, *wait* statements and nested procedure calls should be specifically targeted by the test vectors.

Tests should be developed to verify the functionality of the system completely. This includes testing components of a design as well as testing the design as a whole. It is important to test each component as it is created, and not assume correct operation. As components are connected to form each new level of hierarchy, interdependencies between components may cause unexpected behavior. New simulations should be performed each time a new component is added to the system. Finally, the entire design is integrated and tested as a whole. Most errors should have been found when testing individual components or smaller configurations of components.

11.4.2.1 *Methodology for Developing Test Benches for Executable Specifications*

As stated in the previous section, the main purpose of testing at the behavioral level is to validate the functionality of the executable specification. These known correct responses can then be used to test the MUT at lower levels of abstraction.

The test bench module (TBM) typically performs the following functions:

1. At the start of the simulation, the TBM reads the input image data from an external text file into an internal frame buffer, and then sends it to the MUT in an appropriate format. For example, the input test vectors in the external file could be of type Integer, but the ports of the MUT may be of STD_LOGIC_VECTOR data type. Therefore, TBM must convert the input data to STD_LOGIC_VECTOR data type before sending it to the MUT.

2. Once the data has been processed by the MUT, its output responses are stored in internal frame buffers in the TBM. An important task for the TBM is to decide the time when the output response should be written into the frame buffers. Internal MUT delays and test bench artifacts must be considered.

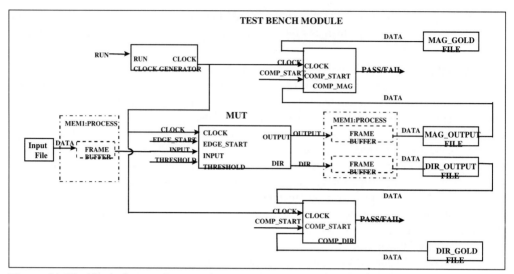

Figure 11.17   Block diagram for a test bench module for the Sobel edge detection system.

3. Finally, the output responses from the frame buffers are written into external text files, which are then read by the comparators to compare the output response of the MUT with the expected response to generate a PASS/FAIL indication.

11.4.2.2   Example Test Bench for the Sobel Edge Detection Executable Specification

To illustrate the methodology, Figure 11.17 shows the block diagram for a TBM that can be used to verify the behavioral implementation of the Sobel edge detection system. It includes the MUT, a clock generator to provide the clock for the entire system, and two comparators to verify the correctness of the outputs.

The stimulus values or test vectors used to test the MUT are provided in a file, specified as the Input File in Figure 11.17. The test bench module uses process MEM1 to read the input image data from the image file and to store it in an internal frame buffer.

The TBM provides four input signals to the MUT. For clarity, the input signals (CLOCK, EDGE_START, INPUT, and THRESHOLD) are given the same names as the MUT input ports to which they are connected. The output signals (OUTPUT and DIR) from the MUT also have the same names as the MUT output ports to which they are connected. Details of these ports are described in an earlier section in this chapter.

The output edge pixel magnitudes and directions are stored in two frame buffers by process MEM1 and, subsequently, moved to two output files (MAG_OUTPUT and DIR_OUT-PUT). Two comparator components (COMP_MAG and COMP_DIR) compare the edge magnitudes and directions produced by the MUT with those stored in the GOLD files MAG_GOLD and DIR_GOLD, respectively. For each test, the GOLD files must be prepared in advance from the written specifications. Separate PASS/FAIL responses are provided for the edge pixel magnitudes and directions. TBM signals RUN and COMP_START are used to enable the clock generator and the two comparators, respectively.

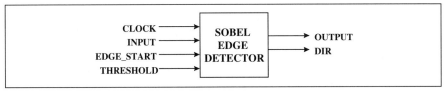

Figure 11.18   Block diagram of the Sobel edge detector.

```
library IEEE;
use ieee.STD_LOGIC_1164.all;
use STD.TEXTIO.all;
library BEH_INT;
use BEH_INT.IMAGE_PROCESSING.all;
---------interface declaration---------
entity TEST is

generic(IN_FILE:  STRING(1 to 11);   -- inut image file
  OUT_FILE:  STRING(1 to 11);--file for output magnitudes
  DIR_FILE:  STRING(1 to 11);--file for output directions
  NUM_ROWS:  NATURAL;-- number of rows in the input image
  NUM_COLS:  NATURAL;-- number of columns in the image
  WAIT_CYCLES:  NATURAL);   -- time required before
               -- Outputs  are written into frame buffers.
end TEST;
```

Figure 11.19   Entity declaration for Sobel test bench module.

Clock generator component.   The clock generator component is described in Chapter 4. Refer to Figure 4.38.

Sobel edge detector component.   Figure 11.18 shows the block diagram of the Sobel edge detector component. It has four input signals: CLOCK, INPUT, EDGE_START, and THRESHOLD. The output signals are OUTPUT and DIR, which represent the magnitude and direction outputs, respectively. This module is the MUT.

Test bench module for the Sobel edge detection system.   The source code for the test bench module has been divided into several sections for easier readability. The code section in Figure 11.19 shows the library declarations and the entity declaration. Since the test bench is self-contained, the entity declaration has no port list. Generic constants are used within the entity to define the input and output text files. These are explained by comments in the Figure 11.19 source code. Because these are generic constants, all file names must have exactly 11 characters, including the file extension. For example, "tesv_i4.dat" is a valid file name. We also use two generics, NUM_ROWS and NUM_COLS, to define the size of the input image.

We use the generic constant WAIT_CYCLES to specify the delay, expressed in clock cycles, between the time at which a pixel arrives at port INPUT and the time at which the MUT generates the corresponding edge information at ports OUTPUT and DIR. This value is defined as a generic instead of a constant since the number of clock cycles varies with the abstraction level. For example, when we simulate the model at the behavioral level, WAIT_CYCLES must

```
--architecture description----
architecture BENCH of TEST is
  type FRAME_IMAGE is array(1 to NUM_ROWS,1 to NUM_COLS)
                           of INTEGER;
  type FRAME_DIRECTION is array(1 to NUM_ROWS,
              1 to NUM_COLS) of BIT_VECTOR(2 downto 0);

  ---component instantiations----
  component CLOCK_GENERATOR1
    port(RUN : in STD_LOGIC;
         CLOCK: out STD_LOGIC);
  end component;

  component EDGE_DETECTOR1
    port (CLOCK: in STD_LOGIC;
          EDGE_START: in STD_LOGIC;
          INPUT: in PIXEL;
          THRESHOLD: in FILTER_OUT;
          OUTPUT: inout PIXEL;
          DIR: inout DIRECTION);
  end component;

  component COMP_MAG1
    port(CLOCK: in STD_LOGIC;
         COMP_START: in STD_LOGIC:='0');
  end component;

  component COMP_DIR1
    port(CLOCK: in STD_LOGIC;
         COMP_START: in STD_LOGIC:='0');
  end component;
```

Figure 11.20   Architecture and component declarations for the Sobel edge detection test bench module.

be 5 clock cycles; however, when simulating the model at the RTL level, WAIT_CYCLES needs to be 13 clock cycles. This is an important generic. Internal delays often change with abstraction level. By providing a similar generic in all components, these components can be used at various levels in the abstraction hierarchy. These generics are particularly useful in multilevel simulations described later in this chapter.

Figure 11.20 shows the declaration section of the architecture body for the test bench module. Two array types, FRAME_IMAGE and FRAME_DIRECTION, have been defined. FRAME_IMAGE represents a frame buffer for Integer data types. FRAME_DIRECTION represents a frame buffer for BIT_VECTOR data types. Figure 11.20 also includes the component declarations for the clock generator, edge detector, and comparators.

The clock generator, edge detector, and comparator components are instantiated, and their port signals are mapped with port map statements as shown in Figure 11.21. The local signals in the architecture body are used to interconnect the components as shown in Figure 11.17. An additional signal, START, has been included to trigger the test bench module. The architecture

```
-----Signal declarations----
  signal RUN:            STD_LOGIC:='0';
  signal CLOCK:          STD_LOGIC:='0';
  signal START:          STD_LOGIC:='0';
  signal INPUT, OUTPUT: PIXEL:=0;
  signal THRESHOLD:      FILTER_OUT:=0;
  signal DIR:            DIRECTION;
  signal EDGE_START:     STD_LOGIC:='0';
  signal COMP_START :    STD_LOGIC:='0';

-- Begin the test bench module.
begin
  START<='0' after 0 ns,'1' after 5 ns,'0' after 105 ns;
  THRESHOLD <= 110 after 0 ns;
  RUN<=transport '1' after 0 ns,'0' after 100000 ns;

---- Component Instantiations ------
P1 : CLOCK_GENERATOR1
  port map(RUN,CLOCK);
P2 : COMP_MAG1
  port map(CLOCK,COMP_START);
P3 : COMP_DIR1
  port map(CLOCK,COMP_START);
P4 : EDGE_DETECTOR1
port map(CLOCK,EDGE_START,INPUT,THRESHOLD,OUTPUT,DIR);

-----Processes------
MEMORY1 : process
  begin
  -- Details in Figures 11.22 through 11.26.
  end process;
end BENCH;
```

Figure 11.21 Signal declarations, component instantiations, and process declarations for Sobel edge detection test bench module.

body contains process MEMORY1 (MEM1 in Figure 11.17), which controls data transfers between internal frame buffers and data files.

Figure 11.22 shows the first section for the source code of the MEMORY1 process. At the rising edge of the clock, if signal START is '1', we initialize the internal variables. These will be explained as they are used. We, then, use a *case* statement for the internal variable BUSY, which specifies the section of the test bench code to be executed during each clock cycle. When START='1', BUSY is initialized to 1. If the condition (BUSY=1) is satisfied, the data from the input image file, IMAGEIN, is read into a frame buffer named INPUT_IMAGE. All elements in the frame buffers OUTPUT_MAG_IMAGE and OUTPUT_DIR_IMAGE are initialized to 0 and "000", respectively, in preparation for storing the magnitude and direction outputs from the MUT. As soon as the input image file is completely read into the frame buffer, BUSY is set to 2. An assertion statement generates a report confirming that the image file has been read into the frame buffer successfully. All these frame buffer operations are completed during the first clock

```
MEMORY1: process
  variable VLINE1,VLINE2:LINE; -- for file I/O
  variable BUSY:INTEGER range 0 to 3;-- internal variable
    -- to trigger the different parts of the test bench

-- Internal frame buffers to store image data
  variable INPUT_IMAGE: FRAME_IMAGE; -- input image  buffer
  variable OUTPUT_MAG_IMAGE:FRAME_IMAGE;-- output magnitude
  variable OUTPUT_DIR_IMAGE:FRAME_DIRECTION;-- directions
  variable Z:INTEGER; -- used for reading the input files
  variable I,J:NATURAL:=1;--frame buffer indexing variables
  variable X,Y:NATURAL:=2; --frame buffer indexing vars
  variable COUNT:INTEGER:=0;-- used to specify when the
  file IMAGEIN: TEXT is in IN_FILE;  -- Input image file
  file IMAGEOUT:TEXT is out OUT_FILE; -- Output magnitudes
  file DIROUT:  TEXT is out DIR_FILE;  -- Output Directions

begin
  wait until rising_edge(CLOCK);
  if START = '1' then    -- Initialize variables
    BUSY :=1; COUNT:=0; I:=1;J:=1; X:=2; Y:=2;
  end if;
  case BUSY is
    when 1 =>
      --load the image to the internal frame buffer---
      for i in 1 to NUM_ROWS loop
        readline(IMAGEIN,VLINE1);
        for j in 1 to NUM_COLS loop
          read(VLINE1,Z);  INPUT_IMAGE(i,j):= Z;
          OUTPUT_MAG_IMAGE(i,j):=0;
          OUTPUT_DIR_IMAGE(i,j):="000";
        end loop;
      end loop;
      BUSY:=2; -- Advance to step 2
      assert (false) report "array read in";
```

Figure 11.22   Sobel test bench module code to read the input image and intitialize output frame buffers.

cycle following a START pulse. This clock cycle, as well as the signal START, do not exist in normal operation of the MUT. They are just artifacts of the test bench module to prepare the system for simulation.

Figure 11.23 continues the test bench module code. When *BUSY=2*, one pixel of the image data from the frame buffer INPUT_IMAGE is transferred to the edge detector component via the signal INPUT during each clock period. Variables I and J, initialized to 1, select the pixel to be transferred from the buffer. The signal EDGE_START is set to '1' for one clock cycle at the beginning of the data transfer to the MUT to initialize the MUT. We use an internal variable COUNT, initialized to 0, to count the number of pixels that have been processed.

Figure 11.24 shows the initial position of the filter window in the upper-left corner of the input image. The first edge filter operation should occur with the window in this position. Since

```
      when 2 =>
            COUNT:=COUNT+1;   -- Update counter

            -- Send next pixel to MUT
            INPUT <= (INPUT_IMAGE(i,j));

            -- Update frame buffer index variables
            if (I=1) and (J=1) then
              EDGE_START<='1';
            else
              EDGE_START<='0';
            end if;
            if (I=NUM_ROWS) and (J=NUM_COLS) then
              I:=1; J:=1;
            elsif J=NUM_COLS then
              J:=1; I:=I+1;
            else
              J:=J+1;
            end if;
```

Figure 11.23   Sobel test bench module code to apply inputs to the MUT.

Row 1	x	x	x	x	x
Row 2	x	x	x	x	x
Row 3	x	x	x		

Figure 11.24   Position of window for first valid application of filters.

the input pixels arrive in raster scan order (one row at a time from top to bottom with the pixels in each row arriving from left to right), the initial filter operation can only occur after at least three pixels in row three have arrived. The total number of pixels that must be received prior to the first filter operation is, therefore, (2*NUM_COLS+3).

Figure 11.25 continues the test bench module code for the Sobel edge detector. The variable WAIT_CYCLES, described in the previous paragraph, defines the time at which the first output edge pixel becomes available at the output of the MUT. As noted in the previous paragraph, WAIT_CYCLES needs to be at least 3 to position the filter window in the upper left-hand corner of the image. In addition, internal delays in the Sobel module require us to wait for additional clock cycles for the edge data to be present at the outputs. The delay varies from level to level. At the systems definition level, we must wait for two additional clock cycles, so WAIT_CYCLES will be 5. When the condition specified for COUNT is satisfied, the magnitude and direction outputs of the edge detector are written into output frame buffers **OUTPUT_MAG_IMAGE** and **OUTPUT_DIR_IMAGE**, respectively. Variables X and Y keep track of the row and column numbers of the pixel being processed. Both X and Y are initialized to 2 because the pixel in the center of the initial window is in the second row and second column of the input image (see Figure 11.24). Also, no output occurs if Y=1 (left boundary pixel) or Y=NUM_COLS (right boundary pixel) because the window is not fully within the

```
    -- This code is a continuation of the "when 2=>" from
    -- Figure 11.23--

    -- write the magnitude and direction outputs into internal
    -- frame buffers --
        if COUNT>=(2*NUM_COLS+WAIT_CYCLES) then
          if not(Y=1) and not(Y=NUM_COLS) then
            OUTPUT_MAG_IMAGE(X,Y):= OUTPUT;
            OUTPUT_DIR_IMAGE(X,Y):= STDLOGIC_TO_BIT(DIR);
          end if;
          Y:=Y+1;
          if Y=NUM_COLS+1 then            --start new row--
            Y:=1;
            X:=X+1;
          end if;
          if (X=NUM_ROWS-1) and Y=NUM_COLS then
          --entire image has been processed--
          BUSY:=3;
          end if;
        end if;
```

Figure 11.25 Code to transfer valid edge information into the output buffers for the Sobel edge detection test bench module.

image boundaries. X and Y are updated to prepare for receiving the next pixel. BUSY is set to 3 when the last valid pixel is processed. We do not process pixels in the last row because the window is not fully within the image boundaries.

Figure 11.26 shows the code that writes the magnitude and direction outputs to text files **IMAGEOUT** and **DIROUT**, respectively. Assertion statements are used to indicate that the outputs have been written successfully to the text files. After the outputs have been written, **COMP_START** is set high to trigger the two comparators. Finally, BUSY is reset to 0 to stop processing data. Data processing is suspended until a new START signal is generated.

Comparator component. Figure 11.27 shows the block diagram of the comparator component used to compare the outputs of the MUT with the expected response and to generate a PASS/FAIL indication. The model has two inputs, CLOCK and START. The START signal is used to trigger the comparator.

Figure 11.28 shows the portion of the source code for the comparator component used to compare the magnitude outputs. When the internal variable BUSY is set high, the data from the text files for the GOLD model (**GOLD_MAGNITUDE**) and the MUT (**TEST_MAGNITUDE**) are read and compared one pixel at a time. If there is an error, we set an internal variable FLAG1 to '1' as shown in Figure 11.29. If FLAG1 is set high, we generate a report indicating that the outputs of the MUT don't match the outputs of the GOLD model. If FLAG1='0', the report states that magnitudes do match.

The comparator for the direction outputs was developed along similar lines.

```
   --Write the outputs to external text files--
     when 3 =>
       RD_START<='0';
       for i in 1 to NUM_ROWS loop
         for j in 1 to NUM_COLS loop
           write(VLINE1,OUTPUT_MAG_IMAGE(i,j));
           write(VLINE1,' ');
           write(VLINE2,OUTPUT_DIR_IMAGE(i,j));
           write(VLINE2,' ');
         end loop;
         writeline(IMAGEOUT,VLINE1);
         writeline(DIROUT,VLINE2);
       end loop;

       -- Processing is complete----
       assert (false) report
        "magnitude and direction outputs written to file";
       COMP_START<='1'; -- Start the compare operation.
       BUSY:=0;         -- Reset the test bench module.
     when others => null;
     end case;
   end process MEMORY1;
end BENCH;
```

Figure 11.26 Code to copy the edge information from the frame buffers to the text files.

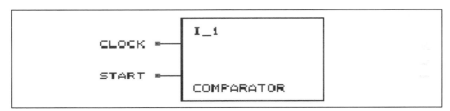

Figure 11.27 Block diagram for the comparator component.

11.4.2.3 Configuration Declarations for an Executable Specification

VHDL allows a great deal of flexibility in configuring a specific model for simulation. Configurations allow selection of different architecture bodies for components and also allow generic constants to be specified in them, rather than in the component instance. Each experiment has a different configuration file. Thus, the experiments can easily be repeated at a later time by simply simulating the configuration. Another use of configurations is to define port maps, particularly to specify type conversion functions. The combination of the selection of architecture bodies and the choice of type conversion functions is a key part in constructing a mixed-level abstraction model, using components from a design database in which components are modeled at multiple levels of abstraction. Effective use of configuration declarations can provide significant assistance in the configuration management of models. However, care must be taken to centralize the late binding decisions in the configuration declaration, rather than distribute this

```
     wait until rising_edge(CLOCK);
   --Set the internal BUSY signal high to start comparison--
    if START ='1' and END_COMP_MAG ='0' then
      BUSY:='1'; FLAG1:='0';
    end if;

    if BUSY='1' then
  ---Read the files and compare the contents----
      for i in 1 to NUM_ROWS loop
        readline(TEST_MAGNITUDE,VLINE1);
        readline(GOLD_MAGNITUDE,VLINE2);
        for j in 1 to NUM_COLS loop
          read(VLINE1,A);
          read(VLINE2,B);
  -- If there is a mismatch in this row, set FLAG1.
          if A /= B then
             FLAG1:='1';
          end if;
        end loop;
      end loop;
    end if;
```

Figure 11.28   Partial code for comparator element.

```
  if FLAG1='1' then
      -- assert message after comparing all the data
        assert (false) report
           "MAGNITUDE VALUES DO NOT MATCH -- FAIL";
      else
        assert (false) report
           "MAGNITUDE VALUES MATCH -- PASS";
      end if;
      END_COMP_MAG<='1';
    end if;
    BUSY:='0';
  end process;
end COMPARE;
```

Figure 11.29   Code for comparator component.

information throughout the different compilation units. For example, if the same test bench can be used with different external data files, the names of the data files should be defined in the configuration declaration, rather than in the architecture body of the design entity that reads the file. A configuration file example is shown in Figures 11.30 and 11.31.

Figure 11.30 shows the declaration for the TEST_BENCH. We declare an entity TB_CONFIG with an architecture containing one component declaration, TEST1, and one instance (con) of TEST1.

```
--declaration of an empty top-level component--
entity TB_CONFIG is
end TB_CONFIG;

architecture TEST_BENCH of TB_CONFIG is
  component TEST1
  end component;

begin
  con: TEST1;
end TEST_BENCH;
```

Figure 11.30   Test bench declaration.

Figure 11.31 shows a configuration declaration that defines the bindings in the following chart:

Component	Entity (Architecture)
con:TEST1	BEH_INT.TEST(BENCH)
P1	BEH_INT.CLOCK_GENERATOR(BEHAVIOR)
P2	BEH_INT.COMP_MAG(COMPARE)
P3	BEH_INT.COMP_DIR(COMPARE)
P4	BEH_INT.EDGE_DETECTOR(BEHAVIOR)

When we need to test the Edge detector model at lower levels of abstraction, we only need to change the name of the configuration and library in the configuration, without having to change anything else in the model. Generic constants have also been assigned values in the configuration. For the top-level entity TEST1, the various input and output file names have been specified. NUM_ROWS and NUM_COLS have been specified as 10 and WAIT_CYCLES has been specified as 5 because the model being simulated is at the behavioral level. For the Clock Generator, HI_TIME and LO_TIME have been specified as 75 ns and 25 ns, respectively.

11.4.2.4   Implementing a Test Plan for the Sobel Edge Detection System

The functionality of the edge detector model was verified effectively by developing the test plan shown in Table 11.2. The test plan lists the various test goals, stimuli source, gold data source, and acceptable outcome.

To illustrate the use of the test plan, we show how to obtain test results for test numbers 1 and 2. These two test goals ensure that the MUT detects horizontal and vertical edges and that appropriate direction values are assigned to each edge. The input test image (file tesv_i1.dat) shown in Figure 11.32(a) is a synthetically generated image that contains horizontal and vertical edges. The expected response for the input image, shown in Figure 11.32(b), is placed in the tesv_g1.dat file. The edge detected output image produced by simulation is shown in Figure 11.32(c). Both the horizontal and vertical edges for the image have been detected correctly.

```
---CONFIGURATION DECLARATION---------
library IEEE;
use ieee.STD_LOGIC_1164.all;

library BEH_INT;
use BEH_INT.IMAGE_PROCESSING.all;
use BEH_INT.all;
use STD.TEXTIO.all;

configuration CONFIG_B_INT of TB_CONFIG is
  for TEST_BENCH
    for con:TEST1 use entity BEH_INT.TEST(BENCH)
      generic map("tesv_i2.dat", "test_o1.dat",
                  "test_d1.dat", 10,10,5);

      for BENCH
        for P1:CLOCK_GENERATOR1 use entity
          BEH_INT.CLOCK_GENERATOR(BEHAVIOR)
          generic map(HI_TIME=>75 ns,LO_TIME=>25 ns);
        end for;

        for P2:COMP_MAG1 use entity
          BEH_INT.COMP_MAG(COMPARE)
          generic map("test_o1.dat","test_gm.dat",10,10);
        end for;

        for P3:COMP_DIR1 use entity BEH_INT.COMP_DIR(COMPARE)
          generic map("test_d1.dat","test_gd.dat",10,10);
        end for;

        for P4:EDGE_DETECTOR1 use entity
          BEH_INT.EDGE_DETECTOR(BEHAVIOR)
          generic map(NUM_ROWS=>10,NUM_COLS=>10);
        end for;
      end for;   -- End of BENCH
    end for;     -- End of con:TEST1
  end for;       -- End TEST_BENCH
end;             -- End of configuration
```

Figure 11.31   Configuration declaration for test bench.

Table 11.3 shows the direction values for the horizontal and the vertical edges that were output by the simulation.

These directions were computed according to Figure 11.4. The results correspond to the acceptable outcomes listed in the table. The Synopsys simulator indicated that the statement coverage was 100 percent. This means that every instruction in the model was executed at least once.

Figures 11.33 and 11.34 show that all edges in a complex artificial image are correctly identified according to test number 12.

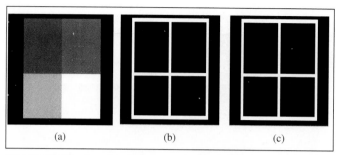

Figure 11.32 Input image, expected response, and output image.

Table 11.2 Test plan.

Test Number	Stimuli Source	Gold Data Source	Acceptable Outcome	Desired Statement Coverage
1	tesv_i1.dat	test_g1.dat	All H edges detected with correct direction.	100%
2	tesv_i1.dat	test_g1.dat	All V edges detected with correct direction.	100%
3	tesv_i2.dat	test_g2.dat	All L edges detected with correct direction.	100%
4	tesv_i2.dat	test_g2.dat	All R edges detected with correct direction.	100%
5	tesv_i1.dat	test_g1.dat	All corners detected with correct direction.	100%
6	tesv_i3.dat	test_g3.dat	Single pixel holes should be filled.	100%
7	tesv_i4.dat	test_g4.dat	Multipixel holes remain unfilled.	100%
8	tesv_i5.dat	test_g5.dat	Single pixel corner holes filled.	100%
9	tesv_i6.dat	test_g6.dat	Multipixel corner holes remain unfilled.	100%
10	tesv_i7.dat	test_g7.dat	Single pixel holes ignored.	100%
11	tesv_i8.dat	test_g8.dat	Noise filtered appropriately and ignored.	100%
12	Tesv_i9.dat	Test_g9.dat	Identify all edges in a complex artificial image.	100%
13	Tesv_i10.dat	Test_g10.dat	Produce a recognizable outline of a real image.	100%

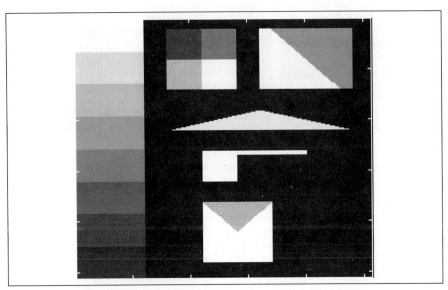

Figure 11.33   Complex artificial image.

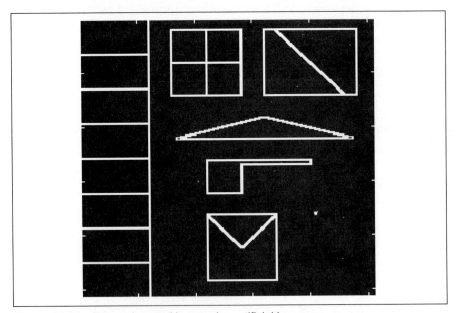

Figure 11.34   Edges detected in complex artificial image.

Table 11.3   Directions of detected edges.

	HORIZONTAL EDGES	VERTICAL EDGES
1	"110" - TOP EDGE	"000" - LEFT EDGE
2	"110" - MIDDLE EDGE	"000" - MIDDLE EDGE
3	"010" - BOTTOM EDGE	"100" - RIGHT EDGE

11.5 ARCHITECTURE DESIGN

During the architecture design, the system is decomposed into subsystems. At the previous design level, the system specifications were refined and captured in a VHDL executable specification model. The designer, working at the architecture design level, can thoroughly document the system being designed by constructing the executable specification model, or by studying the model if it was created by another group. Based on the knowledge obtained from this documentation, the system is decomposed into smaller modules called subsystems. These subsystems can be modeled as components at the same abstraction level or at a lower abstraction level. The interface and the function of each subsystem must be specified. The design methodology of Figure 11.1 must be applied to each subsystem, beginning with a written specification and an executable specification. A VHDL structural model is then developed for the original system that reflects the system decomposition. The structural model specifies an interconnection of the subsystem components. Clearly, the overall interface and the function of the system remain the same as that specified by the original executable specification.

Generally, systems can be decomposed in many different ways. A major goal of architecture design is to investigate alternatives.

11.5.1 System Level Decomposition

To illustrate architecture design, we will develop a structural model of the Sobel edge detection system. By studying the written specification and the VHDL executable specification model, the edge detection algorithm proceeds in three major steps.

Step 1.   During each clock cycle, update the internal memory data structures. The image pixels arrive in raster scan order (rows are scanned from top to bottom, and each row is scanned from left to right) with one pixel arriving at port INPUT every clock period. This pixel value must be stored in the internal 3-row memory array. After enough pixels arrive to construct a full 3x3 filter window, the filter window buffer must be updated by reading a column of three bits from the internal memory as shown in Figure 11.15.

Step 2.   According to Equations 11.1 through 11.4, compute the four Sobel filter functions on the current filter window using the pixels in the window buffer.

Step 3.   Decide whether the center pixel in the current window is an edge pixel. This step requires computing the maximum absolute value for the four filter outputs, the absolute value perpendicular to the maximum value, and computing the magnitude using Equation 11.5. If the magnitude is greater than the specified threshold, the center pixel is declared to be an edge pixel. The appropriate pixel value is assigned, and the direction computed according to Figure 11.4.

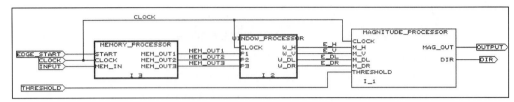

Figure 11.35   Top-level decomposition of the Sobel edge detection system.

System decomposition requires designer insight. Designers might decompose the system in different ways. Figure 11.35 shows one way to decompose the Sobel edge detection system based on the three steps just discussed.

Memory processor component.   The component MEMORY_PROCESSOR contains the 3-row internal memory buffer. During each clock period, it stores the pixel arriving at port MEM_IN into the appropriate location in the internal memory buffer. This component also reads the appropriate 3-pixel column from the internal memory buffer and sends it to the window processor component via ports MEM_OUT1, MEM_OUT2, and MEM_OUT3, so the window processor can update the current filter window. The memory processor component needs access to the system clock, CLOCK, and also to the system start signal (port START) to initialize the memory buffer address pointers when a new image is to be processed.

Window processor component.   The component WINDOW_PROCESSOR contains the 3x3 window buffer. During each clock period, it reads a new column of data from its inputs (P1, P2, P3) and updates the window buffer as shown in Figure 11.15. It then computes the four filter functions and outputs the four intensity gradients at ports W_H (horizontal intensity gradient), W_V (vertical intensity gradient), W_DL (left-diagonal intensity gradient), and W_DR (right-diagonal intensity gradient).

Magnitude processor component.   This component reads the four intensity gradients at its inputs and computes the magnitude gradient according to Equation 11.5 and the direction of the maximum gradient according to Figure 11.4. If the magnitude gradient exceeds the threshold (at input port THRESHOLD), the center pixel in the current window buffer is declared to be an edge pixel. If the center pixel is an edge pixel, MAG_OUT is set to the foreground value, and DIR is set to the direction of the maximum gradient. If the center pixel is not an edge, MAG_0UT is set equal to the background value, and DIR is arbitrary.

The corresponding VHDL structural code can be automatically generated by SGE, or the structural architecture can be easily constructed from the block diagram of Figure 11.35. Figures 11.36 and 11.37 show the VHDL structural architecture for the Sobel edge detection system that is implied by the block diagram in Figure 11.35.

Figure 11.36 shows the component declarations for the structural architecture. Note that delay generics are added to accommodate internal delays, and all outputs are declared to be mode *inout,* so configurations can be used with port type conversion functions. Mixed data type configurations are described in detail later in this chapter.

Figure 11.37 shows the intercomponent signal declarations and the component instantiations. The designer must prepare behavioral models for each of the components. This procedure is similar to that illustrated for the Sobel behavioral model at the systems definition level. We

```
-- structural description of edge detector ---
architecture STRUCTURE of EDGE_DETECTOR is

   -- memory processor component declaration --
   component MEMORY_PROCESSOR1
     generic (NUM_ROWS, NUM_COLS: NATURAL;
              MEM_OUT_DELAY:TIME)
     port(CLOCK: in STD_LOGIC:='0';    -- system CLOCK
          START: in STD_LOGIC:='0';    -- start signal
          MEM_IN: in PIXEL:=0;         -- input pixel
          MEM_OUT1, MEM_OUT2, MEM_OUT3: inout PIXEL:=0);
   end component;

 -- window processor component declaration --
   component WINDOW_PROCESSOR1
     generic(HORIZ_DELAY, VERT_DELAY, LEFT_DIAG_DELAY,
             RIGHT_DIAG_DELAY, WAIT_TIME:TIME)
     port(CLOCK: in STD_LOGIC:='0';   -- system CLOCK
          P1,P2,P3: in PIXEL:=0;       -- input pixels
          W_H: inout FILTER_OUT:=0;   -- horizontal filter
          W_V: inout FILTER_OUT:=0;   -- vertical filter
          W_DL: inout FILTER_OUT:=0; -- left diagonal filter
          W_DR: inout FILTER_OUT:=0);-- right diagonal filter
   end component;

 -- magnitude processor component declaration --
   component MAG_PROCESSOR1
     generic(MAG_DELAY: TIME); -- magnitude processor delay
     port(CLOCK: in STD_LOGIC:='0';   -- system CLOCK
          M_H,M_V,M_DL,M_DR: in FILTER_OUT:=0;   -- filter
             -- values from the window processor
          THRESHOLD: in FILTER_OUT:=0;   -- threshold value
          MAG_OUT: inout PIXEL:= 0; -- edge pixel value
          DIR: inout DIRECTION :="000"); -- edge direction
   end component;
```

Figure 11.36   Structural architecture for Sobel edge detection system, part 1.

now have word descriptions of the three components and must create VHDL behavioral models to serve as executable specifications for each of them. Since this procedure is the same as that described for the top-level Sobel model, we will leave development of these modules to the problems at the end of the chapter.

Binding the instantiations to specific models is done in a configuration, so the bindings can be changed by writing a new configuration. The structural architecture declaration will never change. In this way, we can bind the instances to behavioral models now, and possibly to structural models later, if the component is further decomposed.

Figure 11.38 shows the binding of components to behavioral models.

```
-- intermediate signals between the components --
  signal E_H,E_V,E_DL,E_DR: FILTER_OUT:=0;
  signal MEM_OUT1, MEM_OUT2, MEM_OUT3: PIXEL:=0;

begin
    ---------- component instantiation -------
MEMP1: MEMORY_PROCESSOR1
  generic map(NUM_ROWS => NUM_ROWS, NUM_COLS => NUM_COLS
            MEM_OUT_DELAY => 2 ns)
  port map(CLOCK, EDGE_START, INPUT, MEM_OUT1,
          MEM_OUT2, MEM_OUT3);

WINP: WINDOW_PROCESSOR1
  generic map(HORIZ_DELAY => 3 ns, VERT_DELAY=> 3 ns,
              LEFT_DIAG_DELAY => 3 ns,
              RIGHT_DIAG_DELAY => 3 ns,
              WAIT_TIME => 0 ns)
  port map(CLOCK, MEM_OUT1, MEM_OUT2, MEM_OUT3,
          E_H, E_V, E_DL, E_DR);

MAGP: MAG_PROCESSOR1
  generic map( MAG_DELAY => 3 ns)
  port map(CLOCK, E_H, E_V, E_DL, E_DR,THRESHOLD,
          OUTPUT, DIR)
end STRUCTURE;
```

Figure 11.37   Structural architecture for Sobel edge detection system, part 2.

```
library STRUCT_INT;
use STRUCT_INT.all;

configuration CONFIG_SOBEL_S_L1 of EDGE_DETECTOR is
  for STRUCTURE

    for MEMP1: MEMORY_PROCESSOR1
      use entity STRUC_INT.MEMORY_PROCESSOR(BEHAVIOR);
    end for;

    for WINP: WINDOW_PROCESSOR1
      use entity STRUC_INT.WINDOW_PROCESS(BEHAVIOR);
    end for;

    for MAGP: MAG_PROCESSOR1
      use entity STRUC_INT.MAG_PROCESS(BEHAVIOR);
    end for;
  end for;
end;
```

Figure 11.38   Configuration declaration for top-level decomposition of the Sobel edge detection system.

```
configuration CONFIG_S_INT of TB_CONFIG is
  for TEST_BENCH
    for con:TEST1 use entity BEH_INT.TEST(BENCH)
      generic map("tesv_i2.dat", "test_o2.dat",
                  "test_d2.dat", 10,10,5);
      for BENCH
        for P1:CLOCK_GENERATOR1 use entity
          BEH_INT.CLOCK_GENERATOR(BEHAVIOR)
          generic map(HI_TIME=>75 ns,LO_TIME=>25 ns);
        end for;
        for P2:COMP_MAG1
          use entity BEH_INT.COMP_MAG(COMPARE)
          generic map("test_o2.dat","test_gm.dat",10,10);
        end for;
        for P3:COMP_DIR1
          use entity BEH_INT.COMP_DIR(COMPARE)
          generic map("test_d2.dat","test_gd.dat",10,10);
        end for;
        for P4:EDGE_DETECTOR1
          use entity STRUC_INT.CONFIG_SOBEL_S_L1
          generic map(NUM_ROWS=>10,NUM_COLS=>10);
        end for;
      end for;
    end for;
  end for;
end;
```

Figure 11.39 Configuration declaration for test bench to test the structural architecture for the Sobel edge detection system.

11.5.1.1   *Test Bench for First Level of Structural Decomposition*

The test bench model for the executable specification, BEH_INT.TEST(BENCH), defined in Figures 11.19 to 11.26 can be reused to test the structural architecture for the Sobel edge detection system. It is not necessary to reanalyze all the VHDL code in BEH_INT.TEST(BENCH) if the configuration approach is used. The structural model is tested by analyzing and simulating only the new configuration, CONFIG_S_INT shown in Figure 11.39. The reader should compare Figure 11.39 with Figure 11.31. The same test bench BEH_INT.TEST(BENCH) is reused. The same test vector data file, tesv_i2.dat, and GOLD data files, test_gm.dat and test_gd.dat are also reused. All generic parameter declarations are reused. Only two items are changed. First, different intermediate data files, test_o2.dat and test_d2.dat, are used so that the result files for the behavioral level test are not overwritten. Second, the edge detector component (P4:EDGE_DETECTOR1) is bound to the structural model in Figure 11.38 using configuration STRUC_INT.CONFIG_SOBEL_S_L1 instead of to the behavioral model BEH_INT.EDGE_-DETECTOR(BEHAVIOR).

11.5.2   Hierarchical Decomposition

System-level decomposition is an iterative process that continues until the bottom-level modules are simple enough to synthesize. As an example, the WINDOW_PROCESSOR can be further

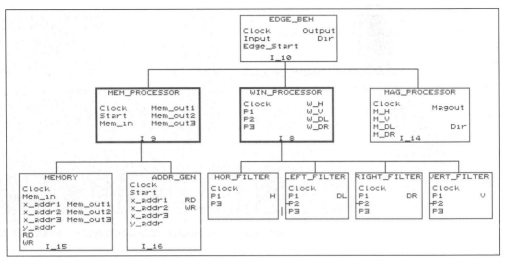

Figure 11.40   Hierarchical structure of Sobel edge detection system design.

decomposed into four filter components (HOR_FILTER, LEFT_FILTER, RIGHT_ FILTER, and VERT_FILTER) as shown in Figure 11.40. Similarly, component MEM_ PROCESSOR can be further decomposed into an address generator component (ADDR_GEN) and a memory component (MEMORY). The written specifications for the second-level components are detailed below.

Address generator component.   Component ADDR_GEN accepts the clock input (Clock) and the start input (Start) to time its operation and to initialize its variables, respectively. This module generates the memory read control signal (RD) and the memory write control signal (WR), which control reads and writes into the internal memory component (MEMORY). If a write operation is to occur during the current clock cycle, ADDR_GEN sends out the row address for the write operation at port x_addr1, the column address for the write operation at port y_addr, and generates a write enable control signal (WR='1') during the low half of the clock cycle. If a window buffer update is to occur during the current clock cycle, ADDR_GEN sends out the column address for all three reads at port y_addr, the row address for reading the top element in the window update column at port x_addr1, the row address for reading the center element in the window update column at port x_addr2, the row address for reading the bottom element in the window update column at port x_addr3, and a read enable control signal (RD='1') during the high half of the clock cycle.

Memory component.   Component MEMORY contains the 3-row memory array of pixels. On the rising edge of input Clock, if WR='1', the memory component writes a new pixel value from input port Mem_in into the memory array at row address x_addr1 and column address y_addr. On the falling edge of input Clock, if RD='1', the memory component reads three pixels in parallel from the internal memory array. The pixel at row address x_addr1 and column address y_addr is sent to output port mem_out1. The pixel at row address x_addr2 and column address y_addr is sent to output port mem_out2. The pixel at row address x_addr3 and column address y_addr is sent to output port mem_out3. These three pixels are sent to the window processor to update the window buffer as shown in Figure 11.15.

Horizontal filter component. Component HOR_FILTER computes the horizontal filter operation of Equation 11.1. The top row of pixels in the window buffer arrives at input port P1 and the bottom row of pixels arrives at input port P3. This filter does not need the pixels in the center row of the window buffer. One pixel arrives at each port every clock period. This component keeps its own internal copy of rows 1 and 3 of the buffer window.

Vertical filter component. Component VERT_FILTER computes the vertical filter operation of Equation 11.2. The top row of pixels in the window buffer arrives at input port P1, the center row of pixels arrives at input port P2, and the bottom row of pixels arrives at input port P3. One pixel arrives at each port every clock period. This component keeps its own internal copy of all three rows of the window buffer.

Left diagonal filter component. Component LEFT_FILTER computes the left diagonal filter operation of Equation 11.3. The top row of pixels in the window buffer arrives at input port P1, the center row of pixels arrives at input port P2, and the bottom row of pixels arrives at input port P3. One pixel arrives at each port every clock period. This component keeps its own internal copy of all three rows of the window buffer.

Right diagonal filter component. Component RIGHT_FILTER computes the right diagonal filter operation of Equation 11.4. The top row of pixels in the window buffer arrives at input port P1, the center row of pixels arrives at input port P2, and the bottom row of pixels arrives at input port P3. One pixel arrives at each port every clock period. This component keeps its own internal copy of all three rows of the window buffer.

Note: No claim is made for this decomposition being optimal. Other possible decompositions might be better.

11.5.3 Methodology for Development of Test Benches for a Hierarchical Structural Model

This section explains a methodology for testing hierarchical structural models. Hierarchical structural models are tested in a bottom-up fashion, even though the models were designed using a top-down methodology.

1. First, we test each individual component at the bottom level of the hierarchy using a separate test bench. This verifies the behavioral performance of each leaf module.
2. After testing all the lower-level modules, modules one level up in the hierarchy are tested. These modules will be structural models using components from the lowest level. If efficient top-down design procedures are used, including using configurations to create test benches at each level, the configurations used to test the behavioral models can be slightly modified to create new configurations to test the structural models. The same test benches with the same test vectors can be reused repeatedly.
3. At the top of the hierarchy, the original test bench for the executable specification can be reused to test the complete hierarchical structural model by simply making a small change in the original configuration for the test bench.

As the test procedure works its way up the hierarchy, the test benches used to test the designs during the top-down design methodology are all reused in the bottom-up testing strategy for the hierarchical structural model—provided configurations were used at each level during the top-down development. Only new configurations need to be analyzed and simulated at each step in the procedure, not the more complex test benches. This can save an enormous amount of computer time and memory space. An additional advantage is an easily followed documentation trail. If changes are made in the future, this same well documented trail can be followed again to reverify the modified design.

To illustrate this methodology, suppose the design group working on the horizontal filter component, HOR_FILTER, has just completed the design but has not yet tested it. Also, suppose the components LEFT_FILTER, RIGHT_FILTER, and VERT_FILTER have already been designed and thoroughly tested using their individual test benches, but that these designs have not yet been tested as structural components of the mid-level component, WIN_PROCESSOR. However, the behavioral version of WIN_PROCESSOR has already been designed and tested, both individually with its behavioral test bench and as a structural component in the top-level EDGE_BEH component. Further, suppose the design group is using configurations to streamline the modeling process. For example, Figure 11.39 shows the configuration used to test the structural architecture at the top level.

The first step is to test the bottom-level component, HOR_FILTER, using its own test bench. Figure 11.41 shows a test bench suitable for the bottom-level test. Because this is a bottom-level component, the configuration specification is included within the model (using *for* statements) instead of having a separate configuration declaration.

The next step is to reuse the test bench originally used to test the behavioral model of the window processor. Figure 11.42 shows the original test bench, TB(WINDOW_TB), used to test the behavioral model of the window processor. Note that no binding is specified for the component WINDOW_PROCESSOR1. Figure 11.43 shows the original configuration (WINDOW_ BEH_INT) used for the behavioral test. The original test bench, TB(WINDOW_TB) of Figure 11.42, can be reused to test the structural model that includes the newly developed filter processors by simulating a new configuration, WINDOW_STRUC_INT, as shown in Figure 11.44. The only difference is that the structural model (defined by configuration STRUCT_INT.- W_INT) instead of the behavioral model (STRUCT_INT.WINDOW_PROCESSOR(BEHAV- IOR)) is bound to the window processor component (L2: WINDOW_PROCES-SOR1).

Finally, the top-level test bench shown in Figures 11.19 through 11.26, can be reused to test the entire system with the new filter components installed, by simply writing a new configuration to replace the old configuration of Figure 11.31. Figure 11.45 shows the new configuration. Entity CONFIG_SOBEL_S_L2 is the configuration that includes the new window architecture. This example shows how easy it is to reuse test benches as the design proceeds down the design hierarchy if configuration declarations are used at each level.

11.6   DETAILED DESIGN AT THE RTL LEVEL

We now move into the detailed design stage. The previous stage was the architecture design stage in which a structural model was created by decomposing the specification into

```
entity TB is
end TB;
architecture HORIZ_TB_INT of TB is
  signal RUN: STD_LOGIC;
  signal CLOCK:STD_LOGIC:='0';
  signal P1,P3: PIXEL:=0;
  signal H: FILTER_OUT:=0;
  component CLOCK_GENERATOR1
    generic(HI_TIME,LO_TIME:TIME);
    port(RUN: in STD_LOGIC;
         CLOCK: out STD_LOGIC);
  end component;
  component  HORIZONTAL_FILTER1
    generic(HORIZ_DELAY: TIME;
            WAIT_TIME:TIME);
    port(CLOCK: in STD_LOGIC:='0';
         P1,P3: in  PIXEL:=0;
         H: inout FILTER_OUT:=0);
  end component;
  for L1: CLOCK_GENERATOR1
    use entity BEH_INT.CLOCK_GENERATOR(BEHAVIOR);
  for L2: HORIZONTAL_FILTER1
    use entity STRUC_INT.HORIZONTAL_FILTER(BEHAVIOR);
begin
  L1: CLOCK_GENERATOR1
    generic map(HI_TIME=>75 ns,LO_TIME=>25 ns)
    port map(RUN,CLOCK);
  L2: HORIZONTAL_FILTER1
    generic map(HORIZ_DELAY=>1 ns,WAIT_TIME=>100 ns)
    port map(CLOCK,P1,P3,H);
  RUN<='0' after 0 ns,'1' after 5 ns,'0' after 1000 ns;
  P1<=11 after 10 ns,10 after 110 ns,9 after 210 ns,
      12 after 310 ns,13 after 410 ns,48 after 510 ns;
  P3<=8 after 10 ns,40 after 110 ns,7 after 210 ns,
      10 after 310 ns,5 after 410 ns,110 after 510 ns;
end HORIZ_TB_INT;
```

Figure 11.41   Independent test bench for horizontal filter.

manageable subsystems. We now need to design each of the bottom-level components in the hierarchical decomposition in detail. The most direct route to gate-level design is to create a register transfer level (RTL) model, and use a synthesis tool to convert it into a gate-level circuit. In this section, we describe a methodology for obtaining the RTL model. In the next section, we discuss the synthesis step.

11.6.1   Register Transfer Level Design

As the level name implies, RTL models describe the transfer of data among various registers. Operations commonly performed on data include arithmetic operations (addition, subtraction, multiplication, etc.), logical operations (*or, and, not,* etc.), shift and rotate operations, and many

```
-------------ENTITY TEST BENCH---------------
entity TB is
end TB;
-------------ARCHITECTURE BODY--------------
architecture WINDOW_TB of TB is
  signal P1,P2,P3: PIXEL:=0;
  signal CLOCK:  STD_LOGIC:='0';
  signal W_H: FILTER_OUT:=0;
  signal W_V: FILTER_OUT:=0;
  signal W_DL: FILTER_OUT:=0;
  signal W_DR: FILTER_OUT:=0;
  signal RUN: STD_LOGIC:='0';
----------COMPONENT DECLARATIONS--------------
  component CLOCK_GENERATOR1
    port(RUN: in STD_LOGIC;
         CLOCK: out STD_LOGIC);
  end component;
  component WINDOW_PROCESSOR1
    port (CLOCK: in STD_LOGIC:='0';
          P1,P2,P3: in   PIXEL:=0;
          W_H: inout FILTER_OUT:=0;
          W_V: inout FILTER_OUT:=0;
          W_DL: inout FILTER_OUT:=0;
          W_DR: inout FILTER_OUT:=0 );
  end component;
begin
  L1: CLOCK_GENERATOR1
    port map(RUN,CLOCK);
  L2:WINDOW_PROCESSOR1
    port map(clock,P1,P2,P3,W_H,W_V,W_DL,W_DR);
  RUN<='0' after 0 ns, '1' after 5 ns, '0' after 1000 ns;
  P1<=  11 after 10 ns, 10 after 110 ns, 9 after 210 ns,
        12 after 310 ns,13 after 410 ns,48 after 510 ns;
  P2<=6 after 10 ns,  5 after 110 ns,14 after 210 ns,
      13 after 310 ns,10 after 410 ns,19 after 510 ns;
  P3<=8 after 10 ns,40 after 110 ns,  7 after 210 ns,
      10 after 310 ns,5 after 410 ns,110 after 510 ns;
end WINDOW_TB;
```

Figure 11.42   Test bench for window processor component.

other similar operations. We will only treat synchronous circuits in this section, although asynchronous designs are certainly possible. One typically performs the following four steps to create an RTL model. Chapter 12 describes formal methods by which this design process might be automated.

Step 1.   The register transfer-level design of a component begins with an analysis of the behavioral model of the component to determine the number and types of registers needed and the types of data operations that need to be performed.

```
configuration WINDOW_BEH_INT of TB is
  for WINDOW_TB
    for L1:CLOCK_GENERATOR1
      use entity BEH_INT.CLOCK_GENERATOR(BEHAVIOR)
        generic map(HI_TIME=>75 ns,LO_TIME=>25 ns);
    end for;
    for L2: WINDOW_PROCESSOR1
      use STRUC_INT.WINDOW_PROCESSOR(BEHAVIOR)
        generic map(HORIZ_DELAY=>3 ns,
                    VERT_DELAY =>3 ns,
                    LEFT_DIAG_DELAY =>3 ns,
                    RIGHT_DIAG_DELAY =>3 ns,
                    WAIT_TIME=> 0 ns);
    end for;
  end for;
end;
```

Figure 11.43   Configuration of window processor test bench to test behavioral model.

```
configuration WINDOW_STRUC_INT of TB is
  for WINDOW_TB
    for L1:CLOCK_GENERATOR1
      use entity BEH_INT.CLOCK_GENERATOR(BEHAVIOR)
        generic map(HI_TIME=>75 ns,LO_TIME=>25 ns);
    end for;
    for L2: WINDOW_PROCESSOR1
      use configuration STRUC_INT.W_INT
        generic map(HORIZ_DELAY=>3 ns,
                    VERT_DELAY =>3 ns,
                    LEFT_DIAG_DELAY =>3 ns,
                    RIGHT_DIAG_DELAY =>3 ns,
                    WAIT_TIME=> 0 ns);
    end for;
  end for;
end;
```

Figure 11.44   Configuration of window processor test bench to test structural model.

Step 2.   Determine the order in which operations must be performed. This step is commonly called *scheduling*.

Step 3.   Map the operations onto the hardware computation components and decide where to place the results. This step is commonly called *allocation*.

Step 4.   Create a VHDL data flow model or structural model to describe the RTL design.

We now illustrate this methodology by showing how to create an RTL structural model for the horizontal filter component at the bottom level of the hierarchical decomposition described in Figure 11.40.

```
configuration CONFIG_S2_INT of TB_CONFIG is
  for TEST_BENCH
    for con:TEST1 use entity BEH_INT.TEST(BENCH)
      generic map("tesv_i2.dat", "test_o2.dat",
                  "test_d2.dat", 10,10,5);
      for BENCH
        for P1:CLOCK_GENERATOR1 use entity
               BEH_INT.CLOCK_GENERATOR(BEHAVIOR)
          generic map(HI_TIME=>75 ns,LO_TIME=>25 ns);
        end for;
        for P2:COMP_MAG1 use entity WORK.COMP_MAG(COMPARE)
          generic map("test_o2.dat","test_gm.dat",10,10);
        end for;
        for P3:COMP_DIR1 use entity WORK.COMP_DIR(COMPARE)
          generic map("test_d2.dat","test_gd.dat",10,10);
        end for;
        for P4:EDGE_DETECTOR1 use entity CONFIG_SOBEL_S_L2
          generic map(NUM_ROWS=>10,NUM_COLS=>10);
        end for;
      end for;
    end for;
  end for;
end;
```

Figure 11.45 Configuration declaration for top-level test bench to test the structural model of the window processor.

11.6.1.1 Design of the Horizontal Filter Component at the RTL

Figure 11.46 shows the behavioral model for the horizontal filter. Pixels arrive in two continuous data streams at input ports P1 and P3. The result of the horizontal filter computation is sent out at port H, also in a continuous stream. The pixels in the top row of the filter window arrive at port P1, and the pixels in the bottom row of the filter window arrive at port P3. The behavioral model stores three pixels from each input data stream in internal buffers FIRST_LINE and THIRD_LINE.

The behavioral model uses two delays (1) HORIZ_DELAY, which represents the internal delay of the filter component, and (2) WAIT_TIME, which is an artificial delay used to balance delays in mixed-level simulations. This delay accounts for differing properties of models at different abstraction levels. By using separate parameters to represent the two distinct effects, it is easier to keep track of the parameters that represent real delays. Otherwise, the two are merged into one parameter, which is confusing.

The horizontal filter performs the WEIGHT function on the pixels in each of the internal buffers and computes the difference between the two values. Figure 11.47 shows the VHDL code for the WEIGHT function. We will denote the pixels in the top row of the filter window as (X1P1, X2P1, X3P1) and the pixels in the bottom row as (X1P3, X2P3, X3P3). After all pixels have arrived, the following computation is inferred by the VHDL code.

$$H = ((X1P1) + 2*(X2P1) + (X3P1)) - ((X1P3) + 2*(X2P3) + X3P3)) \qquad (11.6)$$

```
    library IEEE;
    use ieee.std_logic_1164.all;
    library STRUC_INT;
    use STRUC_RTL.IMAGE_PROCESSING.all;
    -- interface declaration ------------------------
    entity HORIZONTAL_FILTER is
      generic(HORIZ_DELAY: TIME; -- horizontal filter delay
              WAIT_TIME: TIME); -- abstraction level match
      port(CLOCK: in STD_LOGIC:='0';
           P1,P3: in PIXEL:=0;
           H: inout FILTER_OUT:=0);
    end HORIZONTAL_FILTER;
    -- behavioral architecture declaration ------------
    architecture BEHAVIOR of HORIZONTAL_FILTER is
      signal TEMP1: FILTER_OUT:=0;   -- intermediate signal
    begin
      H_FILTER: process
        variable TEMP_H: FILTER_OUT:=0;   -- temporary storage
        variable FIRST_LINE: PIX3:=(0,0,0); -- first scan line
        variable THIRD_LINE: PIX3:=(0,0,0); -- third scan line
      begin
        wait until rising_edge(CLOCK);
    --- store the input pixels in the 3-stage buffers ----
        FIRST_LINE:=SHIFT_LEFT(FIRST_LINE,P1);
        THIRD_LINE:=SHIFT_LEFT(THIRD_LINE,P3);
    TEMP_H:=WEIGHT(FIRST_LINE(1),FIRST_LINE(2),FIRST_LINE(3))
         -WEIGHT(THIRD_LINE(1), THIRD_LINE(2), THIRD_LINE(3));
        TEMP1 <= TEMP_H after HORIZ_DELAY;
      end process H_FILTER;
    ---- make H as a delayed version of temp1 -------------
      H <= TEMP1 after WAIT_TIME; -- Horizontal filter output
    end BEHAVIOR;
```

Figure 11.46 VHDL behavioral model for the horizontal filter.

```
    ----- WEIGHT function -----------------
    function WEIGHT
       (X1,X2,X3: PIXEL)
       return FILTER_OUT is
    begin
       return X1+ 2*X2 + X3;
    end WEIGHT;
```

Figure 11.47 WEIGHT function.

Recall that the pixels arrive in pairs at ports P1 and P3 in this order (X1P1, X1P3), (X2P1, X2P3), and (X3P1, X3P3). By expanding the expression above and collecting together terms that arrive at the same time, we get this expression:

$$H = ((X1P1 - X1P3)) + 2*((X2P1 - X2P3)) + ((X3P1 - X3P3)) \qquad (11.7)$$

We now note that the pair of pixels that arrive at each time step need to be subtracted. Since the pixels are arriving continuously, this observation suggests that a pipelined architecture should be considered. We need hardware to add, subtract, and multiply by 2. Since the data arrives continuously, we cannot share the hardware. This suggests a straight-line scheduling and a simple one-to-one mapping (allocation) of operations to hardware units.

Figure 11.48 shows one way to organize a set of registers and computational units to accomplish the pipelining operation. Recall from the Sobel algorithm description that the pixel values are positive integers in the range from 0 to 255. At the RTL, we will represent all data as binary vectors. Eight bits are required to represent integers in the specified range. Therefore, the inputs at P1 and P2 will be 8-bit binary, positive integers. Internally, since the computation involves subtraction, negative numbers may be generated. We, therefore, need one additional bit to represent the sign of the result. Since we are adding twice in successive steps, the sum requires more than eight bits for an accurate representation. We, therefore, add two additional bits to allow accurate representation of large sums. The total number of bits required is 11. To allow for standard register sizes and standard bus sizes, we decided to use 12-bit data paths inside the filter. So, internally we will use a 12-bit signed 2's complement number representation.

Since the input data are 8-bit positive integers, we need to append four 0's at the left of each data item to convert it into a 12-bit 2's complement representation. Components EXT8_12A and EXT8_12B perform this conversion task for the two data streams. Components DELAYx are 12-bit registers. Components SUM1 and SUM2 are 12-bit 2's complement adders, and Diff1 is a 12-bit 2's complement subtracter. VHDL models for these components can be found in Chapter 4. Filter output H is a 12-bit signed 2's complement number. The complete code for all RTL functions can be found on the CD ROM that accompanies this text.

Tracing one set of data through the pipeline, as the data pairs (X1P1, X1P3), (X2P1, X2P3), and (X3P1, X3P3) arrive in three successive clock periods, the subtracter Diff1 computes (X1P1-X1P3), (X2P1-X2P3), and (X3P1-X3P3). After three clock periods, these are the results:

```
S2=X3P1-X3P3
S3=X2P1-X2P3
S4=X1P1-X1P3
S5=(X1P1-X1P3)+(X3P1-X3P3).
```

After clock period 4, these results are achieved:

```
S6=(X1P1-X1P3)+(X3P1-X3P3)
S4=(X2P1-X2P3)
C=2*(X2P1-X2P3)
S7=((X1P1-X1P3)+(X3P1-X3P3))+2*(X2P1-X2P3)
```

Finally, after clock period 5, we have the desired output at H.

```
H=(X1P1-X1P3)+2*(X2P1-X2P3)+(X3P1-X3P3)
```

The VHDL code for a structural model for the horizontal filter follows from the RTL diagram shown in Figure 11.48 in a straightforward manner. Figure 11.49 shows the structural model and Figures 11.50 and 11.51 show the configuration for the structural model. The reader should study these figures carefully because we have introduced a new concept, the empty com-

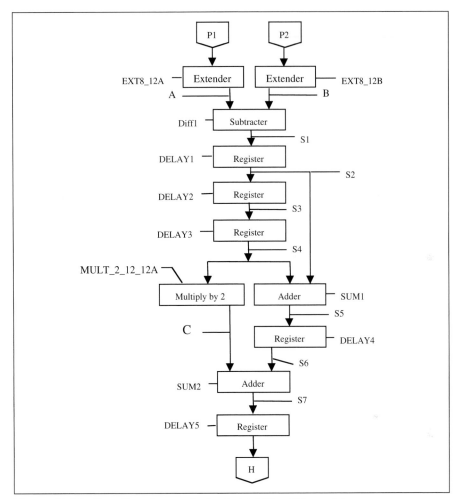

Figure 11.48 Data flow diagram for horizontal filter RTL model.

ponent declaration. Using empty component declarations, the data types on the ports may be deferred to the configurations. This allows us to construct mixed data type structural models with type conversion functions associated with the ports in the configuration declarations. Mixed data type structural models are discussed in the next section.

The RTL structural model of Figure 11.49 contains empty declarations for the 12-bit 2's complement subtracter (SUM12_PM), the 12-bit register (REG12), the 12-bit 2's complement adder (SUM12_PP), the 8- to 12-bit extender (EXT8_12), and the element that multiplies by 2 (MULT_2_12_12). The names in the empty instantiations of the component types correspond to the labels in the data flow diagram of Figure 11.48. These instantiations define the mapping of operations to physical components (allocation). Since this model is a pipelined model, no components can be used to perform multiple operations. Therefore, the mapping is one-to-one,

```
entity HORIZONTAL_FILTER is
  generic(HORIZ_DELAY:TIME);
  port(CLOCK: in STD_LOGIC:='0';
       P1,P3: in PIXEL:="00000000";
       H: inout STD_LOGIC_VECTOR(11 downto 0));
end HORIZONTAL_FILTER;
architecture STRUCTURE of HORIZONTAL_FILTER is
------ Empty component declarations --------
  component SUM12_PM      end component; -- Subtracter
  component REG12         end component; -- Register
  component SUM12_PP      end component; -- Adder
  component EXT8_12       end component; -- Extender
  component MULT_2_12_12 end component; -- Multiply by 2
---- intermediate signal declarations -------
  signal S1,S2,S3,S4,S5,S6,S7,A,B,C:
    STD_LOGIC_VECTOR(11 downto 0);
begin
----- empty component instantiations ---------
  EXT8_12A: EXT8_12;
  EXT8_12B: EXT8_12;
  DIFF1   : SUM12_PM;
  DELAY1 : REG12;
  DELAY2 : REG12;
  DELAY3 : REG12;
  SUM1    : SUM12_PP;
  DELAY4 : REG12;
  MULT_2_12_12A: MULT_2_12_12;
  SUM2    : SUM12_PP;
  DELAY5 : REG12;
end STRUCTURE;
```

Figure 11.49   RTL structural model for horizontal filter.

which means that each operation in Figure 11.48 is mapped to a different component. See Chapter 12 for a discussion of algorithms that can be used to optimize the mapping in more complex situations.

Figures 11.50 and 11.51 show the configuration declaration for the RTL model. Note that the port maps, as well as the generic maps, are included in this declaration since they were absent from the empty component declarations of Figure 11.49. This methodology is used in the next section to expedite the construction of simulation models with mixed data types.

11.6.2 Simulating Structural Models Using Components with Different Data Types

At this point in the design cycle, we have completed the design of the RTL model of the horizontal filter. Suppose we have thoroughly tested the RTL model as a standalone component. We now want to integrate it into the structural model of the window processor and run tests on the window processor with the new RTL model of the horizontal filter as one of the components. Further suppose that the other design groups, working on RTL models of the other filter compo-

```
configuration H_RTL of HORIZONTAL_FILTER is
  for STRUCTURE
--------------- A <= "0000" & P1 ----------------
    for EXT8_12A:EXT8_12 use entity STRUC_RTL.EXT8_12(BEHAVIOR)
      port map(P1,A);
    end for;
--------------- B <= "0000" & P3 ----------------
    for EXT8_12B:EXT8_12 use entity STRUC_RTL.EXT8_12(BEHAVIOR)
      port map(P3,B);
    end for;
--------------- S1 <= A - B --------------------
    for Diff1:SUM12_PM use entity STRUC_RTL.SUM12_PM(PARTS)
      port map(A,B,S1);
    end for;
----------------- S2 <= S1 ----------------
    for DELAY1:REG12 use entity STRUC_RTL.REG12(PARTS)
      generic map(HORIZ_DELAY)
      port map( S1, S2, CLOCK );
    end for;
----------------- S3 <= S2 ----------------
    for DELAY2:REG12 use entity STRUC_RTL.REG12(PARTS)
      generic map(HORIZ_DELAY)
      port map( S2, S3, CLOCK );
    end for;
----------------- S4 <= S3 ----------------
    for DELAY3:REG12 use entity STRUC_RTL.REG12(PARTS)
      generic map(HORIZ_DELAY)
      port map( S3, S4, CLOCK );
    end for;
----------------- S5 <= S4 + S2 ----------------
    for SUM1:SUM12_PP use entity STRUC_RTL.SUM12_PP(PARTS)
      port map( S4, S2, S5 );
    end for;
----------------- S6 <= S5 ----------------
    for DELAY4:REG12 use entity STRUC_RTL.REG12(PARTS)
      generic map(HORIZ_DELAY)
      port map( S5, S6, CLOCK );
    end for;
```

Figure 11.50   Configuration for RTL model for horizontal filter.

nents, have not yet completed their work. We want to substitute the RTL model for the horizontal filter in the window processor structural model but keep the behavioral models for the other filters in place. Ideally, we would like to write a new configuration for the window processor with the new RTL model for the horizontal filter, instead of the behavioral model, being bound to the horizontal filter component in the window processor model. This would work cleanly except for the fact that the ports have different data types in the two models. In the behavioral model of the horizontal filter, the ports have Integer data types. Therefore, the interconnecting signals in the structural model must also have Integer data types. If we substitute a model with

```
    ------------ C <= S4(10 downto 0) & '0' ----------------
        for MULT_2_12_12A:MULT_2_12_12
          use entity STRUC_RTL.MULT_2_12_12(BEHAVIOR)
          port map(S4,C);
        end for;
    ---------------- S7 <= C + S6 ------------------------
        for SUM2:SUM12_PP use entity STRUC_RTL.SUM12_PP(PARTS)
          port map(C, S6, S7);
        end for;
    ---------------- H <= S7 ----------------
        for DELAY5:REG12 use entity STRUC_RTL.REG12(PARTS)
          generic map(HORIZ_DELAY)
          port map( S7, H, CLOCK );
        end for;
      end for; -- End STRUCTURE.
    end; -- End configuration.
```

Figure 11.51   Configuration for RTL model for horizontal filter (continued).

standard logic data types in the structural model, there will be a type mismatch. VHDL does not allow even a hint of type mismatch.

A similar problem exists if we acquire models from outside the company. The data types for those models might be different from the data types our company uses.

Any time we develop a structural model using components with different data types, we experience the same type mismatch problem. One solution is to develop multiple copies of all models with the only difference, being the data types of the ports. This solution uses huge amounts of memory and creates severe maintenance headaches. If the model needs to be updated, we must remember where all the various copies are and update each one. A second solution is to write multiple structural models—one for each combination of component data types. Using this system, the number of structural models will grow exponentially with the number of components and data types. Again, this creates maintenance headaches because if the structural model is to be updated or revised, we must locate all copies and update each one of them. This is clearly not acceptable.

We propose using a common structural model and a separate configuration for each combination of data types needed. If the structural model must be updated, it is easier to find the configurations since they are all configurations of the same entity and architecture. The configuration files are, in general, much smaller than the structural model files, so we use much less memory to store them. Further, we do not need to have multiple copies of the components with different data types if we place type conversion functions in the port association statements in the configuration file. If the configuration happens to be for a test bench, then we have only one copy of the test bench with different configurations for various combinations of components at different levels of abstraction. See the section "Methodology for Development of Test Benches for a Hierarchical Structural Model" earlier in this chapter.

To facilitate developing structural models using components with different data types, we should take advantage of the fact that VHDL allows type conversion functions to appear in the association lists in port map statements. To illustrate this technique, we will integrate the RTL

```
      entity WINDOW_PROCESSOR is
        generic(HORIZ_DELAY,VERT_DELAY,LEFT_DIAG_DELAY,
               RIGHT_DIAG_DELAY,WAIT_TIME: TIME);
        port(CLOCK: in std_logic:='0';   -- system CLOCK
            P1,P2,P3: in PIXEL:=0;        -- input Pixels
            W_H:  inout FILTER_OUT:=0; -- horizontal output
            W_V:  inout FILTER_OUT:=0; -- vertical output
            W_DL: inout FILTER_OUT:=0; -- left diagonal output
            W_DR: inout FILTER_OUT:=0);-- right diagonal output
      end WINDOW_PROCESSOR;
      architecture STRUCTURE of WINDOW_PROCESSOR is
      ------ empty component declarations------------------
        component HORIZONTAL_FILTER1 end component;
        component VERTICAL_FILTER1   end component;
        component LEFT_DIAG_FILTER1  end component;
        component RIGHT_DIAG_FILTER1 end component;
      begin
      ------ empty component instantiations --------------
        HP: HORIZONTAL_FILTER1;
        VP: VERTICAL_FILTER1;
        LDP: LEFT_DIAG_FILTER1;
        RDP: RIGHT_DIAG_FILTER1;
      end STRUCTURE;
```

Figure 11.52   VHDL code for structural model of window processor.

model of the horizontal filter developed in the previous section into a structural model of the window processor. The RTL model requires standard logic data types at its ports; whereas the original structural model was developed at a higher level of abstraction using Integer data types.

11.6.2.1   Case Study for Mixed Data Type Simulation: The Window Processor

Figure 11.52 shows the VHDL code for the structural model of the window processor. It uses empty component declarations and empty instantiations for the components. The generic and port declarations are omitted from the component declarations, and the generic and port maps are omitted from the instantiations. One way to visualize this situation is to use the analogy of chip sockets. The empty component declarations define the set of empty chip sockets and the names of the interconnecting signals. Figure 11.53 shows the block diagram defined by the window processor structural model declarations. Empty sockets represent the four components.

The socket type, which corresponds to the component name specified in the component declaration, appears inside the socket symbol. In this case, each socket has a unique type (HORIZONTAL_FILTER1, VERTICAL_FILTER1, LEFT_DIAG_FILTER1, and RIGHT_DIAG_FILTER1). In general, several sockets could be of the same type. The analogy to chip sockets is easy to visualize. Different types of chips require different types of sockets. For example, a 40-pin microprocessor chip will not fit into the socket designed for a 14-pin chip. With that understanding, the chip socket analogy is very helpful in visualizing the purpose of the structural model declarations. The structural model declarations define the set of chip socket types (corresponding to the set of chip types) and the set of interconnecting signals. The sockets are empty. There are no interconnecting wires at this time.

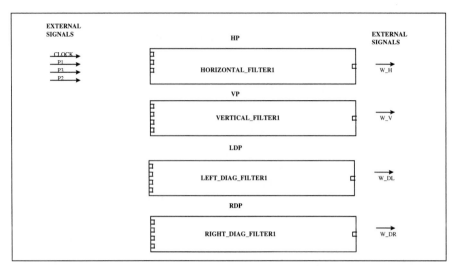

Figure 11.53 Block diagram implied by the structural architecture of Figure 11.52.

The empty instance declarations give a unique name to each socket used. The instance name is written outside the socket symbol. The instance declarations serve to map the set of devices into the set of component types (allocation).

The port and generic maps are included in the configuration declaration with type conversion functions included in the association lists for the port maps. The configuration also specifies the interconnections between chips and the specific chip to be inserted into each socket. Figure 11.54 shows the configuration of the window processor with the RTL model of the horizontal filter inserted into the socket labeled HORIZONTAL_FILTER1. Behavioral models of the other filters are inserted into their respective sockets. Figure 11.55 shows the block diagram implied by the configuration declaration in Figure 11.54. Since the RTL model uses standard logic data types (STD_LOGIC) and the other models use Integer data types, type conversion functions are specified for each pin on the RTL model. The following statements specify the connections and necessary type conversions:

```
P1_H => INT_TO_STDLOGIC8(P1);
P3_H => INT_TO_STDLOGIC8(P3);
```

Function INT_TO_STDLOGIC8 is a user-defined function that converts Integer data types into STD_LOGIC data types. The integer values on the external signals, P1 and P3, are first converted to STD_LOGIC data format; then, sent to the input pins, P1_H and P3_H, respectively, of chip HORIZONTAL_FILTER1.

Similarly, the STD_LOGIC values at output H of HORIZONTAL_FILTER1 need to be converted to integer, and sent to external signal W_H. The following statement accomplishes this task:

```
STDLOGIC_TO_INT(H) => INT_TO_STDLOGIC12(W_H);
```

Function STDLOGIC_TO_INT is a user-defined function that converts values of type STD_LOGIC to type Integer. The function INT_TO_STDLOGIC12 is a user-defined function

```
configuration WINDOW_H_RTL_OTHERS_INTEGER
            of WINDOW_PROCESSOR is
  generic map(HORIZ_DELAY=>3 ns, VERT_DELAY=>3 ns,
         LEFT_DIAG_DELAY=>3 ns,  RIGHT_DIAG_DELAY=>3 ns,
         WAIT_TIME=>0 ns)
    port map(CLOCK, MEM_OUT1, MEM_OUT2, MEM_OUT3,
          E_H, E_V, E_DL, E_DR);
    for structure
      for HP:HORIZONTAL_FILTER1
        use configuration STRUC_RTL.H_RTL
          generic map(HORIZ_DELAY=>HORIZ_DELAY)
          port map (CLOCKH=>CLOCK,
            P1H=>INT_TO_STDLOGIC8(P1),
            P3H=>INT_TO_STDLOGIC8(P3),
            STDLOGIC_TO_INT(H)=>INT_TO_STDLOGIC12(W_H));
      end for;
      for VP: VERTICAL_FILTER1
        use entity STRUC_INT.VERTICAL_FILTER(BEHAVIOR)
          generic map(VERT_DELAY=>VERT_DELAY,
                   WAIT_TIME=>WAIT_TIME)
          port map (CLOCK, P1, P2, P3, W_V);
      end for;
      for LDP: LEFT_DIAG_FILTER1
        use entity STRUC_INT.LEFT_DIAG_FILTER(BEHAVIOR)
          generic map(LEFT_DIAG_DELAY=>LEFT_DIAG_DELAY,
                   WAIT_TIME=>WAIT_TIME)
          port map (CLOCK, P1, P2, P3, W_DL);
      end for;
      for RDP: RIGHT_DIAG_FILTER1
        use entity STRUC_INT.RIGHT_DIAG_FILTER(BEHAVIOR)
          generic map(RIGHT_DIAG_DELAY=>RIGHT_DIAG_DELAY,
                   WAIT_TIME=>WAIT_TIME)
          port map (CLOCK, P1, P2, P3, W_DR);
      end for;
    end for;
end;
```

Figure 11.54 Configuration declaration for window processor structural model with horizontal filter at the RTL and other filters at higher level.

that converts Integer data type to 12-bit STD_LOGIC type. It is needed because the port is mode *inout*. After it is sent to the port, a data value has type Integer. If it needs to be read internally, it must be converted from Integer back into the internal 12-bit format.

The configuration also defines the type of chip to be inserted into each socket. For example, the RTL model of the horizontal filter defined by configuration STRUC_RTL.H_RTL is inserted into the socket labeled HP. Figure 11.55 shows the fully configured structural model of the window processor that is defined by configuration WINDOW_H_RTL_OTHERS_ INTEGER. Compare this with Figure 11.53 to see what the configuration declaration adds to the structural description. The specific chip inserted into each socket is represented by a box inside the socket symbol. The box contains the name of the specific chip to be inserted into the socket.

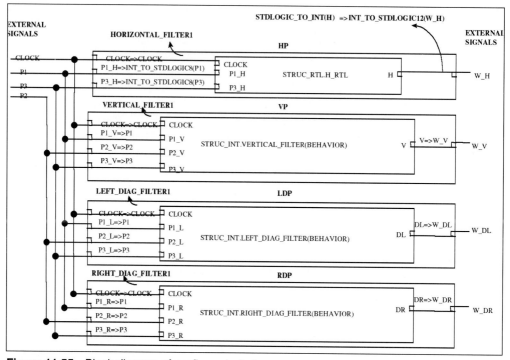

Figure 11.55 Block diagram of configured window processor.

Consider the chip inserted into the vertical filter socket, VP. The chip name includes the entity name (VERTICAL_FILTER), the architecture name (BEHAVIOR), and the name of the library containing the model (STRUC_INT).

11.6.3 Test Bench Development at the RTL

The test bench used for both the executable specification and for the structural model can be reused to test the RTL model. To achieve this we had to change only the architecture and library names for all the filters in the window processor and the magnitude and direction processor in the configuration declaration from STRUC_INT to STRUC_RTL. Also, the generic constant WAIT_CYCLES in the component TEST in the configuration declaration must be changed to 13 since an additional five clock cycles are required for the register delays in comparison with the MUT at the structural level of abstraction. The testing methodology is similar to the one adopted for testing structural-level models. Each of the subcomponents is first tested with its individual test bench. Then, the subcomponents are integrated into upper-level models. If the configuration approach is used, the test benches used to test the original upper-level models can be reused to test the same models with the RTL components replacing the original behavioral components. Since no new principles are involved, we do not show examples for these test benches (see problems at the end of the chapter).

11.7 DETAILED DESIGN AT THE GATE LEVEL

The gate level is the lowest abstraction level at which a model is typically represented in VHDL. Each model at the RTL is individually transformed into a gate-level model. The RTL structural model can be reused by writing a new configuration that binds the structural components to the gate-level models instead of to the RTL models.

The translation of RTL models to gate-level models can be done manually or by using synthesis tools. One motivation for using VHDL is the possibility of automatic synthesis of circuits. At the RTL, a design is represented by a circuit consisting of components, such as adders, subtracters and registers. Packages, if any, and VHDL code for the RTL components are analyzed, and read by the synthesis tool. Next, the RTL structural model of the system is read to generate a gate-level circuit. See Chapter 10 for details on synthesis tools.

The Design Analyser is the synthesis tool for the Synopsis system. In the next section, we show how to use this tool to obtain a gate-level circuit for the horizontal filter.

11.7.1 Gate-Level Design of Horizontal Filter

The synthesis tool does not support all VHDL constructs. In the RTL model for the horizontal filter shown in Figures 11.49, 11.50, and 11.51, the only constructs that are not synthesizable are the timing delays. The following changes must be made to those models to make them synthesizable:

1. Remove the generic HORIZ_DELAY from the entity declaration in Figure 11.49.
2. Remove the generics HORIZ_DELAY from the component bindings in Figures 11.50 and 11.51.
3. Bind all components to entities in library STRUC_GATE, instead of to components in library STRUC_RTL.
4. The components in library STRUC_GATE are modified versions of the components with the same name in library STRUC_RTL. In each model, any generics of type TIME were removed and all *after* clauses were removed from the signal assignment statements.

It should not be surprising that signal delays are not supported by the synthesis tool. After all, the synthesis tool is mapping the design to a library of components that is tied to a specific technology. For example, a CMOS gate has an internal delay that is dictated by the CMOS technology. That delay cannot be changed to accommodate the wishes of a designer. Therefore, it will usually not be possible to design a gate-level circuit using real gates that have a specified overall delay. For this reason, the synthesis tool does not support *after* clauses in signal assignment statements. See Chapter 10 for a thorough discussion of this and other synthesis constraints.

Figure 11.56 shows the gate-level circuit produced by the synthesis tool for the horizontal filter. In the next section, we discuss optimization of this filter with respect to area and timing.

11.7.2 Optimization of Gate-Level Circuits

After the RTL model has been synthesized, the next step is to optimize the circuit. A typical optimization methodology is to optimize for area first and then for timing (see Chapter 10 for

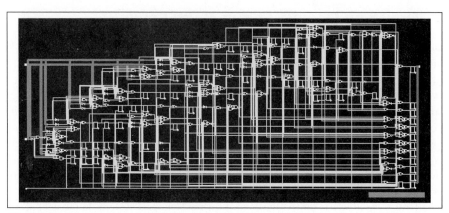

Figure 11.56   Gate-level circuit for horizontal filter produced by the synthesis tool.

other optimization strategies). Hierarchical blocks in a large design are normally optimized starting from the lower-level blocks in a bottom-up process. The synthesis tools carry out the design optimization according to user-defined constraints. Current synthesis tools typically allow constraints to be set for minimal area and minimal timing delay.

The optimization processes use different algorithms to find a circuit implementation that best fits the constraints. A circuit is optimized for minimum area, based on what the optimizer can find. This might not always be the absolute minimum circuit that could be produced if the circuit were carefully designed by hand.

Constraints represent desired circuit characteristics that are specified as part of the design goals. Different constraints cause different optimized circuits to be generated, but with the same functionality. The most common constraints used today are area and timing.

Area: An area constraint is a number corresponding to the desired maximum area of a specified design module. The area number will have units corresponding to the units defined in the target technology library, for example, equivalent gates, cells or transistors.

Timing: Timing constraints put limits on the maximum delay from any input to any output. The static timing analyzer in the synthesis tool extracts timing information in order to compute actual path delays. This includes setup time and hold time of registered elements, and signal delays through combinational logic components.

Signal path delays in the model are computed and compared with specified timing constraints; then, automatic optimization is performed as needed to improve timing characteristics. One strategy is to progressively reduce maximum delay constraints to obtain a series of circuits with various areas, delays, and power consumption. The designer can then select one of the designs, thereby effectively trading off these parameters against each other. The next section applies this strategy to optimize the gate-level circuit for the horizontal filter.

11.7.2.1   Optimization of the Horizontal Filter

Table 11.4 shows the results of a series of synthesis steps for the horizontal filter. The top line in the table shows the initial circuit obtained without any constraints. In line 2, the maximum area is specified as 0, and the clock period is specified as 50 ns with 1 ns skew. By attempting to reach 0 area, the system will obtain the minimum area possible for this timing constraint. Note: The area is the same as for the unconstrained case, but the power consumption is greatly reduced.

We then gradually reduce the clock period from 50 ns to 9 ns and repeat the optimization. The following table gives the series of optimization results. As we decrease the clock period, in general, we get faster circuits that have greater area and consume more power. The designer can trade off circuit speed against area and power by making a selection from the table that is considered optimal. The slack is the difference between the maximum circuit delay and the clock period.

Table 11.4 Series of synthesis steps for the horizontal filter.

Circuit	Maximum area specified	Clock period specified (ns)	Clock skew specified (ns)	Area of circuit generated	Slack in actual circuit	Power consumed in actual circuit (uW)
1	Not specified	Not specified	Not specified	706	Not specified	276.24
2	0	50	1	706	29.02	13.86
3	0	31	1	712	10.53	21.87
4	0	21	1	709	0.81	31.94
5	0	18	1	712	0.3	38.07
6	0	14	1	740	0.05	52.44
7	0	11	1	751	0.02	66.69
8	0	9	1	823	0.01	98.01

11.7.3 Gate-Level Testing

Having synthesized the MUT at the RTL into a netlist, it is important to verify that the functionality of the synthesized netlist matches the functionality of the RTL model. This step is important since a synthesis tool may make interpretations about the VHDL code that might not be identical to those intended by the modeler (see Chapter 10 for details).

As one descends in the hierarchy, the models become increasingly detailed. A major issue is how to compare the outputs of the gate-level model with the outputs of the higher-level models. This introduces the *back annotation* problem. How do we assign delays in the high-level model so that the outputs occur at the same time as the outputs from the gate-level model? Back annotation can be defined as the process of extracting information about timing and circuit characteristics and inserting that information back into the higher-level models, so that the timing of the outputs in the two models matches.

11.7.4 Methodology for Back Annotation

The test bench model developed for the behavioral model of the MUT can be modified to test the model at the gate level. A comparison process is added to the test bench to compare the output of the model at the behavioral level with that at the gate level and to generate a PASS/FAIL indication.

The procedure for obtaining the necessary delay value for back annotation into the behavioral model is:

1. Both the behavioral-level model and the gate-level model are instantiated in the test bench.
2. A clock generator and a comparison process are added. The clock generator drives both the behavioral-level model and the gate-level model. The comparison process compares the outputs of the two models and generates a PASS/FAIL indication.
3. Develop a configuration for the test bench that initially specifies no delay values for the behavioral-level model and simulate the test bench. By observing the outputs of the simulator, determine the relative delay between outputs of the two models.
4. Back annotate the observed relative delay into the behavioral model and simulate again. The outputs from the two models should now agree when the circuits are stable. Due to circuit hazards, the two outputs may disagree shortly after an input change or a clock edge.

The back annotated behavioral model can now be used in system-level simulations with accurate delays.

PROBLEMS

11.1 Write an entity declaration (named WINDOW_PROCESSOR) and a behavioral architecture (named BEHAVIOR) for the window processor component described in Section 11.5.1 and shown in Figure 11.40 as the WIN_PROCESSOR component. Analyze the model and store the analyzed model in library STRUC_INT. Then, use the test bench shown in Figure 11.42 and the configuration shown in Figure 11.43 to test the behavioral architecture.

11.2 Create a behavioral VHDL model for the memory processor component described in Section 11.5.1 and shown in Figure 11.40 as the MEM_PROCESSOR component. Use the following design procedure.

a. Write an entity declaration and a behavioral architecture for the memory processor. Name the entity MEMORY_PROCESSOR and the architecture BEHAVIOR. Analyze the entity and its behavioral architecture and store the analyzed models in library STRUC_INT.

b. Write a test bench for the memory processor component designed in (a). Do not bind the component in the test bench. Use the style in Figure 11.42. Analyze this test bench and store the analyzed model in library STRUC_INT.

c. Write a configuration that binds the memory processor component to its behavioral architecture. Use the style in Figure 11.44. Analyze the configuration and store the analyzed model in library STRUC_INT.

d. Simulate the configuration in (c) to validate the behavioral model of the memory processor.

11.3 Repeat Problem 11.2 for the magnitude processor component described in Section 11.5.1 and shown in Figure 11.40 as the MAG_PROCESSOR component. Name the entity MAG_PRO-CESS and the behavioral architecture BEHAVIOR.

11.4 Write an entity declaration and a behavioral architecture for the horizontal filter component described in Section 11.5.2 and shown in Figure 11.40 as the component HOR_FILTER. Use the independent test bench shown in Figure 11.41 to test the model.

11.5 Write an entity declaration and a behavioral architecture for the vertical filter described in Section 11.5.2 and shown in Figure 11.40 as the component VERT_FILTER. Write an independent test bench to test the vertical filter behavioral model similar to the one in Figure 11.41. Use a configuration to bind the components in the test bench to their behavioral architectures. Simulate the test bench configuration to validate the behavioral model.

11.6 Write an entity declaration and a behavioral architecture for the left diagonal filter described in Section 11.5.2 and shown in Figure 11.40 as the component LEFT_FILTER. Write an independent test bench to test the left diagonal filter behavioral model similar to the one in Figure 11.41. Use a configuration to bind the components in the test bench to their behavioral architectures. Simulate the test bench configuration to validate the behavioral model.

11.7 Write an entity declaration and a behavioral architecture for the right diagonal filter described in Section 11.5.2 and shown in Figure 11.40 as the component RIGHT_FILTER. Write an independent test bench to test the right diagonal filter behavioral model similar to the one in Figure 11.41. Use a configuration to bind the components in the test bench to their behavioral architectures. Simulate the test bench configuration to validate the behavioral model.

11.8 Write an entity declaration and a behavioral architecture for the address generator component described in Section 11.5.2 and shown in Figure 11.40 component ADDR_GEN. Write an independent test bench to test the address generator behavioral model similar to the one in Figure 11.41. Use a configuration to bind the components in the test bench to their behavioral architectures. Simulate the test bench configuration to validate the behavioral model.

11.9 Write an entity declaration and a behavioral architecture for the memory component described in Section 11.5.2 and shown in Figure 11.40 as the component MEMORY. Write an independent test bench to test the memory behavioral model similar to the one in Figure 11.41. Use a configuration to bind the components in the test bench to their behavioral architectures. Simulate the test bench configuration to validate the behavioral model.

11.10 Create a structural VHDL model for the window processor component described in Section 11.5.1 and shown in Figure 11.40 as the WINDOW_PROCESSOR component. Use the following design procedure.

    a. Write an entity declaration and a structural architecture for the window processor. Do not specify any bindings for the four filter components. Use the style illustrated in Figures 11.36 and 11.37. Analyze the entity and its structural architecture and store the analyzed models in library STRUC_INT.

    b. Write a configuration (W_INT) that binds the four filter components to their behavioral models that were created in Problems 11.4, 11.5, 11.6, and 11.7. Analyze the configuration and store the analyzed model in library STRUC_INT.

    c. Reuse the test bench shown in Figure 11.42 to validate your structural model of the window processor. If you have not previously done so, analyze the test bench model of Figure 11.42 and store the analyzed model in library STRUC_INT. Write a configuration using the style illustrated in Figure 11.44 to bind your structural model of the window processor to component WINDOW_PROCESSOR1 in the test bench of Figure 11.42. Simulate the configuration to validate your structural model of the window processor. Note: This configuration binds the window processor component, WINDOW_

PROCESSOR1, to the structural model you created in (a) using the configuration you wrote in (b).

11.11 Create a structural VHDL model for the memory processor component described in Section 11.5.1 and shown in Figure 11.40 as the MEM_PROCESSOR component. Use the following design procedure.

    a. Write an entity declaration and a structural architecture for the memory processor. Do not specify any bindings for the two components. Use the style illustrated in Figures 11.36 and 11.37. Analyze the entity and its structural architecture and store the analyzed models in library STRUC_INT.

    b. Write a configuration (MEM_INT) that binds the two components to their behavioral models that were created in Problems 11.8 and 11.9. Analyze the configuration and store the analyzed model in library STRUC_INT.

    c. Reuse the test bench created in Problem 11.2(b) to validate your structural model of the window processor. If you have not previously done so, analyze the test bench model and store the analyzed model in library STRUC_INT. Write a configuration, using the style illustrated in Figure 11.44, to bind your structural model of the memory processor to the memory component in the test bench. Simulate the configuration to validate your structural model of the memory processor. Note: This configuration binds the memory processor component in the test bench to the structural model you created in (a) using the configuration you wrote in (b).

11.12 Rewrite the code for the structural model of the Sobel edge detector shown in Figures 11.36 and 11.37 in the style of Section 11.6.2. Write configurations using the style of Figure 11.50 for each of the following situations.

    a. All components with Integer data types.

    b. All components with STD_LOGIC data types.

    c. Two components with Integer data types and one with STD_LOGIC data types.

    d. Two components with STD_LOGIC data types and one with Integer data types.

11.13 Rewrite the code for the structural model of the window processor created in Problem 11.10 in the style of Section 11.6.2. Write configurations using the style of Figure 11.50 for each of the following situations.

    a. All components with Integer data types.

    b. All components with STD_LOGIC data types.

    c. Three components with Integer data types and one with STD_LOGIC data types.

    d. Three components with STD_LOGIC data types and one with Integer data types.

    e. Two components with STD_LOGIC data types and one with Integer data types.

11.14 Rewrite the code for the structural model of the memory processor created in Problem 11.11 in the style of Section 11.6.2. Write configurations using the style of Figure 11.50 for each of the following situations.

    a. All components with Integer data types.

    b. All components with STD_LOGIC data types.

    c. One component with Integer data types and one with STD_LOGIC data types.

11.15 Write configurations similar to that in Figure 11.54 for each of the following situations. Draw block diagrams similar to those in Figures 11.53 and 11.55 to illustrate each configuration.

    a. Horizontal and vertical filters both have RTL models with STD_LOGIC data types, but left and right diagonal filters still have only behavioral models with Integer data types.

      b. Horizontal, vertical, and left diagonal filters all have RTL models with STD_LOGIC data types, but the right diagonal filter still only has a behavioral model with Integer data types.

      c. All filters now have RTL models with STD_LOGIC data types.

11.16 Show how to reuse the individual test bench for the horizontal filter (Figure 11.41) to test the RTL model for the filter (Figures 11.49, 11.50, and 11.51).

11.17 Develop an RTL model for the vertical filter VERT_FILTER in Figure 11.40. Use the techniques developed in Section 11.6.1.1. Show how to reuse the test bench created in Problem 11.5 to test the RTL model.

11.18 Develop an RTL model for the left diagonal filter LEFT_FILTER in Figure 11.40. Use the techniques developed in Section 11.6.1.1. Show how to reuse the test bench created in Problem 11.6 to test the RTL model.

11.19 Develop an RTL model for the right diagonal filter RIGHT_FILTER in Figure 11.40. Use the techniques developed in Section 11.6.1.1. how how to reuse the test bench created in Problem 11.7 to test the RTL model.

11.20 Develop an RTL model for the address generator component ADDR_GEN in Figure 11.40. Use the techniques developed in Section 11.6.1.1. Show how to reuse the test bench created in Problem 11.8 to test the RTL model.

11.21 Develop an RTL model for the memory component MEMORY in Figure 11.40. Use the techniques developed in Section 11.6.1.1. Show how to reuse the test bench created in Problem 11.9 to test the RTL model.

11.22 Using a technique similar to that shown in Table 11.4, synthesize the RTL model of the vertical filter created in Problem 11.17.

11.23 Using a technique similar to that shown in Table 11.4, synthesize the RTL model of the left diagonal filter created in Problem 11.18.

11.24 Using a technique similar to that shown in Table 11.4, synthesize the RTL model of the right diagonal filter created in Problem 11.19.

11.25 Using a technique similar to that shown in Table 11.4, synthesize the RTL model of the address generator created in Problem 11.20.

11.26 Using a technique similar to that shown in Table 11.4, synthesize the RTL model of the memory created in Problem 11.21.

11.27 Use the methodology described in Section 11.7.4 to back annotate accurate delays into the behavioral models of each of the following components:

      a. Horizontal filter

      b. Vertical filter

      c. Left diagonal filter

      d. Right diagonal filter

      e. Address generator

      f. Memory component

11.28 PROJECT: Can the magnitude processor MAG_PROCESSOR in Figure 11.40 be logically decomposed? If so, do the decomposition using the configuration techniques discussed in this chapter to specify test benches for testing the components and the integrated system.

11.29 PROJECT: Use the techniques in this chapter to design the calculator described in Problem 10.22.

11.30 PROJECT: Use the techniques in this chapter to design the Booth multiplier described in Problem 10.23.

11.31 Write the configuration CONFIG_SOBEL_S_L2 that is used in Figure 11.45. This configuration should use behavioral models for the memory processor and magnitude processor and a structural model for the window processor.

11.32 Write a configuration CONFIG_SOBEL_S_L3 that uses structural models for both the memory processor and the window processor and a behavioral model for the magnitude processor. Rewrite the test bench configuration shown in Figure 11.45 to accommodate the new configuration.

11.33 Write a configuration CONFIG_SOBEL_S_L4 that uses a structural model for the memory processor and behavioral models for the window processor and magnitude processor. Rewrite the test bench configuration shown in Figure 11.45 to accommodate the new configuration.

Synthesis Algorithms for Design Automation

T he term *algorithmic synthesis* is used in this chapter to describe the automated translation from an abstract behavioral specification at the algorithmic level into a register-level (or gate-level) specification. The target specification can be either a behavioral model or a structural model. Current experimental systems tend to target the structural specification although an intermediate, behavioral finite state machine specification could offer significant advantages by reducing the complexity of some steps in the synthesis. Regardless of the target specification, the steps to be performed are similar.

There are many ways to translate a given behavioral description at the algorithmic level into a description at the register or gate level. In fact, the number of possibilities is generally so large as to be intractable. The synthesis system must select the best design from this very large design space. Exhaustive search is usually not a viable approach. In many cases, where off-the-shelf components are to be used, the design space is constrained by available component types. For custom-designed integrated circuits, commonly referred to as ASICs, the design may be constrained by area or power limitations. The specifications may impose maximum delay times, maximum clock frequencies, or other timing constraints. The synthesis system must search a very large design space for the lowest cost structure to meet all the constraints imposed by the specifications and technology. The task is formidable.

12.1 BENEFITS OF ALGORITHMIC SYNTHESIS

Considering the problem is so difficult, are the benefits worth the effort? These benefits are derived from an algorithmic synthesis capability.

1. **Shorter design cycle.**   By automating the high-level synthesis task, designs may be completed in a more timely manner. This allows a company to be more successful in a highly competitive market.

2. **Lower design cost.**   A shorter design cycle leads naturally to lower design costs. This is particularly important for low-volume production items, where the design cost is the major cost factor.

3. **Lower production cost.**   Because decisions made at the algorithmic level have a much greater impact on the final system complexity (and, therefore, system cost), than decisions made at lower levels of design, substantial reductions in system cost are possible by optimizing decisions at the high level. In current design practices, such optimizations, if achieved at all, are the result of extensive design experience by the design team. An intimate knowledge of lower-level technology details is required. With high-level synthesis tools, the systems designer can take advantage of technology dependent trade-offs without personally possessing extensive knowledge about the details of the technology itself. The result should be better designs.

4. **Fewer design errors.**   Automation removes human error from the design equation. For highly complex designs, the probability of human error increases dramatically. After the synthesis program is debugged (a very difficult task), it should be unsusceptible to a wide range of errors that tend to creep into systems designed by humans.

5. **Easier to determine available design trade-offs.**   The automated synthesizer can quickly and economically produce several designs that meet the behavioral specifications. Using traditional design techniques, the high cost of producing alternative designs frequently prohibits the exploration of a wide range of design trade-offs. Automated design allows the designer to evaluate trade-offs more effectively.

6. **Documentation standards are easier to maintain.**   The automated system can produce the required documentation, including keeping track of design decisions and the reasons for those decisions.

7. **Specification changes are easier to accommodate.**   In traditional design procedures, specification changes are very expensive to implement. Using a high-level synthesizer, the behavioral model can be modified and its function verified. Then, the synthesizer can be applied to redesign the system. The cost and time needed to accomplish this task can be dramatically reduced.

12.2   ALGORITHMIC SYNTHESIS TASKS

Figure 12.1 illustrates the major synthesis tasks. In this book, we begin the high-level synthesis process with a VHDL behavioral description at the algorithmic level. The first task, commonly called *compilation*, is to convert the VHDL description into an intermediate format suitable for automated synthesis. In this book, we will use the data flow graph (DFG) as the intermediate form, although other abstractions are commonly used as well. To demonstrate the synthesis steps, we use the simple VHDL description shown in Figure 12.2. Notice that data types at the high level are often integer or other arithmetic types.

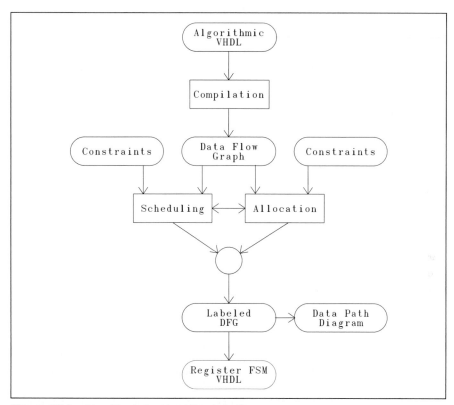

Figure 12.1 Algorithmic-level synthesis tasks.

```
---------------------------------------------------------
-- The entity declaration of synthesis example model.
---------------------------------------------------------
entity SYNEX1 is
  port (A, B, C, D, E: in INTEGER;
        X, Y: out INTEGER);
end SYNEX1;
---------------------------------------------------------
-- The architecture declaration of synthesis example model.
---------------------------------------------------------
architecture HIGH_LEVEL of SYNEX1 is
begin
  X <= E*(A+B+C);
  Y <= (A+C)*(C+D);
end HIGH_LEVEL;
```

Figure 12.2 Example of algorithmic-level VHDL description.

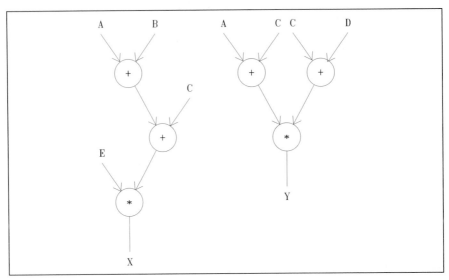

Figure 12.3 Data flow graph example.

12.2.1 Compilation of VHDL Description into an Internal Format

The first major task is to convert the VHDL into a form that is convenient for internal processing. The compilation involves *transformations* and/or *optimizations* that will improve the chances of obtaining a low-cost solution. These can include software-based activities, such as dead code elimination, common subexpression identification, inline expansion of procedures, loop unrolling, and so on. Transformations that are hardware-oriented are also performed, such as integer to binary vector substitution, replacement of multiplication by 2 by a left shift, etc. A data flow graph is frequently used as an intermediate form. Figure 12.3 shows the data flow graph for the example in Figure 12.2. Nodes in the DFG represent operations in the VHDL. Arcs in the DFG represent precedence relations among the operations. For example, an arc from + to * indicates that the addition operation must be performed before the multiplication operation.

12.2.2 Scheduling

Scheduling consists of assigning each operation to a control step. The *control step* is the basic unit of time in a synchronous system and corresponds to one clock cycle. The primary goal of scheduling is to minimize the time needed to complete the global function performed by the device. In computer terms, scheduling attempts to minimize the number of clock cycles needed to execute the algorithm described by the original VHDL program. The scheduling might also be subject to other constraints imposed by the target technology or available functional units. Figure 12.4 shows a straightforward schedule for the example device using only three control steps, assuming there are no other constraints. This schedule uses the ASAP method where each operation is scheduled to occur at the earliest possible time. Although this method achieves maximum throughput when there are no constraints imposed on the scheduling, it generally requires

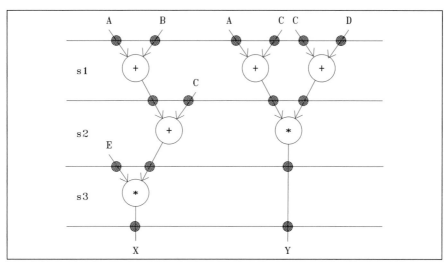

Figure 12.4 ASAP schedule for example data flow graph.

excessive amounts of hardware to implement. For example, this ASAP schedule requires three separate adders to accommodate three parallel add operations during control step s1.

12.2.3 Allocation

Allocation refers to the process of specifying the components in the system and the interconnections among the components. Several specific activities can be identified, although they are not always independent.

- Allocation of register or RAM storage to hold data values.
- Allocation of functional units to perform specified operations.
- Allocation of interconnection paths for the transmission of data among components.

12.2.3.1 Allocation of Registers to Hold Data Values

Figure 12.5 shows one possible allocation of registers and functional units to implement the schedule shown in Figure 12.4. During allocation, some additional knowledge about inputs and outputs is needed. For this allocation, we assumed that the inputs must all be latched before beginning algorithm execution. The addition of state s0 accomplishes the latching operation.

The dark circles in Figure 12.5 indicate data storage requirements. For example, between step s0 and step s1, the device must store five data values corresponding to the five data inputs. Inputs A, B, C, D, and E were assigned to registers R1, R2, R3, R4, and R5, respectively. Between steps s1 and s2, there are again five data values to be stored. The same five registers were reused. Deciding which registers to use for each data item will have a significant effect upon the cost of the system in terms of interconnection complexity, which relates to several important system parameters, such as dollar cost, signal delay, chip area, and fan-in and fan-out loading. Optimizing register assignments is a complex problem.

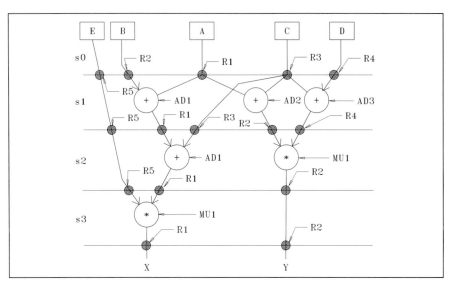

Figure 12.5 Allocation of registers and functional units for the ASAP schedule for example data flow graph.

12.2.3.2   Allocation of Functional Units to Perform Specific Operations

Allocation may be constrained to contain only elements from a given library of components. This form is known as bottom-up allocation because the process is driven by the availability of components in the library. If the designer is free to implement any desired component type, as in ASIC design, or top-down design, then an unconstrained form of allocation can be used. It is clear that several different allocation algorithms might be used for different purposes in the same design group. For this design, we assume that the component library contains adders, multipliers, registers, and multiplexers (MUXes). We will use the bottom-up approach.

During s1, there are three separate additions executing in parallel. Therefore, three separate adders are required, denoted as AD1, AD2, and AD3. During step s2, there is only one addition operation executing. It was allocated to AD1. Since there is at most one multiplication executing during any step, one multiplier is sufficient. All multiply operations are assigned to the same component (MU1).

12.2.3.3   Allocation of Data Paths for Interconnections Among Components

Figure 12.6 shows one possible allocation of data paths for the example system. Table 12.1 shows the component requirements for the data path assignment in Figure 12.6.

12.2.4   Interaction of Scheduling and Allocation

It is apparent that scheduling and allocation are not independent operations. Consider the following observations:

- For two operations to be scheduled in the same control step, it is necessary that the two operations use different functional units. This implies knowledge about the allocation process.

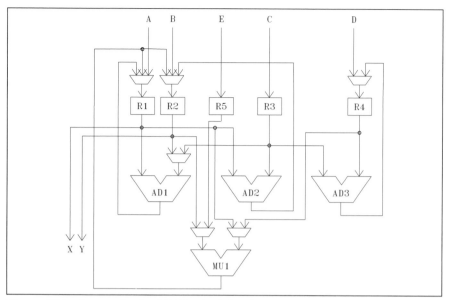

Figure 12.6 Data paths for ASAP scheduling for example data flow graph.

Table 12.1 Components for data path assignment in Figure 12.6.

Component	Quantity
Adder	3
Multiplier	1
Register	5
2x1 MUX	4
3x1 MUX	2

- To obtain an optimum schedule when operations require more than one clock period to execute, we must know how long it takes to execute each operation. This requires knowledge about the speed of the components, which comes from the allocation process.
- To reduce component cost, we want to minimize the number of components. As noted above, the number of components required is influenced by the schedule.

Therefore, it seems that to optimize the allocation of operations to functional units, we must know the schedule. But, in order to optimize the schedule, we must know the allocation of operations to functional devices. The interaction of these two activities presents formidable obstacles to the goal of system optimization.

One approach is to arbitrarily, or systematically, set limits on the allocation process and obtain an optimum schedule based on these limits. Another way is to use iteration by selecting various allocation limits and scheduling for each iteration. Frequently, heuristics are used to

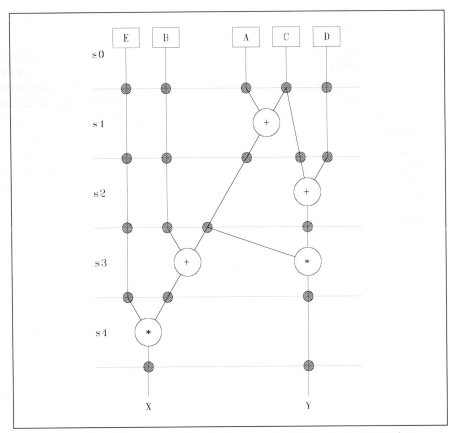

Figure 12.7   Schedule for example data flow graph with hardware constraints.

select the next iteration, in an attempt to have the algorithm converge to an overall optimum design. Other systems develop the schedule and do the allocation simultaneously. Such systems are usually driven by specific goals for the final design, for example, maximum speed or minimal area.

Suppose that the design engineer decides to use only one adder and one multiplier in the example data flow graph shown in Figure 12.3. The schedule shown in Figure 12.7 might be the result. Notice that the hardware constraint mandates one additional control step. The speed of the device is correspondingly reduced. Execution time increases from four clock periods to five clock periods. Also, the given schedule cannot be obtained from the original specification directly. It is necessary to transform the original specification by applying known laws of algebra. The transformed expressions are:

$$X <= E * [(A+C)+B]$$
$$Y <= (A+C) * (C+D)$$

Notice that the common expression (A+C) has been obtained.

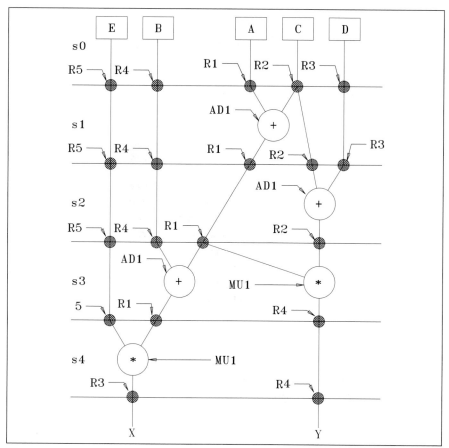

Figure 12.8 Allocation for constrained schedule.

Figure 12.8 shows an allocation of registers and functional units for the constrained schedule, and Figure 12.9 shows one possible allocation of data paths. Table 12.2 shows the component requirements. Relative to the original component count shown in Table 12.1, two adders have been eliminated, two MUXes have been reduced in size from 3x1 to 2x1, and one additional 2x1 MUX has been added. This implementation should require less area than the original design. The issue of interconnect cost has not been addressed.

Table 12.2 Components required for constrained schedule.

Component	Quantity
Adder	1
Multiplier	1
Register	5
2x1 MUX	7

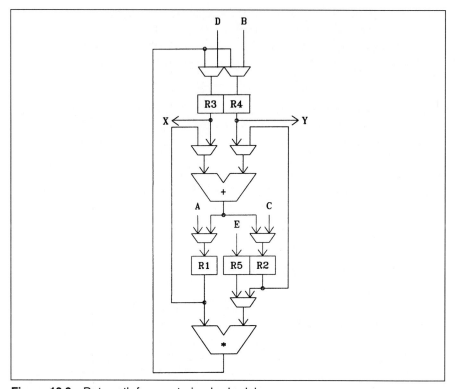

Figure 12.9   Data path for constrained schedule.

	s0	s1	s2	s3	s4
AD1	--	b	b	b	--
MU1	--	--	--	b	b

Figure 12.10   Gantt chart for constrained schedule.

12.2.5   Gantt Charts and Utilization

Before describing scheduling techniques in detail, some basic definitions need to be discussed. After the operations have been assigned to functional units, the Gantt chart is a useful tool for evaluating the results. Figure 12.10 shows the Gantt chart for the schedule of Figure 12.8. In this figure, the symbol "b" indicates that a component is busy, and the symbol "--" indicates that the component is idle. There is a separate row in the Gantt chart for each component in the system, and a separate column for each time step, or control step. This Gantt chart indicates that adder AD1 is being utilized during time steps s1, s2, and s3 and is idle during time steps s0 and s4. Similarly, mutiplier M1 is utilized during time steps s3 and s4 and is idle at all other time steps.

The *execution time* is the total time required for process execution. For the example, the execution time is 5 because five time steps are required. A common goal of scheduling and allocation is to minimize the execution time for a process. When applied to the design methodology, minimizing execution time results in minimum delay in the resulting hardware. As a result, the system operates at maximum speed.

The *utilization* of a component is the ratio of the busy time for the component to the execution time for the process. In the example, the adder utilization is 0.6 (or 60 percent) and the multiplier utilization is 0.4 (or 40 percent). It is sometimes appropriate to maximize utilization because, in some sense, maximum utilization implies optimum use of available resources. If utilization is low, it might be possible to reduce the number of components. Often, however, such reduction results in an increase in execution time. The result is a trade-off of speed for cost. Nevertheless, very low utilization of components may indicate an inefficient design. In other situations, including many VLSI designs, low utilization is of no concern. In the example, the relatively low utilization is an inherent property of the process, and therefore, improving utilization should not be a goal.

12.2.6 Creating FSM VHDL from an Allocation Graph

The FSM VHDL model can be constructed directly from the allocation graph. Figure 12.11 shows the FSM VHDL model for the constrained schedule of Figure 12.8. It was constructed in a straightforward manner by simply declaring the required registers, declaring the set of control steps (called *states* in the VHDL program), and inserting the required data transfers in each state. The Moore state machine template described in Chapter 8 is used. The clock input (CLK) provides timing for state transitions. Notice that we assumed the state following S4 was S0, although the algorithmic level model did not specify any state to follow S4. Since the output is independent of state, the Moore machine template output routine does not need a *case* statement. This simplifies the template somewhat.

To make the device practical, we should add a start signal at the input and provide a completion signal at the output (see Problem 12.1). The model shown in Figure 12.11 should be simulated to verify correct operation of the circuit. Following verification, the hardware can be designed by mapping the integer registers to bit vectors and designing either a hard-wired control unit or a microprogrammed control unit as described in Chapter 8. Finally, the add and multiply functional units with bit vector inputs and outputs must be designed. If they are to be ASIC units, then the techniques described in Chapter 11 can be applied (top-down design). Alternatively, the functional units can be selected from an existing library (bottom-up design).

Using the methods of this section combined with those described in Chapters 5, 8, and 11, we now have a methodology for using VHDL to design systems starting with a word description and finishing with a register-level structural representation, including the design of the control unit. Figure 12.12 summarizes the methodology. The methods in Chapter 5 are employed to create an algorithmic VHDL model from a given word description. Then, the methods in Chapter 12 are used to convert the algorithmic VHDL model to a FSM VHDL model. The methods in Chapter 8 are used to design the control unit for the FSM. The methods in Chapter 12 are used to define a data path diagram. Finally, the individual functional units are either retrieved from a library or are created using the design techniques of Chapters 9, 10, and 11.

```
---------------------------------------------------------
entity FSMEX1 is
  port (A, B, C, D, E: in INTEGER;
        CLK: in BIT;
        X, Y: out INTEGER);
end FSMEX1;
---------------------------------------------------------
architecture FSM of FSMEX1 is
  type STATE_TYPE is (S0, S1, S2, S3, S4);
  signal STATE: STATE_TYPE;
  signal R1, R2, R3, R4, R5: INTEGER;
begin
-- Process to update state and perform register transfers.
STATEP: process (CLK)
  begin
    if CLK'event and CLK='1' then
      case STATE is
        when S0 =>
          -- Data Section
          R5 <= E; R4 <= B; R3 <= D; R2 <= C; R1 <= A;
           -- Control Section
          STATE <= S1;
        when S1 =>
          -- Data Section
          R1 <= R1 + R2;
          -- Control Section
          STATE <= S2;
        when S2 =>
          -- Data Section
          R2 <= R2 + R3;
          -- Control Section
          STATE <= S3;
        when S3 =>
          -- Data Section
          R1 <= R4 + R1;
          R4 <= R1 * R2;
          -- Control Section
          STATE <= S4;
        when S4 =>
          -- Data Section
          R3 <= R5 * R1;
          -- Control Section
          STATE <= S0;
      end case;
    end if;
  end process STATEP;
  -- Assign output.
  X <= R3; Y <= R4;
end FSM;
```

Figure 12.11 FSM VHDL model example.

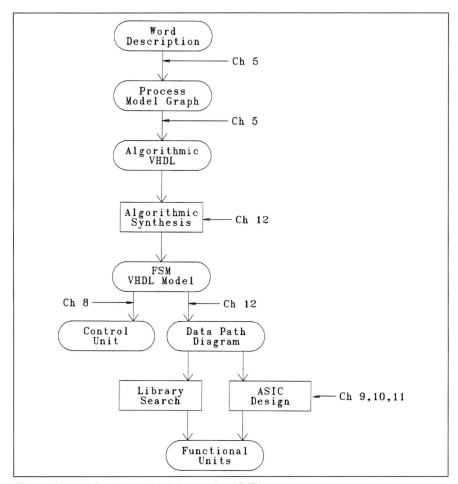

Figure 12.12   Design methodology using VHDL.

12.3   SCHEDULING TECHNIQUES

This section contains descriptions of scheduling techniques. There are two fundamental approaches to scheduling. The first approach, *transformational scheduling*, starts with a default schedule and performs a systematic sequence of modifications to the schedule until it meets the existing constraints. The second approach is *iterative/constructive scheduling*, in which the final schedule is constructed by systematically adding one operation at a time to the schedule until all operations have been scheduled.

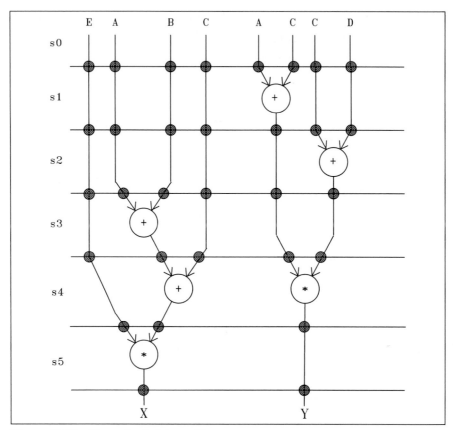

Figure 12.13   Example of state splitting.

12.3.1   Transformational Scheduling

Transformational scheduling usually begins with a schedule that is easy to construct. For example, the ASAP schedule in Figure 12.4 is an example of a *maximally parallel schedule*. One could start with that schedule, and gradually modify it until all constraints are met.

One technique, called *state splitting*, divides any state into one or more substates when hardware constraints or timing constraints are violated. For example, if a limit of one adder and one multiplier is specified, then state S1 would have to be split into three states to accommodate the single-adder constraint. The split can be done in one of several ways. Efficient procedures for optimizing the splitting are not known. Figure 12.13 shows one possible way that state S1 could be split. The final schedule requires six control steps.

Another approach to transformational scheduling, known as *exhaustive search*, is to try all possible known transformations on the initial schedule. The optimum schedule is guaranteed, but the computational burden is prohibitive in all but the simplest situations. It is possible to apply branch and bound techniques to eliminate the search down unproductive paths; however, it is more common to use *heuristics* to direct the search down those paths that have some promise of containing the optimum solution. Although the optimum solution is not guaranteed, "good" designs can often be obtained in reasonable times with a manageable number of computations. This approach is beyond the scope of this text.

12.3.2 Iterative/Constructive Scheduling

With iterative/constructive scheduling, one builds a schedule by adding one operation at a time until all operations are scheduled. Two decisions must be made at each step:

1. Which operation should be scheduled next?
2. In which control step should the operation be scheduled?

Each decision may be governed by either local or global criteria. If local criteria are used, the number of options is reduced, but usually, global optimality is sacrificed. Global criteria usually involve the examination of more alternatives, but provide better solutions than local criteria. The trade-off is between quality of design and complexity of computation.

Techniques in this category differ as to how the next operation to be scheduled is selected and how the control step, in which the operation is to execute, is selected. We now consider several representative iterative/constructive scheduling techniques.

12.3.3 ASAP Scheduling

Perhaps the simplest scheduling algorithm is known as ASAP scheduling. Assuming the number of functional units has already been determined, and the data flow graph has been constructed, the ASAP scheduling technique consists of assigning as many operations in each time step as possible—subject to the availability of data and functional units. That is, each operation is scheduled as soon as possible.

Since there is no attempt to look ahead to any potential scheduling conflicts in later control steps, the selection of which operation is to be scheduled next is, therefore, under the control of local optimization criteria. Because each control step is scheduled in order, starting with the first control step, the selection of which control step to use for a given operation is also local. As many operations as possible are assigned to each control step. All operations for which data is available are scheduled in the current control step. After as many operations as possible are assigned to the current control step, we move on to the next control step and never return to the current control step.

For example, consider the data flow graph in Figure 12.3. Figure 12.14 shows the data flow graph with the nodes numbered in an arbitrary manner for reference purposes only. Assume that a maximum of two adders and two multipliers is specified. In control step s1, the operations available for scheduling are 1, 2, and 3. Since all these are add operations, and we have only two adders available, we arbitrarily select any two of the operations for scheduling. Since we are using only a local criteria for selecting operations for scheduling, we cannot look ahead to anticipate any future scheduling problems. We, therefore, arbitrarily select operations 2 and 3 for scheduling at control step s1. Clearly, there are three possible ways to select the two operations. An exhaustive search technique would try all three possibilities.

Since we assigned operations 2 and 3 to control step s1, the results produced by operations 2 and 3 are available at control step s2. Therefore, the operations that may be scheduled during

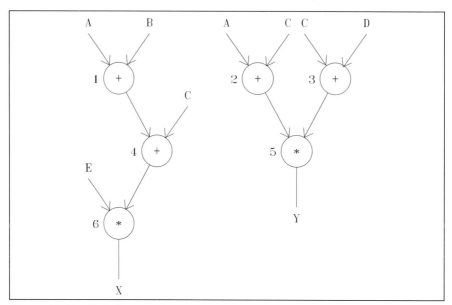

Figure 12.14   Data flow graph with nodes numbered in an arbitrary manner.

s2 are 1 and 5. Because the operations are different, it is possible to schedule both during s2. The only operation available for control step s3 is 4, so we schedule it for s3. Operation 6 is scheduled during control step s4.

Figure 12.15 shows an ASAP schedule, subject to the constraint of two adders and two multipliers. As it turns out, we only used one of the available multipliers, which suggests that a more efficient schedule might be possible. However, in general, an efficient schedule might not require all available components.

Scheduling only assigns time periods for each operation. We must still allocate operations to specific hardware components. Using a greedy approach to functional unit allocation (see Section 12.4.1 Greedy Allocation for details), the Gantt chart shown in Figure 12.16 can be obtained for this ASAP schedule. The basic concept of the greedy approach is to assign each operation to the first available functional unit that can perform the operation.

It is immediately apparent that the functional units have low utilization, which is an indicator that a better schedule might be possible.

12.3.4   ALAP Scheduling

ALAP scheduling is defined as scheduling each operation as late as possible. The algorithm is very similar to the ASAP scheduling algorithm except that we work from the bottom of the data flow graph toward the top, and we start with the last control step and work toward the first control step. Since we do not know initially how many control steps will be required, we do not know what number to give the last step. We, therefore, defer specifying the number of the control steps until we are finished.

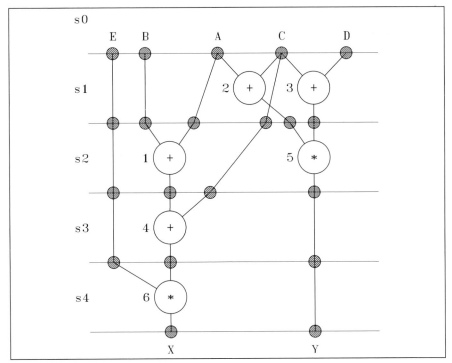

Figure 12.15 ASAP schedule.

	s0	s1	s2	s3	s4
A1	--	2	1	4	--
A2	--	3	--	--	--
M1	--	--	5	--	6
M2	--	--	--	--	--

Figure 12.16 Gantt chart for the ASAP schedule.

Referring to Figure 12.14, the bottom two operations on the data flow graph are both multiplication operations. Since we have two multipliers available, we schedule both operations at the last time step.

For the next to last time step, we have three operations to consider (2, 3, and 4). Since these are all ADD operations, we have only two adders, and we are using only local selection criteria, we arbitrarily select operations 2 and 3 to schedule in the two adders at this step. We do

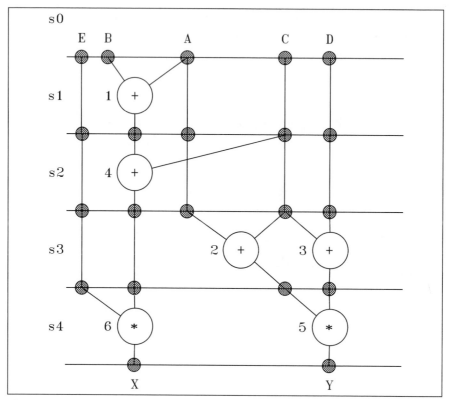

Figure 12.17 ALAP schedule.

not look ahead to check on potential conflicts that might arise from these choices. That leaves operations 1 and 4 to schedule in earlier time steps. However, because the data is not available for both operations, we must schedule operations 1 and 4 in different control steps. Figure 12.17 shows the resulting ALAP schedule. Five time steps are again required.

Again, using a greedy algorithm for functional unit allocation, the Gantt chart shown in Figure 12.18 can be obtained. The utilization of components is again low, indicating that a better schedule might be possible.

12.3.5 List Scheduling

The problem with both the ASAP and ALAP algorithms was that no priority was given to operations on the "critical path." It was, therefore, possible to waste limited resources on less critical items. As a result, at a later time step, resources lay idle waiting for critical items to execute. This problem can be solved by using a more global criteria for selecting the next item to be scheduled.

A *critical path* is a path in the data flow graph of maximum length. For example, the data flow graph in Figure 12.14 has a critical path of length 2, namely, path 1-4-6. Intuitively, operations on the critical path should have priority over other operations, since delaying operations on

	s0	s1	s2	s3	s4
A1	--	1	4	2	--
A2	--	--	--	3	--
M1	--	--	--	--	5
M2	--	--	--	--	6

Figure 12.18 Gantt chart for an ALAP schedule.

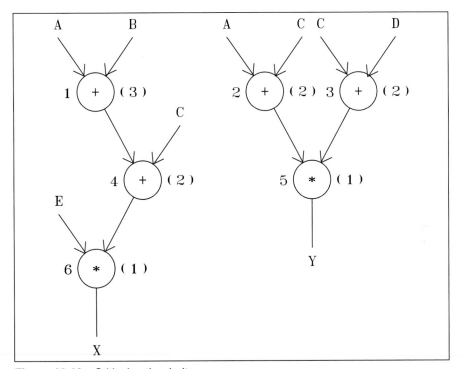

Figure 12.19 Critical path priority.

the critical path tend to increase the total execution time of the graph. However, critical path is a dynamic concept. Once some operations have already been scheduled, the critical path among the remaining operations may differ from the critical path in the original situation.

Various criteria have been proposed to help identify the current critical path. One approach is to define the *critical path priority* of a node as the length of the longest path from the node to

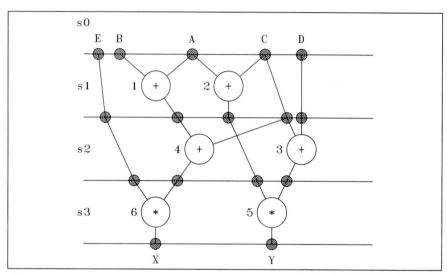

Figure 12.20   List schedule using critical path priority.

the bottom of the graph. Figure 12.19 shows the data flow graph of Figure 12.14 with the critical path priority for each node in parentheses.

A list is then formed of nodes in descending order of priority. The technique attempts to schedule as many items from the head of the list as possible. This gives priority at each time step to those items that are farthest from the bottom of the graph. It is obvious why this method is called *list scheduling*.

For the graph in Figure 12.19, the ordered list is (1,4,2,3,6,5). Nodes with the same critical path priority are placed on the list in arbitrary order. It might be possible to use other criteria to order items with the same critical path priority in such a way as to improve the overall schedule. A list schedule using critical path priority for the graph in Figure 12.19 is shown in Figure 12.20.

The list schedule was obtained as follows. Using critical path scheduling, three add operations are available for control step s1, namely, 1, 2, and 3. Since the nodes on the critical path have priority, operation 1 has priority over operations 2 and 3. Therefore, assign operation 1 and either operation 2 or 3 to step s1. In this case, operation 2 was arbitrarily chosen. At control step s2, operations 3 and 4 are the only ones available. Since we have two adders, both may be assigned to step s2. Operations 5 and 6 are left for step s3. Since we have two multipliers, both may be scheduled for s3.

Figure 12.21 shows the schedule that results from using the greedy algorithm to assign functional units. Notice that, by giving priority to operations on the critical path, we obtained a schedule using only four control steps as compared to the five control steps needed in ASAP or ALAP scheduling. This schedule is an optimum schedule, relative to total execution time because only one operation on the critical path 1-4-6 can be scheduled during each time step. Clearly, we could not do any better than this because the operations on the critical path must be executed in sequence.

Figure 12.21 Gantt chart for list schedule using critical path priority.

Another observation is that the loads on the three components are more balanced for this schedule than for either the ASAP or ALAP schedule. This observation has led some researchers to pursue the concept of load balance as an optimization criteria. The concept of *force directed scheduling* is the result, but we will not describe this technique here.

List scheduling is global in selecting the next operation to be scheduled, but local in selecting the next control step in which to schedule operations.

12.3.6 Freedom-Directed Scheduling

This technique is global in both the selection of the next operation to be scheduled and the selection of the control step in which the operation is placed. The basic principle supporting freedom-directed scheduling is that operations with the least freedom in scheduling should be scheduled first because they are the most difficult to schedule and the most likely to cause blockage at a later time. Operations with more freedom are delayed because they have more options available for scheduling.

The first step is to perform both ASAP and ALAP schedules to obtain the range of control steps into which each operation can be scheduled. An upper bound on the number of control steps must be selected. For the example in Figure 12.14, with an upper bound of 4 (excluding the initialization at s0), the range of control steps into which each operation can be scheduled is shown in Table 12.3.

Table 12.3 Range of control steps for data flow graph in Figure 12.14.

Operation	Earliest ASAP	Latest ALAP	Range
1	1	2	2
2	1	3	3
3	1	3	3
4	2	3	2
5	2	4	3
6	3	4	2

Clearly, the operations with the smallest freedom are 1, 4, and 6. At control step s1, the available operations are 1, 2, and 3, which are all add operations. Since there are only two adders available, only two of the three can be scheduled at time step s1. Since operation 1 has the least freedom, it is scheduled for control step s1. Since 2 and 3 have equal freedom, either one may be selected. Arbitrarily select 2.

The available items for control step s2 are, then, 3 and 4. Since both are add operations and since we have two adders, both can be scheduled. The available operations for control step s3 are 5 and 6. Since both are multiply operations and since we have two multipliers, we can schedule both. Again, the schedule shown in Figure 12.20 is obtained. In this case, freedom-directed scheduling produces an optimum schedule.

12.4 ALLOCATION TECHNIQUES

Data path allocation includes assigning operations to functional units, assigning registers in which to store data values, and designing interconnections among registers and functional units using multiplexers and/or buses. The result is a register-level representation. The allocation process may be constrained by one or more specified limits, such as a maximum time delay, a maximum real estate area, a maximum power consumption, or a maximum total cost. The specification of combinations of goals and limits complicates the design process. For example, the specification might call for a minimum cost solution that does not exceed a given delay time.

Allocation techniques are often classified into two types: iterative/constructive and global. Iterative/constructive techniques use a local criteria to select the next item to be allocated. Items are allocated one at a time until all have been placed. Global techniques examine all remaining items to find the one whose allocation will minimize the design goals while not exceeding any specified limits. Exhaustive search is one example of a global technique. It guarantees an optimum solution but generally requires an excessive amount of computation time. Iterative/constructive techniques are generally more efficient because they search a smaller segment of the design space, but they are less likely to find the optimum solution.

12.4.1 Greedy Allocation

Greedy allocation is an example of an iterative/constructive technique that uses a local selection criterion. The schedule graph is processed from the top to the bottom. Each operation is assigned to the next available functional unit; each data value is assigned to the next available register, and each data path is assigned to the next available bus or multiplexer. At each step, a new component is added only if all current components of the appropriate type are currently busy. The name *greedy* is used to refer to this process because we tend to reuse the first available component we can find, without looking ahead to any potential conflicts that might arise in the future. In other words, we are being greedy by gobbling up all available resources at the present time without any consideration for the future. Some people might claim that an appropriate name for this algorithm would be the Congressional algorithm! Nevertheless, the approach is easy to automate and can give good results for some environments. However, there is little chance of achieving a global optimum result, further supporting the congressional analogy.

12.4.2   Allocation by Exhaustive Search

Global allocation techniques use more comprehensive selection criteria than the iterative/constructive techniques. As a result, more of the design space is searched, and the likelihood of obtaining an optimal solution is increased. The price to pay is greater complexity and an increase in computation time.

Exhaustive search is an example of a global technique. This method searches through the entire allocation space by trying all possible allocations and choosing the best one. Clearly, this approach guarantees finding the optimum solution. However, except in trivial cases, the computational complexity of this approach is unacceptable. Even the fastest computers cannot solve typical problems in a reasonable amount of time. However, some of the approaches discussed in later sections have aspects that are related to exhaustive search, making the concept worthy of consideration.

12.4.3   Left Edge Algorithm

The left edge algorithm is often applied to register allocation after a schedule has been obtained. It is an example of an iterative/constructive technique. The steps in the left edge algorithm are summarized as follows.

1. Give each storage device in the schedule flow chart a unique label. Figure 12.22 shows one possible labeling for the schedule of Figure 12.20.
2. Prepare a chart showing the lifetimes of data storage. Figure 12.23(a) shows how this might be organized. For example, data value E must be held through control steps s1, s2, and s3. During s3, it is used as an input to a multiply unit. Similarly, data value G is created at the end of step s1 and must be held through step s2 until it is used during step s3.
3. Sort the lifetimes by time of origination. All lifetimes with the same starting time are grouped together. The groups are ordered by starting time with an earlier starting time having priority. Within each group having the same starting time, order the lifetimes according to ending time, with earlier ending times having priority. Figure 12.23(b) shows the result of sorting the lifetimes of Figure 12.23(a).
4. Order the available registers from left to right with the highest priority register on the left. Frequently, this ordering is arbitrary. The data values are assigned in the order listed in Figure 12.23(b) from left to right. Each data value is assigned to the left-most register in the list of Figure 12.23(c) that is available at the time of assignment. The final result is shown in Figure 12.23(c).

A data value created at the end of s1 may be stored in the same register as a data value that is used during s1. For example, data value G might be stored in the same register as data value B. On the other hand, data value G cannot be stored in the same register as data value C because both data values must be stored during step s2.

In this example, we must have a minimum of five registers since there are five data items to store during s1. We arbitrarily ordered the registers from R1 to R5. Therefore, data items A, B, C, D, and E are assigned respectively to registers R1, R2, R3, R4, and R5. Data item F is then assigned to R1 and data item G to R2 since these two registers are available during step s2. Sim-

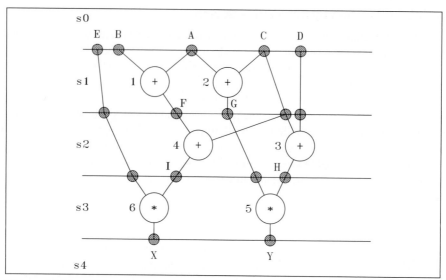

Figure 12.22   DFG with storage devices labeled.

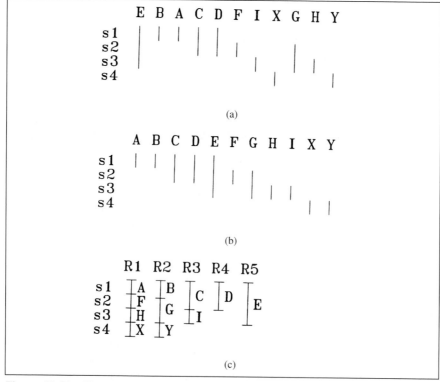

Figure 12.23   Illustration of steps in left edge algorithm.

ilarly, items H and I are assigned to registers R1 and R3, respectively since these registers are available at s3. Items X and Y are assigned to registers R1 and R2 at step s4.

The left edge algorithm does not take into account the cost of data paths. It only minimizes the number of registers needed. We could just as well have assigned item H to R3 and item I to R1 at s3. Assigning a data value to a register is often called *binding* the data value to the register. When there are choices available without adding an additional register, we might want to delay binding of the data value to the register until we assign functional units and interconnect paths. A prudent choice might save on interconnection costs, as illustrated in the next section.

12.4.4 Assigning Functional Units and Interconnection Paths

The assignment of operations to functional units, data values to registers, and interconnection paths to multiplexers or buses are all interrelated. If the left edge algorithm described in the previous section is used to bind data values to registers, without considering functional unit assignment or interconnect costs, the optimum design may be missed.

To illustrate this effect, consider assigning functional units and interconnect paths to the schedule in Figure 12.22. We assume the left edge algorithm has been executed to optimize the register assignment, but we have flagged all binding decisions that might be delayed, as indicated in the previous section.

We have constrained our design to have a maximum of two adders and two multipliers. From execution of the left edge algorithm, we know that five registers are necessary and sufficient to store data values. Figure 12.24 shows the set of functional units and registers that are needed. The assignment of operations to functional units and the final assignment of data values to registers should be made to minimize the cost of the MUXes. In some cases, a MUX might not be necessary at all if only one connection to the input of a register or to the input of a functional unit is needed. If multiple connections are needed, we want to minimize the size of the MUXes by making assignments appropriately.

Although we indicate MUXes for data steering, tristate gates are equally effective.

Minimizing the size of a MUX requires minimizing the number of data sources that have to be provided as inputs to the MUX. Figure 12.25 shows a chart that can be used to record decisions and to help make subsequent decisions. There is one row for each time step and one column for each register or functional unit input. We will record the required connections in the chart as they are developed.

From the register assignment of Figure 12.23(c) data values A, B, C, D, and E are bound to the registers R1, R2, R3, R4, and R5, respectively, during step s1. This is indicated in Figure 12.26 by placing the labels A, B, C, D, and E in row s1, columns R1, R2, R3, R4, and R5, respectively. From the schedule DFG of Figure 12.22, we need to bind operations 1 and 2 to adders AD1 and AD2 during step s1. Since no operations have yet been bound, we arbitrarily decide to bind operation 1 to AD1 and operation 2 to AD2. This decision is indicated in Figure 12.26 by placing a 1 in column AD1-Op and a 2 in column AD2-Op. Because arithmetic addition is commutative, we may connect either A (output of R1) or B (output of R2) to input "a" of AD1. The other will then be connected to input "b." Again, since we have made no connections yet, this decision is arbitrary. We, therefore, arbitrarily connect R1 to AD1b and R2 to AD1a during s1. This is indicated in Figure 12.26 by placing R1 in column AD1-b and R2 in column

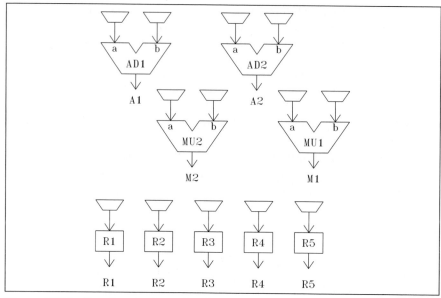

Figure 12.24   Functional units and registers needed in list scheduling example.

	AD1			AD2			MU1			MU2			R1	R2	R3	R4	R5
	Op	a	b	Op	a	b	Op	a	b	Op	a	b					
s1																	
s2																	
s3																	
s4																	

Figure 12.25   Blank connection chart for list scheduling example.

	AD1			AD2			MU1			MU2			R1	R2	R3	R4	R5
	Op	a	b	Op	a	b	Op	a	b	Op	a	b					
s1	1	R2	R1	2	R1	R3	-	-	-	-	-	-	A	B	C	D	E
s2																	
s3																	
s4																	

Figure 12.26   Partial connection chart for list scheduling example.

	AD1			AD2			MU1			MU2			R1	R2	R3	R4	R5
	Op	a	b	Op	a	b	Op	a	b	Op	a	b					
s1	1	R2	R1	2	R1	R3	-	-	-	-	-	-	A	B	C	D	E
s2	4	R3	R1	3	R4	R3	-	-	-	-	-	-	F-A1	G-A2	C	D	E
s3																	
s4																	

Figure 12.27   Partial connection chart for list scheduling example.

AD1-a of row s1. Similarly, since no connections have yet been assigned to AD2, we can arbitrarily connect A (from R1) and C (from R3) to the AD2a and AD2b inputs, respectively. The first row of the connection chart in Figure 12.26 is now completed.

At time step s2, we must bind operations 3 and 4 to AD1 and AD2, and data values F and G to registers R1 and R2. The left edge algorithm bound F to R1 and G to R2, but we could reverse these if it would lower the cost.

Operation 3 requires data from R3(C) and R4(D). By assigning it to AD2, we can use the connection from R3 that already exists. Operation 4 requires data from F and C. By binding F to R1, G to R2, and operation 4 to AD1, as indicated in Figure 12.27, we can use the connection from R1 to AD1b that already exist from s1. Thus, the binding of operations to functional units and data values to registers at s2 will have a significant effect on the interconnection cost. Notice that this binding requires the output of AD1 (A1) to be connected to the input of R1 at the end of s1, and the output of AD2 (A2) to be connected to the input of R2 at the end of s1. The notation F-A1 in column R1 of row s2 indicates that data value F is stored in register R1 during control step s2 and that the output of adder AD1 (A1) must be connected to the input of register R1 in order to transfer F to R1. Register R1 and R2 now have two input connections each. The input of R1 is connected to A and A1, and the input of R2 is connected to B and A2. Row s2 of Figure 12.27 shows the connections that result from this assignment. Note that the "b" column under AD1 has both entries equal to R1, but the "a" column has entries R2 and R3. As a result, through step s2, AD1 needs no MUX at the "b" input, but needs a 2-input MUX at the "a" input. Similarly, AD2 needs no MUX at input "b", but needs a 2-input MUX at input "a."

At s3, we must assign operations 5 and 6 to the two multipliers (M1) and (M2). Since no operations have been assigned to the multipliers yet, this decision is arbitrary. However, from the left edge chart (Figure 12.23), we notice that data values H and I could have been bound to register R1, R3, or R4 at s3. This choice provides a new kind of decision.

Since the value of I comes from AD1 at s2, we must provide a connection from the output of AD1 to the register bound to I. Since we already have a connection from the output of AD1 to register R1, it would be better to bind I to register R1 because we would not have to add any new connections to the input of R1. The left edge algorithm did not have this information available, and therefore, made an arbitrary choice to bind I to R3. However, with the new information, we can decide to bind I to R1 instead.

Since the value of H comes from AD2 at s2, we must provide a connection from the output of AD2 to the register bound to H. Since the output of AD2 is not yet connected to any of the registers, R1, R3, or R4, we will have to add a new connection in either case. We, therefore, arbi-

	AD1			AD2			MU1			MU2			R1	R2	R3	R4	R5
	Op	a	b	Op	a	b	Op	a	b	Op	a	b	R1	R2	R3	R4	R5
s1	1	R2	R1	2	R1	R3	-	-	-	-	-	-	A	B	C	D	E
s2	4	R3	R1	3	R4	R3	-	-	-	-	-	-	F-A1	G-A2	C	D	E
s3	-	-	-	-	-	-	6	R1	R5	5	R3	R2	I-A1	G-A2	H-A2	-	E
s4																	

Figure 12.28   Partial connection chart for list scheduling example.

	AD1			AD2			MU1			MU2			R1	R2	R3	R4	R5
	Op	a	b	Op	a	b	Op	a	b	Op	a	b	R1	R2	R3	R4	R5
s1	1	R2	R1	2	R1	R3	-	-	-	-	-	-	A	B	C	D	E
s2	4	R3	R1	3	R4	R3	-	-	-	-	-	-	F-A1	G-A2	C	D	E
s3	-	-	-	-	-	-	6	R1	R5	5	R3	R2	I-A1	G-A2	H-A2	-	E
s4	-	-	-	-	-	-	-	-	-	-	-	-	X-M1	Y-M2			
mux inps		2	1		2	1		1	1		1	1	3	3	2	1	1

Figure 12.29   Final connection chart for list scheduling example.

trarily bind H to R3 since we have already decided to bind I to R1. The result thus far is shown in Figure 12.28.

At the end of s3, we must store the multiply results in registers. Although s4 was not needed to compute the results, it is during s4 that the multiply results will become available. We have, therefore, included s4 in the chart. The binding of X and Y to registers provides yet another new kind of choice. Any of the registers R1 through R5 are available. Registers R1, R2, and R3 have two connections at their inputs, so a 2-input MUX would be sufficient. Registers R4 and R5 have only one connection, so no MUX is needed at all. Since X and Y come from the multipliers, and since there is no path from either multiplier to any register, we must add new paths for both X and Y. If we add the path to R3 or R4, we must add a 2-input MUX to accommodate the new path. If we add the path to R1, R2, or R3, we must expand the 2-input MUX to accommodate at least three inputs. Sometimes this means expanding to a 4-input MUX. The decision then depends upon which costs less. This decision may be technology dependent, and we may want to delay the binding until a decision on technology is made.

To complete our example, we decide to bind X to R1 (requiring a connection from the output of M1 to R1, and Y to R2 (requiring a connection from the output of M2 to R2). Figure 12.29 shows the complete connection table and Figure 12.30 shows the resulting connection diagram.

The connection diagram is constructed directly from the final connection chart in a straightforward manner. For example, because the only entry in column "b" of AD1 is R1, we can connect the output of register R1 directly to the "b" input of AD1. Since both R1 and R4

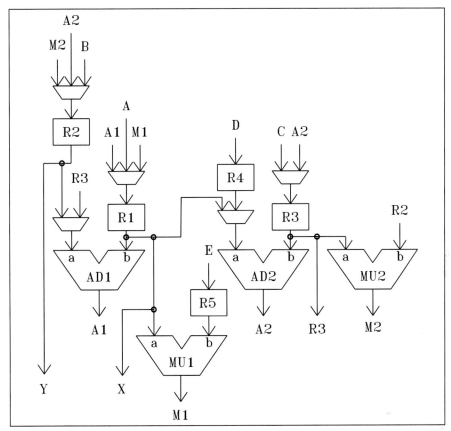

Figure 12.30 Connection diagram for list scheduling example.

appear in the "a" column of AD2a, we need a 2-input MUX to select either R1 or R4 to connect to the "a" input of AD2.

In the R1 column, we see entries A, F-A1, I-A1, and X-M1. The entry "A" means that data item A is assigned to register R1. Since A is a primary input, there is a connection from the primary input to R1. The entries F-A1 and I-A1 indicate that data values F and I are assigned to register R1. Since F and I are produced at the output of AD1, there must be a connection from the output of AD1 to R1. Similarly, entry X-M1 indicates that data value X is assigned to register R1, which requires a connection from the output of M to R1. A MUX can be used to select among the three data paths, or alternatively, the three data sources can be gated onto a bus using tristate gates that are disjointly enabled and a connection established from the bus to the input circuit of register R1.

Table 12.4 shows the parts list for this implementation. If standard parts are being used, 4x1 MUXes may be used in place of the 3x1 MUXes specified.

Table 12.4   Parts list for the example.

Type	Quantity
adder	2
multiplier	2
register	5
2x1 MUX	3
3x1 MUX	2

Note that many choices still remain at each step, and the long-term effect of choices is not always evident. Therefore, we need some techniques for optimizing the choices. The next few sections describe formal mathematical techniques to aid in the allocation process.

12.4.5   Analysis of the Allocation Process

Graphical methods can lend useful insights into the allocation process. Whether we are allocating registers, functional units, or interconnection paths, two items can be assigned to the same device only if there are no usage conflicts. Items to be assigned are called *variables*. The variables are assigned to *resources*. When assigning data values to registers, the variables are the data values, and the registers are the resources. When assigning operators to functional units, the operators are the variables, and the functional units are the resources. When defining a connection, we assign the connection to either a multiplexer or data bus. In this case, the connections are the variables, and the multiplexers or data buses are the resources. In this way, all three allocation activities are modeled as an assignment of a list of variables to a list of resources. The assignment, in each case, is limited by potential conflicts defined by the schedule of activities. In this section, we assume that a schedule has already been determined by some means. We will call the assignment of variables to resources in such a way as to minimize the number of resources required, the *resource allocation problem*.

Now we define a graphical representation for the resource allocation problem. Each variable is represented by a node in an undirected graph. Two nodes are joined by an arc if the two variables can be assigned to the same resource. The resulting graph is called the *compatibility graph*. Two nodes that are joined by an edge are *compatible*. The graph in Figure 12.31(a) is the compatibility graph for functional unit allocation for the schedule in Figure 12.15. Two operators are compatible only if they can be assigned to the same functional unit and are active in different time periods. The compatibility graph is frequently constructed by starting with a complete graph (one with an edge between every pair of nodes), and gradually eliminating the edges that connect incompatible nodes.

To construct the compatibility graph for the operators in Figure 12.15, we could start with the complete graph with six nodes. Assume we have separate functional units to perform the add and multiply operations. Operator 5 is incompatible with operators 1, 2, 3, and 4 because they are different operators. From the schedule, operators 2 and 3 are both add operators but are incompatible because they occur in the same time period. On the other hand, add operators 1 and 4 are compatible operators because they occur in different time periods. Therefore, it should be

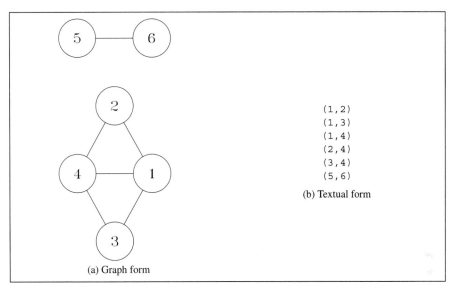

(b) Textual form

(a) Graph form

Figure 12.31 Compatibility graph for functional unit allocation for ASAP schedule.

possible to use the same adder to perform both operations without conflicts. We, therefore, include an edge between nodes 1 and 4 in the compatibility graph.

It is convenient to label each node with a unique integer and to represent the graph with a set of node pairs, where the node pair (i,j) $(i < j)$ represents the undirected edge between nodes i and j. By sorting the nodes in increasing order of i, followed by increasing order of j, the graph is represented by a unique ordered list of integer pairs as shown in Figure 12.31(b).

The goal is to use as few functional units as possible, and at least three functional units are required. Minimally, two adders are required because 2 and 3 are to be executed in parallel. At least one multiplier is also required. From an examination of the graph in Figure 12.31(a), it is clear that operators 5 and 6 can be performed by the same multiplier. Also, operations 1, 2, and 4 can be performed by the same adder. Table 12.5 shows some possible ways to map the operators onto three functional units.

Table 12.5 Possible mappings of operators to functional units.

	Mapping			
Unit	1	2	3	4
MUL	5,6	5,6	5,6	5,6
ADD1	2,4	1,2	1,2,4	2
ADD2	1,3	3,4	3	1,3,4

A set of operators that can be mapped to the same functional unit is called a *compatible set*. A set of operators forms a compatible set if and only if each pair of operators in the set is compatible. For example, operators {1,3,4} form a compatible set because (1,3), (1,4), and (3,4) are all compatible pairs. However, {1,2,3} is not a compatible set even though (1,2) and (1,3) are both compatible pairs because (2,3) is not a compatible pair. Notice that 2 and 3 are not compatible because they represent operations that must be done in the same time period and, therefore, require separate functional units. A set of nodes with the property that each pair of nodes in the subset is connected by an edge is called a *complete subgraph*. There is a one-to-one correspondence between complete subgraphs and compatible sets.

Since each operator must be assigned to a unique functional unit, the collection of compatible sets used to assign functional units must be disjoint. Also, each operator must be assigned to some functional unit. A collection of subsets of nodes that includes each node exactly once is called a *partition* of the set of nodes. We define a *cluster* to be a complete subgraph of a graph. The problem to be solved can then be stated in terms of graph theory as:

Minimum cluster partition problem (MCPP).   Find the minimum number of clusters that constitute a partition of the nodes of a graph.

The next section describes a mathematical formulation for this problem.

12.4.6   Nearly Minimal Cluster Partitioning Algorithm

A *clique* of graph G is a complete subgraph of G that is not contained in any other complete subgraph of G. For the graph of Figure 12.31(a), complete subgraph {1,2,4} is a clique, but complete subgraph {2,4} is not. The set {2,4} is a complete subgraph, but it is contained in a bigger complete subgraph, namely {1,2,4}. Searching for cliques in a graph is known to be NP-complete. Therefore, an efficient algorithm for finding cliques, which are simply maximal clusters, probably cannot be found. The following algorithm, which is of polynomial complexity, finds nearly minimal cluster partitions for the nodes of a graph. It is assumed that the nodes of a graph are labeled with unique integers. This formulation is adapted from a paper by Tseng and Siewiorek, published in 1986.

Nearly Minimal Cluster Partitioning Algorithm (CPA).

C1. For each edge, (i,j),
   a. Compute the number of common neighbors of i and j. A node (r) is a *common neighbor* of i and j if r is connected to i by an edge, and r is connected to j by an edge.
   b. Compute the number of edges that will be deleted from the graph if i and j are merged into a single node. When i and j are merged, the edge (i,j) will be deleted along with all edges r that are neighbors of either i or j but not both. If node r is a common neighbor of both i and j, then one of the edges from r to i and j will be deleted, and one will be kept.
C2. Start collecting nodes for the next cluster.
   a. If there are no edges in the graph, stop. Otherwise, continue.
   b. Select an edge (p,q) that has the maximum number of common neighbors. If there is more than one such edge, select one that will result in the fewest edge deletions when p and q are merged. If there is still more than one choice, select an arbitrary representative.
   c. Merge p and q into a cluster. We will list the elements of the cluster in order of increasing label and will call the lowest labeled element in the cluster the head

of the cluster. At this time, the current cluster contains only the two elements p and q.

    d. Call the graph update subroutine, shown at the end of this description, to merge nodes p and q of the graph into a single node. Label the combined node p, where p is the head of the cluster. At this time, p represents the cluster $\{p,q\}$.

C3. Node p in the graph represents the current cluster. Add an additional node to the current cluster if possible by performing this step.

    a. If there are no edges connected to node p, then node p represents the full cluster. Go to step 2 to start the next cluster. Otherwise, continue in order to add another node to the current cluster.

    b. Consider the nodes connected to node p. Select edge (i,p) or (p,i), such that nodes i and p have the most common neighbors. If more than one choice is available, select the node i, so the number of edges removed when nodes i and p are merged is minimal. If there is still more than one choice, arbitrarily select any choice for node i.

    c. Add node i to the current cluster. If $i<p$, i becomes the new head of the cluster; otherwise, p remains as the cluster head.

    d. Call the graph update subroutine to merge nodes i and p.

    e. Set $p=\min\{i, p\}$. Node p is the new cluster head.

    f. Repeat step 3.

Graph Update Subroutine for Merging Nodes x and y $(x<y)$.

S1. Merge nodes x and y into a single node labeled x.

S2. Update the edges of the graph by:

    a. Deleting all edges involving node y.

    b. Deleting any edge between node r and x, unless there was also an edge between r and y before merger. Note: The effect of edge removal is to retain only those arcs that represent common neighbors of nodes x and y.

S3. Recompute the number of common neighbors and number of edges deleted for each edge in the merged graph.

12.4.6.1 *Example of Cluster Partitioning*

Figure 12.32 shows an example of a compatibility graph, with an ordered list of edges, number of common neighbors for each edge (CN), and number of edges deleted (ED) if the nodes at each end of the edge are merged. To illustrate the calculations in step C1, consider edge (7,8). Nodes 7 and 8 have only one common neighbor, namely node 6. If nodes 7 and 8 are merged into a new node labeled 7, edge (6,7) is retained because it is a common neighbor of nodes 7 and 8. Edge (5,7) is removed because node 5 is not a common neighbor of nodes 7 and 8. Edges (7,8), (6,8), and (8,9) are removed because node 8 vanishes. Therefore, a total of four edges will be removed. The result of merging nodes 7 and 8 is shown in Figure 12.33. Note: This result is for illustration of c1 calculations only, we do not actually merge nodes 7 and 8 as part of the algorithm.

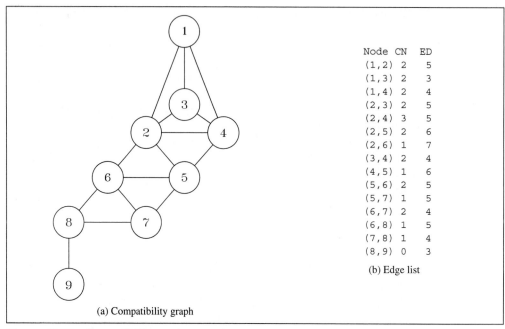

Node	CN	ED
(1,2)	2	5
(1,3)	2	3
(1,4)	2	4
(2,3)	2	5
(2,4)	3	5
(2,5)	2	6
(2,6)	1	7
(3,4)	2	4
(4,5)	1	6
(5,6)	2	5
(5,7)	1	5
(6,7)	2	4
(6,8)	1	5
(7,8)	1	4
(8,9)	0	3

(b) Edge list

(a) Compatibility graph

Figure 12.32   Compatibility graph example.

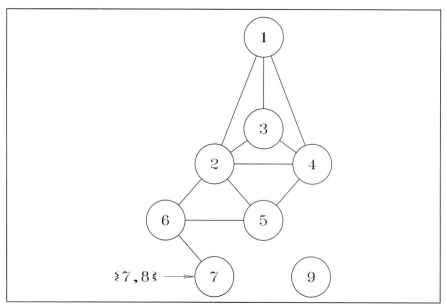

Figure 12.33   Result of merging nodes 7 and 8.

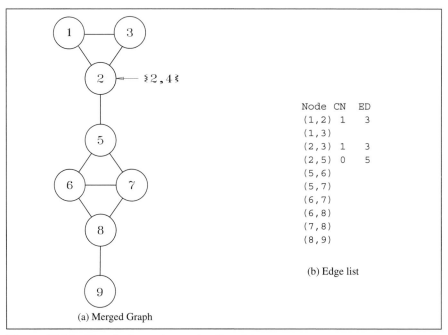

Node CN   ED
(1,2) 1   3
(1,3)
(2,3) 1   3
(2,5) 0   5
(5,6)
(5,7)
(6,7)
(6,8)
(7,8)
(8,9)

(b) Edge list

(a) Merged Graph

Figure 12.34 Result of merging nodes 2 and 4.

In step C2(b), we select edge (2,4) because it has the most common neighbors (three). In this case, there was only one node with three neighbors so no tie breaking was necessary. At this time, $p=2$ and $q=4$. In step C2(c), we start a new cluster with nodes 2 and 4 as the only elements. Node 2 is the current head of the cluster.

In step C2(d), we now call the graph update subroutine to merge nodes 2 and 4. In the subroutine, set $x=2$ and $y=4$ because $2<4$. In step S1, we merge nodes 2 and 4 into a new node labeled 2. In step S2(a), we delete edges (1,4), (2,4), (3,4), and (4,5). In step S2(b), we delete edge (2,6) because there is no edge (4,6) in the original graph. We retain edges (1,2), (2,3), and (2,5) because nodes 1, 3, and 5 are all common neighbors of nodes 2 and 4. Figure 12.34 shows the merged graph and list of edges.

Note: In step C2(b) of the main algorithm, we only computed the number of common neighbors for edges incident on node 2 since we had not yet completed the current cluster. It is only necessary to compute the common neighbors and number of edges deleted for all nodes when starting a new cluster. We have now finished updating the graph as a result of the merge of nodes 2 and 4. This completes step C2 of the main algorithm.

Following step C3(a), we continue on to step C3(b) because there are still edges incident on node 2. In step C3(b), there are three nodes connected to node 2 by edges (1,3,5). Therefore, we must consider edges (1,2), (2,3), and (2,5). Both (1,2) and (2,3) have one common neighbor. Since the number of edges deleted is 3 in both cases, we can make an arbitrary choice. We arbitrarily decided to merge nodes 1 and 2. In step C3(c), we add node 1 to the current cluster, which now consists of the set {1,2,4}. The new cluster head is node 1. In step C3(d), we call the graph

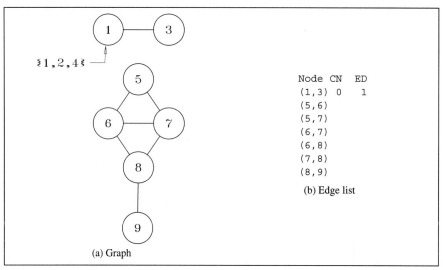

Figure 12.35   Result of merging nodes 1 and 2.

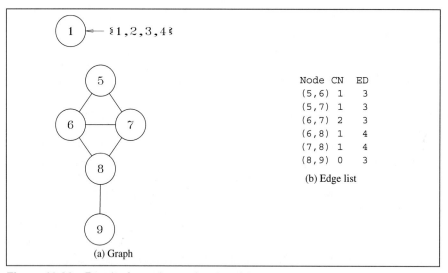

Figure 12.36   Result of merging nodes 1 and 3.

update subroutine with *x*=1 and *y*=2. Figure 12.35 shows the new graph and list of graph edges. In step C3(e), *p* is set equal to 1 to indicate the new cluster head.

After one more pass through step C3, we decide to merge nodes 1 and 3. The current cluster is then {1,2,3,4} with the graph and edge list shown in Figure 12.36. Node 1 now represents the cluster {1,2,3,4}. Because there are no edges incident on node 1, at step C3(a) we return to step C2 to start the next cluster. We now must compute the number of common neighbors and arcs deleted for each node in the graph, in order to execute step C2(b).

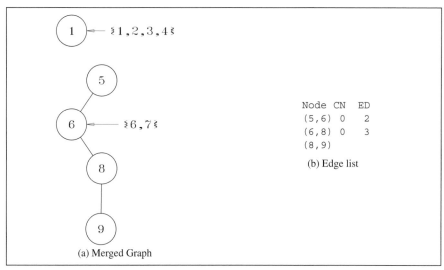

Node CN   ED
(5,6) 0   2
(6,8) 0   3
(8,9)

(b) Edge list

(a) Merged Graph

Figure 12.37   Result of merging nodes 6 and 7.

At step C2(a), we continue since there are edges left in the graph. At step C2(b), we select edge (6,7) because it is the only edge with two common neighbors. After combining nodes 6 and 7, the graph and edge list shown in Figure 12.37 is obtained.

At step C3(b), the current cluster is {6,7} with node 6 remaining in the graph to represent the cluster. All nodes connected to node 6 have 0 common nodes. However, this time edge (6,8) results in three edges being deleted, while edge (5,6) results in only two edges being deleted. Therefore, we must select edge (5,6) since that results in fewer edges being deleted. Notice that it is very important to make this selection, because selecting (6,8) will isolate node 5 and result in an extra cluster being added that is not needed.

We have now illustrated all types of situations and will state the final results. After the algorithm is completed, the final clusters are {1,2,3,4}, {5,6,7}, and {8.9}. We conclude that all nine variables can be covered with only three resources.

CPA can be applied to any of the three resource allocation problems individually. The next section describes an extension of CPA that allows additional problem dependent information to be used to improve the quality of the solutions obtained.

12.4.7   Profit Directed Cluster Partitioning Algorithm (PDCPA)

Although the cluster partitioning algorithm previously described can find good solutions, it is possible to improve performance if additional information is available regarding the potential profits to be obtained at each merge opportunity. If we can identify certain merge opportunities as being more profitable than others, we can direct the selection process to take advantage of this knowledge.

We will assume that it is possible to partition the edges of the original graph into a set of categories, based on expected profit from combining the nodes connected by each edge. We will number the categories, starting at 1, in increasing profitability. For example, if there are two cat-

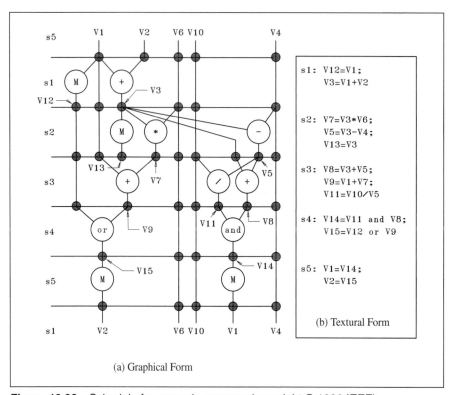

s1: V12=V1;
    V3=V1+V2

s2: V7=V3*V6;
    V5=V3-V4;
    V13=V3

s3: V8=V3+V5;
    V9=V1+V7;
    V11=V10/V5

s4: V14=V11 and V8;
    V15=V12 or V9

s5: V1=V14;
    V2=V15

(b) Textural Form

(a) Graphical Form

Figure 12.38 Schedule for example program (copyright © 1986 IEEE).

egories, category 1 will be less profitable than category 2. If there are m categories, we will reduce the graph by starting with edges of category m and working our way down to edges of category 1. For each category, $k>1$, create a subgraph of G called G_k that consists of all edges in G of category k. Starting with the subgraph G_m consisting of the most profitable edges, simultaneously reduce G and G_m until there are no edges left in G_m. At this time, we will have performed the most profitable reductions. Then, simultaneously reduce G and G_{m-1} until G_{m-1} has no edges. Continue in this manner until the lowest category is reached. At that time, G will be equal to G_1. The algorithm in the previous section will then be applied to graph G.

The two graphs G and G_k will be used at iteration k to merge nodes of category k according to the following algorithm. This algorithm was adapted from the paper by Tseng and Siewiorek, 1986.

Nearly Minimal Profit Directed Cluster Partitioning Algorithm (PDCPA). Given a graph G and a subgraph G_k of G that specifies edges with the highest potential profitability, simultaneously reduce G and G_k. Apply the nearly minimal cluster partitioning algorithm of the previous section to subgraph G_k, except use the following graph update subroutine.

```
(1,8)    (1,9)    (1,11)   (1,13)   (1,14)   (1,15)   (2,3)
(2,5)    (2,7)    (2,8)    (2.9)    (2,11)   (2,12)   (2,13)
(2,14)   (2,15)   (3,8)    (3,9)    (3,11)   (3,13)   (3,14)
(3,15)   (4,13)   (5,8)    (5,9)    (5,11)   (5,13)   (5,14)
(5,15)   (6,13)   (7,8)    (7,9)    (7,11)   (7,13)   (7,14)
(7,15)   (8,13)   (8,14)   (8,15)   (9,13)   (9,14)   (9,15)
(10,13)  (11,13)  (11,14)  (11,15)  (12,13)  (12,14)  (12,15)
(13,14)  (13,15)
```

Figure 12.39   Compatibility graph for register allocation (copyright © 1986 IEEE).

Profit Directed Graph Update Subroutine for Merging Nodes x and y ($x<y$).

1. Merge nodes x and y into a single node labeled x in both G and G_k.
2. Update the edges of graphs G and G_k as follows:
   a. Delete all edges involving node y in both G and G_k.
   b. In graph G delete any edge between node r and x, unless there was also an edge between r and y before merger. If the deleted edge in G is also in G_k, then delete it from G_k.
   c. If there was an edge between r and x and also an edge between r and y in graph G before merger, then keep the edge in graph G. If the edge was also in G_k before merger, then keep the edge in graph G_k. If the edge that is kept in G was not in before merger, it might need to be added to G_k. As a result of merger, the edge might change category. If the new category of the retained edge in graph G is greater than or equal to k, and the edge was not previously in G_k, then add it to G_k.
   d. Recompute the number of common neighbors and number of edges deleted for each edge remaining in graph G_k.

12.4.7.1   *Application of PDCPA to Register Allocation*

Registers can either be independent or can be located in RAM memory. The goal of register allocation is usually to minimize the number of registers. For two values to share a storage element, their lifetimes must not overlap. For example, assume the schedule of Figure 12.38(a) is obtained by some scheduling algorithm. Figure 12.38(b) shows a textual representation of the schedule. The schedule is part of a program loop, so step s1 follows step s5. Values V7 and V9 can share a register because their active lifetimes do not overlap. V7 is active only during s3, and V9 is active during s4. The values V4, V6, and V10 are used repeatedly during the loop without being updated; that is, they are initialized at s0 only, but are used during each pass through the loop. Since they are all active for the duration of the loop, no two may share a register.

The profit directed cluster partitioning algorithm can be used to find a register assignment using a nearly minimal number of registers. First, construct a compatibility graph (G) in which each value is represented by a node. There is an edge between nodes i and j if and only if nodes i and j represent values that can be merged into a single register, that is, their lifetimes do not overlap. Figure 12.39 shows the compatibility graph in textual form for the schedule of Figure 12.38 where value Vi is represented by integer i.

```
(1,14)   (2,15)   (3,13)
```

Figure 12.40   Subgraph G_2, which consists of edges that correspond to pure data transfers.

```
(1,8)    (1,9)    (1,11)   (2,3)    (2,5)    (2,7)    (2,8)
(2,9)    (2,11)   (2,12)   (3,8)    (3,9)    (3,11)   (5,8)
(5,9)    (5,11)   (7,8)    (7,9)    (7,11)
```

Figure 12.41   Compatibility graph after reduction relative to subgraph G_2.

Let us define a *pure data transfer* as a move operation in which data is simply moved from one variable to another. The example contains four pure data transfers represented by the symbol M. Consider the transfer from V15 to V2 at step s5. If we can store V15 and V2 in the same register, the entire operation will be eliminated, which reduces the number of control functions, registers, and data paths needed in the system. Furthermore, if a time step consists of only pure data transfers, and all the data transfers can be eliminated, then the entire control step can be eliminated, which speeds up the operation of the circuit. In the example, step s5 consists of only pure data transfers; therefore, it might be possible to eliminate step s5 entirely.

A data value that is never used, such as V13, is compatible with all other data values and will be automatically eliminated by PDCPA. To account for the additional profits that could be realized if pure data transfers are eliminated, we will partition the arcs in the graph into two categories. Edges that represent pure data transfers will be placed in the higher category (category 2), and all other edges will be placed in category 1. Let G_2 be the subgraph of G that consists of all pure data transfers, and let G_1 be the subgraph of G that includes all other transfers. Edge (i,j) is in G_2 if and only if (i,j) represents a pure data transfer in G. Figure 12.40 shows subgraph G_2. It is never really necessary to construct subgraph G_1.

The number of registers can be minimized by solving PDCPA for graph G with respect to subgraphs G_2 and G_1. The result of applying PDCPA to graphs G and G_2 is the following cluster partition:

$$(\{1,14\}, \{2,15\}, \{3,13\}, \{4\}, \{5\}, \{6\}, \{7\}, \{8\}, \{9\}, \{10\}, \{11\}, \{12\})$$

Figure 12.41 shows the reduced graph with node 1 representing cluster {1,14}; node 2 represents cluster {2,15}, and node 3 represents cluster {3,13}.

The result of applying CPA to the reduced graph of Figure 12.41 is the following cluster partition:

$$(\{1,14\}, \{2,7,9,15\} \{3,8,13\} \{4\} \{5,11\}, \{6\}, \{10\}, \{12\})$$

The values (variables) in each of these clusters can be assigned to the same storage location, which may be either a register or RAM location. The fifteen variables in the original problem are reduced to eight. In addition, since all the data transfers in step s5 are eliminated, step 5 is no longer needed. Figure 12.42(a) shows the reduced schedule in graphical format, and Figure 12.42(b) shows the reduced schedule in textual format.

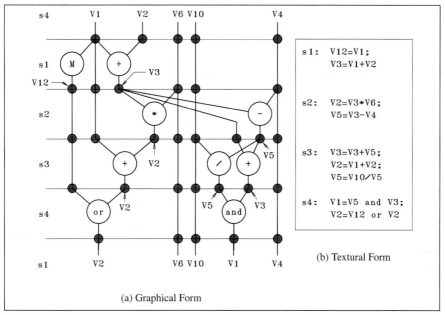

s1:   V12=V1;
      V3=V1+V2

s2:   V2=V3*V6;
      V5=V3−V4

s3:   V3=V3+V5;
      V2=V1+V2;
      V5=V10/V5

s4:   V1=V5 and V3;
      V2=V12 or V2

(b) Textural Form

(a) Graphical Form

Figure 12.42   Representation of the schedule after register allocation.

12.4.7.2   *Application of PDCPA to Functional Unit Allocation*

The goal of functional unit allocation is to minimize the number of computational units and the amount of gating logic needed to move data to and from the computational units. In this application, the amount of profit to be gained by assigning two data operators to the same functional unit varies according to the number of operands that are common. Figure 12.43 illustrates these concepts.

Figure 12.43(a) shows the hardware required if separate adders are used to implement the two add operations. Figure 12.43(b) shows the hardware required if the two add operations are performed by a single adder but each pair of data values is different. Notice that a multiplexor and select signal are required for each input to the adder, and two separate control signals are required to store the results. The cost reduction is the cost of the saved adder minus the cost of the two MUXes and two additional control signals. This design is better only if the cost of the two MUXes and two MUX select signals are less than the cost of the additional adder. However, if each of the three pairs of data items are common, then the hardware shown in Figure 12.43(c) is sufficient. Notice that the complexity is reduced by two multiplexers and three control signals, relative to the cost of using one adder when all pairs of operators are different, and is reduced by one adder and one control signal, relative to the design using separate adders.

Functional unit allocation is also accomplished by using PDCPA. The compatibility graph has a node for each data operation and an edge from node *i* to node *j* if and only if the data operators represented by nodes *i* and *j* can both be assigned to the same functional unit. Clearly, the profit is different, depending on the number of common data operators. Table 12.6 shows eight subgraphs, taken from Tseng and Siewiorek, 1986, that represent eight profit categories, where

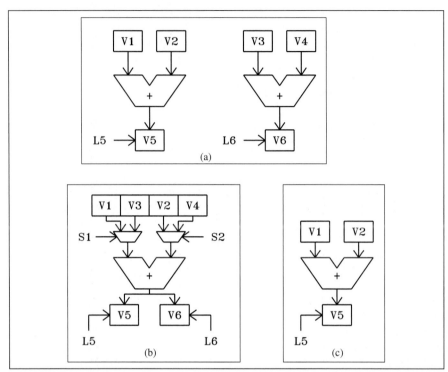

Figure 12.43 Variation in profit as a function of the number of common operands.

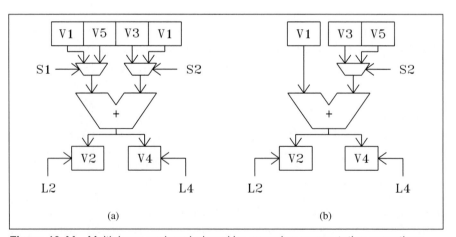

Figure 12.44 Multiplexer savings induced by reversing commutative operations.

the profits for category 8 are the greatest, and the profits for category 1 are the lowest. Be advised that these are only suggested rankings of categories. The actual rankings in a specific design environment might be different. Nevertheless, this ranking serves to illustrate the solution techniques. Notice that we allow for the possibility that two different data operators may be mapped to the same functional unit. An ALU is an example of such a component.

Table 12.6 Subgraphs for profit categories in functional unit allocation.

G_8	The two operations and the three pairs of variables are all the same.
G_7	The operations are different but the three pairs of variables are all the same.
G_6	The operations and two pairs of variables are the same. The third pair of variables is different.
G_5	Two pairs of variables are the same. The operations are different, and the third pair of variables is different.
G_4	The two operations and one pair of variables are the same. The other two pairs of variables are different.
G_3	One pair of variables is the same. The two operations are different, and the other two pairs of variables are different.
G_2	The operations are the same. All three pairs of variables are different.
G_1	The two operations and all three pairs of variables are different.

To solve this problem, seven applications of PDCPA followed by an application of CPA are required.

The quality of the solution obtained may be substantially improved by using a pre-processing step that checks the possibility of swapping the variables in commutative operations. Frequently, by reversing the order of the operands, it becomes apparent that sources can be shared, which were previously incompatible. For example, consider this situation:

```
s2:  V2 = V1 + V3
s3:  V4 = V5 + V1
```

If the two add operations are performed by the same adder, and the equations are used as given, then the network shown in Figure 12.44(a) results. However, if the commutative property of addition is invoked, the statements can be written as:

```
s2:  V2 = V1 + V3
s3:  V4 = V1 + V5
```

Figure 12.44(b) shows that one multiplexer can be saved in the resulting implementation.

As an example, consider the reduced schedule of Figure 12.42. Each operator is assigned an integer in an arbitrary manner as shown in Figure 12.45. Before beginning the application of PDCPA, we will check the commutative operations. In this example, the only commutative operations are +, *, "and", and "or". For the "and" operation, we notice that V3 is the right operand but that all operations in other time steps that could be assigned to the same functional unit as the "and" operation have V3 as the left operand. Therefore, we decide to reverse the operands for the "and" operation in anticipation of possible savings. This change is already incorporated into Figure 12.45.

Pure data transfers are not assigned an integer because a functional unit is not required. Only a data path is needed. In some systems, this assumption will not be valid. For example, it

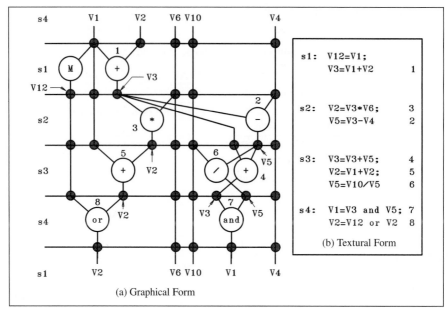

Figure 12.45 **Figure 12.45**   Assignment of integers for functional unit allocation.

```
(1,2)1   (1,3)1   (1,4)4   (1,5)6   (1,6)1   (1,7)1   (1,8)3
                  (2,4)3   (2,5)1   (2,6)3   (2,7)3   (2,8)1
                  (3,4)3   (3,5)3   (3,6)1   (3,7)3   (3,8)3
                                             (4,7)5   (4,8)1
                                             (5,7)1   (5,8)5
                                             (6,7)3   (6,8)1
```

Figure 12.46   Compatibility graph for functional unit allocation (copyright © 1986 IEEE).

```
(1,2)1   (1,3)3                           (1,7)1   (1,8)5
                  (2,4)3          (2,6)3   (2,7)3   (2,8)1
                  (3,4)3          (3,6)1   (3,7)3   (3,8)3
                                           (4,7)5   (4,8)1

                                           (6,7)3   (6,8)1
```

Figure 12.47   Reduced compatibility graph after processing G_6.

might be possible to use an existing path through an ALU to implement a pure data transfer. In that case, we would assign integers to pure data transfers as well.

Let us assume that ALUs are available as functional units and that the ALUs can perform any of the required operations except pure data transfers. Figure 12.46 shows the compatibility graph (G) used for functional unit allocation taken from Tseng and Siewiorek, 1986. Operations

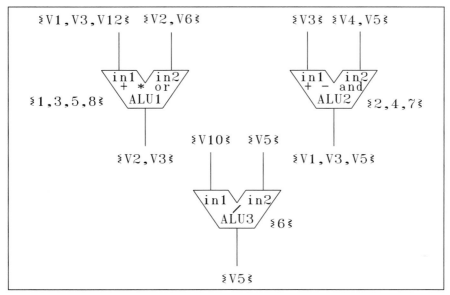

Figure 12.48 ALU configuration obtained from functional unit allocation.

are incompatible only if they occur in the same time step. The number outside the parentheses represents the category number for each edge. For example, (1,4)4 indicates that operations 1 and 4 are compatible (i.e., they occur in different control steps) and that the profit category is G_4, that is, the operations are the same (both add) and one pair of operands are common (the output is V3 in both cases).

Subgraphs G_8 and G_7 are empty. Subgraph G_6 consists of only one edge, namely (1,5). Executing PDCPA on graphs G and G_6 results in combining nodes 1 and 5. In the reduced graph, edges (1,3) and (1,8) are upgraded to categories 3 and 5, respectively. Figure 12.47 shows the reduced graph with node 1 representing the cluster {1,5}.

In a similar manner, the graph is reduced relative to subgraphs G_5 and G_3 using PDCPA to yield this partition:

$$(\{1,3,5,8\}, \{2,4,7\}, \{6\})$$

Figure 12.48 shows the ALU configuration implied by this partition. ALU1 must perform three different functions (+, *, "or"). Connections to each ALU are indicated by labels on the inputs and outputs of the ALU. For example, ALU1.in1 must be connected to three registers (V1, V3, and V12).

12.4.7.3   Application of PDCPA to Data Path Allocation

Interconnection variables that are never used simultaneously can be transmitted over a common data path. Combining interconnection variables into data paths can also be solved using PDCPA. The number of required bus drivers and receivers is used to create categories.

Less profit is gained from combining two interconnection variables, which originate at different sources and connect to different destinations, because the number of drivers required is

Source Name	Destination Name	Index
V1	V12	1
V1	ALU1.in1	2
V2	ALU1.in2	3
V3	ALU1.in1	4
V3	ALU2.in1	5
V4	ALU2.in2	6
V5	ALU2.in2	7
V5	ALU2.in2	8
V6	ALU1.in2	9
V10	ALU3.in1	10
V12	ALU1.in1	11
ALU1.out	V2	12
ALU1.out	V3	13
ALU2.out	V1	14
ALU2.out	V3	15
ALU2.out	V5	16
ALU3.out	V5	17

Figure 12.49 Assignment of indices to interconnections (copyright © 1986 IEEE).

very high. However, if two interconnection variables share the same source, they can also share the same bus drivers or the same multiplexers. Similarly, if two interconnection variables share the same destination, they can share the same termination lines and control signal.

Create a compatibility graph with a node for every interconnection variable and an edge between nodes i and j if and only if the two variables are never transferred at the same time. Again, a convenient method for constructing the graph is to begin with a complete graph with nodes for the interconnection variables. Edges that represent data transfers that occur at the same time are systematically removed by examining the schedule graph or the textual representation of the schedule.

To allow profit considerations, we will partition the connections into two categories as defined by the subgraphs in Table 12.7.

Table 12.7 Subgraphs for profit categories in data path allocation.

G_2	Connections share either a common source or a common destination.
G_1	Connections share neither a common source nor a common destination.

PDCPA can be used to find a nearly minimal clustering of interconnection variables into busses.

As an example, consider the reduced schedule in Figure 12.42 along with the ALU allocation as obtained in the previous section:

```
ALU1: {1,3,5,8}
ALU2: {2,4,7}
ALU3: {6}
```

We arbitrarily assign indices to the interconnections as shown in Figures 12.49 and 12.50. The figures for this example were reprinted from Tseng and Siewiorek, 1986.

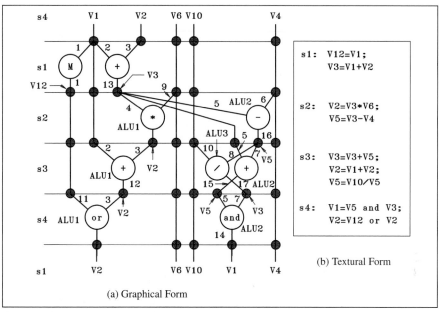

Figure 12.50 Assignment of integers for data path allocation.

```
(1,2)2    (1,4)      (1,5)       (1,6)      (1,7)       (1,8)
(1,9)     (1,10)     (1,11)      (1,12)     (1,14)      (1,15)
(1,16)    (1,17)     (2,4)2      (2,6)      (2,9)       (2,11)2
(2,14)    (2,16)     (3,4)       (3,6)      (3,9)2      (3,16)
(4,5)2    (4,7)      (4,8)       (4,10)     (4,11)2     (4,13)
(4,14)    (4,15)     (4,17)      (5,13)     (6,7)2      (6,8)
(6,10)    (6,11)     (6,13)      (6,14)     (6,15)      (6,17)
(7,8)2    (7,9)      (7,13)      (7,16)     (8,9)       (8,11)
(8,13)    (8,14)     (8,16)      (9,10)     (9,11)      (9,13)
(9,14)    (9,15)     (9,17)      (10,11)    (10,13)     (10,14)
(10,16)   (11,13)    (11,15)     (11,16)    (11,17)     (12,13)2
(13,14)   (13,15)2   (13,16)     (13,17)    (14,15)2    (14,16)2
(14,17)   (15,16)2   (16,17)2
```

Figure 12.51 Compatibility graph for interconnection unit allocation (copyright © 1986 IEEE).

Using the schedule in Figure 12.42 and the assignment of indices shown in Figure 12.49, construct the compatibility graph in Figure 12.51.

Application of the profit directed cluster partitioning algorithm to the compatibility graph with two categories of profit yields the following partition.

$$(\{1,2,4,11\}, \{3,9\}, \{5\}, \{6,7,8\}, \{10\}, \{12\}, \{13,14,15,16\}, \{17\})$$

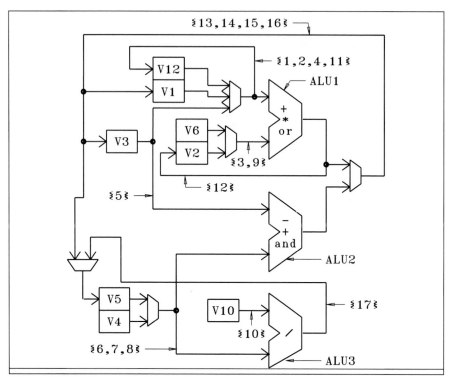

Figure 12.52   Data path diagram for the example using multiplexers.

Figure 12.52 shows the block diagram using multiplexers that results from the three applications of the profit directed cluster partitioning algorithm to register allocation, to functional unit allocation, and to data path allocation. The multiplexers could be replaced by tristate buffers if desired.

12.5 STATE OF THE ART IN HIGH-LEVEL SYNTHESIS

The algorithms given in previous sections perform well for each synthesis task taken in isolation. However, as mentioned in the introduction to this chapter, the decisions of the various synthesis tasks are not independent. The problem of how to find good solutions to all tasks simultaneously is still an open problem.

None of the algorithms produce guaranteed, optimal solutions even to an isolated task, which is probably impossible because the tasks are NP-complete problems. Therefore, the best that we can hope for are efficient algorithms that give near-minimal or "good" solutions or efficient algorithms that solve special cases of the general problem. Much ongoing research is directed toward reducing the number of computations required during the synthesis task.

Two new approaches seem promising. The first is to use expert knowledge to guide the reduction process. The second approach is to narrow the problem domain so that more specific knowledge can be used. For example, the digital signal processing domain, microprocessor

```
      package TYPES is
        attribute ENCODING: STRING;
        type ENUM is (A, B, C, D);
        attribute ENCODING of ENUM: type is "00 01 10 11";
      end TYPES;
      -----------------------------------------------------------
      use work.TYPES.all;
      entity MUX is
        port (X, Y: in BIT;  VECT: in BIT_VECTOR(3 downto 0);
              CHOICE: in ENUM; INDEX: in INTEGER range 3 downto 0;
              Z1, Z2, Z3: out BIT);
      end MUX;
      -----------------------------------------------------------
      architecture MUX_CONSTRUCTS of MUX is
      begin
      MUX1: process (CHOICE, X, Y)
        begin
          case CHOICE is
            when A => Z1 <= X;
            when B => Z1 <= Y;
            when C => Z1 <= not X;
            when D => Z1 <= not Y;
          end case;
        end process MUX1;
      -----------------------------------------------------------
      MUX2: process (X, Y, VECT)
        begin
          if X = '1' then
            Z2 <= VECT(3);
          elsif Y = '1' then
            Z2 <= VECT(2);
          else
            Z2 <= VECT(1) and VECT(0);
          end if;
        end process MUX2;
      -----------------------------------------------------------
      MUX3: process (VECT, INDEX)
        begin
          Z3 <= VECT(INDEX);
        end process MUX3;
      end MUX_CONSTRUCTS;
```

Figure 12.53 VHDL constructs that map to multiplexer elements.

domain, and pipelined domain have all been investigated with varying degrees of success. There remain a number of open problems in the synthesis area.

Human interaction with the synthesis programs will probably be necessary. However, none of the existing programs allow very much human interaction, and no definitive study has been done to explore the amount of human interaction that might be needed or useful.

Design verification refers to a proof that the design produced actually satisfies the given specifications. This might be done by simulation in which the output of the final design produced by a low-level simulator is compared with the output produced by a high-level simulation of the design specifications. The problem is knowing when to stop the simulation. How much is enough? Another approach is to develop a mathematical proof that the design steps preserve the behavior of the initial specifications.

Integration of design levels so that low-level information about layout area, power requirements, etc., can be used in making high-level decisions, is still in its infancy. For example, to make intelligent decisions at the algorithmic level, we must know something about how the lower-level tools will process the models that we pass to them.

A number of more specific synthesis tasks remain to be solved. Interface design, timing constraints, interaction between scheduling and allocation, interaction between the various allocation tasks, and the effects of high-level activities on lower-level designs are all areas that need further investigation.

In summary, high-level synthesis, as an abstract problem involving scheduling and allocation, is well understood, and there are a variety of techniques that have been applied to the solution of isolated tasks. However, a systematic integrated approach to the complete problem is beyond the current state of the art.

12.6 AUTOMATED SYNTHESIS OF VHDL CONSTRUCTS

In this section, we discuss the possibility of automatically translating a representative sample of VHDL constructs into hardware. We will concentrate on translations that are application independent. Other sections of the book discuss translations from specified programming styles into restricted sets of hardware.

12.6.1 Constructs that Involve Selection

There are a variety of VHDL constructs that imply selection of a specific element from a specified set. The *case* statement implies selection of one case from a specified set of cases. The *if...then...else* statement implies selection of the highest priority condition that is TRUE from a prioritized list of conditions. Also, one element of a vector may be selected by specifying an index value. All these statements involve selection and can, therefore, be implemented in hardware using a multiplexer element. Figure 12.53 shows several examples of VHDL constructs that can be mapped to multiplexer elements. A hypothetical entity called MUX is declared to illustrate the concepts. The purpose of entity MUX is to provide an environment in which three independent processes can be specified. MUX has inputs X and Y of type BIT, input VECT of type BIT_VECTOR, input CHOICE of type ENUM, which is an enumerative data type, and input INDEX of type INTEGER. MUX has three independent outputs Z1, Z2, and Z3. Each process (MUX1, MUX2, and MUX3) involves selection in some form and can be realized with a multiplexer element. Each process is discussed in detail in the next three sections.

12.6.2 Mapping *case* Statements to Multiplexers

Process MUX1 in Figure 12.53 is an example of a *case* statement that can be implemented with a multiplexer element. The *case* statement selects a value to assign to output Z1, based on the value of enumerative data type CHOICE. In a logic circuit, elements of an enumerative data type

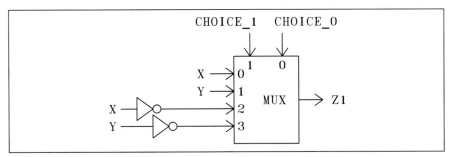

Figure 12.54 Hardware implementation for process MUX1.

must be represented by binary vectors. When CHOICE represents the states in a finite state machine, the assignment of binary values to states is called the *state assignment problem*. In an automated system, we might allow the program to make the assignment. However, there is no known efficient algorithm for finding the best assignment. Therefore, we might want to allow the designer to specify the binary value to be associated with each element in the enumerative data type. One way to specify the assignment is to use the concept of an *attribute*. In package TYPES, we declare an attribute ENCODING of type STRING. We then associate the attribute with the type ENUM with this statement:

```
attribute ENCODING of ENUM: type is "00 01 10 11"
```

The attribute ENCODING directs the automated design tool to assign binary codes to elements of type ENUM as shown in Table 12.8.

Table 12.8 Binary codes assigned to ENUM by attribute ENCODING.

ENUM element	Binary value
A	00
B	01
C	10
D	11

With this information, the automated design tool can translate process MUX1 into the hardware circuit shown in Figure 12.54. In this diagram, signal pair (CHOICE_1, CHOICE_0) represents the binary code for elements of input CHOICE. This binary code is connected to the address inputs of the multiplexer and, therefore, selects the appropriate data input to transfer to the output Z1. For example, when CHOICE = (00), Z1 = X as implied by the VHDL code for MUX1. This example illustrates that *case* statements can be translated directly into multiplexer elements. The CHOICE expression defines the address inputs to the MUX. The values assigned to Z1 become data inputs to the MUX.

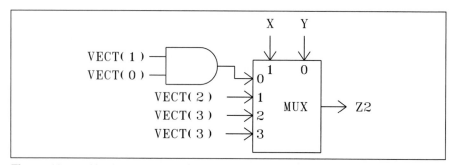

Figure 12.55   Hardware implementation for process MUX2.

12.6.3   Mapping *if...then...else* Statements to Multiplexers

The VHDL construct *if...then...else* also involves selection among several alternative actions. Therefore, multiplexer elements can be used to implement *if...then...else* constructs. Process MUX2 in Figure 12.53 shows an example of this construct, which involves inputs X, Y, and VECT. By scanning the *if...then...else* clause, the automatic design tool can produce the information in Table 12.9.

Table 12.9   Information contained in *if...then...else* clause.

X	Y	Z2
0	0	VECT(1) and VECT(0)
0	1	VECT(2)
1	0	VECT(3)
1	1	VECT(3)

Notice that the first *if* clause that is TRUE selects the action to be performed. It is possible that more than one *if* clause is TRUE. For example, if X=Y=1, then two of the clauses are TRUE. However, in this case, Z2 is assigned the value VECT(3) because the clause (*if* X='1') takes precedence over the clause (*elsif* Y='1').

The preceding table is constructed as follows. The clause (*if* X = '1' *then* Z2 <= VECT(3)) implies that table entries for the two input combinations XY=10 and XY=11 should both be VECT(3). The clause (*elsif* Y = '1' *then* Z2 <= VECT(2)) implies that the vacant table entry for input combination XY=01 should be VECT(2). Since the table entry for input combination XY=11 has already been filled, the *elsif* clause has no effect on this entry, even though Y='1' for this combination. The clause (*else* Z2 <= VECT(1) and VECT(0)) requires that all entries not yet filled be assigned to the function (VECT(1) and VECT(0)). In this case, the only entry not yet filled is for the combination XY=00.

The table directly implies the multiplexer implementation in Figure 12.55, where signals X and Y are connected to the address inputs to the multiplexer, and the data inputs for each XY combination are specified by the table entries.

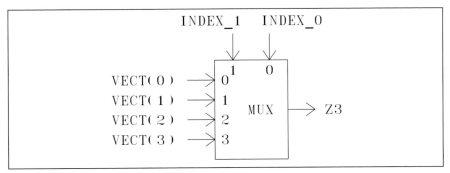

Figure 12.56 Hardware implementation for process MUX3.

12.6.4 Mapping Indexed Vector References to Multiplexers

If VECT is a vector and INDEX is an integer, then a variable assignment of the form:

```
Z <= VECT(INDEX)
```

is also a selection activity. In this case, one of the elements of the vector VECT is being assigned to the variable Z. The specific element is selected, based on the value of the variable INDEX. This type of statement also maps directly to a multiplexer device. Process MUX3 in Figure 12.53 shows an example of this type of selection activity. Variable Z3 is assigned the value of one of the elements of vector VECT. An automated design tool could map this VHDL construct to the hardware circuit of Figure 12.56. In this case, since the vector VECT has four elements with index values ranging from 3 down to 0 (see the declaration for VECT in entity MUX), the automated translator would need to convert variable INDEX into a 2-bit binary representation. In the absence of an attribute specifying a specific code, the automatic translator can use a straight binary number representation for the integer values. This code leads directly to the circuit in Figure 12.56.

12.6.5 Loop Constructs

The general algorithmic modeling style recommended in Chapter 10 suggested that loops should be avoided unless there was an obvious hardware implementation. In this section, we discuss a type of program loop that leads naturally to iterative network implementations.

We will start with a variation on the EXCLUSIVE-OR network first introduced at the end of Chapter 1. Consider the entity XOR4 declared in Figure 12.57. The input to entity XOR4 is a 4-bit vector, A, with index values ranging from 3 down to 0. The output is the variable X of type BIT. We want X to be equal to the EXCLUSIVE-OR of the four bits in the input vector A.

Architecture XOR4_LOOP describes this relationship by using a simple loop structure. An internal variable, X_INT, is declared inside a process. X_INT is initialized to '0'. In the initial pass through the loop (I=0), X_INT is replaced by the EXCLUSIVE-OR of A(0) with the constant '0', which produces a result equal to A(0). During the next pass through the loop (I=1), X_INT is replaced by the EXCLUSIVE-OR of the current value of X_INT (A(0)) with A(1). The new value of X_INT is (A(0) xor A(1)). Each pass through the loop adds another element of A to the calculation until the value of X_INT, after four passes, becomes:

```
-------------------------------------------------
-- Entity declaration for 4-bit XOR circuit --
-------------------------------------------------
entity XOR4 is
  port (A: in BIT_VECTOR (3 downto 0);
        X: out BIT);
end XOR4;
-------------------------------------------------------
-- Standard loop definition for circuit behavior.--
-------------------------------------------------------
architecture XOR4_LOOP of XOR4 is
begin
  process (A)
    variable X_INT: BIT;
  begin
    X_INT := '0';
    for I in 0 to 3 loop
      X_INT := X_INT xor A(I);
    end loop;
    X <= X_INT;
  end process;
end XOR4_LOOP;
```

Figure 12.57   VHDL program with a simple program loop.

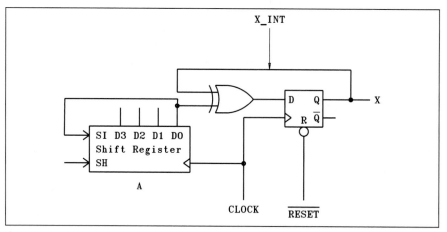

Figure 12.58   Hardware implementation for architecture XOR4_LOOP of entity XOR4.

$$X\_INT = A(0) \text{ xor } A(1) \text{ xor } A(2) \text{ xor } A(3).$$

Finally, the value of X_INT is assigned to the signal X.

This simple loop can be implemented as a sequential circuit using a shift register as shown in Figure 12.58. It is assumed that the process in architecture XOR4_LOOP is embedded in a larger system not shown. Vector A is stored in a shift register. How the storage occurs is not

```
----------------------------------------------
-- Entity declaration for 4-bit XOR circuit--
----------------------------------------------
entity XOR4 is
  port (A: in BIT_VECTOR (3 downto 0);
        X: out BIT);
end XOR4;
--------------------------------------------------------------
-- Space_Time transformation of standard loop definition.--
--------------------------------------------------------------
architecture XOR4_SPACE of XOR4 is
begin
  process (A)
    variable X_INT: BIT_VECTOR (4 downto 0);
  begin
    X_INT(0) := '0';
    for I in 0 to 3 loop
      X_INT(I+1) := X_INT(I) xor A(I);
    end loop;
    X <= X_INT(4);
  end process;
end XOR4_SPACE;
```

Figure 12.59 VHDL program obtained by applying a space-time transformation to architecture XOR4_LOOP.

specified. A flip-flop is used to implement the internal variable X_INT. The flip-flop must be initialized to '0'. The bits of the vector A are shifted in the register, so the bit needed at each instance of time is present at terminal D0. After four shifts, the desired result will be present at the FF output (X). The source of the control signals needed to time the shift operation are not included.

An iterative combinational logic network can be obtained by applying *a space-time transformation* to the architecture XOR4_LOOP. The internal variable X_INT that is assigned a time series of values in Figure 12.57 is replaced by a 5-bit spatial vector with index values ranging from 4 down to 0. Values are assigned to the elements of the spatial vector by the modifed loop in Figure 12.59. Element X_INT(0) is initialized to the value of '0', which corresponds to the initialization of variable X_INT in the original loop. Inside the loop, each result of applying the *xor* operation is assigned to a different element of the spatial vector X_INT.

A similar space-time transformation can be applied to any loop of the form shown in architecture XOR4_LOOP. The transformation is easy to describe in an algorithmic fashion that can be easily automated.

Architecture XOR4_SPACE can be mapped directly onto the iterative combinational network in Figure 12.60. The successive values of vector X_INT are generated by individual XOR gates. The network corresponds to the loop calculations in architecture XOR4_SPACE in a very natural way.

This *xor* example is representative of a large class of iterative combinational networks. All members of the class can be implemented either as shift register-based sequential networks or as iterative combinational logic networks. The networks are all of the same general form

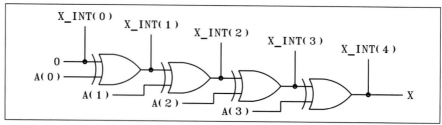

Figure 12.60 Hardware implementation for architecture XOR4_SPACE of entity XOR4.

```
-----------------------------------------------------
-- Entity declaration for a 4-bit binary adder --
-----------------------------------------------------
entity ADD4 is
  port (A,B: in BIT_VECTOR (3 downto 0);
        CIN: in BIT;
        S: out BIT_VECTOR (3 downto 0);
        COUT: out BIT);
end ADD4;
-----------------------------------------------------
-- Typical definition of adder using a program loop --
-----------------------------------------------------
architecture LOOP_ADDER of ADD4 is
begin
  process (A, B, CIN)
    variable CARRY: BIT := '0';
    variable SUM: BIT_VECTOR (3 downto 0);
  begin
    CARRY := CIN;
    for I in 0 to 3 loop
      SUM(I) := A(I) xor B(I) xor CARRY;
      CARRY := (A(I) and B(I)) or (A(I) and CARRY) or
               (B(I) and CARRY);
    end loop;
    S <= SUM;
    COUT <= CARRY;
  end process;
end LOOP_ADDER;
```

Figure 12.61 VHDL program for a ripple carry adder.

except that the function performed is, in general, much more complex than a simple *xor* function.

We will illustrate the general case with the classic ripple carry adder circuit. Figure 12.61 shows the VHDL code written as a simple loop. This code can be directly translated into a shift register-based sequential network as shown in Figure 12.62. The control signals and source of data for registers A and B are not included. It is assumed that the process in architecture

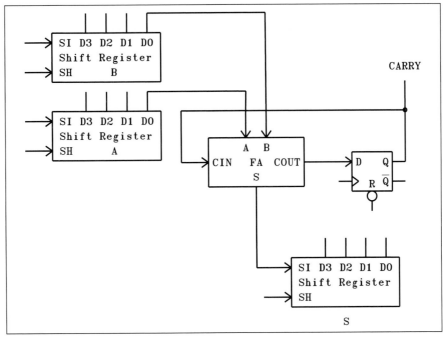

Figure 12.62 Hardware implementation for architecture LOOP_ADDER of entity ADD4.

LOOP_ADDER is embedded in a larger system that is not shown. The values of input vectors A and B are shifted into the Full Adder (FA) network one bit at a time. The sum is created at the output of the FA, one bit at a time, and stored in the shift register S. The value of the internal variable CARRY is stored in a flip-flop for use during the next loop iteration. The result is called a *serial adder.*

Applying a space-time transformation to the VHDL code of Figure 12.61 results in the VHDL code in Figure 12.63. Notice that the internal variable CARRY in architecture LOOP_ADDER is replaced by a vector CARRY with index values in the range from 4 down to 0 in architecture SPACE_ADDER. The logic equations in the original loop (denoted by the symbol FA), are used to assign values to successive outputs and to successive elements of the CARRY vector. The CARRY flip-flop has been eliminated. The new loop in architecture SPACE_ADDER directly maps to the iterative combinational logic network of Figure 12.64.

12.6.6 Functions and Procedures

In this section, we illustrate possible mappings of functions and procedures to hardware.

Figure 12.65 shows VHDL code for architecture FUNCTION_ADDER of entity ADD4 obtained from the code for architecture SPACE_ADDER by replacing the logic equations for the full adder by function declarations. There are separate declarations for each output of the FA. The logic equation for SUM(I) is declared to be function FA_S; the logic equation for CARRY(I+1) is declared as function FA_C. Inside the program loop, the assignments to SUM(I)

```
----------------------------------------------------------
-- Applying space-time transformation to LOOP_ADDER-------
-- architecture --
----------------------------------------------------------
architecture SPACE_ADDER of ADD4 is
begin
  process (A, B, CIN)
    variable CARRY: BIT_VECTOR (4 downto 0) := "00000";
    variable SUM: BIT_VECTOR (3 downto 0);
  begin
    CARRY(0) := CIN;
    for I in 0 to 3 loop
      SUM(I) := A(I) xor B(I) xor CARRY(I);
      CARRY(I+1) := (A(I) and B(I)) or (A(I) and CARRY(I))
               or (B(I) and CARRY(I));
    end loop;
    S <= SUM;
    COUT <= CARRY(4);
  end process;
end SPACE_ADDER;
```

Figure 12.63 Result of applying a space-time transformation to architecture LOOP_ADDER.

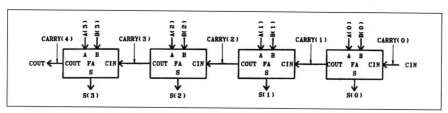

Figure 12.64 Hardware implementation for architecture SPACE_ADDER of entity ADD4.

and to CARRY(I+1) are replaced by function calls. Since we have only changed the notation, not the basic operation of the algorithm, it is clear that architecture FUNCTION_ADDER can be mapped to the same hardware as architecture SPACE_ADDER.

The general conclusion is that functions should be mapped to combinational logic circuits.

Similarly, Figure 12.66 shows how the ripple carry adder might be described in VHDL, using procedures. First, a procedure FA is defined for the full adder. The connections in the iterative adder are then represented by procedure calls within the program loop. This example illustrates that procedures are used mainly as a convenience for ease of programming. In general, any VHDL code that uses procedures can be mapped to the same hardware as equivalent code without procedures. The main difference between procedures and functions is that functions always map to combinational logic; whereas, procedures can map to sequential logic. This example only illustrates the case where a procedure maps to combinational logic. Other examples throughout the book show a variety of procedure definitions.

```
---------------------------------------------
-- Use of functions to define the adder --
---------------------------------------------
architecture FUNCTION_ADDER of ADD4 is
  -- Function to compute the sum output
  function FA_S (AIN, BIN, CIN: BIT) return BIT is
  begin
    return AIN xor BIN xor CIN;
  end FA_S;
  -- Function to compute the carry output
  function FA_C (AIN, BIN, CIN: BIT) return BIT is
  begin
    return (AIN and BIN) or (AIN and CIN) or (BIN and CIN);
  end FA_C;
begin
  process (A, B, CIN)
    variable CARRY: BIT_VECTOR (4 downto 0) := "00000";
    variable SUM: BIT_VECTOR (3 downto 0) := "0000";
  begin
    CARRY(0) := CIN;
    for I in 0 to 3 loop
      SUM(I) := FA_S(A(I), B(I), CARRY(I));
      CARRY(I+1) := FA_C(A(I), B(I), CARRY(I));
    end loop;
    S <= SUM;
    COUT <= CARRY(4);
  end process;
end FUNCTION_ADDER;
```

Figure 12.65   Using functions to represent combinational logic.

PROBLEMS

12.1   Add a start signal and a completion signal to the FSM VHDL model in Figure 12.11 to make the device more useful. Simulate the resulting model to verify correct operation.

12.2   Consider this VHDL code:

```
----------------------------------------------------------
   -- Entity declaration --
----------------------------------------------------------
entity SCHED1 is
  port (A, B, C, D, E, F: in INTEGER;
        CLK: in BIT;
        X, Y: out INTEGER);
end SCHED1;
----------------------------------------------------------
   -- Achitecture declaration --
----------------------------------------------------------
architecture HIGH_LEVEL of SCHED1 is
begin
    X <= F*(A+B+C*D+D*E);
    Y <= (A*B+E)*D*C;
end HIGH_LEVEL;
```

```
-------------------------------------------
-- Use of a procedure to define the adder. --
-------------------------------------------
architecture PROCEDURE_ADDER of ADD4 is
  procedure FA(AIN, BIN, CIN: in BIT;
               SOUT, COUT: out BIT) is
  begin
    SOUT := AIN xor BIN xor CIN;
    COUT := (AIN and BIN) or (AIN and CIN) or
            (BIN and CIN);
  end FA;
begin
  process (A, B, CIN)
    variable CARRY: BIT_VECTOR (4 downto 0) := "00000";
    variable SUM: BIT_VECTOR (3 downto 0) := "0000";
  begin
    CARRY(0) := CIN;
    for I in 0 to 3 loop
      FA(A(I), B(I), CARRY(I), SUM(I), CARRY(I+1));
    end loop;
    S <= SUM;
    COUT <= CARRY(4);
  end process;
end PROCEDURE_ADDER;
```

Figure 12.66   Using procedures to represent combinational logic.

The following tasks refer to the VHDL code above. Assume there are no hardware constraints.
   a. Draw a data flow graph.
   b. Derive an ASAP schedule.
   c. Derive an ALAP schedule.
   d. Derive a list schedule using the critical path priority metric.
   e. Derive a schedule using the freedom-directed method.

12.3 Repeat Problem 12.2 using each of these hardware constraints:
   a. Hardware is limited to one adder and one multiplier.
   b. Hardware is limited to one adder and two multipliers.
   c. Hardware is limited to two adders and one multiplier.
   d. Hardware is limited to two adders and two multipliers.

12.4 Can any of the schedules in Problem 12.2 or Problem 12.3 be improved by applying basic mathematical transformations? In particular, note that the product C*D is common to both X and Y. There might be other simplifications as well.

12.5 Consider this VHDL code:

```
-------------------------------------------------------
  -- Entity declaration
-------------------------------------------------------
entity SCHED2 is
  port (A, B, C, D, E, F: in INTEGER;
        CLK: in BIT;
        X, Y: out INTEGER);
end SCHED2;
```

```
--------------------------------------------------------
-- Architecture declaration
--------------------------------------------------------
architecture HIGH_LEVEL of SCHED2 is
  signal Z: INTEGER;
begin
  X <= (A-B)*Z;
  Y <= (A*B)+Z;
  Z <= C*D + (E+F)/D;
end HIGH_LEVEL;
```

The following tasks refer to the VHDL code above. Assume there are no hardware constraints.

    a. Draw a data flow graph.

    b. Derive an ASAP schedule.

    c. Derive an ALAP schedule.

    d. Derive a list schedule using the critical path priority metric.

    e. Derive a schedule using the freedom-directed method.

12.6 Repeat Problem 12.5 using each of these hardware constraints:

    a. Hardware is limited to one adder, one subtracter, one multiplier, and one divider.

    b. Assume addition and subtraction are done in the same hardware module (ADD_S), and multiplication and division are done in a single module (MD). Also, assume that only one module of each type is available.

    c. Assume addition and subtraction are done in the same hardware module (ADD_S), and multiplication and division are done in a single module (MD). Also, assume that two modules of each type are available.

    d. Assume all operations are done in an ALU module. Assume that two ALU modules are available.

12.7 Add address control signals to the MUXes and load signals to the registers in Figure 12.9, and use the FSM architecture of Figure 12.11 to design the control unit for device FSMEX1. Use the techniques developed in Chapter 8.

12.8 Consider the best of the schedules obtained in Problem 12.2. Use a simple greedy algorithm to allocate registers, functional units, and data paths.

    a. Show an allocation diagram for registers and functional units similar to Figure 12.5.

    b. Show the data path allocation by constructing a hardware diagram similar to Figure 12.6.

    c. Construct a Gantt chart and discuss utilization.

    d. Develop and simulate a FSM model for the control unit similar to Figure 12.11.

    e. Add address control signals to the MUXes and load signals to the registers. Use the FSM architecture to design the control unit. Use the techniques developed in Chapter 8.

12.9 Repeat Problem 12.8 using the best schedule obtained in each of the following problems. Compare and contrast the schedules.

    a. Problem 12.3(a)

    b. Problem 12.3(b)

    c. Problem 12.3(c)

    d. Problem 12.3(d)

    e. Problem 12.5

    f. Problem 12.6(a)

    g. Problem 12.6(b)

    h. Problem 12.6(c)

    i. Problem 12.6(d)

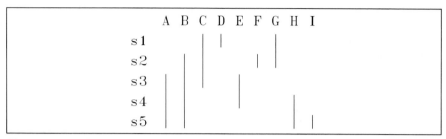

Figure 12.67   Data lifetime chart for Problem 12.10.

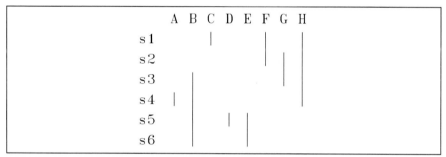

Figure 12.68   Data lifetime chart for Problem 12.11.

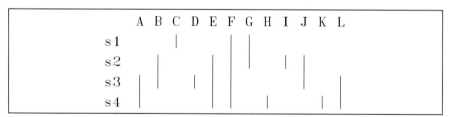

Figure 12.69   Data lifetime chart for Problem 12.12.

12.10 For the data lifetime chart shown in Figure 12.67, use the left edge algorithm to obtain an efficient register allocation.

12.11 For the data lifetime chart shown in Figure 12.68, use the left edge algorithm to obtain an efficient register allocation.

12.12 For the data lifetime chart shown in Figure 12.69, use the left edge algorithm to obtain an efficient register allocation.

12.13 For the data lifetime chart shown in Figure 12.70, use the left edge algorithm to obtain an efficient register allocation.

12.14 For the data lifetime chart shown in Figure 12.71, use the left edge algorithm to obtain an efficient register allocation.

12.15 For the data lifetime chart shown in Figure 12.72, use the left edge algorithm to obtain an efficient register allocation.

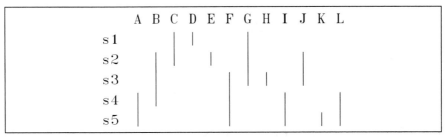

Figure 12.70 Data lifetime chart for Problem 12.13.

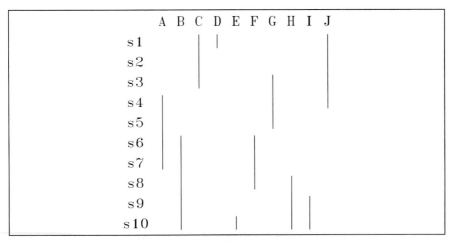

Figure 12.71 Data lifetime chart for Problem 12.14.

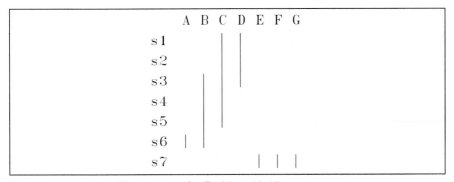

Figure 12.72 Data lifetime chart for Problem 12.15.

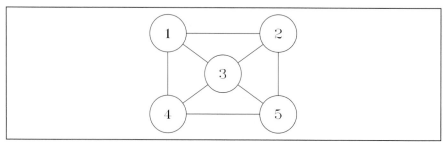

Figure 12.73   Compatibility graph for Problem 12.19.

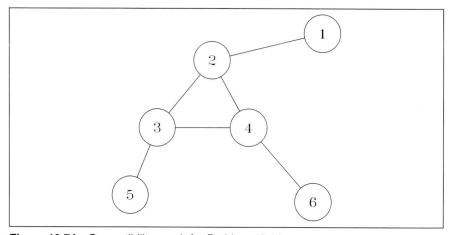

Figure 12.74   Compatibility graph for Problem 12.20.

12.16 For the schedule shown in Figure 12.38, use the left edge algorithm to obtain an efficient register allocation. Compare the results with the results generated by the graphical technique, PDCPA, shown in Figure 12.42. Did the left edge algorithm provide all the information generated by PDCPA? If not, could the left edge algorithm be modified to generate the missing information?

12.17 Consider the best of the schedules obtained in Problem 12.2. Use the left edge algorithm to allocate registers and the connection chart method (see Figure 12.29) to allocate functional units and data paths.
    a. Show an allocation chart similar to the one in Figure 12.29.
    b. Construct a hardware diagram similar to the one in Figure 12.30.
    c. Construct a Gantt chart and discuss utilization.
    d. Develop and simulate an FSM model for the control unit similar to the one in Figure 12.11.
    e. Add address control signals to the MUXes, load signals to the registers, and any other control signals that are needed to the hardware diagram constructed above. Use the FSM architecture developed above to design the control unit. Use the techniques developed in Chapter 8.

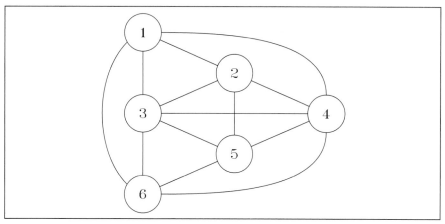

Figure 12.75 Compatibility graph for Problem 12.21.

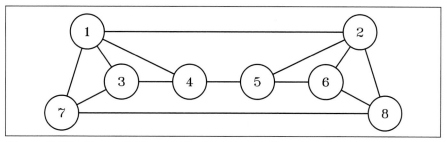

Figure 12.76 Compatibility graph for Problem 12.22.

12.18 Repeat Problem 12.17. Use the best schedule obtained in each of these problems:
  a. Problem 12.3(a)
  b. Problem 12.3(b)
  c. Problem 12.3(c)
  d. Problem 12.3(d)
  e. Problem 12.5
  f. Problem 12.6(a)
  g. Problem 12.6(b)
  h. Problem 12.6(c)
  i. Problem 12.6(d)

12.19 For the compatibility graph shown in Figure 12.73, execute the cluster partitioning algorithm (CPA) to determine a nearly minimal number of clusters that partitions the nodes of the graph.

12.20 For the compatibility graph shown in Figure 12.74, execute the cluster partitioning algorithm (CPA) to determine a nearly minimal number of clusters that partitions the nodes of the graph.

12.21 For the compatibility graph shown in Figure 12.75, execute the cluster partitioning algorithm (CPA) to determine a nearly minimal number of clusters that partitions the nodes of the graph.

12.22 For the compatibility graph shown in Figure 12.76, execute the cluster partitioning algorithm (CPA) to determine a nearly minimal number of clusters that partitions the nodes of the graph.

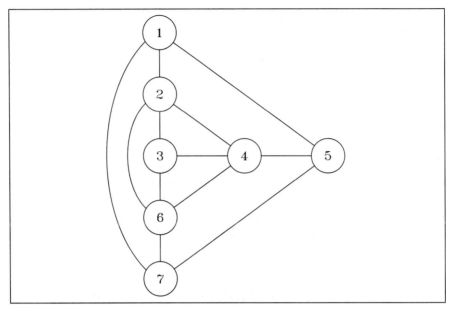

Figure 12.77   Compatibility graph for Problem 12.23.

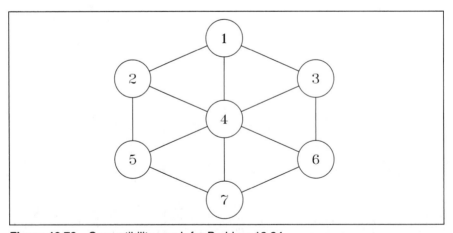

Figure 12.78   Compatibility graph for Problem 12.24.

12.23 For the compatibility graph shown in Figure 12.77, execute the cluster partitioning algorithm (CPA) to determine a nearly minimal number of clusters that partitions the nodes of the graph.

12.24 For the compatibility graph shown in Figure 12.78, execute the cluster partitioning algorithm (CPA) to determine a nearly minimal number of clusters that partitions the nodes of the graph.

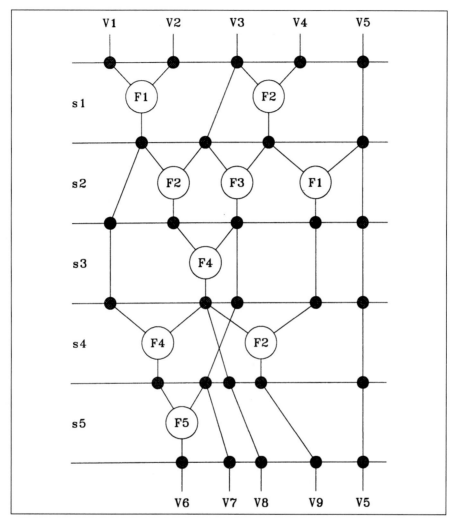

Figure 12.79 Typical schedule.

12.25 For the list schedule of Figure 12.20, use the profit directed cluster partitioning algorithm (PDCPA) to allocate registers, functional units, and data paths. Compare the results obtained with PDCPA with the results obtained using the left edge algorithm and the connection chart in the text.

12.26 Design a control unit for the device designed in Problem 12.25. Start by developing and simulating a VHDL model similar to the one in the one in Figure 12.11. Use the methods in Chapter 8 to design the control unit. Discuss the possibilities for design automation.

12.27 Consider the best of the schedules obtained in Problem 12.2. Use the profit directed cluster partitioning algorithm (PDCPA) to allocate registers, functional units, and data paths.

    a. Show a diagram similar to the one in Figure 12.50 that documents register and functional unit allocation.

    b. Construct a hardware diagram similar to the one in Figure 12.52 to document data path allocation.

    c. Construct a Gantt chart and discuss utilization.

    d. Develop and simulate an FSM model for the control unit similar to the one in Figure 12.11.

    e. Add address control signals to the MUXes, function select signals to the functional units, and load signals to the registers in the hardware diagram developed above. Use the FSM architecture developed above to design the control unit. Use the techniques developed in Chapter 8.

12.28 Repeat Problem 12.27. Use the best schedule developed in each of these problems:

    a. Problem 3(a)

    b. Problem 3(b)

    c. Problem 3(c)

    d. Problem 3(d)

    e. Problem 5

    f. Problem 6(a)

    g. Problem 6(b)

    h. Problem 6(c)

    i. Problem 6(d)

12.29 Use the profit directed cluster partitioning algorithm (PDCPA) to allocate registers, functional units, and data paths for the schedule shown in Figure 12.79. Assume each function is commutative and ALU functional units are available that can perform any of the functions.

    a. Show a diagram similar to Figure 12.50 that documents register and functional unit allocation.

    b. Construct a hardware diagram similar to Figure 12.52 to document data path allocation.

    c. Construct a Gantt chart and discuss utilization.

    d. Develop and simulate an FSM model for the control unit similar to Figure 12.11.

    e. Add address control signals to the MUXes, function select signals to the functional units, and load signals to the registers in the hardware diagram developed above. Use the FSM architecture developed above to design the control unit. Use the techniques developed in Chapter 8.

12.30 Design a PRIORITY ANALYSIS device. Assume there are eight devices contending for access to a data bus. Each device generates an access request signal that is logic 1 when the device desires access to the data bus. The inputs to the PRIORITY ANALYSIS device are the eight request signals. The PRIORITY ANALYSIS device produces eight acknowledge output signals, one for each contending device. At any time, exactly one of the eight acknowledge outputs will be logic 1. All other outputs will be logic 0. The active output will correspond to the contending device with highest priority that is currently requesting access. Consider the eight request signals to be 8-bit vector REQ(7 downto 0). REQ(7) has highest priority and

REG(0) lowest priority. Let the eight acknowledge outputs be 8-bit vector ACK(7 downto 0). For an example, if REQ = 00101100, then ACK should be 00100000.

   a. Write an algorithmic-level VHDL description of the PRIORITY ANALYSIS device using a program loop and simulate it to verify correct input-output behavior.

   b. Translate the algorithmic-level VHDL description into a hardware implementation. Document the translation procedure used.

   c. Simulate the hardware implementation using a hardware simulator. Compare the behavior of this simulation with the algorithmic VHDL simulation.

12.31 Write an algorithmic-level VHDL description for a PRIORITY ANALYSIS device similar to that in the previous problem except use generics to make the module accept any number of inputs greater than or equal to 2.

12.32 Design a binary COMPARATOR device that will accept two, 8-bit binary vectors, A(7 downto 0) and B(7 downto 0), as inputs and produce three binary output signals as follows. The inputs represent positive integers.

   AGTB   Signal is logic 1 if and only if integer A is strictly greater than integer B.

   AEQB   Signal is logic 1 if and only if integer A is equal to integer B.

   ALTB   Signal is logic 1 if and only if integer A is strictly less than B.

   a. Write an algorithmic-level VHDL description for the COMPARATOR device using a program loop and simulate it to verify correct behavior.

   b. Translate the algorithmic-level VHDL description into a hardware implementation. Document the translation procedure used.

   c. Simulate the hardware implementation using a hardware simulator. Compare the behavior of this simulation with the algorithmic VHDL simulation.

12.33 Add generics to the algorithmic-level VHDL description of the COMPARATOR device described in the previous problem, so it will work for input vectors of any size.

12.34 Design a device, COMPARATOR2, that will accept two, 8-bit binary vectors, A(7 downto 0) and B(7 downto 0), as inputs and produce three binary output signals as follows. The inputs represent positive integers.

   AGEB   Signal is logic 1 if and only if integer A is greater than or equal to integer B.

   AEQB   Signal is logic 1 if and only if integer A is equal to integer B.

   ALEB   Signal is logic 1 if and only if integer A is less than or equal to B.

   a. Write an algorithmic-level VHDL description for the COMPARATOR2 device using a program loop and simulate it to verify correct behavior.

   b. Translate the algorithmic-level VHDL description into a hardware implementation. Document the translation procedure used.

   c. Simulate the hardware implementation using a hardware simulator. Compare the behavior of this simulation with the algorithmic VHDL simulation.

12.35 Add generics to the algorithmic-level VHDL description of the COMPARATOR2 device described in the previous problem, so it will work for input vectors of any size.

12.36 Shown below is the code for a VHDL model. The ADD8 function performs an 8-bit add. The INC8 function performs an 8-bit increment.

   a. Explain in words what the model does.

   b. Draw a schematic of the hardware implied by the description. Your schematic should be at the register level of abstraction.

```
use work.funcs.all;
entity SYS is
  port(C: in BIT; COM: BIT_VECTOR(0 to 1);
       INP: in BIT_VECTOR(0 to 7));
end SYS;
architecture CODE of SYS is
  signal X,Y: BIT_VECTOR(0 to 7);
begin
  process(C)
  begin
    if C='1' then
      case COM is
        when "00" => X <= INP;
        when "01" => Y <= INP;
        when "10" => X <= ADD8(X,Y);
        when "11" => X <= ADD8(X,INC8(not(Y)));
      end case;
    end if;
  end process;
end CODE;
```

12.37 Synthesize the VHDL shown below into hardware. Use a block diagram where the function, inputs, and outputs of each block are clearly identified.

```
use work.PRIMS.all;
entity TEST_QUESTION is
  port(CLK,W,X,Y,Z: in BIT; O: out BIT);
end TEST_QUESTION;
architecture BEHAV of TEST_QUESTION is
  signal  C: BIT_VECTOR(0 to 1);
begin
  BLK: block(not CLK'STABLE and CLK='1')
  begin
    C <= guarded INC(C);   --INC is an increment function
  end block BLK;
  O <= W when C="00" else
       X when C="01" else
       Y when C="10" else
       Z;
end BEHAV;
```

References

1. Acock, S.J.B, Dimond, K.R., "Automatic mapping of algorithms onto multiple FPGA-SRAM modules," *Field-programmable Logic and Applications. 7th International Workshop, FPL '97 Proceedings*. Springer-Verlag, Berlin, Germany, 1997, pp. 255–64.

2. Airiau, R., Berge, J., Olive, V., *Circuit Synthesis with VHDL*, Kluwer Academic Publishers, Boston, MA, 1994.

3. Arato, P., Visegrady, T., "Effective graph generation from VHDL descriptions," *Microelectronics Journal*. vol. 29, no. 3, March 1998, pp. 113–21.

4. Armstrong, J. R., Gray, F. G., *Structured Logic Design With VHDL*, Prentice Hall PTR, Englewood Cliffs, N.J., 1993.

5. Armstrong, J. R., "Process-level modeling with VHDL," *Proceedings International Verilog HDL Conference and VHDL International Users Forum* (Cat. No.98TB100230), IEEE Computer Society Press, Los Alamitos, CA, USA, 1998, p.72–6.

6. Armstrong, J., Cho, C., Shah, S., Kosaraju, C. "The VHDL validation suite," *Proceedings-27th ACM/IEEE Design Automation Conference* (Cat. #90CH2894-4), IEEE, Piscataway, NJ, 1990, pp. 2–7.

7. Armstrong, J. R., *Chip-Level Modeling with VHDL*, Prentice Hall PTR, Englewood Cliffs, NJ, 1989.

8. Armstrong, J. R., "The use of the process model graph in defining the structure of behavioral models," *SIGDA Newsletter*, vol. 18, no. 4, pp. 65–70.

9. Armstrong, J. R., Gray, F. G., Lin, M. W., "VHDL modeling and model testing for DSP applications," *IEEE Transactions on Industrial Electronics*, vol. 46, no. 1, February 1999, pp. 13–22.

10. Armstrong, J. R., Burnette, D. G., "Automated assists to the behavioral modeling process," *First International Workshop on Rapid System Prototyping. Shortening the Path from Specification to Prototype* (Cat. #91TH0380-6), IEEE, Piscataway, NJ, pp. 187–95.

11. Armstrong, J. R., Burnette, D. G., "A systematic approach to chip level modeling with VHDL," *Wescon/89 Conference Record*, Electronic Conventions Management, Ventura, CA, pp. 333–338.

12. Armstrong, J. R., Gray, F. G., Lin, M. W., "VHDL modeling and model testing for DSP applications," *IEEE Transactions on Industrial Electronics*, vol. 46, no. 1, February 1999, pp. 13–22.

13. Ashenden, P., *The Designer's Guide to VHDL*, Morgan Kaufman Publishers, Inc., San Francisco, 1996.

14. Ashenden, P., *The Student's Guide to VHDL*, Morgan Kaufman Publishers, Inc., San Francisco, 1998.

15. Ashenden, P., "Modeling digital systems using VHDL," *IEEE-Potentials*. vol. 17, no. 2, April–May 1998, pp. 27–30.

16. Austin, S. M., "Automated translation of ASIC designs," *1998 IEEE AUTOTESTCON Proceedings, IEEE Systems Readiness Technology Conference. Test Technology for the 21st Century* (Cat. No.98CH36179), IEEE, New York, NY, USA, 1998, p. 667.

17. Barton, D. L., "Behavioral descriptions in VHDL," *VLSI SystemsDesign*, vol. 9, no. 6, pp. 28–31, 33.

18. Baumgartner, K. M., *Computer Scheduling Algorithms: Past, Present and Future*, Elsevier Science Publishing Co., Inc., 1991.

19. Berenyi, A., et al, "Continuously live image processor for drift chamber track segment triggering," *IEEE Transactions on Nuclear Science*, vol. 46, no. 3, pt.1, June 1999, pp. 348–53.

20. Berge, J., Fonkura, A., Maginot, S., Rouillard, J., *VHDL 92: The New Features of the VHDL Language*, Kluwer Academic Publishers, Boston, MA, 1993.

21. Berge, J., Fonkura, A., Maginot, S., Rouillard, J., *VHDL Designer's Reference*, Kluwer Academic Publishers, Boston, MA, 1992.

22. Bhasker, J., "Process-graph analyzer: a front-end tool for VHDL behavioral synthesis," *Software-Practice and Experience*, vol. 18, no. 5, pp. 469–83.

23. Bhasker, J., *A Guide to VHDL Syntax*, Prentice Hall, Englewood Cliffs, NJ, 1995.

24. Bhasker, J., *A VHDL Primer*, Prentice Hall, Englewood Cliffs, NJ, 1995.

25. Bhasker, J., *A VHDL Synthesis Primer*, Star Galaxy Publishing, Allentown, PA, 1996.

26. Bonk, J., Stone, A., Manolakos, E. S., "Synthesis of array architectures for block matching motion estimation," *1999 IEEE International Conference on Acoustics, Speech, and Signal Processing. Proceedings, ICASSP99,* vol. 4 (Cat. No.99CH36258), IEEE, Piscataway, NJ, USA, 1999, pp. 1925–8.

27. Borriello, G., Detjens, E., "High-level synthesis: Current status and future directions," *Proceedings–25th Design Sutomation Conference*, pp. 477–82.

28. Brown, S. D., Francis, R. J., Rose J., Vranesic Z. G., *Field Programmable Gate Arrays*, Kluwer Academic Publishers, Boston, MA, 1992.

29. Buhler, M., Baitinger, U. G., "VHDL-based development of a 32-bit pipelined RISC processor for educational purposes," *9th Mediterranean Electrotechnical Conference Proceedings*, vol. 1 (Cat. No.98CH36056), IEEE, New York, NY, USA, 1998, pp. 138–42.

30. Buzzoni, M., Cardini, D., Gallino, R., Romagnese, R., "ATM traffic management systems: ASIC fast prototyping," *Proceedings Tenth IEEE International Workshop on Rapid System*

Prototyping. Shortening the Path from Specification to Prototype (Cat. No.PR00246), IEEE Computer Society Press, Los Alamitos, CA, USA, 1999, pp. 74–80.

31. Campasono, R., Wolf, W., *High Level Synthesis*, Kluwer Academic Publishers, Boston, MA, 1991.

32. Camposano, R., Wolf, W., *High-Level VLSI Synthesis*, Kluwer Academic Publishers, Dordrecht, Netherlands, 1991.

33. Carter, J. W., *Digital Designing with Programmable Logic Devices*, Prentice Hall, Upper Saddle River, NJ, 1997.

34. Chang, K. C., *Digital Design and Modeling with VHDL and Synthesis*, IEEE Computer Society Press, Los Alamitos, CA, 1997.

35. Cleaver, C., Derr, M., "Design automation through synthesis of VHDL," *Design Automation*, vol. 2, no. 1, pp. 20–4.

36. Coelho, D. R., *The VHDL Handbook*, Kluwer Academic Publishers, Netherlands, 1989.

37. Compass Design Automation, *VHDL Scout*, Compass Design Automation, San Jose, CA, 1994.

38. Dalcolmo, J., Lauwereins, R., Ade, M., "Code generation of data dominated DSP applications for FPGA targets," *Proceedings. Ninth International Workshop on Rapid System Prototyping* (Cat. No.98TB100237), IEEE Computer Society Press, Los Alamitos, CA, USA, 1998, pp. 162–7.

39. DeMicheli, G., Ku, D. C., "HERCULES—A system for high-level synthesis," *Proceedings–25th Design Automation Conference*, 1988, pp. 483–8.

40. Devadas, S., Newton, A. R., "Algorithms for hardware allocation in data path synthesis," *IEEE Transactions on Computer-Aided Design*, vol. 8 (July 1989), pp. 768–81.

41. Dewey, A., *Analysis and Design of Digital Systems With VHDL*, PWS Publishing Co., Boston, MA, 1997.

42. Dinu A., Cirstea, M. N., McCormick, M., "Virtual prototyping of a digital neural current controller," *Proceedings Ninth International Workshop on Rapid System Prototyping* (Cat. No.98TB100237), IEEE Computer Society Press, Los Alamitos, CA, USA, 1998, pp. 176–81.

43. Dipert, B., "Getting a handle on HDLs," *EDN-(US-Edition)*, vol. 43, no. 10, May 1998, pp. 71–2, 75–6, 79–80, 83–4, 86, 90.

44. Dutt, N. D., Gajski, D. D. "Designer controlled behavioral synthesis," *26th ACM/IEEE Design Automation Conference* (ACM #477890 and IEEE Cat. #89CH2734-2), ACM, New York, pp. 754–7.

45. Filippi, E., Licciardi, L., Montanaro, A., Paolini, M., Turolla, M., Taliercio, M., *Proceedings of the IEEE 1998 Custom Integrated Circuits Conference* (Cat. No.98CH36143), IEEE, New York, NY, USA, 1998, pp. 97–100.

46. Frank, G. A., Gray, F. G., Gopalakrishnan, S., Song, W., "Reuse of models and test benches at different levels of abstraction," *Proceedings International Verilog HDL Conference and VHDL International Users Forum* (Cat. No.98TB100230), IEEE Computer Society Press, Los Alamitos, CA, USA, 1998, pp. 130–7.

47. Frenkil, J., "The practical engineer [IC design, power reduction]," *IEEE-Spectrum,* vol. 35, no. 2, February 1998, pp. 54–60.

48. Gajski, D. D., Dutt, N. D., Wu, A., Lin, S. Y-L., *High-Level Synthesis: Introduction to Chip and System Design*, Kluwer Academic Publishers, Dordrecht, Netherlands, 1992.

49. Gajski, D. D., Vahid F., Narayan S., Gong J., *Specification and Design of Embedded Systems*, Prentice Hall PTR, Englewood Cliffs, N.J., 1994.

50. Gajski, D. D., Dutt, N. D., Wu, A., Lin, S., *High Level Synthesis: Introduction to Chip and System Design*, Kluwer Academic Publishers, Boston, MA, 1992.

51. Garbergs, B., Sohlberg, B., "Implementation of a state space controller in a FPGA," *MELECON '98 9th Mediterranean Electrotechnical Conference Proceedings,* vol. 1 (Cat. No.98CH36056), IEEE, New York, NY, USA, 1998, pp. 566–9.

52. Gray, F. G., Frank, G. A., Ziegenbein, D., Vuppala, S., Balasubramanian, P., "Tools for rapid construction of VHDL performance models for DSP systems," *Proceedings International Verilog HDL Conference and VHDL International Users Forum* (Cat. No.98TB100230), IEEE Computer Society Press, Los Alamitos, CA, USA, 1998, pp. 77–82.

53. Hamblen, J., "A VHDL synthesis model of the MIPS processor for use in computer architecture laboratories," *IEEE-Transactions-on-Education*, vol. 40, no. 4, November 1997, pp. 10.

54. Harr, R. E., Stanculescu, A. G., *Applications of VHDL to Circuit Design*, Kluwer Academic Publishers, Dordrecht, Netherlands.

55. Hines, J., "Where VHDL fits within the CAD environment," *ConferenceProceedings-24th ACM/IEEE Design Automation Conference*, 1987, IEEE, Piscataway, NJ, pp. 491–4.

56. Hong, Y. S., Park, K. H., Kim, M., "Automatic synthesis of data paths based on the path-search algorithm," *IEEE Int. Conf. on Computer-Aided Design*, 1987, pp. 270–3.

57. *IEEE Standard VHDL Language Reference Manual*. IEEE, Piscataway, NJ, 1987.

58. *IEEE Standard VHDL Language Reference Manual*. IEEE, Piscataway, NJ, 1993.

59. Jenkins, J. H., *Designing With FPGAs and CPLDs*, Prentice Hall PTR, Englewood Cliffs, NJ, 1994.

60. Jong, C. C., Lam, Y. H., Ng, L. S., "FPGAs implementation of a digital IQ demodulator using VHDL, field-programmable logic and applications," *7th International Workshop, FPL '97 Proceedings*, Springer-Verlag, Berlin, Germany, 1997, pp. 410–17.

61. Kapusta, R., "Writing reusable VHDL," *Electronic-Product-Design*, vol. 19, no. 8, Aug. 1998, pp. 17–18, 20.

62. Kelly, M., Hsu, K. W., "A flexible pipelined image processor," *Proceedings Eleventh Annual IEEE International ASIC Conference* (Cat. No.98TH8372), IEEE, New York, NY, USA, 1998, pp. 325–32.

63. Kelly, M., Hsu, K. W., "VHDL implementation of an image processor," *Proceedings of the SPIE The International Society for Optical Engineering*, vol. 3422, 1998, pp. 120–31.

64. Kivioja, M., Isoaho, J., Vanska, L., "Design and implementation of Viterbi decoder with FPGAs," *Journal of VLSI Signal Processing Systems for Signal, Image, and Video Technology*, vol. 21, no. 1, May 1999, pp. 5–14.

65. Komanec, R., Vrba, R., "Considerable advantages of VHDL over classic design approach," *Proceedings, Vol. 1, AMSE*, Assoc. Advancement of Modeling & Simulation Tech. Enterprises, Tassin la Demi Lune, France, 1995, pp. 144–7.

66. Krup, P., Abbasi, T., *Logic Synthesis Using Synopsys*, Kluwer Academic Publishers, Boston, MA, 1997.

67. Kurdahi, F. J., Parker, A.C., "REAL: A program for register allocation," *24th ACM/IEEE Design Automation Conference*, 1987, pp. 210–5.

68. Kuusilinna, K., Hamaainen, T., Saarinen, J., "Field programmable gate array-based PCI interface for a coprocessor system," *Microprocessors-and-Microsystems*, vol. 22, no. 7, January 1999, pp. 373–88.

69. Leung, S. S., Shanblatt, M. *ASIC System Design with VHDL : A Paradigm*, Kluwer Academic Publishers, 1989.

70. Lin, M. W., Armstrong, J. R., Frank, G. A, Concha, L., "A functional test planning system for validation of DSP circuits modeled in VHDL," *Proceedings International Verilog HDL Conference and VHDL International Users Forum* (Cat. No.98TB100230), IEEE Computer Society Press, Los Alamitos, CA, USA, 1998, pp. 172–7.

71. Lin, M. W., Armstrong, J. R., Gray, F. G., "A goal tree based high-level test planning system for DSP real number models," *Proceedings International Test Conference 1998* (IEEE Cat. No. 98CH36270), Washington DC, USA, 1998, pp. 1000–9.

72. Lipman, J., "Covering your HDL chip-design bets," *EDN-(US-Edition)*, vol. 43, no. 22, October 1998, pp. 65–6, 68–70, 72, 74.

73. Lipsett, R., Schaefer, C., Ussery, C. *VHDL: Hardware Description and Design*, Kluwer Academic Publishers, 1989.

74. Lis, J. S., Gajski, D. D. "Synthesis from VHDL," *Proceedings-1988 IEEE International Conference on Computer Design, VLSI Computing Process ICCD 88* (Cat. #88CH2643-5), IEEE, Piscataway, NJ, pp. 378–81.

75. Lis, J. S., Gajski, D. D. "VHDL synthesis using structured modeling," *Proceedings-Design Automation Conference* (Cat. #89CH2734-2), IEEE, Piscataway, NJ, pp. 606–9.

76. Mazor, S., Langstraat, P., *A Guide to VHDL*, Kluwer Academic Publishers, Dordrecht, Netherlands, 1992.

77. McCanny, J. V., Trainor, D., Hu, Y., Ding, T. J., "Rapid design of complex DSP cores," *ESSCIRC '97 Proceedings of the 23rd European Solid-State Circuits Conference*, Editions Frontieres, Paris, France, 1997, pp. 284–7.

78. McCloskey, J., "Application of VHDL to software radio technology," *Proceedings International Verilog HDL Conference and VHDL International Users Forum* (Cat. No.98TB100230), IEEE Computer Society Press, Los Alamitos, CA, USA, 1998, pp. 90–5.

79. McLeod, J. "'Top-down' design is the watchword at DAC," *Electronics*, vol. 63, no. 6, June 1990, pp. 78–80.

80. McLeod, J., "New kind of engineering fuels CAD," *Electronics*, vol. 63, no. 4, April 1990, pp. 50–3.

81. Menchini, P. J., "A minimalist approach to VHDL logic modeling," *IEEE Design & Test of Computers*, vol. 7, pp. 12–23.

82. Meyer, E., "VHDL opens the road to top-down design," *ComputerDesign*, vol. 28, pp. 57–62.

83. Meyer, E., "VHDL strives to cover both synthesis and modeling," *Computer Design*, vol. 28, no. 19, pp. 42–5.

84. Mora, F., Sebastia, A., Muller, H., Fernandes, C., Ermoline, Y., "Design of a high-performance PCI interface for an SCI network," *Computing & Control Engineering Journal*, vol. 9, no. 6, December 1998, pp. 275–82.

85. More, M., Vidal, J., Lecha, E., Rincon, F., Teres, L., "Experiences on VHDL based methodologies on industrial ASIC design," *1998 International Semiconductor Conference. CAS'98 Proceedings* vol. 1 (Cat. No.98TH8351), IEEE, New York, NY, USA, 1998, pp. 167–70.

86. Munch, M., When, N., Glesner, M., "An efficient ILP-based scheduling algorithm for control-dominated VHDL descriptions," *ACM Transactions on Design Automation of Electronic Systems*, vol. 2, no. 4, October 1997, pp. 344–64.

87. Narayan, S., Vahid, F., Gajski, D. D., "Translating system specifications to VHDL," *EDAC Proceedings of the European Conference on Design Automation*, 1991, IEEE Computer Society Press, Los Alamitos CA, pp. 390–4.

88. Navabi, Z. *VHDL: Analysis and Modeling of Digital Systems*, McGraw-Hill, Inc., USA, 1993.

89. Nebhrajani, V. A., Suthar, N., "Finite state machines: a deeper look into synthesis optimization for VHDL," *Proceedings Eleventh International Conference on VLSI Design* (Cat. No. 98TB100217), IEEE Computer Society Press, Los Alamitos, CA, USA, 1997, pp. 516–21.

90. O'Neill, M. D., January, D. D., Cho, C. H., Armstrong, J. R., "BTG: A behavioral test generator," *Computer Hardware Description Languages and Their Applications. Proceedings of the IFIP WG 10.2 Ninth InternationalSymposium*, North-Holland, Amsterdam, Netherlands, pp. 347–61.

91. Olcoz, S., Castellvi, A., Garcia, M., "Improving VHDL soft-cores reuse with software-like reviews and audit procedures," *Proceedings. International Verilog HDL Conference and VHDLInternational Users Forum* (Cat. No. 98TB100230), IEEE Computer Society Press, Los Alamitos, CA, USA, 1998, pp. 143–6.

92. Oldfield, J. V., Dorf, R. C., *Field Programmable Gate Arrays*, John Wiley and Sons, New York, NY, 1995.

93. Ott, D. E., Wilderotter, T. J., *A Designer's Guide to VHDL Synthesis*, Kluwer Academic Publishers, Boston, MA, 1994.

94. Park, N., Parker, A., "Sehwa: A software package for synthesis of pipelines from behavioral specifications," *IEEE Transactions on Computer-Aided Design* (March 1988), vol. 7, no. 3.

95. Paulin, P. G., Kight, J. P., "Algorithms for high-level synthesis," *IEEE Design and Test of Computers* (December 1989), pp. 18–31.

96. Perez, M. A. J, Luque, W. M., Damiani, F., "Biologically-inspired digital circuit for a self-organising neural network," *Proceedings of the 1998 Second IEEE International Caracas Conference onDevices, Circuits and Systems*, ICCDCS 98, On the 70th Anniversary of the MOSFET and 50th of the BJT (Cat. No.98TH8350), IEEE, New York, NY, USA, 1998, pp. 172–7.

97. Perry, D. L., *VHDL*, McGraw-Hill, Inc, New York, 1991.

98. Popp, R. L., Montana, D. J., Gassner, R. R. Vidaver, G., Iyer, S., "Automated hardware design using genetic programming," *VHDL and FPGAs, SMC'98 Conference Proceedings*, 1998 IEEE International Conference on Systems, Man, and Cybernetics, vol. 3 (Cat. No. 98CH36218), IEEE, New York, NY, USA, 1998, pp. 2184–9.

99. Renner, F. M., Becker, J., Glesner, M., "An FPGA implementation of a magnetic bearing controller for mechatronic applications," *Field-Programmable Logic and Applications. From FPGAs to Computing Paradigm. 8th International Workshop, FPL'98 Proceedings*, Springer-Verlag, Berlin, Germany, 1998, pp. 179–88.

100. Rosenstiel, W., Camposano, R. "Synthesizing circuits from behavioral level specifications," *Computer Hardware Description Languages and their Applications*, Elsevier Science Publishers B.V., North-Holland.

101. Ruiz, P. L., Riesgo, T., Uceda, J., "Design and prototyping of DSP custom circuits based on a library of arithmetic components," *Proceedings of the IECON'97 23rd International Conference on Industrial Electronics, Control, and Instrumentation,* vol. 1 (Cat. No. 97CH36066), IEEE, New York, NY, USA, 1997, pp. 191–6.

102. Ryan, R., "X-state handling in VHDL," *Design Automation*, vol. 2, no. 4, pp. 32–5.

103. Salcic, Z. and Smailagic, A., *Digital Systems Design and Prototyping Using Field Programmable Logic*, Kluwer Academic Publishers, Boston, MA, 1997.

104. Sargunaraj, J. J., Rao, S. S., "An optimal implementation approach for discrete wavelet transform using FIR filter banks on FPGAs," *1998 International Semiconductor Conference. CAS'98 Proceedings*, vol. 1 (Cat. No. 98TH8351), IEEE, New York, NY, USA, 1998, pp. 167–70.

105. Schoen, J., *Performance and Fault Modeling with VHDL*, Prentice Hall, Englewood Cliffs, NJ, 1989.

106. Schroeter, J., *Surviving the ASIC Experience*, Prentice Hall, Englewood Cliffs, NJ, 1992.

107. Schutti, M., Pfaff, M., Hagelauer, R., "VHDL design of embedded processor cores: the industry-standard microcontroller 8051 and 68HC11," *Proceedings Eleventh Annual IEEE International ASIC Conference* (Cat. No.98TH8372), IEEE, New York, NY, USA, 1998, pp. 265–9.

108. Shaditalab, M., Bois, G., Sawan, M., "Self-sorting radix-2 FFT on FPGAs using parallel pipelined distributed arithmetic blocks," *Proceedings IEEE Symposium on FPGAs for Custom Computing Machines* (Cat. No. 98TB100251), IEEE Computer Society Press, Los Alamitos, CA, USA, 1998, pp. 337–8.

109. Shahdad, M., "An overview of VHDL language and technology," *Proceedings-23rdACM/ IEEE Design Automation Conference, 1986* (Cat. #86CH2288-9), IEEE Computer Society Press, Washington, D.C., pp. 320–6.

110. Shahdad, M., Lipsett, R., Marchschner, E., Sheehan, K., Cohen, H., Waxman, R., Ackley, D. "VHSIC hardware description language," *Computer*, vol. 18, no. 2, pp. 94–103.

111. Skahill, K., *VHDL for Programmable Logic*, Adddison-Wesley, Menlo Park, CA, 1996.

112. Slorach, C. G., Sharman, K. C., "A novel single chip evolutionary hardware design using FPGAs," *Proceedings of the SPIE -The International Society for Optical Engineering*, vol. 3526, 1998, pp. 114–23.

113. Smith, D. J., "To create successful designs, know your HDL simulation and synthesis issues," *EDN-Europe*, November 1997, pp. 135–6, 138, 140, 142, 144.

114. Smith, S. P., Larson, J. "A high performance VHDL simulator with integrated switch and primitive modeling," *Computer Hardware Description Languages and their Applications. Proceedings of the IFIPWG 10.2 Ninth International Symposium.* North-Holland, Amsterdam, Netherlands, pp. 299–313.

115. Smith, S. P., Acosta, R. D., "Value system for switch-level modeling," *IEEE Design & Test of Computers*, vol. 7, no. 3, June 1990, pp. 33–41.

116. Smith, S., Taylor, D, Benaissa, M., "Design automation of Reed-Solomon codecs using VHDL," *Microelectronics-Journal*, vol. 29, no. 12, December 1998, pp. 977–82.

117. Stone, A., Manolakos, E. S., "Using DG2VHDL to synthesize an FPGA implementation of the 1-D discrete wavelet transform," *1998 IEEE Workshop on Signal Processing Systems. SIPS 98. Design and Implementation* (Cat. No.98TH8374), IEEE, New York, NY, USA, 1998, pp. 489–98.

118. Sullivan, R., Asher, L. R., "VHDL for ASIC design and verification," *VLSI Systems Design, Semicustom Design Guide*, 1998, pp. 64–72.

119. Tan. S., Furber, S. B., Wen-Fang-Yen, "The design of an asynchronous VHDL synthesizer," *Proceeding. Design, Automation and Test in Europe* (Cat. No. 98EX123), IEEE Computer Society Press, Los Alamitos, CA, USA, 1998, pp. 44–51.

120. Thomas, D. E., Moorby, P. R. *The Verilog Hardware DescriptionLanguage*, Kluwer Academic Publishers, Dordrecht, Netherlands, 1991.

121. Trimberger, S., *Field-Programmable Gate Array Technology*, Kluwer Academic Publishers, Boston, MA, 1994.

122. Tseng, C.-J., Siewiorek, D. P., "Facet: A procedure for the automated synthesis of digital systems," *20th Design Automation Conference*, 1983, pp. 490–6.

123. Tseng, C.-J., Siewiorek, D. P., "Automated synthesis of data paths in digital systems," *IEEE Transactions on Computer-Aided Design* (July 1986), pp. 379–95.

124. Uht, A. K., Ying-Sun, "The laboratory environment of the URI Integrated Computer Engineering Design (ICED) curriculum," *FIE '98. 28th Annual Frontiers in Education Conference. Moving from `Teacher-Centered' to `Learner-Centered' Education. Conference Proceedings,* vol. 1 (Cat. No.98CH36214). IEEE, Piscataway, NJ, USA, 1998, pp. 331–6.

125. Van den Bout, D., *The Practical Xilinx Designer Lab Book*, Prentice Hall PTR, Upper Saddle River, NJ, 1999.

126. Vassileva, T., Tchoumatchenko, V., Shishkov, V., Guyot, A., "High performance adder's synthesis using efficient macro generator," *ECCTD '97 Proceedings of the 1997 European Conference on Circuit Theory and Design*, vol. 3, Tech. Univ. Budapest, Budapest, Hungary, 1997, pp. 1343–6.

127. Walker, P. A., Ghosh, S., "On the nature and inadequacies of transport timing delay constructs in VHDL descriptions," *IEEE Transactions on Computer Aided Design of Integrated Circuits and Systems*, vol. 16, no. 8, August 1997, pp. 894–915.

128. Ward, P. C., Armstrong, J. R., "Behavioral fault simulation inVHDL," *Proceedings-27th ACM/IEEE Design Automation Conference,1990* (Cat. #90CH2894-4), IEEE, Piscataway, NJ, pp. 587–93.

129. Wicks, J. A., Armstrong, J. R., "Efficiency ratings for VHDL behavioral models," *Proceedings IEEE Southeastcon '98. 'Engineering for a New Era'* (Cat.No.98CH36170), IEEE, New York, NY, USA, pp. 401–4.

130. Wilsey, P. A., Martin, D. E., Subramani, K., "SAVANT/TyVIS/WARPED: components for the analysis and simulation of VHDL," *Proceedings International Verilog HDL Conference and VHDL International Users Forum* (Cat. No. 98TB100230), IEEE Computer Society Press, Los Alamitos, CA, USA, 1998, pp. 195–201.

131. Yin, T. H., Yuan, H. W., Jer, S. H., "Rapid prototyping of hardware/software codesign for embedded signal processing," *Journal of Information Science and Engineering*, vol. 4, no. 3, September 1998, pp. 605–32.

132. Zeidman, B., *Verilog Designer's Library*, Prentice Hall PTR, Upper Saddle River, NJ, 1997.

133. Zhang, D., Liu, M., "VHDL high level design of digital systems with HLS/BIT," *Journal of Computer Science and Technology (English Language Edition)*, vol. 13, supplemental issue, December 1998, pp. 82–8.

Index

About the Authors

Dr. James R. Armstrong teaches graduate and undergraduate courses in computer architecture, HDLs, and logic design. He was a member of the original VHDL IEEE standardization committee; authored *Chip Level Modeling With VHDL*, and co-authored *Structured Logic Design With VHDL*, both from Prentice Hall. He has done extensive research on the use of HDLs. His work has been published in IEEE journals and presented at international symposia.

Dr. F. Gail Gray teaches graduate and undergraduate courses in computer engineering, logic design, hardware description languages, coding theory, fault tolerant computing, testing, and microprocessor system design. His work has been published by *IEEE Transactions on Computers; Journal of VLSI Signal Processing for Signal, Image, and Video Technolocy; Design Automation Conference; the VHDL International Users Forum,* and many other leading journals and conferences.

LICENSE AGREEMENT AND LIMITED WARRANTY

READ THE FOLLOWING TERMS AND CONDITIONS CAREFULLY BEFORE OPENING THIS SOFTWARE MEDIA PACKAGE. THIS LEGAL DOCUMENT IS AN AGREEMENT BETWEEN YOU AND PRENTICE-HALL, INC. (THE "COMPANY"). BY OPENING THIS SEALED SOFTWARE MEDIA PACKAGE, YOU ARE AGREEING TO BE BOUND BY THESE TERMS AND CONDITIONS. IF YOU DO NOT AGREE WITH THESE TERMS AND CONDITIONS, DO NOT OPEN THE SOFTWARE MEDIA PACKAGE. PROMPTLY RETURN THE UNOPENED SOFTWARE MEDIA PACKAGE AND ALL ACCOMPANYING ITEMS TO THE PLACE YOU OBTAINED THEM FOR A FULL REFUND OF ANY SUMS YOU HAVE PAID.

1.   **GRANT OF LICENSE:** In consideration of your payment of the license fee, which is part of the price you paid for this product, and your agreement to abide by the terms and conditions of this Agreement, the Company grants to you a nonexclusive right to use and display the copy of the enclosed software program (hereinafter the "SOFTWARE") on a single computer (i.e., with a single CPU) at a single location so long as you comply with the terms of this Agreement. The Company reserves all rights not expressly granted to you under this Agreement.

2.   **OWNERSHIP OF SOFTWARE:** You own only the magnetic or physical media (the enclosed software media) on which the SOFTWARE is recorded or fixed, but the Company retains all the rights, title, and ownership to the SOFTWARE recorded on the original software media copy(ies) and all subsequent copies of the SOFTWARE, regardless of the form or media on which the original or other copies may exist. This license is not a sale of the original SOFTWARE or any copy to you.

3.   **COPY RESTRICTIONS:** This SOFTWARE and the accompanying printed materials and user manual (the "Documentation") are the subject of copyright. You may not copy the Documentation or the SOFTWARE, except that you may make a single copy of the SOFTWARE for backup or archival purposes only. You may be held legally responsible for any copying or copyright infringement which is caused or encouraged by your failure to abide by the terms of this restriction.

4.   **USE RESTRICTIONS:** You may not network the SOFTWARE or otherwise use it on more than one computer or computer terminal at the same time. You may physically transfer the SOFTWARE from one computer to another provided that the SOFTWARE is used on only one computer at a time. You may not distribute copies of the SOFTWARE or Documentation to others. You may not reverse engineer, disassemble, decompile, modify, adapt, translate, or create derivative works based on the SOFTWARE or the Documentation without the prior written consent of the Company.

5.   **TRANSFER RESTRICTIONS:** The enclosed SOFTWARE is licensed only to you and may not be transferred to any one else without the prior written consent of the Company. Any unauthorized transfer of the SOFTWARE shall result in the immediate termination of this Agreement.

6.   **TERMINATION:** This license is effective until terminated. This license will terminate automatically without notice from the Company and become null and void if you fail to comply with any provisions or limitations of this license. Upon termination, you shall destroy the Documentation and all copies of the SOFTWARE. All provisions of this Agreement as to warranties, limitation of liability, remedies or damages, and our ownership rights shall survive termination.

7.   **MISCELLANEOUS:** This Agreement shall be construed in accordance with the laws of the United States of America and the State of New York and shall benefit the Company, its affiliates, and assignees.

8.   **LIMITED WARRANTY AND DISCLAIMER OF WARRANTY:** The Company warrants that the SOFTWARE, when properly used in accordance with the Documentation, will operate in substantial conformity with the description of the SOFTWARE set forth in the Documentation. The Company does not

warrant that the SOFTWARE will meet your requirements or that the operation of the SOFTWARE will be uninterrupted or error-free. The Company warrants that the media on which the SOFTWARE is delivered shall be free from defects in materials and workmanship under normal use for a period of thirty (30) days from the date of your purchase. Your only remedy and the Company's only obligation under these limited warranties is, at the Company's option, return of the warranted item for a refund of any amounts paid by you or replacement of the item. Any replacement of SOFTWARE or media under the warranties shall not extend the original warranty period. The limited warranty set forth above shall not apply to any SOFTWARE which the Company determines in good faith has been subject to misuse, neglect, improper installation, repair, alteration, or damage by you. EXCEPT FOR THE EXPRESSED WARRANTIES SET FORTH ABOVE, THE COMPANY DISCLAIMS ALL WARRANTIES, EXPRESS OR IMPLIED, INCLUDING WITHOUT LIMITATION, THE IMPLIED WARRANTIES OF MERCHANTABILITY AND FITNESS FOR A PARTICULAR PURPOSE. EXCEPT FOR THE EXPRESS WARRANTY SET FORTH ABOVE, THE COMPANY DOES NOT WARRANT, GUARANTEE, OR MAKE ANY REPRESENTATION REGARDING THE USE OR THE RESULTS OF THE USE OF THE SOFTWARE IN TERMS OF ITS CORRECTNESS, ACCURACY, RELIABILITY, CURRENTNESS, OR OTHERWISE.

IN NO EVENT, SHALL THE COMPANY OR ITS EMPLOYEES, AGENTS, SUPPLIERS, OR CONTRACTORS BE LIABLE FOR ANY INCIDENTAL, INDIRECT, SPECIAL, OR CONSEQUENTIAL DAMAGES ARISING OUT OF OR IN CONNECTION WITH THE LICENSE GRANTED UNDER THIS AGREEMENT, OR FOR LOSS OF USE, LOSS OF DATA, LOSS OF INCOME OR PROFIT, OR OTHER LOSSES, SUSTAINED AS A RESULT OF INJURY TO ANY PERSON, OR LOSS OF OR DAMAGE TO PROPERTY, OR CLAIMS OF THIRD PARTIES, EVEN IF THE COMPANY OR AN AUTHORIZED REPRESENTATIVE OF THE COMPANY HAS BEEN ADVISED OF THE POSSIBILITY OF SUCH DAMAGES. IN NO EVENT SHALL LIABILITY OF THE COMPANY FOR DAMAGES WITH RESPECT TO THE SOFTWARE EXCEED THE AMOUNTS ACTUALLY PAID BY YOU, IF ANY, FOR THE SOFTWARE.

SOME JURISDICTIONS DO NOT ALLOW THE LIMITATION OF IMPLIED WARRANTIES OR LIABILITY FOR INCIDENTAL, INDIRECT, SPECIAL, OR CONSEQUENTIAL DAMAGES, SO THE ABOVE LIMITATIONS MAY NOT ALWAYS APPLY. THE WARRANTIES IN THIS AGREEMENT GIVE YOU SPECIFIC LEGAL RIGHTS AND YOU MAY ALSO HAVE OTHER RIGHTS WHICH VARY IN ACCORDANCE WITH LOCAL LAW.

ACKNOWLEDGMENT

YOU ACKNOWLEDGE THAT YOU HAVE READ THIS AGREEMENT, UNDERSTAND IT, AND AGREE TO BE BOUND BY ITS TERMS AND CONDITIONS. YOU ALSO AGREE THAT THIS AGREEMENT IS THE COMPLETE AND EXCLUSIVE STATEMENT OF THE AGREEMENT BETWEEN YOU AND THE COMPANY AND SUPERSEDES ALL PROPOSALS OR PRIOR AGREEMENTS, ORAL, OR WRITTEN, AND ANY OTHER COMMUNICATIONS BETWEEN YOU AND THE COMPANY OR ANY REPRESENTATIVE OF THE COMPANY RELATING TO THE SUBJECT MATTER OF THIS AGREEMENT.

Should you have any questions concerning this Agreement or if you wish to contact the Company for any reason, please contact in writing at the address below.

Robin Short
Prentice Hall PTR
One Lake Street
Upper Saddle River, New Jersey 07458

About the CD

The CD-ROM contains four subdirectories:

Figure_Models: This directory contains source code for all models in the book. There is a file for each chapter. Within each chapter, the figures are organized by figure number. They are text files that can be analyzed. This directory also contains all image processing packages mentioned in the text. Components of some models are not included because they are proprietary.

Miscellaneous_Models: This directory source code for various models collected by the authors over the years. They are text files that can be analyzed. The files are listed alphabetically in the log file: Misc_Log.txt. The four-valued logic package described in the text is located in this directory.

Project_Sequences: This directory contains files for a six-project sequence for one semester. It contains: project descriptions, input files, and solutions.

Other_Projects: This directory contains a collection of other projects that the authors have used. The projects are listed in the log file: Project_Log.txt.

TECHNICAL SUPPORT

Prentice Hall does not offer technical support for this CD-ROM. However, if there is a problem with the media, you may obtain a replacement copy by emailing us with your problem at: disc_exchange@prenhall.com